Advanced Engineering Cementitious Composites and Concrete Sustainability

Advanced Engineering Cementitious Composites and Concrete Sustainability

Editor

Dumitru Doru Burduhos Nergis

MDPI • Basel • Beijing • Wuhan • Barcelona • Belgrade • Manchester • Tokyo • Cluj • Tianjin

Editor
Dumitru Doru Burduhos
Nergis
Faculty of Materials Science
and Engineering
"Gheorghe Asachi" Technical
University
Iasi
Romania

Editorial Office
MDPI
St. Alban-Anlage 66
4052 Basel, Switzerland

This is a reprint of articles from the Special Issue published online in the open access journal *Materials* (ISSN 1996-1944) (available at: www.mdpi.com/journal/materials/special_issues/Cem_Compos_Concr_Sustain).

For citation purposes, cite each article independently as indicated on the article page online and as indicated below:

LastName, A.A.; LastName, B.B.; LastName, C.C. Article Title. *Journal Name* **Year**, *Volume Number*, Page Range.

ISBN 978-3-0365-7627-5 (Hbk)
ISBN 978-3-0365-7626-8 (PDF)

© 2023 by the authors. Articles in this book are Open Access and distributed under the Creative Commons Attribution (CC BY) license, which allows users to download, copy and build upon published articles, as long as the author and publisher are properly credited, which ensures maximum dissemination and a wider impact of our publications.
The book as a whole is distributed by MDPI under the terms and conditions of the Creative Commons license CC BY-NC-ND.

Contents

Dumitru Doru Burduhos-Nergis
Special Issue "Advanced Engineering Cementitious Composites and Concrete Sustainability"
Reprinted from: *Materials* **2023**, *16*, 2582, doi:10.3390/ma16072582 1

Allice Tan Mun Yin, Shayfull Zamree Abd Rahim, Mohd Mustafa Al Bakri Abdullah, Marcin Nabialek, Abdellah El-hadj Abdellah and Allan Rennie et al.
Potential of New Sustainable Green Geopolymer Metal Composite (GGMC) Material as Mould Insert for Rapid Tooling (RT) in Injection Moulding Process
Reprinted from: *Materials* **2023**, *16*, 1724, doi:10.3390/ma16041724 5

Zhishuan Lv, Yang Han, Guoqi Han, Xueyu Ge and Hao Wang
Experimental Study on Toughness of Engineered Cementitious Composites with Desert Sand
Reprinted from: *Materials* **2023**, *16*, 697, doi:10.3390/ma16020697 41

Muhd Hafizuddin Yazid, Meor Ahmad Faris, Mohd Mustafa Al Bakri Abdullah, Muhammad Shazril I. Ibrahim, Rafiza Abdul Razak and Dumitru Doru Burduhos Nergis et al.
Mechanical Properties of Fly Ash-Based Geopolymer Concrete Incorporation Nylon66 Fiber
Reprinted from: *Materials* **2022**, *15*, 9050, doi:10.3390/ma15249050 57

Ali İhsan Çelik, Yasin Onuralp Özkılıç, Özer Zeybek, Memduh Karalar, Shaker Qaidi and Jawad Ahmad et al.
Mechanical Behavior of Crushed Waste Glass as Replacement of Aggregates
Reprinted from: *Materials* **2022**, *15*, 8093, doi:10.3390/ma15228093 77

Moslih Amer Salih, Shamil Kamil Ahmed, Shaymaa Alsafi, Mohd Mustafa Al Bakri Abullah, Ramadhansyah Putra Jaya and Shayfull Zamree Abd Rahim et al.
Strength and Durability of Sustainable Self-Consolidating Concrete with High Levels of Supplementary Cementitious Materials
Reprinted from: *Materials* **2022**, *15*, 7991, doi:10.3390/ma15227991 95

Nurul Aida Mohd Mortar, Mohd Mustafa Al Bakri Abdullah, Rafiza Abdul Razak, Shayfull Zamree Abd Rahim, Ikmal Hakem Aziz and Marcin Nabiałek et al.
Geopolymer Ceramic Application: A Review on Mix Design, Properties and Reinforcement Enhancement
Reprinted from: *Materials* **2022**, *15*, 7567, doi:10.3390/ma15217567 115

Özer Zeybek, Yasin Onuralp Özkılıç, Memduh Karalar, Ali İhsan Çelik, Shaker Qaidi and Jawad Ahmad et al.
Influence of Replacing Cement with Waste Glass on Mechanical Properties of Concrete
Reprinted from: *Materials* **2022**, *15*, 7513, doi:10.3390/ma15217513 143

Kaffayatullah Khan, Muhammad Arif Aziz, Mukarram Zubair and Muhammad Nasir Amin
Biochar Produced from Saudi Agriculture Waste as a Cement Additive for Improved Mechanical and Durability Properties—SWOT Analysis and Techno-Economic Assessment
Reprinted from: *Materials* **2022**, *15*, 5345, doi:10.3390/ma15155345 159

Rebeca Martínez-García, P. Jagadesh, Osama Zaid, Adrian A. Șerbănoiu, Fernando J. Fraile-Fernández and Jesús de Prado-Gil et al.
The Present State of the Use of Waste Wood Ash as an Eco-Efficient Construction Material: A Review
Reprinted from: *Materials* **2022**, *15*, 5349, doi:10.3390/ma15155349 175

Dickson Ling Chuan Hao, Rafiza Abd Razak, Marwan Kheimi, Zarina Yahya, Mohd Mustafa Al Bakri Abdullah and Dumitru Doru Burduhos Nergis et al.
Artificial Lightweight Aggregates Made from Pozzolanic Material: A Review on the Method, Physical and Mechanical Properties, Thermal and Microstructure
Reprinted from: *Materials* 2022, 15, 3929, doi:10.3390/ma15113929 . . . . **195**

Muhammad Nasir Amin, Waqas Ahmad, Kaffayatullah Khan and Mohamed Mahmoud Sayed
Mapping Research Knowledge on Rice Husk Ash Application in Concrete: A Scientometric Review
Reprinted from: *Materials* 2022, 15, 3431, doi:10.3390/ma15103431 . . . . **217**

Marwan Kheimi, Ikmal Hakem Aziz, Mohd Mustafa Al Bakri Abdullah, Mohammad Almadani and Rafiza Abd Razak
Waste Material via Geopolymerization for Heavy-Duty Application: A Review
Reprinted from: *Materials* 2022, 15, 3205, doi:10.3390/ma15093205 . . . . **239**

Kaffayatullah Khan, Muhammad Ishfaq, Muhammad Nasir Amin, Khan Shahzada, Nauman Wahab and Muhammad Iftikhar Faraz
Evaluation of Mechanical and Microstructural Properties and Global Warming Potential of Green Concrete with Wheat Straw Ash and Silica Fume
Reprinted from: *Materials* 2022, 15, 3177, doi:10.3390/ma15093177 . . . . **257**

Mohd Izrul Izwan Ramli, Mohd Arif Anuar Mohd Salleh, Mohd Mustafa Al Bakri Abdullah, Ikmal Hakem Aziz, Tan Chi Ying and Noor Fifinatasha Shahedan et al.
The Influence of Sintering Temperature on the Pore Structure of an Alkali-Activated Kaolin-Based Geopolymer Ceramic
Reprinted from: *Materials* 2022, 15, 2667, doi:10.3390/ma15072667 . . . . **285**

Dumitru Doru Burduhos Nergis, Petrica Vizureanu, Andrei Victor Sandu, Diana Petronela Burduhos Nergis and Costica Bejinariu
XRD and TG-DTA Study of New Phosphate-Based Geopolymers with Coal Ash or Metakaolin as Aluminosilicate Source and Mine Tailings Addition
Reprinted from: *Materials* 2021, 15, 202, doi:10.3390/ma15010202 . . . . **297**

Bogdan Bolborea, Cornelia Baera, Sorin Dan, Aurelian Gruin, Dumitru-Doru Burduhos-Nergis and Vasilica Vasile
Concrete Compressive Strength by Means of Ultrasonic Pulse Velocity and Moduli of Elasticity
Reprinted from: *Materials* 2021, 14, 7018, doi:10.3390/ma14227018 . . . . **311**

Zarina Yahya, Mohd Mustafa Al Bakri Abdullah, Long-yuan Li, Dumitru Doru Burduhos Nergis, Muhammad Aiman Asyraf Zainal Hakimi and Andrei Victor Sandu et al.
Behavior of Alkali-Activated Fly Ash through Underwater Placement
Reprinted from: *Materials* 2021, 14, 6865, doi:10.3390/ma14226865 . . . . **327**

Editorial

Special Issue "Advanced Engineering Cementitious Composites and Concrete Sustainability"

Dumitru Doru Burduhos-Nergis

Faculty of Materials Science and Engineering, Gheorghe Asachi Technical University of Iasi, 700050 Iasi, Romania; doru.burduhos@tuiasi.ro

Citation: Burduhos-Nergis, D.D. Special Issue "Advanced Engineering Cementitious Composites and Concrete Sustainability". *Materials* **2023**, *16*, 2582. https://doi.org/10.3390/ma16072582

Received: 15 March 2023
Accepted: 19 March 2023
Published: 24 March 2023

Copyright: © 2023 by the author. Licensee MDPI, Basel, Switzerland. This article is an open access article distributed under the terms and conditions of the Creative Commons Attribution (CC BY) license (https://creativecommons.org/licenses/by/4.0/).

Concrete, one of the most often-used building materials today, is the cornerstone of modern buildings all over the world, being used for foundations, pavements, building walls, architectural structures, highways, bridges, overpasses, and so on. Because of its adaptability, concrete may be found in practically every construction, in some form or another. Yet, the diverse nature of its components, their combinations, and their doses result in a very wide range of concrete kinds with varying properties. As a result, concrete is a material that is always evolving and is popular even now, especially when it comes to circular economy.

Other ways of concrete manufacturing are now being researched to lessen or remove the limits of this material, which are connected to its brittleness and poor environmental effects. As a result, the development of engineering cementitious composites has resulted in a significant reduction in flexibility issues, while the introduction of new additives and the optimization of the manufacturing process has resulted in a significant reduction in the negative effects of virgin raw material exploitation. In-depth research is still required to optimize and increase the sustainability of these advanced engineering cementitious composites or alternative concretes.

In this Special Issue (SI), state-of-the-art research and review articles on the emerging material systems for AM are collected, with a focus on the process–structure–properties relationships. In total, eleven research papers and six reviews have been collected. Considering the high interest in this field for finding alternatives for virgin raw materials, in the research article conducted by Lv, Z. et al. [1], an interesting experimental study was conducted on the effect of replacing ordinary sand with desert sand on the obtainment and characterization of engineered cementitious materials. Additionally, A. İ. Çelik et al. [2] observed that a 20% replacement of fine aggregates and coarse aggregates with recycled crushed glass resulted in a significant increase in the mechanical properties of concrete. In another study, Ö. Zeybek et al. [3] evaluated the effect of replacing cement with fine glass microparticles on the tensile and flexural strengths of concrete, and showed that a 10% replacement would result in better mechanical properties. Burduhos Nergis, D.D. et al. [4] evaluated the possibility of obtaining acid-activated geopolymers, using mine tailings as a substitute for fine aggregates. In their article, M.I.I. Ramli et al. [5] aimed to obtain alkali-activated ceramics and determined the influence of high curing temperatures on the morphology of kaolin-based geopolymers. To improve the main characteristics of these cementitious composites, some researchers designed and obtained engineered materials by integrating different types of reinforcing elements, or by involving advanced techniques to characterize them. M.H. Yazid et al. [6] obtained geopolymer concrete with improved mechanical performances and water absorption by introducing low amounts of diamond-shaped nylon66 fibers. M.A. Salih et al. [7] incorporated high amounts of supplementary cementitious materials, such as fly ash, ground-granulated blast furnace slag, and microsilica, into self-consolidating concrete, in order to improve the durability and properties of fresh and cured Ordinary Portland Cement (OPC)-based concrete. Z. Yahya et al. [8] developed another self-consolidating concrete for underwater structures and showed that a class

C fly ash, activated with a mixture of sodium silicate and sodium hydroxide, could achieve more than 70 MPa when cured in seawater, river water, or lake water. K. Khan et al. [9] showed that ecofriendly concrete could be obtained by replacing OPC with wheat straw ash and/or silica fume. According to their study, this differently engineered composition could achieve better mechanical performances at lower CO_2-eq. In another study, K. Khan et al. [10] showed that biochar could be used to obtain advanced concrete by performing a SWOT analysis and a techno-economic assessment on the introduction of this by-product as substitute for OPC. Because concrete durability is difficult to assess, particularly for in situ applications, B. Bolborea et al. [11] conducted an experimental investigation on the forecasting of the mechanical properties of concrete, using a non-destructive approach, namely ultrasonic pulse velocity.

In the review articles published in this SI, A.T.M. Yin et al. [12] discussed the potential of producing mold inserts for rapid tooling, using geopolymer composites that were reinforced with recycled metal particles, while N.A.M. Mortar [13] conducted a comprehensive literature analysis on the obtainment and characterization of kaolin-based geopolymers for ceramic applications. R. Martínez-García et al. [14] reviewed the recent developments of the effect produced by the addition of waste wood ash on the composition of different types of concrete. D.L.C. Hao et al. [15] assessed the previous studies on the characterization of artificial aggregates that were manufactured by sintering, cold bonding, or autoclaving, and concluded that the last two methods were suitable for producing lightweight aggregates for industrial use. M. Kheimi et al. [16] presented an overview of the research that was conducted on the parameters that influence the performances of geopolymers that are used in heavy-duty applications, and observed that the mixing design, curing conditions, alkali activator, and binder type are the key factors that define the properties of the final product. A scientometric analysis, considering the publications that are indexed in the Scopus database, was conducted by M.N. Amin et al. [17], in order to establish the statical overview and mapping of the research on rice husk ash utilization in concrete compositions. According to their study, despite the high number of papers published in this field, the lack of standardization in the preparation, process, and use of geopolymers, is the main limitation toward the industrial use of this material.

Conflicts of Interest: The authors declare no conflict of interest.

References

1. Lv, Z.; Han, Y.; Han, G.; Ge, X.; Wang, H. Experimental Study on Toughness of Engineered Cementitious Composites with Desert Sand. *Materials* **2023**, *16*, 697. [CrossRef] [PubMed]
2. Çelik, A.İ.; Özkılıç, Y.O.; Zeybek, Ö.; Karalar, M.; Qaidi, S.; Ahmad, J.; Burduhos-Nergis, D.D.; Bejinariu, C. Mechanical Behavior of Crushed Waste Glass as Replacement of Aggregates. *Materials* **2022**, *15*, 8093. [CrossRef] [PubMed]
3. Zeybek, Ö.; Özkılıç, Y.O.; Karalar, M.; Çelik, A.İ.; Qaidi, S.; Ahmad, J.; Burduhos-Nergis, D.D.; Burduhos-Nergis, D.P. Influence of Replacing Cement with Waste Glass on Mechanical Properties of Concrete. *Materials* **2022**, *15*, 7513. [CrossRef] [PubMed]
4. Burduhos-Nergis, D.D.; Vizureanu, P.; Sandu, A.V.; Burduhos-Nergis, D.P.; Bejinariu, C. XRD and TG-DTA Study of New Phosphate-Based Geopolymers with Coal Ash or Metakaolin as Aluminosilicate Source and Mine Tailings Addition. *Materials* **2022**, *15*, 202. [CrossRef] [PubMed]
5. Ramli, M.I.I.; Salleh, M.A.A.M.; Abdullah, M.M.A.B.; Aziz, I.H.; Ying, T.C.; Shahedan, N.F.; Kockelmann, W.; Fedrigo, A.; Sandu, A.V.; Vizureanu, P.; et al. The Influence of Sintering Temperature on the Pore Structure of an Alkali-Activated Kaolin-Based Geopolymer Ceramic. *Materials* **2022**, *15*, 2667. [CrossRef] [PubMed]
6. Yazid, M.H.; Faris, M.A.; Abdullah, M.M.A.B.; Ibrahim, M.S.I.; Razak, R.A.; Burduhos Nergis, D.D.; Burduhos Nergis, D.P.; Benjeddou, O.; Nguyen, K.S. Mechanical Properties of Fly Ash-Based Geopolymer Concrete Incorporation Nylon66 Fiber. *Materials* **2022**, *15*, 9050. [CrossRef] [PubMed]
7. Salih, M.A.; Ahmed, S.K.; Alsafi, S.; Abullah, M.M.A.B.; Jaya, R.P.; Abd Rahim, S.Z.; Aziz, I.H.; Thanaya, I.N.A. Strength and Durability of Sustainable Self-Consolidating Concrete with High Levels of Supplementary Cementitious Materials. *Materials* **2022**, *15*, 7991. [CrossRef] [PubMed]
8. Yahya, Z.; Abdullah, M.M.A.B.; Li, L.Y.; Nergis, D.D.B.; Hakimi, M.A.A.Z.; Sandu, A.V.; Vizureanu, P.; Razak, R.A. Behavior of Alkali-Activated Fly Ash through Underwater Placement. *Materials* **2021**, *14*, 6865. [CrossRef] [PubMed]

9. Khan, K.; Ishfaq, M.; Amin, M.N.; Shahzada, K.; Wahab, N.; Faraz, M.I. Evaluation of Mechanical and Microstructural Properties and Global Warming Potential of Green Concrete with Wheat Straw Ash and Silica Fume. *Materials* **2022**, *15*, 3177. [CrossRef] [PubMed]
10. Khan, K.; Aziz, M.A.; Zubair, M.; Amin, M.N. Biochar Produced from Saudi Agriculture Waste as a Cement Additive for Improved Mechanical and Durability Properties—SWOT Analysis and Techno-Economic Assessment. *Materials* **2022**, *15*, 5345. [CrossRef] [PubMed]
11. Bolborea, B.; Baera, C.; Dan, S.; Gruin, A.; Burduhos-Nergis, D.D.; Vasile, V. Concrete Compressive Strength by Means of Ultrasonic Pulse Velocity and Moduli of Elasticity. *Materials* **2021**, *14*, 7018. [CrossRef] [PubMed]
12. Yin, A.T.M.; Rahim, S.Z.A.; Al Bakri Abdullah, M.M.; Nabialek, M.; Abdellah, A.E.; Rennie, A.; Tahir, M.F.M.; Titu, A.M. Potential of New Sustainable Green Geopolymer Metal Composite (GGMC) Material as Mould Insert for Rapid Tooling (RT) in Injection Moulding Process. *Materials* **2023**, *16*, 1724. [CrossRef] [PubMed]
13. Mohd Mortar, N.A.; Abdullah, M.M.A.B.; Abdul Razak, R.; Abd Rahim, S.Z.; Aziz, I.H.; Nabiałek, M.; Jaya, R.P.; Semenescu, A.; Mohamed, R.; Ghazali, M.F. Geopolymer Ceramic Application: A Review on Mix Design, Properties and Reinforcement Enhancement. *Materials* **2022**, *15*, 7567. [CrossRef] [PubMed]
14. Martínez-García, R.; Jagadesh, P.; Zaid, O.; Șerbănoiu, A.A.; Fraile-Fernández, F.J.; de Prado-Gil, J.; Qaidi, S.M.A.; Grădinaru, C.M. The Present State of the Use of Waste Wood Ash as an Eco-Efficient Construction Material: A Review. *Materials* **2022**, *15*, 5349. [CrossRef] [PubMed]
15. Hao, D.L.C.; Razak, R.A.; Kheimi, M.; Yahya, Z.; Abdullah, M.M.A.B.; Nergis, D.D.B.; Fansuri, H.; Ediati, R.; Mohamed, R.; Abdullah, A. Artificial Lightweight Aggregates Made from Pozzolanic Material: A Review on the Method, Physical and Mechanical Properties, Thermal and Microstructure. *Materials* **2022**, *15*, 3929. [CrossRef] [PubMed]
16. Kheimi, M.; Aziz, I.H.; Abdullah, M.M.A.B.; Almadani, M.; Razak, R.A. Waste Material via Geopolymerization for Heavy-Duty Application: A Review. *Materials* **2022**, *15*, 3205. [CrossRef] [PubMed]
17. Amin, M.N.; Ahmad, W.; Khan, K.; Sayed, M.M. Mapping Research Knowledge on Rice Husk Ash Application in Concrete: A Scientometric Review. *Materials* **2022**, *15*, 3431. [CrossRef] [PubMed]

Disclaimer/Publisher's Note: The statements, opinions and data contained in all publications are solely those of the individual author(s) and contributor(s) and not of MDPI and/or the editor(s). MDPI and/or the editor(s) disclaim responsibility for any injury to people or property resulting from any ideas, methods, instructions or products referred to in the content.

Review

Potential of New Sustainable Green Geopolymer Metal Composite (GGMC) Material as Mould Insert for Rapid Tooling (RT) in Injection Moulding Process

Allice Tan Mun Yin [1], Shayfull Zamree Abd Rahim [1,2,*], Mohd Mustafa Al Bakri Abdullah [2,3], Marcin Nabialek [4], Abdellah El-hadj Abdellah [5], Allan Rennie [6], Muhammad Faheem Mohd Tahir [2,3] and Aurel Mihail Titu [7]

1. Faculty of Mechanical Engineering & Technology, Universiti Malaysia Perlis, Arau 02600, Malaysia
2. Center of Excellence Geopolymer and Green Technology (CEGeoGTech), Universiti Malaysia Perlis, Kangar 01000, Malaysia
3. Faculty of Chemical Engineering & Technology, Universiti Malaysia Perlis, Kangar 01000, Malaysia
4. Department of Physics, Faculty of Production Engineering and Materials Technology, Częstochowa University of Technology, 42-201 Czestochowa, Poland
5. Laboratory of Mechanics, Physics and Mathematical Modelling (LMP2M), University of Medea, Medea 26000, Algeria
6. Lancaster Product Development Unit, Engineering Department, Lancaster University, Lancaster LA1 4YW, UK
7. Industrial Engineering and Management Department, Faculty of Engineering, "Lucian Blaga" University of Sibiu, 10 Victoriei Street, 550024 Sibiu, Romania
* Correspondence: shayfull@unimap.edu.my

Abstract: The investigation of mould inserts in the injection moulding process using metal epoxy composite (MEC) with pure metal filler particles is gaining popularity among researchers. Therefore, to attain zero emissions, the idea of recycling metal waste from industries and workshops must be investigated (waste free) because metal recycling conserves natural resources while requiring less energy to manufacture new products than virgin raw materials would. The utilisation of metal scrap for rapid tooling (RT) in the injection moulding industry is a fascinating and potentially viable approach. On the other hand, epoxy that can endure high temperatures (>220 °C) is challenging to find and expensive. Meanwhile, industrial scrap from coal-fired power plants can be a precursor to creating geopolymer materials with desired physical and mechanical qualities for RT applications. One intriguing attribute of geopolymer is its ability to endure temperatures up to 1000 °C. Nonetheless, geopolymer has a higher compressive strength of 60–80 MPa (8700–11,600 psi) than epoxy (68.95 MPa) (10,000 psi). Aside from its low cost, geopolymer offers superior resilience to harsh environments and high compressive and flexural strength. This research aims to investigate the possibility of generating a new sustainable material by integrating several types of metals in green geopolymer metal composite (GGMC) mould inserts for RT in the injection moulding process. It is necessary to examine and investigate the optimal formulation of GGMC as mould inserts for RT in the injection moulding process. With less expensive and more ecologically friendly components, the GGMC is expected to be a superior choice as a mould insert for RT. This research substantially impacts environmental preservation, cost reduction, and maintaining and sustaining the metal waste management system. As a result of the lower cost of recycled metals, sectors such as mould-making and machining will profit the most.

Keywords: rapid tooling; geopolymer metal composite; additive manufacturing; injection moulding process

1. Introduction

Time to market is a crucial aspect of a product development strategy, and speed is frequently compared to other factors such as functionality, creativity, or performance [1–3]. With numerous new technologies, worldwide rivalry for product creation is soaring. Furthermore, companies are always looking for cutting-edge technologies that are cost-effective, capable of manufacturing goods in tiny quantities while maintaining excellent performance, and able to meet sustainability goals. This has driven the development of rapid tooling (RT) techniques, which are needed in today's market to replace traditional techniques with rapid product innovation and improve manufacturing processes, particularly mould-making [4–6].

As shown in Figure 1, RT provides quicker manufacturing for completing tests and starting final production, minimises costs, and reduces project time [7].

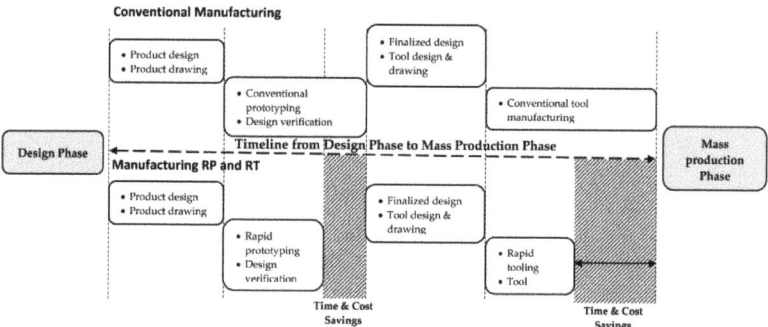

Figure 1. A Review of the Application of Rapid Tooling in Manufacturing [8].

Every industry, regardless of size, experiences a time when rapid tooling is required to address particular problems. Additionally, an improved tooling system is required for creating a limited number of functional prototypes to assess the product development cycle [8–10]. A small quantity of items is often utilised as a marketplace trial, evaluation need, and manufacturing process design [9,10].

Before mass manufacturing, functioning tools or prototypes must be launched for every scientific study [11–14]. These are not made available in large quantities to consumers but rather in limited amounts to researchers. RT is highly advantageous in this circumstance since it allows for the rapid introduction of items. Furthermore, the uses of production tools allow mass production to be obtained at a lower price because manufacturing costs are cheap. For this reason, many brand-new businesses and even big organisations prefer this technology to boost their profits and obtain a market advantage over their rivals [1–3].

Prototype companies or mould producers typically employ mild steel or aluminium for the mould inserts in RT. Production toolmaking is time-consuming and costly, and machining involves the same computer numerical control (CNC), electrical discharge machining (EDM), and electric discharge machining (wire EDM) procedures [15,16]. Recently, additive manufacturing (AM) has been employed to create mould inserts for RT [13,16]. For a limited number of prototypes, RT often uses models or prototypes made by AM as templates for manufacturing mould inserts or uses the AM process directly [4–6]. Numerous RT technologies are available on the market, such as a hybrid technique combining RT and AM to shorten RT production time.

RT can be categorised as either an indirect or direct technique and differs from traditional tooling in that the amount of time needed to create the tooling is significantly reduced [17,18]. Automated manufacturing methods use the AM process to generate mould inserts without the requirement for values to be predicted. Direct tooling includes processes such as additive manufacturing (AM), stereolithography (SLA), jet photopolymerisation (PolyJet), fused deposition moulding (FDM), and selective laser sintering (SLS) [5,6,19]. Alternatively, the AM project as a master model is considered a secondary approach to create moulds for casting or plastic moulding processes. This technique combines the 3D KelTool process, metal casting, plastic casting, elastic moulding, and other comparable procedures to create injection moulding inserts [5,6,15,20]. Inserts constructed from epoxy-acrylate and utilising material for injection moulding by homo-polypropylene are used for quick tooling applications for 3D-printed injection moulds [21]. The mould insert constructed from steel and copper for hybrid prototype mould applications is created using a mix of laser powder bed fusion (L-PBF) and casting. In a study, L-PBF printed the steel shell with conformal cooling channels, and the shell was cast with copper [22]. For a material jetting (PolyJet) mould, a mould insert constructed from epoxy-acrylate resin is used and its in-mould behaviour is compared to a guidance mould insert fabricated from aluminium [23].

Failure in the moulding process in generating RT mould inserts is common for technologies with poor thermal and mechanical quality, such as SLS, FDM, SLA, PolyJet, and AM [5,6,19]. Zakzewski et al. [24] presented the bucking π-theorem, which was modified to analyse and characterise the poor surface quality Ra issue of SLS/SLM-processed samples, as well as the existence of porosity, a material structure defect. Furthermore, the use of SLA models is physically restricted. These constraints can be solved by developing procedures that use SLA parts as the "master blueprint" for the silicone mould process. In comparison to mechanical techniques, the DMLS process is inefficient for the design of basic plastic components. Furthermore, DMLS imposes a few constraints for a specific feature design for complicated components. According to previous studies on RT mould inserts, the stress applied to the mould insert during the injection cycle has a significant influence on the mould life [11,15,19,20]. Nowadays, a combination of RP technique and production tooling helps produce RT more quickly but faces dimensional accuracy and surface finish issues [17,18]. Moreover, the injection moulding process faces a cooling time issue where most of the mould inserts fabricated using RP techniques have very low thermal conductivity; thus, increasing the cooling rate will undoubtedly influence the cycle time for producing the components [4,16,25,26]. One of the RT options to increase competitiveness is using metal epoxy composite (MEC), which provides greater heat conductivity as mould inserts in RT application and lowers tooling production costs and lead times by 25% and 50%, respectively [3,6]. Using optimisation methods to determine the optimal composition for materials, as recommended in the linked literature, can be considered for future research, such as determining the best amount of Al or Cu to mix with epoxy resin for desirable mechanical properties [27–37]. The use of MEC mould inserts for RT in the injection moulding process, which uses pure metal filler particles combined with epoxy resin, has attracted the attention of many researchers [20,29,38–40].

The use of MEC materials as mould inserts offers better thermal and mechanical properties as compared to mould inserts produced using AM technologies [40–43]. However, dimensional accuracy and surface quality still need to be improved, so after the fabrication of mould inserts using MEC material and casting techniques, the mould inserts need to go through a secondary process (machining) to improve mould dimensional precision and surface quality in terms of cavities, especially for precision of plastic products (± 0.05 mm).

Secondary (recycled) materials compete with primary materials in the metals business. Primary materials require the use of finite resources. Producing scrap materials or processed secondary metals can sometimes be more cost-effective than producing new primary materials, provided that the cost of collecting the waste is not prohibitively expensive [42]. Due to its spherical morphology and manageable particle size dispersion, gas-atomised (GA) powder is the most frequent feedstock for AM. However, much energy and inert gas are required to make GA powders [43]. Water-atomised powder is another option when increased powder solidification rates and reduced manufacturing costs are the priorities [29,44,45]. In contrast, melting the metal before ejection from the atomisation nozzles is energy-intensive because of the significant enthalpy difference between the liquid and solid states [45]. In metal AM, the feedstock powder is often remelted. This repeated melting is a costly and inefficient process [46]. As it can reduce materials of varying sizes to powder, mechanical milling offers a chance for environmentally friendly powder production [31,47–49]. For mechanical milling, ambient or cryogenic temperatures are typically used [49,50]. The aforementioned considerable energy input is no longer required to reach atomisation temperatures [50]. Due to these potential benefits, mechanical milling is being used to reduce metal machining chips to powders that can be used in AM [46].

On the other hand, Davidovits' geopolymer technology is one of the groundbreaking innovations resulting in an affordable and greener binder alternative. The silica and aluminium in geosource materials such as metakaolin (calcined kaolin), and maybe techniques such as fly ash and bottom ash, are combined with the alkaline liquid to generate a geopolymer, an alkali-activated binder [1,11]. As a result, it reduces not only CO_2 emissions but also recycles industrial waste, specifically using an aluminium–silicate mix to create products of higher value [9,10]. MEC using pure metal filler particles is beginning to be used by some researchers to investigate mould inserts in the injection moulding process [4,5,7]. However, a type of epoxy that can withstand high temperatures (>220 °C) is hard to find and still costly.

Additionally, besides municipal solid waste, coal combustion production (CCP) has been identified as the second-largest pollutant in the world. In 2011, about 130 metric tonnes (MT) of CCP were generated, with only 56.57 MT (43.50%) effectively used [51].

The four forms of solid waste created in substantial amounts by the CCP are boiler slag, bottom ash, fly ash, and flue gas desulphurisation (FGD) material [40–42]. One hundred and thirty metric tonnes of CCP included around 59.9 MT of fly ash. Fly ash was disposed of in surface impoundments covered with compacted clay soil, a plastic sheet, or both for the remaining 22.9 MT (38.36%) in landfills or surface impoundments [51–53]. The Environmental Protection Agency (EPA) of the United States (US) is now investigating the positive uses of fly ash [53–57]. This eliminates major health concerns associated with heavy metals and radioactive elements accumulated from fly ash disposal over time. Geopolymers derived from environmentally friendly materials, such as slag, or industrial by-products and used as a binding material are known as "green material".

One interesting property of geopolymer is that it can withstand temperatures up to 1000 °C. Nevertheless, geopolymer only has a compressive strength of 60–80 MPa (8700–11,600 psi), while epoxy has a compressive strength of 68.95 MPa (10,000 psi) [29,58]. However, employing geopolymer material has similar issues to using epoxy resin, which necessitates determining the optimal strength, accuracy, acceptable surface finish, and good thermal characteristics.

Early strength of geopolymer can be obtained as early as 1 day, with compressive strength up to 15 MPa, and continues increase up to 40–50MPa within 7 days, which is comparable with the strength offered by epoxy. Nevertheless, the optimum strength of geopolymer material can be obtained by 28 days (80 MPa) and the strength will keep on increasing over time [59].

It was recognised that the filler's interlaminar strength controls the bond strength of geopolymer reinforced with filler. The fact that filler with a bigger particle size has a lower binding strength is also well known. In addition, compared to epoxy resin, geopolymer showed high bond strength for both wet and dry interface surface conditions [59].

On the other hand, as the need for an environmentally friendly society grows, the quantity of waste material must be continually decreased. Hence, in order to achieve zero emissions, the idea of recycling metal waste from factories and workshops needs to be examined (waste free) [60–63]. Metal recycling helps to conserve natural resources while requiring less energy for manufacturing new products than would be required for virgin raw materials. Waste-free recycling reduces the emission of carbon dioxide and certain other harmful gases while also saving money and enabling industrial companies to reduce their production costs [64,65].

Through a Google Patents search (https://patents.google.com/ accessed on 10 February 2023), six patents granted/published that make use of (1) metal composite and composites made of (2) geopolymer and (3) metal were located. Table 1 lists the search terms for this review's related field, rapid tooling. Fibre-reinforced metal composites (aluminium matrix composites) were developed by Yamamoto et al. [65] using aluminium alloy with 6–11 wt. % nickel as the metal matrix and reinforcing fibres. To mass produce complex parts with near-net shapes, Behi et al. [66] proposed using steel tooling in an injection moulding machine. In comparison to more traditional methods, this approach to producing complicated metal tooling is relatively cost-effective, making it possible to rapidly fabricate complex shaped parts using normal metal, ceramic, and plastic processing. According to the metal matrix composite introduced by Shaikh et al. [67], the fibre to metal or alloy ratio ranges from about 9:1 to less than about 1:1, and the fibres have an average diameter of approximately eight micrometres with a coating. Amaya and Crounse [68] discovered rapid manufacturing of mould inserts by employing blank die inserts formed from material typically used in the metal injection moulding process of complex shaped components to achieve high machinability rates, time and cost savings, extended tool life, and material savings. The dry-mix composition, as proposed by Nematollahi and Sanjayan [69], includes (a) an aluminosilicate material rich in silica and alumina and (b) a powdered alkali activator. Moreover, the dry-mix composition is chosen so that (i) the SHGC may be generated at ambient temperature without liquid activator, and (ii) strain-hardening behaviour and multiple cracking behaviours are observed. A strain-hardened, ambient temperature-cured geopolymer composite (SHGC) is generated by adding water and using a method of manufacturing an ambient temperature-cured SHGC. Qiang et al. [70] proposed a geopolymer composite material that is a type of 3D print as well as their preparation technique and applications, which included blast furnace slag powder accounting for 20~25% of the total composition weight, steel-making slag powder accounting for 10~15%, fly ash accounting for 0~5%, mine tailing machine-made sand accounting for 33~45%, exciting composite agent accounting for 3~5%, high molecular weight polymer accounting for 2.5~3%, volume stabiliser accounting for 1~3%, thixotropic agent accounting for 1~2%, defoamer accounting for 0.05~0.1%, and mixing water accounting for 13.9~12.45%. Each component is stirred, and subsequently pumped into the 3D printer applications for construction. The present invention's geopolymer composite material demonstrates good caking property, strong stability, good go-out pump from holding capacity and adhesive property, excellent form, and volume stability, resulting in the construction of buildings with good overall stability and safety during use. The six patents granted/published from 1990 to February 2023 are listed in Table 1.

RT is a cost-effective solution in the transition phase from new product development to mass production in the manufacturing industry [71,72]. RT, often referred to as bridge tooling, prototype tooling, or soft tooling, is a fast way to preproduce hundreds or even thousands of plastic parts prior to mass production, for design optimisation, functional testing, or preproduction verification, which can be a bridge between rapid prototyping (RP) and mass production. Shape, fit, and function prototype components are frequently made using RP technology, such as additive manufacturing [71,73,74]. Recycled metal waste such as mild steel, aluminium, copper, and brass after machining processes are as shown in Figure 2.

However, since 3D material qualities vary from those used in injection moulding, 3D-printed samples cannot provide a thorough evaluation of an injection-moulded part's functional performance [18,59], making RT extremely crucial for the manufacturing industry.

Table 1. Six patents granted/published from 1990 to February 2023.

No.	Patent Number	Title	Inventor/s	Granted/Publication Date	Patent Summary
1	US4980242A	Fibre-reinforced metal composite	Tadashi Yamamoto, Michiyuki Suzuki, Yoshiharu Waku, Masahiro Tokuse [65]	25 December 1990	• Aluminium matrix composite is a fibre-reinforced metal composite containing 6–11% nickel.
2	US6056915A	Rapid manufacture of metal and ceramic tooling	Mohammad Behi, Mike Zedalis, James M. Schoonover [66]	2 May 2000	• Steel tooling is needed to produce near-net form, complex items in high volume. • The technology is economical to make complex metal tooling for quick fabrication of complex shaped parts using conventional metal, ceramic, and plastic processes.
3	US6376098B1	Low-temperature, high-strength metal-matrix composite for rapid-prototyping and rapid-tooling	Furqan Zafar Shaikh, Howard Douglas Blair, Tsung-Yu Pan [67]	23 April 2002	• Fibre to metal or alloy ratio can vary from 9:1 to 1:1. • Fibres have an average diameter of 8 micrometres, and metal or alloy is distributed within them.
4	US20020187065A1	Method for the rapid fabrication of mould inserts	Herman Amaya, Dennis Crounse [68]	12 December 2002	• Mould inserts manufactured from metal injection moulding material provide high machinability rates, time and cost savings, extended tool life, and material savings. • The process involves developing cutting path programmes from CAD files, machining cavity and core inserts to predefined sizes, and processing them to transform the soft material into a dense, hardenable material.
5	WO2017070748A1	Geopolymer composite and geopolymer matrix composition	Behzad Nematollahi, Jay Sanjayan [69]	4 May 2017	• The dry-mix composition allows for the formation of ambient temperature-cured SHGC without the need for a liquid activator.
6	CN106082898A	3D printed geopolymer composite material, its production and applications	Lin Xi Qiang, Li Jing Fang, Zhang Tao, Huo Liang, Li Guo You, Zhang Nan, Liao Juan, Wang Bao Hua, Ji Wen Zhan [70]	31 July 2018	• Slag powder composition includes blast furnace slag, steel-making slag powder, fly ash, mine tailing sand, composite exciting agent, volume stabiliser, thixotropic agent, defoamer, and mixing water. • The invention's geopolymer composite material has good caking properties, stability, form and volume stability, providing good stability and safety for building construction.

Figure 2. (**a**) Metal scraps from turning; (**b**) metal scraps from grinding; (**c**) metal scraps from milling.

Using fly ash (waste from coal combustion) as the raw material, the metal scraps from the machining process are ground using a ball mill machine into a small and uniform size and mixed with geopolymer material to create green geopolymer metal composite (GGMC) material as in Figure 3. Then, this material can be used as mould inserts for RT applications which is expected to reduce tooling production costs and lead periods by up to 25% and 50%, respectively. The effect of GGMC material as mould inserts for RT in an injection moulding process and its relationship with compressive strength and thermal conductivity should be examined accordingly. Therefore, this research aims to determine whether geopolymer material may be used as RT mould inserts in the injection moulding process. The process by which GMCs are used as a new material for mould inserts is depicted in Figure 4.

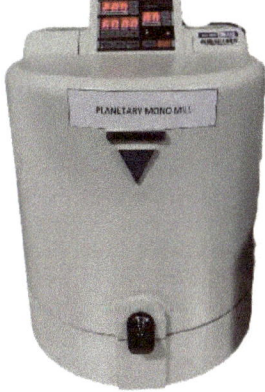

Figure 3. Planetary Mono Mill Pulverisette 6 can be used in the ball mill process.

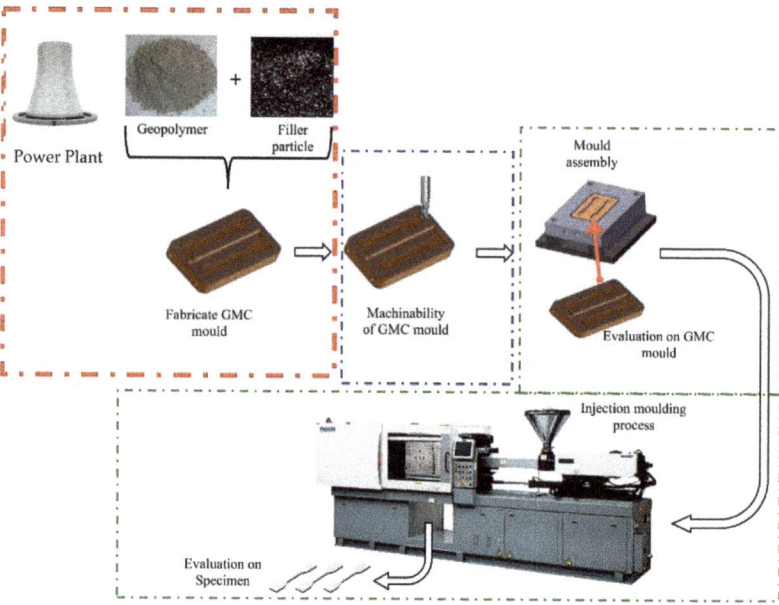

Figure 4. Graphical representation of GGMC as new material for mould inserts.

A power plant is a structure that produces waste geopolymer and generates electric energy from another form of energy. The geopolymer material is then combined with filler particles (waste from machining after a ball milling process to form a powder filler). The ratio of geopolymer and powder filler is evaluated accordingly in terms of thermal conductivity and compressive strength. Next, the optimised ratio is used to fabricate the GMC mould inserts. Then, GMC mould inserts are machined accordingly to fit the insert size and assembled in the mould base. Following the examination of the GMC mould inserts, the GMC mould is assembled in the injection moulding machine to mould out the specimen for further evaluation of the mould parts' quality in terms of shrinkage and warpage, including the cooling time required, which is definitely influenced by the thermal conductivity of the GMC mould inserts. The reliability of the GMC mould inserts is evaluated accordingly in terms of the number of shots (specimens) that can be produced before the mould starts to crack or wear.

2. Injection Moulding Process

2.1. Important Processing Parameters in the Injection Moulding Process

Processing parameters are essential to produce good-quality moulded parts in the injection moulding process. Previously, the trial-and-error approach to determine processing parameters relied upon a plastic injection moulding process. However, the trial-and-error approach is ineffective for complex manufacturing processes [75–77]. Therefore, many studies had been carried out over the years to minimise shrinkage and warpage defects by optimising the processing parameters [77–80]. In addition, it has also been observed that various critical processing factors, including packing pressure, melt temperature, packing shrinkage duration, mould temperature, and cooling time, have an impact on the quality of the moulded components produced (warpage) [77,81–83].

2.1.1. Melt Temperature

Melt temperature is the temperature required to melt the plastic material in a pellet formed in the screw barrel of the injection moulding machine before the injection stage to fill the mould cavities [84,85]. Some researchers reported that melt temperature is a significant processing parameter that causes warpage defects on the moulded parts produced. The relationship between melt temperature and the flow of molten plastic into the mould cavities through feeding system has been studied and it was reported that the amount of material flow into the cavities is affected by the melt temperature [84,85].

2.1.2. Cooling Time

When the molten plastic hits the walls of the mould cavities, it starts to cool down and continues to solidify. The mould stays closed until the moulded part reaches the ejection temperature. The part is ejected out from the injection mould once it becomes rigid enough [85,86]. When the cooling time, including that needed for the moulded component to achieve the injection temperature, is increased, shrinkage and warp issues are reduced [87,88]. However, the appropriate cooling time needs to be determined in order to produce moulded parts with good quality within the optimal cycle time.

2.1.3. Packing Pressure

Packing pressure is the pressure used to inject and compress the molten plastic material into mould cavities until the gate freezes [85]. According to previous research, packing pressure is a crucial processing parameter that impacts the accuracy and quality of the moulded components produced. In addition, packing pressure is also a significant processing parameter after packing time which has a significant impact on shrinkage and flexural strength of the moulded parts produced [80–82]. Any changes in packing pressure will cause degradation of the mechanical properties of the parts moulded from virgin and recycled plastic material in various compositions. Inappropriate settings of packing pressure may result in high shrinkage defects in the moulded parts [85].

2.1.4. Mould Temperature

Mould temperature is known as the temperature of the mould that needs to be controlled in order to solidify the molten plastic material that flows into the mould cavities towards the ejection temperature. Previous studies showed that mould temperature is one of the significant processing parameters that affects warpage and shrinkage defects [83,87]. Kamaruddin et al. [86] examined mould temperature using the Taguchi methods, and reported that the shrinkage of moulded parts affected by mould temperature is a critical factor. This supports the findings of a study by Chen et al. [89] which found that the temperature of the mould plays a role in the shrinkage of the resulting moulded products in both the transverse and longitudinal axes. In addition, mould temperature cannot be set directly but it can be controlled by controlling the temperature of coolant used in the injection moulding process.

2.1.5. Packing Time

The packing time is known as the time required to fill the mould cavities without pressing the mould or flashing the finished parts entirely with additional material [90]. The packing time is generally determined by the freeze time of the gate [91]. When gates freeze, the material is not permitted to flow into the mould cavities. Nevertheless, if the packing time is shorter, the molten material returns to the feeding system and causes a backflow phenomenon [89,92].

It can be seen that, in terms of material used as mould inserts for injection moulding, the thermal conductivity (which influences the melt temperature, mould temperature, packing time, and definitely cooling time) and compressive strength (which influences packing pressure and reliability of mould inserts) are important parameters that require the attention of the mould fabrication industries.

2.2. Mould Base Material

The selection of material for mould base parts depends on the product that needs to be manufactured. Choosing suitable materials can help a company to save costs and time. The materials of the mould base are divided into four types, which are mild steel, high-alloy steel, stainless steel, and tool steel, as tabulated in Table 2 [83–87].

Table 2. Types of mould base material with examples [81–87].

Mould Base Material	Example of Material
Carbon steel	1018
	1050
Alloy steel	AISI 4130
	AISI M2
Stainless steel	420
	316L
	17-4 PH
Tool steel	O-1
	A-6
	S-7
	D-2
	P-20
	H13

Mild steel is a type of iron that has varied levels of carbon added to it and no addition of other elements. There are different percentages of carbon where the carbon content ranges from mild, to medium, to high. Examples of carbon steel are carbon steel 1018 and 1050 [83,86]. High-alloy steel is a variety of steel that is alloyed with additional components ranging from 1 wt. % to 50 wt. % through the addition of carbon to enhance the material's different qualities.

High-alloy steel is therefore made of iron that has been alloyed with additional elements including copper, chromium, and aluminium. It can also alloy more than two metals. Examples of alloy steel are alloy steel AISI 4130 and AISI M2 [35]. Stainless steel provides excellent corrosion resistance and machinability. Stainless steel is a class of iron-based alloys notable for their corrosion and heat resistance.

Furthermore, stainless steel is produced by adding chromium at a rate of about 11% and the use of stainless steel is selected because it does not corrode or oxidise. Stainless steel does not require stress relief because its material qualities are stable. Examples of stainless steel are stainless steel 420, 316L, and 17-4 PH [88,92,93]. Tool steel refers to a range of carbon and alloy steels that are especially well-suited to be produced into tools.

In addition, tool steel contains elements such as tungsten, vanadium, cobalt, and molybdenum [94]. These elements are used to improve hardenability and generate harder and more thermally stable carbides. Examples of tool steels are tool steel O-1, A-6, S-7, D-2, P-20, and H13 [83,88]. RT is the AM technology that refers to the manufacturing methods of tooling [94–96].

Injection mould bases can be made from a wide variety of materials. However, selecting the right mould base material is essential for making high-quality components, since different materials have different properties.

Selecting Mould Base Material

Material selection for the mould base is important because it will affect the performance of the mould. Selecting the suitable material during the tool-making stage can reduce cost. Several factors need to be considered, which are strength, good wear resistance, excellent surface finish, dimensional stability, machinability, and corrosion resistance. First, highly compressive loads must be able to be absorbed by the material without cracking or splitting. Next, good wear resistance is needed so that the mould can be used longer. Good surface finish is also vital to be considered because it will affect the product surface. Other parameters also need to be considered so that the product can be used longer, and to save cost and time. An example of this consideration is the use of H13 which is selected because it can perform well at high temperatures, and has high dimensional stability, hardness, and wear resistance [97]. The recommended mould material for transparent products is stainless steel AISI 420, which has a hardness of up to 54HRC [98].

On the other hand, mould inserts are assembled in a mould base and form the cavities where the molten plastic will be injected to form the products. Therefore, the material of the mould insert is an important aspect that will have a direct impact on the defects of the moulded parts produced.

2.3. Mould Insert Material

The material of a mould insert will affect the cooling time of a product as it influences the overall cycle time of the injection moulding process [36]. Other than that, improving cooling time can also reduce defects such as shrinkage and warpage [90,99,100]. Tool steel material takes longer to achieve the ejection temperature than pure copper (Cu) and beryllium copper (BeCu) as tabulated in Table 3 [86]. This is because Cu and BeCu have higher thermal conductivities which can remove more heat than tool steel material. It is important because the temperature needs to be evenly distributed from the cavity to the core of the mould [84]. Although pure copper is proven to be the best according to simulation results, other factors need to be considered in choosing the mould insert material, including properties such as hardness. The hardness of BeCu is higher compared to pure copper and other properties that need to be considered are, namely, durability and resistance to non-oxidising acids.

Table 3. Simulation results of mould inserts by researchers [100].

Material	Time to Reach Ejection Temperature (s)	Mould Core Insert Temperature (°C)	Volumetric Shrinkage (%)	Warpage (mm)
Pure copper	8.804	28.10	1.605	0.1602
Tool steel	12.400	76.82	1.759	0.1700
Beryllium copper	9.483	41.62	1.160	0.1614

However, the materials used to fabricate mould inserts for the product designed in the development stage do not have to be the same as materials used for the hard tooling (mould used for mass production) because the product design is not yet finalised and there are still some tests and evaluations to be carried out, as well as a need to improve the product's features in terms of ease of assembly and reliability tests in order to ensure the high quality of product. An alternative material of mould inserts for low production in the product development industry is in high demand, especially in the effort to reduce the expenses in the research and development stage.

2.3.1. Alternative Materials for the Mould Insert

Small numbers of functional plastic parts that range from five to one thousand units are usually needed during the product development stage to confirm the development stage before mass manufacturing. An alternative material is required for mould inserts to reduce the cost, time, part quality, and production number [48]. Currently, the alternative material that is used in mould insert fabrication is epoxy resin. The different types of epoxy resin with their properties are listed in Table 4 [101].

Table 4. Different types of resin and their properties [102].

Name	EP250	NeuKadur VGSP5	EPO 752	XD4532 or XD4533	Reshape-Express 2000™
Resin Producer	MCP HEK Tooling GmbH, Lubeck, Germany	Altropol Kunststoff GmbH, Stockelsdorf, Germany	Axson Technologies (Shanghai) Co., Ltd., Shanghai, China	Ciba Specialty Chemicals Holding Inc., Basel, Switzerland	
Density (kg/m^3)	2	2.8	1.7–1.78	1.7 ± 0.02	1.8
Tensile strength (MPa)	67	50	49	38 ± 4	62
Compressive strength (MPa)	260	180	NA	145 ± 5	251
Flexural strength (MPa)	120	NA	88	90 ± 5	82
Deflexion temperature (°C)	250	150	195	220	234
Linear expansion ($\times 10^6$ mm/K)	30–35	30–35	50	NA	42
Hardness	112 (Rc)	90 (Shore D)	90 (Shore D15)	90 (Shore D)	91 (Shore D)

Nevertheless, there are some restrictions when using epoxy as a mould insert in RT for injection moulding. Epoxy has limitations that must be overcome, such as its low hardness and strength [103]. Geopolymer can be used to replace epoxy since it is robust and strong and is now utilised in building concrete. In addition, it preserves the environment, reduces cost, and supports sustainability of waste management systems [81,94,95,104,105]. As an implication, industries related to mould-making will benefit the most due to the reduced cost when using recycled materials.

2.3.2. Rapid Tooling (RT) Mould Inserts

Rapid tooling is an example of how rapid prototyping is used in the manufacturing industry. It enables the rapid and low-cost construction of moulds for small batches of manufacturing goods. Tooling may be either harsh or soft, and can be classified as direct or indirect [9]. An efficient method of direct tooling involves the use of soft materials in a rapid prototype process such as stereolithography material [96,106–108]. Numerous different tools, such those made of powder metal [96,103], are made of tough materials, and in the indirect tooling method, a casting pattern is made by rapid prototyping and then used to manufacture the proper tool. Due to its simplicity of usage in producing mould inserts, aluminium-filled epoxy resin [109,110] is becoming a popular soft material. Silicone rubber is mostly utilised in the manufacture of indirect tools [107].

The most significant factor for injection moulds made utilising the RT process is tool life. RP technology has improved to the point where tools directly generated by RP machines should represent all of the model's various elements and features accurately and precisely. On the other hand, many soft rapid prototyping materials are unable to tolerate sufficiently high pressure and melt temperature owing to their poor heat conductivity [107], resulting in shortened useful life of the instrument. Although other methods, including metal laser sintering, may be employed to apply metal coatings on pliable materials [108] to enhance their hardness, it will raise the manufacturing process's difficulties. Alternatively, epoxy resin is often used in indirect moulds due to its plasticity or compatibility with casting models. The use of metal powder may greatly boost its hardness and heat conductivity, prolonging tool life even more. However, this does not prevent the epoxy resin from hardening within the mould chamber and becoming brittle. Indirectly crafted tools thus wear very quickly [104]. Some of the studies concentrating on RT are listed in Table 5.

Tomori et al. [110] investigated how changing the material formulation and determining the validity of composite tooling boards affected mould efficiency and component quality. An example of the method for setting up a tooling board is illustrated in Figure 5. The boards were constructed using three materials: RP4037 (fluid), RP4037 hardener, and silicon carbide (SiC) filler (powder). For the six moulds, two cutting speeds (1.00 and 1.66 m/s) and three tooling board formulations (28.5%, 34.75%, and 39.9% wt. % SiC filler) were used. The surface roughness of the moulded components served as the study's response variable, while cutting speed served as the study's independent parameter. As there was no visible mould damage, the physical structure of the mould was unchanged by SiC concentration and cutting speed. This discovery indicated that the SiC content in the mould has a significant impact on the surface roughness of the moulded items. Additionally, the flexural strength rose with the SiC filler concentration (from 58.75 to 66.49 MPa), following a pattern comparable to the heat conductivity of the mould material. The influence of filler concentration primarily on the direction of welding for moulded components was not examined in this research.

Senthilkumar et al. [111] studied the effects of epoxy resin on the mechanical characteristics of aluminium (Al) particles. The sample was cast utilising Al filler mixed into epoxy resin at various concentrations. Optical microscopy revealed that the Al particles were uniformly dispersed throughout the epoxy resin matrix. These results show that increasing the amount of Al particles inside the epoxy resin matrix significantly raises both the thermal conductivity (3.97 to 5.39 W/mK) and the hardness value of the composite (69 to 89 RHL). The sample's fatigue life decreased from 15,786 cycles to 734 cycles as the Al content of the epoxy resin increased. The best percentage of Al filler particle for enhancing mould performance and durability was found to be between 45 and 55 wt.% There was an improvement of 72 RHL in durability, 10,011 cycles in fatigue resistance, and 4.06 W/mK in thermal conductivity. However, the hardness value increased by 4.34% for every 5% increase in Al filler particles, which might reduce the fatigue life by 36.58%. Nevertheless, there has been no further research on the moulded components' flexural strength, compressive strength, tensile strength, or surface appearance.

Srivastava and Verma [27] attempted to determine how the addition of Cu and Al particles to epoxy resin composites altered their mechanical properties. Epoxy resin was mixed with Cu and Al particles (1, 5, 8, and 10 wt. %) to create a variety of filler compositions. The results of the mechanical tests showed that the epoxy resin with Al reinforcement has excellent tensile properties, with a tensile strength of 104.5 MPa at 1 wt. %, while the epoxy resin composites with Cu filler was optimal in the hardness test (22.4 kgF/mm^2 at 8 wt. %) and had a compressive strength of 65 MPa at 10 wt. %. In addition, epoxy resin composites filled with Cu demonstrated better performance than those filled with Al despite having a lower hardness. This finding demonstrated that the tensile strength, wear loss, and hardness of the material all decreased steadily with increasing filler content, whereas the compressive strength, friction coefficient, and hardness all showed an increase. However, the impact of the welding direction on the surface of the moulded components is yet to be determined.

Table 5. Research on epoxy materials as mould inserts for RT.

	Researchers	Epoxy Resin/Hardener	Particles/Fillers Used	Weight Percentage of Filler (wt. %)	Particle Size	Arithmetic Mean Roughness (Ra) (μm)	Flexural Strength (MPa)	Hardness Test (R_H)	Thermal Conductivity (W/m·K)	Fatigue Test	Tensile Strength (MPa)	Compressive Strength (MPa)	Vickers Hardness (kgF/mm²)	Shore D Hardness Test	Density (g/cm³)	Thermal Diffusivity (mm²/s)	Surface Roughness
1.	Tomori et al. (2004) [110]	RP4037 (resin); RP4037 (hardener)	SiC	28.5; 34.7; 39.9	N/A	1.03 to 1.35	58.75 to 66.49	N/A	N/A	N/A	N/A	N/A	N/A	N/A	N/A	N/A	N/A
2.	Senthilkumar et al. (2012) [111]	Araldite LY 556 (resin)	Al	40; 45; 50; 55; 60	45–150 μm	N/A	N/A	69 to 89	3.97 to 5.39	15,786 to 734	N/A	N/A	N/A	N/A	N/A	N/A	N/A
3.	Srivastava and Verma (2015) [27]	PL411 (resin); PH-861 (hardener)	Cu; Al	1; 5; 8; 10	N/A	N/A	N/A	N/A	N/A	N/A	≈85 (pure epoxy)	Cu = 65 at 10 wt. %	Cu = 22.4 at 8 wt. %	N/A	N/A	N/A	N/A
4.	Fernandes et al. (2016) [26]	RenCast 436 (resin with Al filler); Ren HY 150 (hardener)	Al	21.4	N/A	N/A	N/A	N/A	N/A	N/A	Steel AISI P20 inserts = 20.0 ± 4.5; Epoxy resin/Al inserts = 22.0 ± 5.0	N/A	N/A	Steel AISI P20 inserts = 66 ± 3.2; Epoxy resin/Al inserts = 61 ± 1.6	N/A	N/A	N/A
5.	Khushairi et al. (2017) [112]	RenCast CW 47 (resin with Al filler); Ren HY 33 (hardener)	Brass; Cu	10; 20; 30	N/A	N/A	N/A	N/A	Brass: 10% = 1.18, 20% = 1.21, 30% = 1.37; Cu: 10% = 1.66, 20% = 1.73, 30% = 1.87	N/A	N/A	Brass: 10% = 95.61, 20% = 93.23, 30% = 92.69; Cu: 10% = 80.83, 20% = 81.51, 30% = 73.17	N/A	N/A	Brass: 10% = 1.85, 20% = 2.01, 30% = 2.22; Cu: 10% = 1.83, 20% = 1.96, 30% = 2.08	Brass: 10% = 0.644, 20% = 0.657, 30% = 0.740; Cu: 10% = 0.837, 20% = 0.923, 30% = 1.112	N/A
6.	Kuo and Lin (2019) [113]	TE-375 (Al filled epoxy resin)	N/A	N/A	N/A	N/A	N/A	N/A	N/A	N/A	N/A	N/A	N/A	N/A	N/A	N/A	Average microgroove depth of Al-filled epoxy resin was 90.5%; Average microgroove width of Al-filled epoxy resin was 98.9%

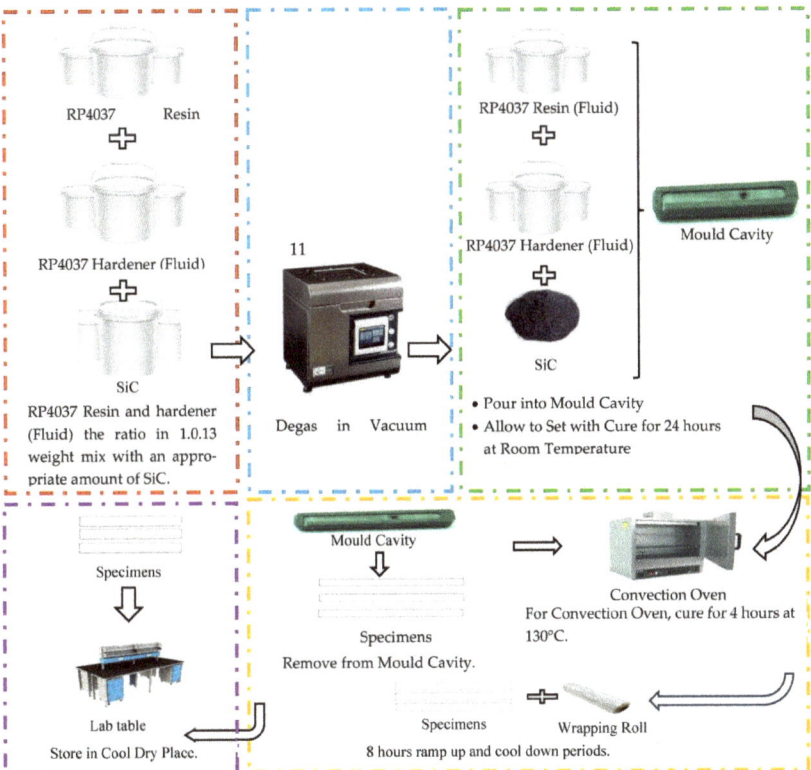

Figure 5. Method for setting up a tooling board.

Fernandes et al. [26] studied the dimensions and mechanical characteristics of epoxy resin/Al insert-moulded PP injection components for RT. A 140 mm diameter sphere was made up of five chambers with 2 mm thick walls that formed the work's central geometrical component. The length of the test was 60 mm, the diameter of the entrance was 6.5 mm, and the draught angle was 2°. To test the suggested mould, a novel hybrid mould comprising epoxy resin and Al was employed in this work to insert polypropylene (PP) pieces. In addition, comparable pieces were inserted to use an AISI P20 (conventional) steel mould, the same as in the genuine application. Epoxy resin/Al insert-filled components had slightly higher tensile strength at yield (22.0 ± 5.0 MPa) than steel AISI P20 insert-filled components (20.0 ± 4.5 MPa), but the difference was not statistically significant. Epoxy resin/Al-injected parts had lower values for ultimate tensile strength, elongation at break, and modulus of elasticity than steel AISI P20-injected parts. Furthermore, the Shore D hardness of objects formed by AISI P20 steel inserts increased by 8.5% in comparison to goods moulded by epoxy/Al inserts. When compared to components injected using an epoxy/Al mould, those injected using an AISI P20 steel mould showed less shrinkage. Based on these findings, epoxy/Al moulding blocks may be a high-quality alternative to fast tooling for producing single units or small series. Furthermore, this research did not investigate whether the orientation of welding on the moulded components was affected by the impact.

Khushairi et al. [112] investigated various epoxy compositions using Al, Cu, and brass fillers which were tested for their mechanical and thermal properties. In Al-filled epoxy, different combinations of brass and Cu filler (10, 20, and 30% wt. %) were used. Brass and Cu densities were 2.22 g/cm^3 and 2.08 g/cm^3 at the optimum filler content, respectively. When 30% Cu fillers were added to an epoxy matrix, the total thermal diffusivity (1.12 mm^2/s) and thermal conductivity (1.87 W/mK) were the maximum, but adding brass had no effect on thermal properties. When 20% brass filler was added, compressive strength increased from 76.8 MPa to 93.2 MPa, whereas 10% Cu filler raised compressive strength from 76.8 MPa to 80.8 MPa. As a result of porosity, multiple metal fillers diminished the compressive strength. According to this research, fillers boost mechanical, thermal, and density properties of Al-filled epoxy. Nonetheless, a careful evaluation of the surface characteristics, notably the welding line of the moulded components, is necessary to determine the moulded parts' quality.

Kuo and Lin [113] examined the quick injection moulding of Fresnel lenses from liquid silicone rubber. The experiment was conducted utilising RT and liquid silicone rubber (LSR) parts to build a horizontal LSR moulding machine (Allrounder 370S 700–290, ARBURG, Loßburg, Germany). Injection moulds for LSR injection moulding could be manufactured using Al-filled epoxy resin. The total microgroove depth and width of the Al-filled epoxy resin mould were 90.5% and 98.9%, respectively. LSR-moulded components exhibited typical microgroove depth and width transcription rates of roughly 91.5% and 99.2%, respectively. LSR-moulded components' microgroove depth as well as width may be modified to within 1 m. The mean surface polish of the Al-filled epoxy resins increased by around 12.5 nm following 200 LSR injector test cycles. However, further testing on tensile strength, compressive strength, hardness, and density, as well as weld line observations, is essential to understand the impact of quick injection moulding on the recommended mould in terms of moulded component quality.

From the review that has been carried out, it can be seen that numerous elements such as flexural strength, hardness, thermal conductivity, tensile strength, compressive strength, density, thermal diffusivity, and surface roughness of the new material introduced are important factors that need to be considered prior to its use as mould inserts for RT in the injection moulding process.

2.4. Geopolymer

A geopolymer is formed by combining a dry solid containing high aluminosilicate content, called a precursor, with alkaline solution and other ingredients if needed [114]. It is a semicrystalline, three-dimensional structure made of the tetrahedral structures of silica and alumina that share oxygen [115]. Geopolymer precursor can be obtained in two ways: from geological origin or industrial by-products. Examples of geological origins are kaolinite and clay, while industrial by-products are fly ash (FA), wheat straw ash, and furnace ash. Geopolymers are activated using high-alkali solution for the polymeric reaction to occur by using sodium hydroxide (NaOH), potassium hydroxide (KOH), or a mixture of sodium oxide (N_2O) and silicon monoxide (SiO) [116].

The geopolymer concrete curing process has a significant impact on mechanical characteristics and microstructure development [117,118]. Excellent mechanical strength, reduced creep, improved acid resistance, and minimal danger of shrinkage are all characteristics of geopolymer concrete [41,119,120]. The durability of waste pozzolan-based geopolymer concrete that is cured at high temperatures has been extensively studied [121–124]. By curing the geopolymer at a higher temperature, one may enhance the geopolymer's mechanical properties, polymerisation level, microstructure density, and overall strength [117,125–127].

Geopolymers come in a variety of unique shapes, and each type has certain properties. Geopolymers are an alternative material in the tooling industry. However, changing the geopolymer composition will change the qualities of the geopolymer, where selecting the correct geopolymer precursor will give the tooling industry greater advantages.

2.4.1. Effect of Different Geopolymer Precursors on Mechanical Properties of Geopolymer

Concrete for building uses geopolymer because of its great compressive strength. Its mechanical qualities, however, can vary depending on the type of geopolymer used [128–131]. Previous studies employing various geopolymer precursors are presented in Table 6.

Table 6. Research on the effects of different geopolymer compositions on mechanical properties.

No.	Researchers	Curing Days	Curing Temperature	Material Composition	Mechanical Properties	Result
1.	Girish et al. (2017) [131]	• 7, 14, 28	• 60 °C	• NaOH solution from 8 M to 14 M	• Compressive strength	• The greatest strength attained was 62.15 MPa at 28 days. • Compressive strength ratings suggest an increase in the strength of all combinations. • At 28 days, the compressive strength of the cement concrete surpassed the stiff pavement's minimum compressive strength requirement (40 MPa).
2.	Girish et al. (2018) [132]	• 7, 28, 56	• 30 °C	• SiO_2/Al_2O_3 ratio of 3.0–3.8 • Na_2O/Al_2O_3 ratio of 1	• Compressive strength • Flexural strength • Split tensile strength • Modulus of elasticity • Flexural strength of beams sliced from slab	• The highest strength achieved was 71.78 MPa after ambient curing at 56 days. • Compressive strength values indicate an increase in the strength of all mixes. • At 28 days, the compressive strength of the cement concrete exceeded the rigid pavement's minimum compressive strength requirement (40 MPa).
3.	Izzati et al. (2020) [133]	• 3	• FA and slag at 27 °C • Kaolin at 80 °C	• 1.0 wt. % of either FA, kaolin, slag geopolymer particles in Sn-0.7Cu	• Hardness	• Slag geopolymer in SnCu solder paste impacts on the microhardness values. • Slag geopolymer particles enhanced hardness by up to 7.84 Hv.
4.	Hussein and Fawzi (2021) [134]	• 7, 28	• 40 °C	• Cement: fine agg.: coarse agg. with 0% • 5% copper fibre and fly ash and slag: fine agg. • Coarse agg. with 5% copper fibre	• Compressive strength • Splitting tensile strength • Flexural strength	• The greatest improvement in compressive strength, splitting tensile strength, and flexural strength. • Copper wire fibre increases splitting tensile strength and flexural strength, and when the age of the concreate increases, the MPa increases.
5.	Hussein and Fawzi (2021) [135]	• 2	• 40 °C	• MR0 and MR1 cement: fine aggregate • MR2, MR3, MR4—fly ash in slag at 0.75:0.25, 0.65:0.35, and 0.55:0.45	• Compressive strength • Splitting tensile strength • Flexural strength	• MR1 has the greatest preliminary compressive strength. • Geopolymer mix MR4 has the highest mechanical properties. • In splitting tensile strength and bending strength tests, fibre addition produces better results.

Girish et al. [131] investigated the feasibility of employing geopolymer concrete as fine aggregate in stiff paving-grade concrete comprising quarry dust and sand. The 60/40 mixture consisted of fly ash and ground granulated blast furnace slag (GGBS), had different solid–liquid ratios, and was examined at 3, 14, and 28 days. Increasing the molar ratio of the NaOH solution from 8M to 14M increased the strength of the resulting concrete but reduced the solution's workability. The experiment used a 12M NaOH solution, and the fine aggregates included both quarry dust and sand. The maximum strength was 62.15 MPa, and it was reached after 28 days. The results of the compressive strength test as depicted in Figure 6 showed that the strength of all the mixtures had increased. The achieved compressive strength at 28 days was more than the 40 MPa minimum required for stiff pavement cement concrete. However, research needs to be undertaken to investigate whether the compressive strength of geopolymer concrete is affected by the substitution of quarry dust for sand.

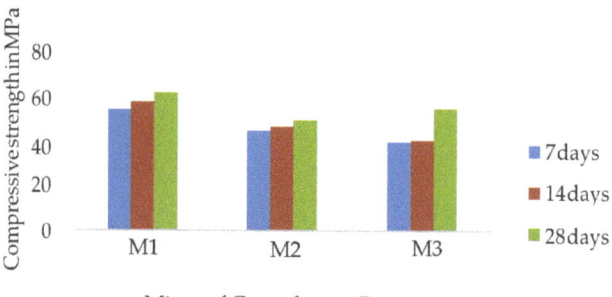

Figure 6. Compressive strength of geopolymer concrete with different mixture compositions [131].

Girish et al. [132] investigated self-consolidating geopolymer concrete for fixed-form pavement. Optimal strength geopolymer concrete is produced with a SiO_2/Al_2O_3 ratio between 3.0 and 3.8 and a Na_2O/Al_2O_3 ratio of 1. Compressive strength of 40 MPa was targeted for this mixture, which also included class F fly ash, ground blast furnace slag (GGBS), NaOH particles and solution form (molar concentration: 10 and 12), Na_2SiO_3 (A-53 grade), fine aggregate (quarry dust and river sand), coarse aggregate (below −20 mm), retarder (Conplast SP500), sugar solution, and water. The average compressive strength of the ambient-cured M10 mix after 28 days was 56.47 MPa, which is 40% higher than the intended compressive strength. At day 56, the compressive strength had increased to a peak of 71.78 MPa. However, as highlighted in Table 7, the proposed combination lacks considerable green strength, which is essential for slip-form paving applications, due to its low viscosity and yield stress. To make the SGC more environmentally friendly and appropriate for sliding mould applications, it might be beneficial to include nanoclays and/or fibres in the material.

Table 7. Hardened properties of M10 mix [132].

Curing Period in Days	Compressive Strength (MPA)	Flexural Strength (MPa)	Split Tensile Strength (MPa)	Modulus of Elasticity (MPa)	Flexural Strength of Beams Sliced From Slab
7	45.22	3.85	-	-	4.05
28	56.41	4.63	3.96	37,471.44	4.95
56	71.78	5.42	4.96	38,197.20	5.22

Izzati et al. [133] evaluated the use of different levels of geopolymer. No geopolymers, 1.0 wt. % fly ash, kaolin, or slag geopolymer particles were added to Sn-0.7Cu. All the mix designs were cured for 3 days and the temperature of curing for fly ash and slag was 27 °C and that for kaolin was 80 °C. As illustrated in Figure 7, using slag geopolymer is more challenging compared to not using geopolymer and using other geopolymers. Future research can attempt at using a higher percentage of geopolymer to test the composition's hardness. This may result in higher hardness compared to 1% geopolymer. To be comparable to other geopolymers, future research needs improve its preparation procedure in terms of curing temperature.

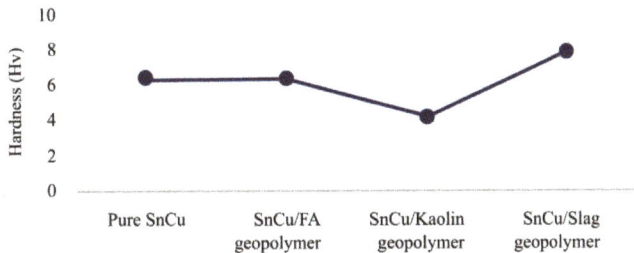

Figure 7. Different compositions of composite solder hardness value [133].

Hussein and Fawzi [134] tested various geopolymer contents in mix composition. The normal composition was cement with fine aggregate and coarse aggregate and 0% and 5% copper fibre, while the geopolymer composition had varied amounts of fly ash (FA) and slag with fine aggregate and coarse aggregate and 0% and 5% copper fibre. The preparation was cured at 40 °C for seven to twenty-eight days to evaluate compressive strength, splitting tensile strength, and bending strength. Figure 8 demonstrates that the maximum compressive strength, splitting tensile strength, and bending strength increase when the FA to ground granulated blast furnace slag (GGBFS) ratio is 0.55:0.45 with 0.5% copper wire fibre. It indicates that the compressive strength increases as the GGBFS level rises. The maximum strength of the geopolymer content can be determined by employing longer curing times and greater FA to GGBFS ratios.

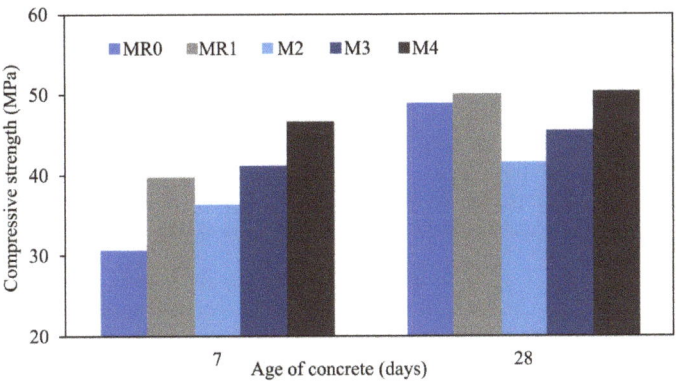

Figure 8. Compressive strength of geopolymer with different mixtures [134].

Hussein and Fawzi [135] analysed different contents of geopolymer by using different ratios of fly ash (FA) to ground granulated blast furnace slag (GGBFS). Cement, fine aggregate, and coarse aggregate were used in the preparation of MR0 and MR1, while fly ash to slag ratios for MG0, MG1, MG2, and MG3 were 0.75:0.25, 0.65:0.35, and 0.55:0.45 and mixed with fine aggregate and coarse aggregate in MR1, MG1, MG2, and MG3 with 0.5% copper fibre added. The preparation was cured at 40 °C for seven and twenty-eight days. As depicted in Figure 9, the larger the proportion of GGBFS, the greater the compressive strength and, at ninety days, 45% GGBFS had the highest compressive strength. MG3 with a content of 45% GGBFS shows the highest split tensile strength and flexural strength. To determine the ideal fly ash to slag ratio for assessing hardness, an analysis with a higher fly ash to slag ratio could be carried out.

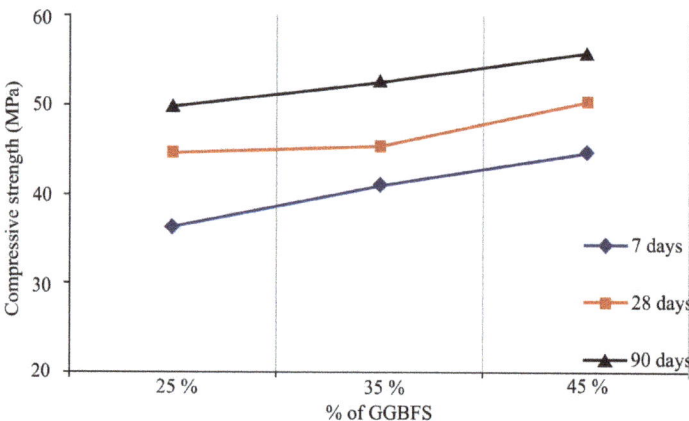

Figure 9. Different content percentages of GGBFS show different compressive strengths [135].

According to the review, mechanical qualities can be improved by utilising slag geopolymer. Research is necessary to determine whether a particular geopolymer can enhance mechanical properties. Furthermore, according to the studies mentioned, there are several preparations that would affect the strength, therefore the sample preparation procedure should be fixed, such as curing at the same temperature, to ensure that the results are unaffected. Varied drying times will result in different compressive strengths.

The mechanical characteristics of geosynthetics are affected by several geosynthetic precursors. The strength of geosynthetic polymers is improved by using varied ratios of sodium silicate/sodium hydroxide and fly ash/alkaline activators.

2.4.2. Effect of Different Ratios of Sodium Silicate/Sodium Hydroxide and Fly Ash/Alkaline Activators on the Mechanical Properties of Geopolymer

The current investigation looks into the influence of sodium silicate/sodium hydroxide ratios on geopolymer feasibility. Different studies showing the various proportions of sodium silicate/sodium hydroxide and fly ash/alkaline activator to improve geopolymer properties are listed in Table 8 [27,111]. The ideal preparation of fly ash can be determined by testing varying concentrations of sodium silicate, sodium hydroxide, fly ash, and alkaline activator.

Morsy et al. [136] evaluated the influence of sodium silicate/sodium hydroxide ratios on the viability of fly ash-based geopolymer synthesis at 80 °C. In this study, 10 M NaOH was combined with Na_2SiO_3 and alkaline activator ratios of 0.5, 1.0, 1.5, 2.0, and 2.5. The compressive strength of fly ash geopolymer mortars increased with age at 3, 7, 28, and 60 days. The compressive strengths of fly ash geopolymer mortars M1, M2, M3, M4, and M5 after three days were 34.7, 61.6, 40.4, 40.5, and 22.3 MPa, respectively. The S/N ratio of alkali activator had a significant impact on the strength of low-calcium fly ash geopolymer cured at 80 °C. Maximum strength was achieved when the ratio of sodium silicate to sodium hydroxide (S/N) was equal to 1. Other than that, future research should investigate preparation methods for mixtures and ensuring homogeneity so that they are comparable with other geopolymers.

Table 8. Research on the effects of various proportions of sodium silicate/sodium hydroxide and fly ash/alkaline activators on the mechanical properties.

No.	Researchers	The Ratio of Sodium Silicate/Sodium Hydroxide	The Ratio of Fly Ash/Alkaline Activator	Curing Temperature and Days	Mechanical Properties	Result
1.	Morsy et al. (2014) [136]	• 0.5, 1.0, 1.5, 2.0, 2.5	• 2.5	• 80 °C	• Compressive strength • Flexural strength	• Curing time has a direct correlation with the increase in compressive and flexural strength.
2.	Liyana et al. (2014) [137]	• 1.0, 1.5, 2.0, 2.5	• 1.0, 1.5, 2.0, 2.5	• Room temperature for 24 h	• Flexural strength	• Fly ash/alkaline activator yielded the highest flexural strength at ratio 2.0. • Maximum flexural strength is achieved with a 2.5 sodium silicate to sodium hydroxide ratio.
3.	Bakri et al. (2011) [138]	• 0.5, 1.0, 1.5, 2.0, 2.5, 3.0	• 1.5, 2.0, 2.5	• 70 °C for 24 h	• Compressive strength	• When combined with sodium silicate and sodium hydroxide, fly ash and alkaline activator may boost concrete's compressive strength.
4.	Nis (2019) [139]	• 1, 1.5, 2, 2.5	• 180	• Ambient curing 6 ± 4 °C for 26 days • Delayed oven-curing at 70 °C for 48 h	• Compressive strength	• As the ratio of alkali activators increased, the compressive strength of the specimens dropped at 14 M.
5.	Abdullah et al. (2021) [140]	• 0.5, 1.0, 1.5, 2.0, 2.5, 3.0	• 1.5, 2.0, 2.5	• 40–80 °C for 24 h	• Compressive strength	• The maximum compressive strength at 60 °C is achieved with a ratio of 2.0 fly ash to alkaline activator, and a ratio of 2.5 sodium silicate to sodium hydroxide.

According to Liyana et al. [137], in their study, the proportions of Na_2SiO_3/NaOH solution and fly ash to alkaline activator were synthesised in four different ratios: 1.0, 1.5, 2.0, and 2.5, in a 24 h period during which curing was carried out at room temperature. According to the results, the fly ash/alkaline activator ratio of 2.0 had the highest results compared to other ratios, and the sodium silicate/sodium hydroxide ratio of 2.5 had the highest results compared to other ratios. The best mechanical properties can be obtained through research using various molarities and curing temperatures.

The study by Bakri et al. [138] used a 12 M concentration of NaOH and fly ash to alkaline activator ratios of 0.5, 1.0, 1.5, 2.0, 2.5, and 3.0. Only the three ratios of 1.5, 2.0, and 2.5 were employed. Due to the geopolymer paste's high workability, which makes it challenging to handle, the ratios of 0.5 and 1.0 could not be used, and the ratio 3.0 could not be used due to the paste's low workability. Five different ratios of Na_2SiO_3/NaOH were used: 0.5, 1.0, 1.5, 2.0, and 2.5. The sample was cured for 24 h at 70 °C before being tested for compressive strength on the seventh day. The fly ash/alkaline activator ratio of 2.0 and the sodium silicate/sodium hydroxide ratio of 2.5 had the maximum compressive strength. Future studies could examine various curing temperatures to achieve the best compressive strength.

Nis [139] investigated geopolymer content using various NaOH concentrations and sodium silicate to sodium hydroxide ratios. The sodium silicate (Na_2SiO_3) and sodium hydroxide (NaOH) solutions were used with four different sodium silicate to sodium hydroxide ratios (1, 1.5, 2, and 2.5) and three different molarities (6 M, 10 M, and 14 M) for alkali activation to evaluate the impact of these parameters on the compressive strength of the alkali-activated fly ash/slag concrete under ambient-curing (AC) and delayed oven-curing (OC) conditions. The specimens' compressive strengths varied greatly with molarity concentration; those with the greatest NaOH molarity (14 M) concentration had the greatest compressive strength, as depicted in Figure 10. Other than that, more research can consider investigating the impact of oven-curing conditions on compressive strength.

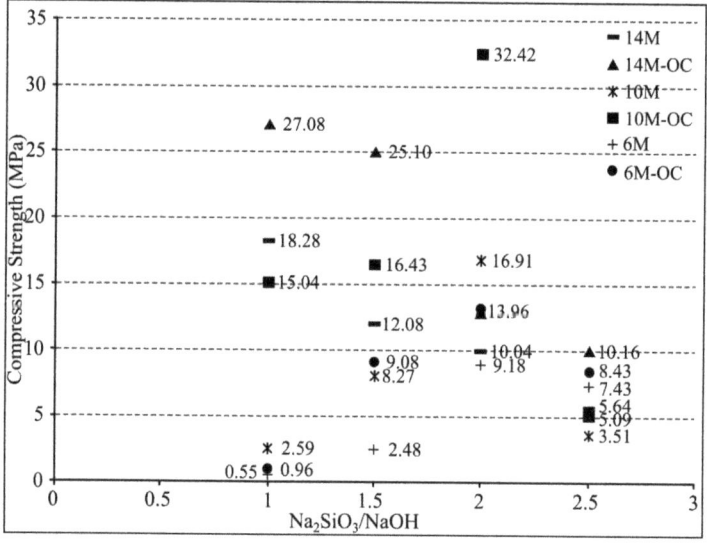

Figure 10. Compressive strengths of alkali-activated fly ash/slag specimens based on alkaline activator ratio type [139].

Abdullah et al. [140] investigated several curing temperatures with a constant NaOH concentration of 12 M using different fly ash/alkaline activator ratios and Na_2SiO_3/NaOH ratios. The samples were cured at different temperatures from 40 °C to 80 °C for 24 h and compressive strength was tested on the seventh day. The fly ash/alkaline activator ratio of 2.0, sodium silicate/sodium hydroxide ratio of 2.5, and curing temperature of 60 °C resulted in the maximum compressive strength. Different curing days may be investigated in order to enhance compressive strength.

Based on various studies [137–141], the mechanical characteristics may be affected by the use of different ratios of sodium silicate, sodium hydroxide, and fly ash/alkaline activator. The strongest strength resulted from the fly ash/alkaline activator ratio of 2.0 and the sodium silicate/sodium hydroxide ratio of 2.5, which were used as the ideal ratio for sample preparation. However, from the previous investigation, more improvements can be made, which are optimising the curing temperature and curing durations as using various curing durations can potentially increase the geopolymer's mechanical qualities.

Although the mechanical characteristics are influenced by the ratio of sodium silicate/sodium hydroxide and fly ash/alkaline activator, the preparation of different molarities of sodium hydroxide is another key aspect that influences overall mechanical characteristics of geopolymer.

2.4.3. Effect of Sodium Hydroxide Molarity on the Mechanical Properties of Geopolymer

The molarity of the alkali activator, the curing temperature, the number of days, and other parameters all have an impact on the sample's characteristics during the creation of the geopolymer [27,111–113]. The different molarities of sodium hydroxide, that acts as an alkali activator and affects the mechanical properties of geopolymer, are listed in Table 9.

Table 9. Research on the effect of the molarity of NaOH on mechanical properties.

No.	Researchers	Curing Days	Curing Temperature	NaOH Molarity	Material Composition	Mechanical Properties	Result
1.	Bakri et al. (2011) [142]	• 1, 2, 3, 7	• 70°C	• 6 M • 8 M • 10 M • 12 M • 14 M • 16 M	• Fly ash • Sodium hydroxide • Sodium silicate	• Compressive Strength (MPa)	• 12 M shows the highest compressive strength reached on the seventh day. • The highest compressive strength was achieved on the third day of curing.
2.	Gum et al. (2013) [143]	• 1, 3, 7, 14, 28, 56, 91 (7 classes)	• Oven: 60 °C for 24 h • Air: 20 °C for 24 h	• 6 M • 9 M • 12 M	• Fly ash • Sodium hydroxide • Sodium silicate	• Compressive Strength (MPa)	• Compressive strength and early strength both seemed to improve with increased NaOH molarity, which was employed as the alkaline activator. • 9 M and 12 M NaOH increased the strength by 45 MPa and 46 MPa after curing for 56 days.
3.	Lee et al. (2013) [144]	• 3, 7, 14, 28, 56	• 15 °C, 28 °C	• 4 M • 6 M • 8 M	• Fly ash, slag • Sodium hydroxide • Sodium silicate • Water glass • Sand	• Compressive Strength (MPa)	• The molarity of NaOH was increased while alkali activator duration was decreased due to the amount of slag and water glass. • The amount of slag was increased 25% and 30% at 28 days while the amount of slag decreased after 56 days of curing due to crack evolution.
4.	Rathanasalam et al. (2020) [145]	• 3, 7, 28	• 60 °C	• 10 M • 12 M • 14 M	• 5%, 10%, and 15% UFGGBFS replaced fly ash, with crushed stone or copper slag	• Compressive Strength (MPa)	• 14 M NaOH concentration has the maximum compressive strength.
5.	Khan et al. (2021) [146]	• 28	• 40 °C, 50 °C, 60 °C, 70 °C	• 8 M • 10 M • 12 M • 14 M	• Fly ash • Copper slag • Crusher dust	• Compressive Strength (MPa)	• To achieve maximum strength, the SS/SH was maintained at 142.4, and the molarity of NaOH was maintained at 14.14 M.

Gum et al. [143] studied the impact of making geopolymer concrete with an alkaline activator on the compressive strength of mortars using fly ash as a binder and different curing temperatures and moles of sodium hydroxide. Fly ash was combined with a mixture of 6, 9, and 12 M NaOH, and the curing conditions were 60 °C in the oven and 20 °C outside for 7 classes of curing days. After the chemicals were mixed, it was poured into moulds with dimensions of 50 mm × 50 mm × 50 mm and measured for compressive strength according to ASTM C 109. An alkaline activator that used NaOH at a higher molarity demonstrated increased compressive strength. The compressive strength decreased as the SiO_2/Na_2O and Al_2O_3/Na_2O ratios increased. When the SiO_2/Na_2O ratio exceeded 8.01 and the Al_2O_3/Na_2O ratio exceeded 1.94, the strength decrease rate appeared to accelerate sharply at 28 days. Based on these findings, the strength at 28 days for series 1 appeared to have increased by more than 1.7 times at a NaOH molarity of 9 M when compared to a molarity of 6 M. However, the 9 M and 12 M results showed nearly identical strengths. This highlights the significance of the SiO_2/Al_2O_3, SiO_2/Na_2O, and Al_2O_3/Na_2O ratios in alkali-activated geopolymer based on fly ash. As SiO_2/Al_2O_3 was constant in this investigation, the values of 8.01 and 1.94 for SiO_2/Na_2O and Al_2O_2/Na_2O ratios yielded the best strength development. The use of NaOH and sodium silicate (SiO_2/Na_2O = 8) in a 1:1 ratio demonstrated that it is possible to activate the geopolymerisation of fly ash and create a significant increase in strength, with a compressive strength of around 47 MPa. The evaluations of the impacts of the SiO_2/Na_2O and Al_2O_3/Na_2O ratios on strength under equal SiO_2/Al_2O_3 ratios are illustrated in Figure 11. The requirement for high-strength concrete is over 40 MPa, demonstrating the possibility of employing fly ash as a cement substitute. Future research can evaluate whether increasing the molarity and pH of NaOH during the curing process will increase compressive strength, including multiple curing temperatures.

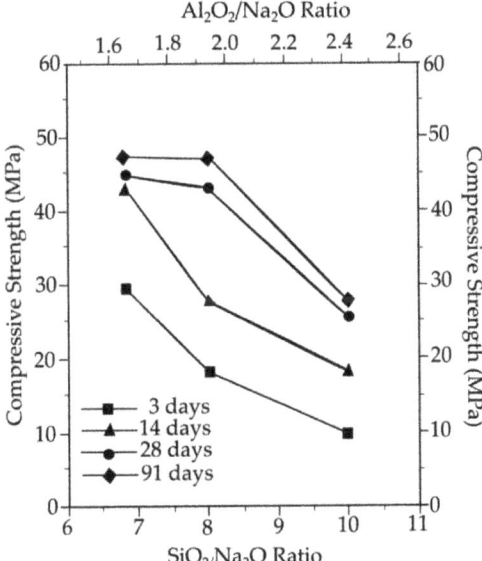

Figure 11. Compressive strength versus ages for molarity of NaOH [143].

Lee et al. [144] analysed the effects of increasing amounts of slag, water glass, and varying curing temperatures and NaOH molarities on curing time reduction. In the preparation, the alkali activators were water glass (Korean Industrial Standards, KS 3-grade; SiO_2 (29%), Na_2O (10%), H_2O (61%, specific gravity 1.38 g/mL), and 98% pure NaOH. The room temperature for the combined alkali-activated fly ash/slag paste was between 17 °C and 28 °C. For setting time tests, the molarity of NaOH was 4 M and 6 M, and the mass ratio of NaOH was 0.5, 1.0, and 1.5. Then, 8 M NaOH was used to accelerate the setting of alkali-activated fly ash/slag paste. For each mixed sample, a 100 mm × 200 mm cylinder mould was employed. The compressive strength and setting times of ASTM C 191-08 [139] were evaluated at 3, 7, 14, 28, and 56 days of curing. At 17 °C, the alkali-activated fly ash/slag paste took 55 min to start and 160 min to finish when the NaOH solution was 4 M and the water glass to NaOH solution by weight ratio was 0.5, as illustrated in Figure 12. Due to the presence of slag and water glass, the molarity of NaOH rose while the alkali activator's duration shortened. The quantity of slag grew by 25% and 30% after 28 days, respectively, but reduced after 56 days due to crack growth. Future research can examine different NaOH molarities to determine whether they can boost compressive strength.

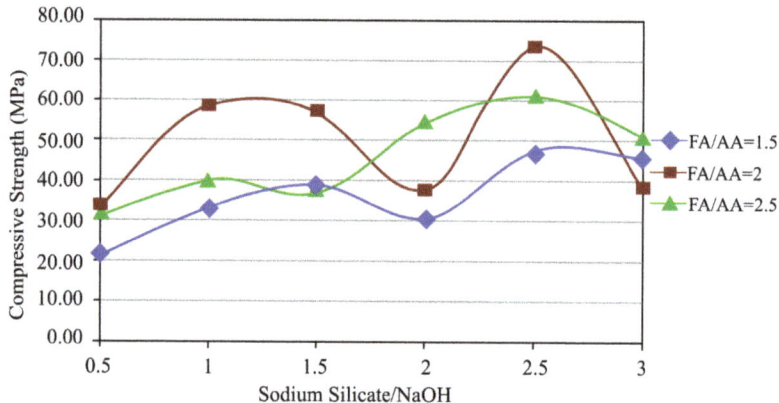

Figure 12. Time versus NaOH solution molarity [144].

Khan et al. [146] examined the material properties of fly ash, copper slag, and crusher dust at different curing temperatures and NaOH concentrations. There were 16 different mix designs that used varying curing temperatures and NaOH concentrations. The design was cured for 28 days before testing, and the analysis revealed that the sodium silicate/sodium hydroxide (SS/SH) ratio should be maintained at 2.4. The molarity of NaOH should be kept at 14 M to produce maximum strength and dotted line was an average region, as shown in Figure 13. The setting time was found to decrease from 449.8 min to 340.8 min. There are some limitations, such as the fact that the greater the molarity, the greater the compressive strength, and this could be due to secondary parameters that may affect the performance of geopolymer, including mixing time and other parameters that can influence the complexity of the mix design; therefore, additional research is required to determine their characteristics.

Rathanasalam et al. [145] investigated different sodium hydroxide (NaOH) molarities of 10 M, 12 M, and 14 M and developed a mixture utilising 5%, 10%, and 15% ultrafine ground granulated blast furnace slag (UFGGBFS) replacing fly ash, with crushed stone or copper slag. After curing for 3, 7, and 28 days at 60 °C, the compressive strength was evaluated. The compressive strength of all mix designs was tested using a 150 mm × 150 mm × 150 mm cube. From the different types of design with different curing days depicted in Figures 14–16, it can be concluded that all the mixtures with 14 M NaOH concentration have the maximum compressive strength. Future studies can look into using higher NaOH molarities to determine the ideal NaOH molarity to make the mix design with the maximum compressive strength.

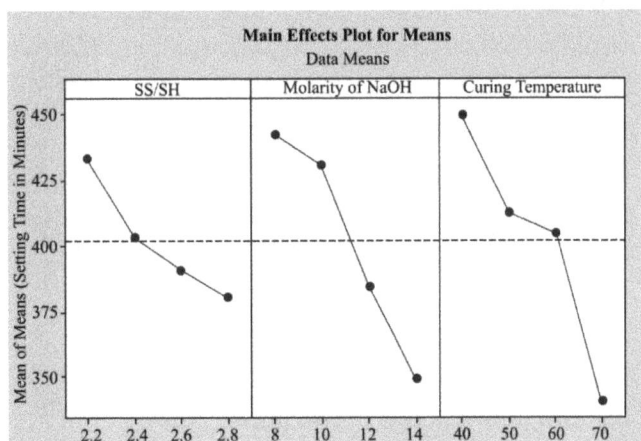

Figure 13. Main effects plot for S/N ratios based on compressive strength [146].

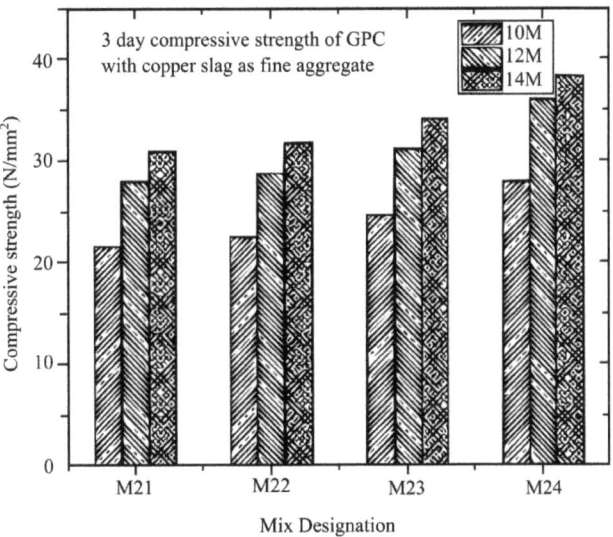

Figure 14. Different molarities in the mix design of GPC and copper slag at 3 days [145].

Bakri et al. [142] investigated the compressive strength of fly ash at various sodium hydroxide molarities. The sodium hydroxide molarities of 6, 8, 10, 12, 14, and 16 M and 1, 2, 3, and 7 curing days were used for the mix design samples. The proportion of fly ash to alkali activator was maintained constant at 2.50, as was the proportion of sodium silicate to sodium hydroxide. Prior to testing, all mixtures were cured at 70 °C, and the results indicated that for sodium hydroxide with molarity of 12 M, the compressive strength result was the highest among the other molarities on the third day, and on the seventh day, it demonstrated the highest compressive strength, as illustrated in Figure 17. Future research can examine whether increasing the curing temperature will increase compressive strength.

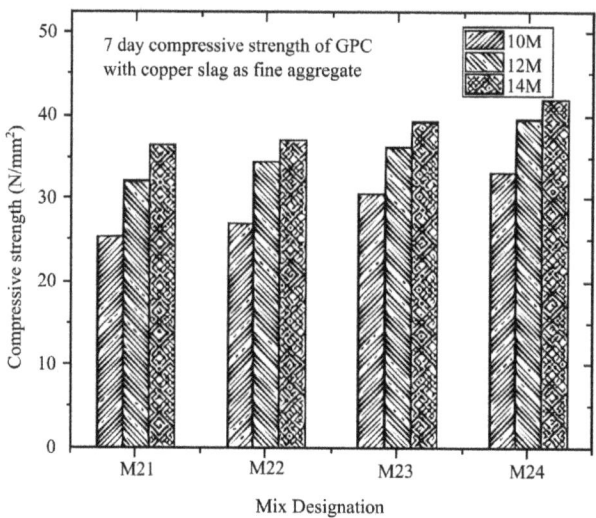

Figure 15. Different molarities in the mix design of GPC and copper slag at 7 days [145].

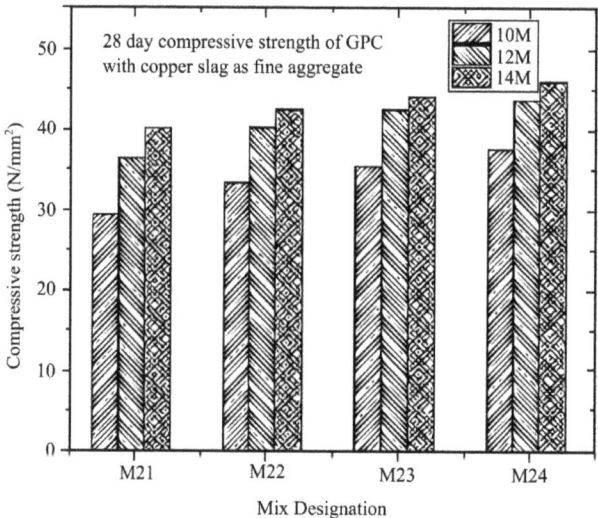

Figure 16. Different molarities in the mix design of GPC and copper slag at 28 days [145].

Previous studies have investigated the effect of molarity of sodium hydroxide on mechanical properties. According to the majority of the studies, the higher the sodium hydroxide molarity, the higher the mechanical characteristics of the geopolymer. Although lower compressive strength is seen in mix designs when sodium hydroxide molarity is 15 M, according to research by Khan et al. [146], this may not be due to the influence of sodium hydroxide. First, it might be affected by the addition of other materials such as copper slag and crusher dust, as well as other aspects that lower compressive strength such as the SS/SH ratio and curing temperature. Although increasing the molarity improves the mechanical properties of the geopolymer, research by Fakhrabadi et al. [147] shows that when the sodium hydroxide molarity is 15 M, unconfined compressive strength is lower than when the sodium hydroxide molarity is 11 M, while research by Bakri et al. [142]

suggested that molarity of 14 M is optimal for improving the mechanical properties of fly ash. The development of sodium aluminate silicate hydrate was caused by an increase in the molarity of sodium hydroxide (NASH) [148]. The use of sodium hydroxide with a high molarity may enhance the geopolymerisation reaction and the dissolution of initially solid materials, leading to better compressive strength [149].

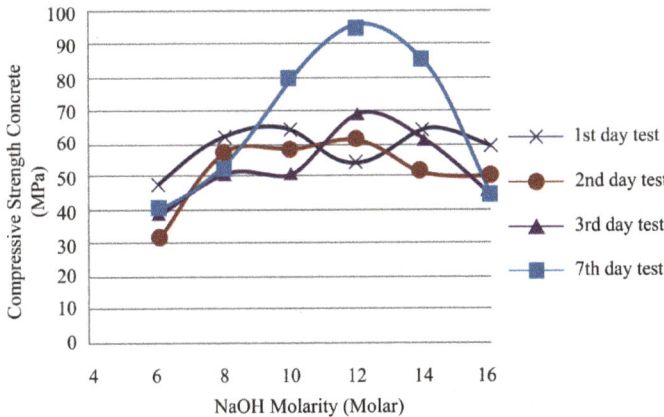

Figure 17. Compressive strength of all mixtures with different molarities and curing days [142].

The success of the geopolymer preparation demonstrates that it is possible to increase the material's strength through geopolymer preparation. However, the low thermal conductivity of geopolymer can be improved by adding metal filler to the mould insert.

3. Summary and Future Works

A combination of RP technique with production tooling helps carry out RT more quickly but faces dimensional accuracy and surface finish issues. Moreover, the injection moulding process faces an issue with cooling time where most mould inserts fabricated using RP techniques have very low thermal conductivity, thus increasing the cooling time, which will definitely affect the cycle time to produce the parts.

Many researchers have started to explore the use of metal epoxy composite (MEC) as mould inserts for RT in the injection moulding process by using pure metal filler particles. However, epoxy that can withstand high temperatures (>220 °C) is hard to find and costly. Therefore, there is a potential opportunity for epoxy to be replaced by geopolymer materials, especially fly ash as raw material. Geopolymer material can withstand temperatures up to 1000 °C. Similarly, the compressive strength of epoxy is 68.95 MPa (10,000 psi) as compared to geopolymer which has strength of 60–80 MPa (8700–11,600 psi). The challenges of using geopolymer material are similar to those of epoxy resin in that optimal strength, good accuracy, acceptable surface finish, and good thermal characteristics must be determined. Based on the gaps found from the literature, recommendations for future studies are as follows:

i. The mechanical and metallurgical properties of GGMC mould inserts should be evaluated to provide significant information and benefits to mould-making and rapid tooling industries.
ii. The size precision and surface integrity of the GGMC mould inserts after the casting process should be evaluated accordingly and compared to the GGMC mould inserts after machining in order to produce precision plastic product with a high-quality surface finish.

iii. To enhance the qualities of the outcomes, various geopolymers filled with scrap metal fillers should be mixed to increase thermal conductivity, or two or more kinds of filler materials can be added to improve thermal conductivity.

iv. The purpose of carrying out RT before production tooling for mass production is to evaluate the part performance and mostly requires modification of the mould inserts. Thus, an investigation on the effects of dimensional accuracy and surface quality in the machining process is definitely required.

This review has provided a clear reference for future development of mould inserts for RT using GGMC material. Thus, initiative needs to be taken to conduct an analysis on the effect of incorporating metal particles in geopolymer material as mould inserts for RT and its relationship with compressive strength and thermal conductivity. Moreover, the integration of metal scraps from machining with geopolymer formed from waste makes this research more interesting. GGMC material should be examined for metallurgical parameters such as corrosion rate, coefficient of expansion, surface roughness, and additive manufacturability. Furthermore, the machinability and the reliability of GGMC mould inserts should be explored and evaluated accordingly. At the end of this research, the discovery of new sustainable green material will benefit moulding and rapid prototyping industries, including with its environmentally friendly attributes.

Author Contributions: Conceptualisation, A.T.M.Y., S.Z.A.R., M.M.A.B.A., A.E.-h.A. and A.R.; data curation, A.T.M.Y., S.Z.A.R., M.M.A.B.A., M.N., A.E.-h.A. and A.R.; formal analysis, M.F.M.T. and A.M.T.; investigation, M.F.M.T. and A.M.T.; methodology, S.Z.A.R., M.M.A.B.A., A.E.-h.A. and A.R.; project administration, S.Z.A.R., M.M.A.B.A., A.R., M.N., A.E.-h.A. and A.M.T.; validation, A.T.M.Y., M.N., M.F.M.T. and A.M.T.; writing—review and editing, A.T.M.Y., S.Z.A.R., M.M.A.B.A., A.E.-h.A., A.R., M.F.M.T. and A.M.T. All authors have read and agreed to the published version of the manuscript.

Funding: This study was supported by the Center of Excellence Geopolymer and Green Technology (CEGeoGTech) UniMAP and Faculty of Technology Mechanical Engineering, UniMAP. The authors wish to thank the Ministry of Education, Malaysia, for their financial support of this study through the Fundamental Research Grant Scheme (FRGS), FRGS/1/2020/TK0/UNIMAP/03/19.

Institutional Review Board Statement: Not applicable.

Informed Consent Statement: Not applicable.

Data Availability Statement: Not applicable.

Acknowledgments: We would like to acknowledge the reviewers for the helpful advice and comments provided.

Conflicts of Interest: The authors declare no conflict of interest.

References

1. Shamsaei, E.; Bolt, O.; de Souza, F.B.; Benhelal, E.; Sagoe-Crentsil, K.; Sanjayan, J. Pathways to commercialisation for brown coal fly ash-based geopolymer concrete in Australia. *Sustainability* **2021**, *13*, 4350. [CrossRef]
2. Islam, A.; Alengaram, U.J.; Jumaat, M.Z.; Bashar, I.I.; Kabir, S.A. Engineering properties and carbon footprint of ground granulated blast-furnace slag-palm oil fuel ash-based structural geopolymer concrete. *Constr. Build. Mater.* **2015**, *101*, 503–521. [CrossRef]
3. Malenab, R.A.J.; Ngo, J.P.S.; Promentilla, M.A.B. Chemical treatment of waste abaca for natural fibre-reinforced geopolymer composite. *Materials* **2017**, *10*, 579. [CrossRef]
4. Costabile, G.; Fera, M.; Fruggiero, F.; Lambiase, A.; Pham, D. Cost models of additive manufacturing: A literature review. *Int. J. Ind. Eng. Comput.* **2016**, *8*, 263–282. [CrossRef]
5. Kuo, C.-C.; Li, M.-R. Development of sheet metal forming dies with excellent mechanical properties using additive manufacturing and rapid tooling technologies. *Int. J. Adv. Manuf. Technol.* **2017**, *90*, 21–25. [CrossRef]
6. Equbal, A.; Sood, A.K.; Shamim, M. Rapid tooling: A major shift in tooling practice. *Manuf. Ind. Eng.* **2015**, *14*, 3–4. [CrossRef]
7. Nee, A.Y.C. *Handbook of Manufacturing Engineering and Technology*; Springer: London, UK, 2015.
8. Huzaim, N.H.M.; Rahim, S.Z.A.; Musa, L.; Abdellah, A.E.-H.; Abdullah, M.M.A.B.; Rennie, A.; Rahman, R.; Garus, S.; Błoch, K.; Sandu, A.V.; et al. Potential of Rapid Tooling in Rapid Heat Cycle Moulding: A Review. *Materials* **2022**, *15*, 3725. [CrossRef]
9. Afonso, D.; Pires, L.; de Sousa, R.A.; Torcato, R. Direct rapid tooling for polymer processing using sheet metal tools. *Procedia Manuf.* **2017**, *13*, 102–108. [CrossRef]

10. Rayna, T.; Striukova, L. From rapid prototyping to home fabrication: How 3D printing is changing business model innovation. *Technol. Forecast. Soc. Change* **2016**, *102*, 214–224. [CrossRef]
11. Mellor, S.; Hao, L.; Zhang, D. Additive manufacturing: A framework for implementation. *Int. J. Prod. Econ.* **2014**, *149*, 194–201. [CrossRef]
12. Achillas, C.; Aidonis, D.; Iakovou, E.; Thymianidis, M.; Tzetzis, D. A methodological framework for the inclusion of modern additive manufacturing into the production portfolio of a focused factory. *J. Manuf. Syst.* **2015**, *37*, 328–339. [CrossRef]
13. Ciurana, J. Designing, prototyping and manufacturing medical devices: An overview. *Int. J. Comput. Integr. Manuf.* **2014**, *27*, 901–918. [CrossRef]
14. Faludi, J.; Bayley, C.; Bhogal, S.; Iribarne, M. Comparing environmental impacts of additive manufacturing vs traditional machining via life-cycle assessment. *Rapid Prototyp. J.* **2015**, *21*, 14–33. [CrossRef]
15. Tofail, S.A.M.; Koumoulos, E.P.; Bandyopadhyay, A.; Bose, S.; O'Donoghue, L.; Charitidis, C. Additive manufacturing: Scientific and technological challenges, market uptake and opportunities. *Mater. Today* **2018**, *21*, 22–37. [CrossRef]
16. Mendible, G.A.; Rulander, J.A.; Johnston, S.P. Comparative study of rapid and conventional tooling for plastics injection moulding. *Rapid Prototyp. J.* **2017**, *23*, 344–352. [CrossRef]
17. Altaf, K.; Rani, A.A.M.; Ahmad, F.; Baharom, M.; Raghavan, V.R. Determining the effects of thermal conductivity on epoxy moulds using profiled cooling channels with metal inserts. *J. Mech. Sci. Technol.* **2016**, *30*, 4901–4907. [CrossRef]
18. Altaf, K.; Qayyum, J.A.; Rani, A.M.A.; Ahmad, F.; Megat-Yusoff, P.S.M.; Baharom, M.; Aziz, A.R.A.; Jahanzaib, M.; German, R.M. Performance analysis of enhanced 3D printed polymer moulds for metal injection moulding process. *Metals* **2018**, *8*, 433. [CrossRef]
19. Jain, P.; Kuthe, A. Feasibility study of manufacturing using rapid prototyping: FDM approach. *Procedia Eng.* **2013**, *63*, 4–11. [CrossRef]
20. Thomas, P.A.; Aahlada, P.K.; Kiran, N.S.; Ivvala, J. A Review on Transition in the Manufacturing of Mechanical Components from Conventional Techniques to Rapid Casting Using Rapid Prototyping. *Mater. Today Proc.* **2018**, *5*, 11990–12002. [CrossRef]
21. Krizsma, S.; Kovács, N.; Kovács, J.; Suplicz, A. In-situ monitoring of deformation in rapid prototyped injection moulds. *Addit. Manuf.* **2021**, *42*, 102001. [CrossRef]
22. Török, D.; Zink, B.; Ageyeva, T.; Hatos, I.; Zobač, M.; Fekete, I.; Boros, R.; Hargitai, H.; Kovács, J.G. Laser powder bed fusion and casting for an advanced hybrid prototype mould. *J. Manuf. Process.* **2022**, *81*, 748–758. [CrossRef]
23. Krizsma, S.; Suplicz, A. Comprehensive in-mould state monitoring of Material Jetting additively manufactured and machined aluminium injection moulds. *J. Manuf. Process.* **2022**, *84*, 1298–1309. [CrossRef]
24. Zakrzewski, T.; Kozak, J.; Witt, M.; Dębowska-Wąsak, M. Dimensional analysis of the effect of SLM parameters on surface roughness and material density. *Procedia CIRP* **2020**, *95*, 115–120. [CrossRef]
25. Sarkar, P.; Modak, N.; Sahoo, P. Mechanical and Tribological Characteristics of Aluminium Powder filled Glass Epoxy Composites. *Mater. Today Proc.* **2018**, *5*, 5496–5505. [CrossRef]
26. De Carvalho Fernandes, A.; De Souza, A.F.; Howarth, J.L.L. Mechanical and dimensional characterisation of polypropylene injection moulded parts in epoxy resin/aluminium inserts for rapid tooling. *Int. J. Mater. Prod. Technol.* **2016**, *52*, 37–52. [CrossRef]
27. Srivastava, V.K.; Verma, A. Mechanical Behaviour of Copper and Aluminium Particles Reinforced Epoxy Resin Composites. *Am. J. Mater. Sci.* **2015**, *5*, 84–89. [CrossRef]
28. Kalami, H.; Urbanic, R.J. Design and fabrication of a low-volume, high-temperature injection mould leveraging a 'rapid tooling' approach. *Int. J. Adv. Manuf. Technol.* **2019**, *105*, 3797–3813. [CrossRef]
29. Tan, J.H.; Wong, W.L.E.; Dalgarno, K.W. An overview of powder granulometry on feedstock and part performance in the selective laser melting process. *Addit. Manuf.* **2017**, *18*, 228–255. [CrossRef]
30. Enayati, M.H.; Bafandeh, M.R.; Nosohian, S. Ball milling of stainless steel scrap chips to produce nanocrystalline powder. *J. Mater. Sci.* **2007**, *42*, 2844–2848. [CrossRef]
31. Radhwan, H.; Sharif, S.; Shayfull, Z.; Suhaimi, M.A.; l-hadj, A.; Khushairi, M.T.M. Thermal-Transient Analysis for Cooling Time on New Formulation of Metal Epoxy Composite (MEC) as Mould Inserts. *Arab. J. Sci. Eng.* **2021**, *46*, 7483–7494. [CrossRef]
32. Kim, K.; Kim, K.; Kim, M. Characterization of municipal solid-waste incinerator fly ash, vitrified using only end-waste glass. *J. Clean Prod.* **2021**, *318*, 128557. [CrossRef]
33. Jian, C.Y. The role of green manufacturing in reducing carbon dioxide emissions. In Proceedings of the 2013 5th Conference on Measuring Technology and Mechatronics Automation, ICMTMA, Hong Kong, China, 16–17 January 2013; Volume 2013, pp. 1223–1226. [CrossRef]
34. Gilmer, E.L.; Miller, D.; Chatham, C.A.; Zawaski, C.; Fallon, J.J.; Pekkanen, A.; Long, T.E.; Williams, C.B.; Bortner, M.J. Model analysis of feedstock behavior in fused filament fabrication: Enabling rapid materials screening. *Polymer* **2018**, *152*, 51–61. [CrossRef]
35. Colangelo, F.; Farina, I.; Travaglioni, M.; Salzano, C.; Cioffi, R.; Petrillo, A. Eco-efficient industrial waste recycling for the manufacturing of fibre reinforced innovative geopolymer mortars: Integrated waste management and green product development through LCA. *J. Clean. Prod.* **2021**, *312*, 127777. [CrossRef]
36. Ahmad, A.; Leman, Z.; Azmir, M.; Muhamad, K.; Harun, W.; Juliawati, A.; Alias, A. Optimisation of warpage defect in in-jection moulding process using ABS material. In Proceedings of the 2009 3rd Asia International Conference on Modelling and Simulation, AMS, Bandung, Indonesia, 25–29 May 2009; Volume 2009, pp. 470–474. [CrossRef]

37. Bajpai, R.; Choudhary, K.; Srivastava, A.; Sangwan, K.S.; Singh, M. Environmental impact assessment of fly ash and silica fume based geopolymer concrete. *J. Clean. Prod.* **2020**, *254*, 120147. [CrossRef]
38. Hussin, R.; Sharif, S.; Nabiałek, M.; Rahim, S.Z.A.; Khushairi, M.; Suhaimi, M.; Abdullah, M.; Hanid, M.; Wysłocki, J.; Błoch, K. Hybrid mould: Comparative study of rapid and hard tooling for injection moulding application using metal epoxy composite (MEC). *Materials* **2021**, *14*, 665. [CrossRef]
39. Altaf, K.; Rani, A.M.A.; Raghavan, V.R. Prototype production and experimental analysis for circular and profiled conformal cooling channels in aluminium filled epoxy injection mould tools. *Rapid Prototyp. J.* **2013**, *19*, 220–229. [CrossRef]
40. Bin Hussin, R.; Bin Sharif, S.; Rahim, S.Z.B.A.; Bin Suhaimi, M.A.; Khushairi, M.T.B.M.; El-Hadj, A.A.; Bin Shuaib, N.A. The potential of metal epoxy composite (MEC) as hybrid mould inserts in rapid tooling application: A review. *Rapid Prototyp. J.* **2021**, *27*, 1069–1100. [CrossRef]
41. Zulkifly, K.; Cheng-Yong, H.; Yun-Ming, L.; Abdullah, M.M.A.B.; Shee-Ween, O.; Bin Khalid, M.S. Effect of phosphate addition on room-temperature-cured fly ash-metakaolin blend geopolymers. *Constr. Build. Mater.* **2021**, *270*, 121486. [CrossRef]
42. Söderholm, P.; Ekvall, T. Metal markets and recycling policies: Impacts and challenges. *Miner. Econ.* **2020**, *33*, 257–272. [CrossRef]
43. Anderson, I.E.; White, E.M.; Dehoff, R. Feedstock powder processing research needs for additive manufacturing development. *Curr. Opin. Solid State Mater. Sci.* **2018**, *22*, 8–15. [CrossRef]
44. Cacace, S.; Demir, A.G.; Semeraro, Q. Densification Mechanism for Different Types of Stainless Steel Powders in Selective Laser Melting. *Procedia CIRP* **2017**, *62*, 475–480. [CrossRef]
45. Jacobson, L.A.; Mckittrick, J. A Review Journal Rapid solidification processing. *Mater. Sci. Eng. R Rep.* **1994**, *11*, 355–408. [CrossRef]
46. Fullenwider, B.; Kiani, P.; Schoenung, J.; Ma, K. Two-stage ball milling of recycled machining chips to create an alternative feedstock powder for metal additive manufacturing. *Powder Technol.* **2019**, *342*, 562–571. [CrossRef]
47. da Costa, C.E.; Zapata, W.C.; Parucker, M.L. Characterization of casting iron powder from recycled swarf. *J. Mater. Process. Technol.* **2003**, *143–144*, 138–143. [CrossRef]
48. Afshari, E.; Ghambari, M. Characterization of pre-alloyed tin bronze powder prepared by recycling machining chips using jet milling. *Mater. Des.* **2016**, *103*, 201–208. [CrossRef]
49. Witkin, D.; Lavernia, E. Synthesis and mechanical behavior of nanostructured materials via cryomilling. *Prog. Mater. Sci.* **2006**, *51*, 1–60. [CrossRef]
50. Ma, K.; Smith, T.; Lavernia, E.J.; Schoenung, J.M. Environmental Sustainability of Laser Metal Deposition: The Role of Feedstock Powder and Feedstock Utilization Factor. *Procedia Manuf.* **2017**, *7*, 198–204. [CrossRef]
51. Park, Y.; Abolmaali, A.; Kim, Y.H.; Ghahremannejad, M. Compressive strength of fly ash-based geopolymer concrete with crumb rubber partially replacing sand. *Constr. Build. Mater.* **2016**, *118*, 43–51. [CrossRef]
52. Martinho, P.G.; Pouzada, A.S. Alternative materials in moulding elements of hybrid moulds: Structural integrity and tribo-logical aspects. *Int. J. Adv. Manuf. Technol.* **2021**, *113*, 351–363. [CrossRef]
53. Skousen, J.; Ziemkiewicz, P.F.; Yang, J.E. Use of coal combustion by-products in mine reclamation:Review of case studies in the USA. *Geosystem Eng.* **2012**, *15*, 71–83. [CrossRef]
54. Li, J.; Zhuang, X.; Leiva, C.; Cornejo, A.; Font, O.; Querol, X.; Moeno, N.; Arenas, C.; Fernández-Pereira, C. Potential utilization of FGD gypsum and fly ash from a Chinese power plant for manufacturing fire-resistant panels. *Constr. Build. Mater.* **2015**, *95*, 910–921. [CrossRef]
55. Shrestha, P. Development of Geopolymer Concrete for Precast Structures. Master's Theses, University of Texas Arlington, Arlington, TX, USA, 12 March 2014.
56. Turan, C.; Javadi, A.A.; Vinai, R. Effects of Class C and Class F Fly Ash on Mechanical and Microstructural Behavior of Clay Soil—A Comparative Study. *Materials* **2022**, *15*, 1845. [CrossRef]
57. Sun, X.; Li, J.; Zhao, X.; Zhu, B.; Zhang, G. A Review on the Management of Municipal Solid Waste Fly Ash in American. *Procedia Environ. Sci.* **2016**, *31*, 535–540. [CrossRef]
58. Sugiyama, S.; Mera, T.; Yanagimoto, J. Recycling of minute metal scraps by semisolid processing: Manufacturing of design materials. *Trans. Nonferrous Met. Soc. China* **2010**, *20*, 1567–1571. [CrossRef]
59. Sharkawi, A.; Taman, M.; Afefy, H.M.; Hegazy, Y. Efficiency of geopolymer vs. high-strength grout as repairing material for reinforced cementitious elements. *Structures* **2020**, *27*, 330–342. [CrossRef]
60. Diaz-Loya, I.; Juenger, M.; Seraj, S.; Minkara, R. Extending supplementary cementitious material resources: Reclaimed and remediated fly ash and natural pozzolans. *Cem. Concr. Compos.* **2019**, *101*, 44–51. [CrossRef]
61. Evans, J.C.; Piuzzi, G.P.; Ruffing, D.G. Assessment of Key Properties of Solidified Fly Ash with and without Sodium Sulfate. *Grouting 2017* **2017**, 197–206. [CrossRef]
62. Bakshi, R.; Ghassemi, A. Injection Experiments on Basaltic Tuffs under Triaxial and Heated Conditions with Acoustic Emissions Monitoring. Rock Properties of Crystalline Basement and Control on Seismicity View Project Poroelasticity View Project. Master's Thesis, Mewbourne School of Petroleum and Geological Engineering, Norman, OK, USA, 26 June 2016.
63. Gorman, M.R.; Dzombak, D.A.; Frischmann, C. Potential global GHG emissions reduction from increased adoption of metals recycling. *Resour. Conserv. Recycl.* **2022**, *184*, 106424. [CrossRef]

64. Kurniawan, T.A.; Maiurova, A.; Kustikova, M.; Bykovskaia, E.; Othman, M.H.D.; Goh, H.H. Accelerating sustainability transition in St. Petersburg (Russia) through digitalization-based circular economy in waste recycling industry: A strategy to promote carbon neutrality in era of Industry 4.0. *J. Clean. Prod.* **2022**, *363*, 132452. [CrossRef]
65. Yamamoto, T.; Suzuki, M.; Waku, Y.; Tokuse, M. Fibre-Reinforced Metal Composite. U.S. Patent 4,980,242, 25 December 1990.
66. Behi, M.; Zedalis, M.; Schoonover, J.M. Rapid Manufacture of Metal and Ceramic Tooling. U.S. Patent 6,056,915, 2 May 2000.
67. Shaikh, F.Z.; Howard, T.; Blair, D.; Tsung-Yu, R. Low-Temperature, High-Strength Metal-Matrix Composite for Rapid-Prototyping and Rapid-Tooling. U.S. Patent 6,376,098 B1, 23 April 2002.
68. Amaya, H.E.; Hills, V. Method for the Rapid Fabrication of Mould Inserts. U.S. Patent 2002/0187065 A1, 12 December 2002.
69. Nematollahi, B.; Sanjayan, J. Geopolymer Composite and Geopolymer Matrix Composition. U.S. Patent WO2017070748 A1, 5 April 2017.
70. Lin, X.Q.; Li, J.F.; Zhang, T.; Huo, L.; Li, G.Y.; Zhang, N.; Liao, J.; Wang, B.H.; Ji, W.Z. 3D Printed Geopolymer Composite Material, Its Production and Applications. U.S. Patent CN106082898A, 31 July 2018.
71. Yang, Y.; Okonkwo, E.G.; Huang, G.; Xu, S.; Sun, W.; He, Y. On the sustainability of lithium ion battery industry—A review and perspective. *Energy Storage Mater.* **2021**, *36*, 186–212. [CrossRef]
72. Griffin, P.W.; Hammond, G.P.; Norman, J.B. Industrial energy use and carbon emissions reduction: A UK perspective. *Wiley Interdiscip. Rev. Energy Environ.* **2016**, *5*, 684–714. [CrossRef]
73. Allwood, J.M.; Azevedo, J.; Clare, A.; Cleaver, J.; Cullen, J.; Dunant, C.; Fellin, T.; Hawkins, W.; Horrocks, I.; Horton, P.; et al. Absolute Zero: Delivering the UK's Climate Change Commitment with Incremental Changes to Today's Technologies. Absolute Zero 2019. Available online: https://www.repository.cam.ac.uk/bitstream/handle/1810/299414/REP_Absolute_Zero_V3_20200505.pdf?sequence=9Allowed=y (accessed on 18 November 2022).
74. Vijayakumar, S.R.; Gajendran, S. Improvement of Overall Equipment Effectiveness (Oee). *Inject. Mould. Process Ind.* **2014**, *2*, 47–60.
75. Chen, W.-C.; Kurniawan, D. Process parameters optimisation for multiple quality characteristics in plastic injection moulding using Taguchi method, BPNN, GA, and hybrid PSO-GA. *Int. J. Precis. Eng. Manuf.* **2014**, *15*, 1583–1593. [CrossRef]
76. Singh, G.; Pradhan, M.; Verma, A. Multi Response optimisation of injection moulding Process parameters to reduce cycle time and warpage. *Mater. Today Proc.* **2018**, *5*, 8398–8405. [CrossRef]
77. Zhao, N.-Y.; Lian, J.-Y.; Wang, P.-F.; Xu, Z.-B. Recent progress in minimizing the warpage and shrinkage deformations by the optimisation of process parameters in plastic injection moulding: A review. *Int. J. Adv. Manuf. Technol.* **2022**, *120*, 85–101. [CrossRef]
78. Zhao, J.; Cheng, G.; Ruan, S.; Li, Z. Multi-objective optimisation design of injection moulding process parameters based on the improved efficient global optimisation algorithm and non-dominated sorting-based genetic algorithm. *Int. J. Adv. Manuf. Technol.* **2015**, *78*, 1813–1826. [CrossRef]
79. Kashyap, S.; Datta, D. Process parameter optimisation of plastic injection moulding: A review. *Int. J. Plast. Technol.* **2015**, *19*, 1–18. [CrossRef]
80. Mehat, N.M.; Kamaruddin, S. Multi-Response Optimisation of Injection Moulding Processing Parameters Using the Taguchi Method. *Polym. Plast. Technol. Eng.* **2011**, *50*, 1519–1526. [CrossRef]
81. Shi, F.; Lou, Z.; Zhang, Y.; Lu, J. Optimisation of Plastic Injection Moulding Process with Soft Computing. *Int. J. Adv. Manuf. Technol.* **2003**, *21*, 656–661. [CrossRef]
82. Dimla, D.E.; Camilotto, M.; Miani, F. Design and optimisation of conformal cooling channels in injection moulding tools. *J. Mater. Process. Technol.* **2005**, *164–165*, 1294–1300. [CrossRef]
83. Hatta, N.M. An Improved Grey Wolf Optimiser Sine Cosine Algorithm for Minimisation of Injection Moulding Shrinkage. Ph.D. Thesis, Universiti Teknologi Malaysia, Skudai, Malaysia, 2019.
84. Fu, J.; Ma, Y. A method to predict early-ejected plastic part air-cooling behavior towards quality mould design and less moulding cycle time. *Robot Comput. Integr. Manuf.* **2019**, *56*, 66–74. [CrossRef]
85. Annicchiarico, D.; Alcock, J.R. Review of factors that affect shrinkage of moulded part in injection moulding. *Mater. Manuf. Process.* **2014**, *29*, 662–682. [CrossRef]
86. Mancini, S.D.; Zanin, M. Recyclability of PET from virgin resin. *Mater. Res.* **1999**, *2*, 33–38. [CrossRef]
87. Abbasalizadeh, M.; Hasanzadeh, R.; Mohamadian, Z.; Azdast, T.; Rostami, M. Experimental study to optimize shrinkage behavior of semi-crystalline and amorphous thermoplastics. *Iran. J. Mater. Sci. Eng.* **2018**, *15*, 41–45. [CrossRef]
88. Chen, C.-P.; Chuang, M.-T.; Hsiao, Y.-H.; Yang, Y.-K.; Tsai, C.-H. Simulation and experimental study in determining injection moulding process parameters for thin-shell plastic parts via design of experiments analysis. *Expert Syst. Appl.* **2009**, *36*, 10752–10759. [CrossRef]
89. Tang, C.; Tan, J.; Wong, C. A numerical investigation on the physical mechanisms of single track defects in selective laser melting. *Int. J. Heat Mass Transf.* **2018**, *126*, 957–968. [CrossRef]
90. Shoemaker, J. *Mouldflow Design Guide: A Resource for Plastics Engineers*; Hanser: Munich, Germany, 2006; p. 409.
91. Ibrahim, A.A.; Khalil, T.; Tawfeek, T. Study the influence of a new ball burnishing technique on the surface roughness of AISI 1018 low carbon steel. *Int. J. Eng. Technol.* **2015**, *4*, 227. [CrossRef]
92. Selvaraj, S.; Venkataramaiah, P. Design and fabrication of an injection moulding tool for cam bush with baffle cooling channel and submarine gate. *Procedia Eng.* **2013**, *64*, 1310–1319. [CrossRef]
93. Højerslev, C.; Risø, F. *Tool Steels*; Risø National Laboratory: Roskilde, Denmark, 2001.

94. Sandanayake, M.; Gunasekara, C.; Law, D.; Zhang, G.; Setunge, S.; Wanijuru, D. Sustainable criterion selection framework for green building materials—An optimisation based study of fly-ash Geopolymer concrete. *Sustain. Mater. Technol.* **2020**, *25*, e00178. [CrossRef]
95. Rahmati, S.; Dickens, P. Rapid tooling analysis of Stereolithography injection mould tooling. *Int. J. Mach. Tools Manuf.* **2007**, *47*, 740–747. [CrossRef]
96. Beal, V.E.; Erasenthiran, P.; Ahrens, C.H.; Dickens, P. Evaluating the use of functionally graded materials inserts produced by selective laser melting on the injection moulding of plastics parts. *Proc. Inst. Mech. Eng. B J. Eng. Manuf.* **2007**, *221*, 945–954. [CrossRef]
97. Yadroitsev, I.; Krakhmalev, P.; Yadroitsava, I. Hierarchical design principles of selective laser melting for high quality metallic objects. *Addit. Manuf.* **2015**, *7*, 45–56. [CrossRef]
98. Reddy, K.P.; Panitapu, B. High thermal conductivity mould insert materials for cooling time reduction in thermoplastic injection moulds. *Mater. Today Proc.* **2017**, *4*, 519–526. [CrossRef]
99. Sanap, P.; Dharmadhikari, H.M.; Keche, A.J. Optimisation of Plastic Moulding by Reducing Warpage with the Application of Taguchi Optimisation Technique & Addition of Ribs in Washing Machine Wash Lid Component. *IOSR J. Mech. Civ. Eng.* **2016**, *13*, 61–68. [CrossRef]
100. Pontes, A.J.; Queirós, M.P.; Martinho, P.G.; Bártolo, P.J.; Pouzada, A.S. Experimental assessment of hybrid mould performance. *Int. J. Adv. Manuf. Technol.* **2010**, *50*, 441–448. [CrossRef]
101. Ferreira, J.; Mateus, A. Studies of rapid soft tooling with conformal cooling channels for plastic injection moulding. *J. Mater. Process. Technol.* **2003**, *142*, 508–516. [CrossRef]
102. Ong, H.; Chua, C.; Cheah, C. Rapid Moulding Using Epoxy Tooling Resin. *Int. J. Adv. Manuf. Technol.* **2002**, *20*, 368–374. [CrossRef]
103. Ma, S.; Gibson, I.; Balaji, G.; Hu, Q. Development of epoxy matrix composites for rapid tooling applications. *J. Mater. Process. Technol.* **2007**, *192–193*, 75–82. [CrossRef]
104. Pozzo, A.D.; Carabba, L.; Bignozzi, M.C.; Tugnoli, A. Life cycle assessment of a geopolymer mixture for fireproofing applica-tions. *Int. J. Life Cycle Assess.* **2019**, *24*, 1743–1757. [CrossRef]
105. Kuo, C.-C.; Tasi, Q.-Z.; Hunag, S.-H. Development of an Epoxy-Based Rapid Tool with Low Vulcanization Energy Con-sumption Channels for Liquid Silicone Rubber Injection Moulding. *Polymers* **2022**, *14*, 4534. [CrossRef]
106. Hopkinson, N.; Dickens, P. A comparison between stereolithography and aluminium injection moulding tooling. *Rapid Prototyp. J.* **2000**, *6*, 253–258. [CrossRef]
107. Ferreira, J.C. Manufacturing core-boxes for foundry with rapid tooling technology. *J. Mater. Process. Technol.* **2004**, *155–156*, 1118–1123. [CrossRef]
108. Polytechnica, P.; Eng, S.M. Influence of Mould Properties on the Quality of Injection Moulded Parts. *Period. Polytech. Mech. Eng.* **2005**, *49*, 115–122.
109. Rossi, S.; Deflorian, F.; Venturini, F. Improvement of surface finishing and corrosion resistance of prototypes produced by direct metal laser sintering. *J. Mater. Process. Technol.* **2004**, *148*, 301–309. [CrossRef]
110. Tomori, T.; Melkote, S.; Kotnis, M. Injection mould performance of machined ceramic filled epoxy tooling boards. *J. Mater. Process. Technol.* **2004**, *145*, 126–133. [CrossRef]
111. Senthilkumar, N.; Kalaichelvan, K.; Elangovan, K. Mechanical behaviour of aluminum particulate epoxy compo-site-experimental study and numerical simulation. *Int. J. Mech. Mater. Eng.* **2012**, *7*, 214–221.
112. Khushairi, M.T.M.; Sharif, S.; Jamaludin, K.R.; Mohruni, A.S. Effects of Metal Fillers on Properties of Epoxy for Rapid Tooling Inserts. *Int. J. Adv. Sci. Eng. Inf. Technol.* **2017**, *7*, 1155. [CrossRef]
113. Kuo, C.-C.; Lin, J.-X. Fabrication of the Fresnel lens with liquid silicone rubber using rapid injection mould. *Int. J. Adv. Manuf. Technol.* **2019**, *101*, 615–625. [CrossRef]
114. Burduhos Nergis, D.D.; Abdullah, M.M.A.B.; Vizureanu, P.; Faheem, M.T.M. Geopolymers and Their Uses: Review. *IOP Conf. Ser. Mater. Sci. Eng.* **2018**, *374*, 012019. [CrossRef]
115. Amritphale, S.S.; Bhardwaj, P.; Gupta, R. Advanced Geopolymerization Technology. In *Geopolymers and Other Geosynthetics*; IntechOpen: London, UK, 2019. [CrossRef]
116. Mehta, A.; Siddique, R. An overview of geopolymers derived from industrial by-products. *Constr. Build. Mater.* **2016**, *127*, 183–198. [CrossRef]
117. Mehta, A.; Siddique, R.; Ozbakkaloglu, T.; Shaikh, F.U.A.; Belarbi, R. Fly ash and ground granulated blast furnace slag-based alkali-activated concrete: Mechanical, transport and microstructural properties. *Constr. Build. Mater.* **2020**, *257*, 119548. [CrossRef]
118. Nguyen, T.T.; Goodier, C.I.; Austin, S.A. Factors affecting the slump and strength development of geopolymer concrete. *Constr. Build. Mater.* **2020**, *261*, 119945. [CrossRef]
119. Wardhono, A.; Gunasekara, C.; Law, D.W.; Setunge, S. Comparison of long term performance between alkali activated slag and fly ash geopolymer concretes. *Constr. Build. Mater.* **2017**, *143*, 272–279. [CrossRef]
120. Cao, V.D.; Bui, T.Q.; Kjøniksen, A.-L. Thermal analysis of multi-layer walls containing geopolymer concrete and phase change materials for building applications. *Energy* **2019**, *186*, 115792. [CrossRef]
121. Nurruddin, E.A. Methods of curing geopolymer concrete: A review. *Int. J. Adv. Appl. Sci.* **2018**, *5*, 31–36. [CrossRef]

122. Ridzuan, A.R.M.; Abdullah, M.M.A.B.; Arshad, M.F.; Tahir, M.F.M.; Khairulniza, A. The effect of NaOH concentration and curing condition to the strength and shrinkage performance of recycled geopolymer concrete. *Mater. Sci. Forum* **2015**, *803*, 194–200. [CrossRef]
123. Riahi, S.; Nazari, A. The effects of nanoparticles on early age compressive strength of ash-based geopolymers. *Ceram. Int.* **2012**, *38*, 4467–4476. [CrossRef]
124. Castel, A.; Foster, S.J. Bond strength between blended slag and Class F fly ash geopolymer concrete with steel reinforcement. *Cem. Concr. Res.* **2015**, *72*, 48–53. [CrossRef]
125. Deb, P.S.; Nath, P.; Sarker, P.K. The effects of ground granulated blast-furnace slag blending with fly ash and activator content on the workability and strength properties of geopolymer concrete cured at ambient temperature. *Mater. Des.* **2014**, *62*, 32–39. [CrossRef]
126. Krishnan, T.; Purushothaman, R. Optimisation and influence of parameter affecting the compressive strength of geopolymer concrete containing recycled concrete aggregate: Using full factorial design approach. *IOP Conf. Ser. Earth Environ. Sci.* **2017**, *80*, 1. [CrossRef]
127. Vora, P.R.; Dave, U.V. Parametric studies on compressive strength of geopolymer concrete. *Procedia Eng.* **2013**, *51*, 210–219. [CrossRef]
128. Dao, D.V.; Ly, H.-B.; Trinh, S.H.; Le, T.-T.; Pham, B.T. Artificial intelligence approaches for prediction of compressive strength of geopolymer concrete. *Materials* **2019**, *12*, 983. [CrossRef]
129. Aredes, F.; Campos, T.; Machado, J.; Sakane, K.; Thim, G.; Brunelli, D. Effect of cure temperature on the formation of metakaolinite-based geopolymer. *Ceram. Int.* **2015**, *41*, 7302–7311. [CrossRef]
130. Nguyen, K.T.; Nguyen, Q.D.; Le, T.A.; Shin, J.; Lee, K. Analyzing the compressive strength of green fly ash based geopolymer concrete using experiment and machine learning approaches. *Constr. Build. Mater.* **2020**, *247*, 118581. [CrossRef]
131. Girish, M.G.; Shetty, K.K.; Rao, A.R. Geopolymer Concrete an Eco-Friendly Alternative to Portland Cement Paving Grade Concrete. *Int. J. Civ. Eng. Technol.* **2017**, *8*, 886–892.
132. Girish, M.G.; Shetty, K.K.; Raja, A.R. Self-Consolidating Paving Grade Geopolymer Concrete. *IOP Conf. Ser. Mater. Sci. Eng.* **2018**, *431*, 092006. [CrossRef]
133. Izzati, Z.N.; Al-Bakri, A.M.; Salleh, M.A.A.M.; Ahmad, R.; Mortar, N.A.M.; Ramasamy, S. Microstructure and Mechanical Properties of Geopolymer Ceramic Reinforced Sn-0. 7Cu Solder. *IOP Conf. Ser. Mater. Sci. Eng.* **2020**, *864*, 012041. [CrossRef]
134. Hussein, S.S.; Fawzi, N.M. Behavior of Geopolymer Concrete Reinforced by Sustainable Copper Fibre. *IOP Conf. Ser. Earth Environ. Sci.* **2021**, *856*, 012022. [CrossRef]
135. Hussein, S.S.; Fawzi, N.M. Influence of Using Various Percentages of Slag on Mechanical Properties of Fly Ash-based Geo-polymer Concrete. *J. Eng.* **2021**, *27*, 50–67. [CrossRef]
136. Morsy, M.S.; Alsayed, S.H.; Al-Salloum, Y.; Almusallam, T.H. Effect of Sodium Silicate to Sodium Hydroxide Ratios on Strength and Microstructure of Fly Ash Geopolymer Binder. *Arab. J. Sci. Eng.* **2014**, *39*, 4333–4339. [CrossRef]
137. Liyana, J.; Al Bakri, A.M.M.; Hussin, K.; Ruzaidi, C.; Azura, A.R. Effect of fly ash/alkaline activator ratio and sodium silicate/NaOH ratio on fly ash geopolymer coating strength. *Key Eng. Mater.* **2014**, *594–595*, 146–150. [CrossRef]
138. Abdullah, M.M.A.B.; Kamarudin, H.; Nizar, I.K.; Bnhussain, M.; Zarina, Y.; Rafiza, A. Correlation between Na2SiO3/NaOH ratio and fly ash/alkaline activator ratio to the strength of geopolymer. *Adv. Mater. Res.* **2012**, *341–342*, 189–193. [CrossRef]
139. Niş, A. Compressive strength variation of alkali activated fly ash/slag concrete with different NaOH concentrations and so-dium silicate to sodium hydroxide ratios. *J. Sustain. Constr. Mater. Technol.* **2019**, *4*, 351–360. [CrossRef]
140. Abdullah, M.M.A.; Kamarudin, H.; Bnhussain, M.; Nizar, I.K.; Rafiza, A.; Zarina, Y. The relationship of NaOH molarity, Na2SiO3/NaOH ratio, fly ash/alkaline activator ratio, and curing temperature to the strength of fly ash-based geopolymer. *Adv. Mater. Res.* **2011**, *328–330*, 1475–1482. [CrossRef]
141. Das, S.K.; Shrivastava, S. Influence of molarity and alkali mixture ratio on ambient temperature cured waste cement concrete based geopolymer mortar. *Constr. Build. Mater.* **2021**, *301*, 124380. [CrossRef]
142. Parthasarathy, P.; Srinivasula, R.M.; Dinakar, P.; Rao, B.K.; Satpathy, B.N.; Mohanty, A. Effect of Na2 SiO3/NaOH Ratios and NaOH Molarities on Compressive Strength of Fly-Ash-Based Geopolymer. *Geo-Chicago 2016* **2012**, *109*, 503. [CrossRef]
143. Ryu, G.S.; Lee, Y.B.; Koh, K.T.; Chung, Y.S. The mechanical properties of fly ash-based geopolymer concrete with alkaline activators. *Constr. Build. Mater.* **2013**, *47*, 409–418. [CrossRef]
144. Lee, N.; Lee, H. Setting and mechanical properties of alkali-activated fly ash/slag concrete manufactured at room temperature. *Constr. Build. Mater.* **2013**, *47*, 1201–1209. [CrossRef]
145. Rathanasalam, V.; Perumalsami, J.; Jayakumar, K. Characteristics of blended geopolymer concrete using ultrafine ground granulated blast furnace slag and copper slag. *Ann. Chim. Sci. Des. Mater.* **2020**, *44*, 433–439. [CrossRef]
146. Rathanasalam, V.; Perumalsami, J.; Jayakumar, K. Design and development of sustainable geopolymer using industrial copper byproduct. *J. Clean. Prod.* **2021**, *278*, 123565. [CrossRef]
147. Fakhrabadi, A.; Ghadakpour, M.; Choobbasti, A.J.; Kutanaei, S.S. Evaluating the durability, microstructure and mechanical properties of a clayey-sandy soil stabilized with copper slag-based geopolymer against wetting-drying cycles. *Bull. Eng. Geol. Environ.* **2021**, *80*, 5031–5051. [CrossRef]

148. Saha, S.; Rajasekaran, C. Enhancement of the properties of fly ash based geopolymer paste by incorporating ground granulated blast furnace slag. *Constr. Build. Mater.* **2017**, *146*, 615–620. [CrossRef]
149. Temuujin, J.; Williams, R.; van Riessen, A. Effect of mechanical activation of fly ash on the properties of geopolymer cured at ambient temperature. *J. Mater. Process. Technol.* **2009**, *209*, 5276–5280. [CrossRef]

Disclaimer/Publisher's Note: The statements, opinions and data contained in all publications are solely those of the individual author(s) and contributor(s) and not of MDPI and/or the editor(s). MDPI and/or the editor(s) disclaim responsibility for any injury to people or property resulting from any ideas, methods, instructions or products referred to in the content.

Article

Experimental Study on Toughness of Engineered Cementitious Composites with Desert Sand

Zhishuan Lv [1], Yang Han [1,2,*], Guoqi Han [1], Xueyu Ge [1] and Hao Wang [1]

1. College of Civil Engineering, Kashi University, Kashi 844008, China; ksdxlzs@163.com (Z.L.); 2210577@tongji.edu.cn (G.H.); g15553669878@163.com (X.G.); 18399340386@163.com (H.W.)
2. Institute of Engineering Disaster Prevention and Mitigation, Henan University of Technology, Zhengzhou 450001, China
* Correspondence: hanyang@haut.edu.cn; Tel.: +86-186-2371-7360

Abstract: In this paper, engineered cementitious composites (ECCs) were prepared with desert sand instead of ordinary sand, and the toughness properties of the ECCs were studied. The particle size of the desert sand was 0.075–0.3 mm, which is defined as ultrafine sand. The ordinary sand was sieved into one control group with a size of 0.075–0.3 mm and three other reference groups. Together with the desert sand group, a total of five groups of ECC specimens were created. Through a uniaxial tensile test, three-point bending test and single-seam tensile test on the ECC specimens, the influence of aggregate particle size and sand type on the ECC tensile strength, deformation capacity, initial crack strength, cement-matrix-fracture toughness, multiple cracking characteristics and strain-hardening properties were studied. The experimental results show that the 28d tensile strain of the four groups of the ordinary sand specimens was 8.13%, 4.37%, 4.51% and 4.23%, respectively, which exceeded 2% and satisfied the requirements for the minimum strain of the ECCs. It is easier to achieve the ECC strain hardening with sand with a fine particle size; thus, a particle size below 0.3 mm is preferred when preparing the ECCs to achieve a high toughness. The multiple cracking performance (MCP) and the pseudostrain hardening (PSH) of desert sand and ordinary sand with a 0.075–0.3 mm grain size were 2.88 and 2.33, and 8.76 and 8.17, respectively, all of which meet the strength criteria and energy criteria and have similar properties. The tensile strength and tensile deformation of the desert sand group were 4.97 MPa and 6.78%, respectively, and the deformation capacity and strain–strengthening performance were outstanding. It is verified that it is feasible to use desert sand instead of ordinary sand to prepare the ECCs.

Keywords: desert sand; engineered cementitious composites (ECCs); particle size; uniaxial tension; toughness

1. Introduction

Since its first appearance in the middle of the 18th century, concrete has been widely used in various industrial and civil buildings, roads, bridges and other infrastructures due to its high strength, extensive sources of raw materials, strong applicability and low cost, in addition to other advantages. However, because of its low tensile strength, poor ductility, easy-cracking characteristics and deterioration under environmental effects, the problem of concrete durability is prominent, which increases the cost of maintenance and protection and restricts the development of concrete materials [1,2]. Almost all infrastructures that are damaged due to a lack of toughness, durability and sustainability can be traced back to tensile cracking and the fracture of concrete.

To overcome the defects of traditional concrete materials, such as high energy consumption, a high degree of brittleness failure and a poor crack-control ability, engineered cementitious composites (ECCs) have been developed and the strength criteria and energy criteria based on micromechanics were proposed by Victor C. Li. A theoretical basis based on these two criteria was provided for the engineering application of the ECCs [3–5]. If both the strength criteria and the energy criteria are satisfied, the characteristics of the multiple cracking and strain hardening of the ECCs under tension can be realized. The uniaxial tensile strain of the ECCs can exceed 2% [6–8], which is more than 200 times that of ordinary cement-based materials [9]. In the failure process for the ECCs, many fine cracks are produced with a crack width that is generally within 100 μm [10–12], and the invasion of harmful substances can effectively be prevented and the impermeability [13–15], self-healing ability [16–18] and durability [19–22] of concrete can be improved. The research on the ECCs from different perspectives and in combination with different factors has been conducted by scholars all over the world [23–26].

A large number of industrial wastes are used in the ECC preparation process to replace some or all of the cement, such as fly ash, slag, lithium slag and red-mud slag [27–31], and the energy consumption of the material-production process is greatly reduced, which conforms to the goal of achieving environmental sustainability [32]. Aggregates are an important component of the ECCs, which account for a large volume proportion and significantly affect the workability, strength, elastic modulus, ductility and other properties of the ECCs. In addition, aggregates can also reduce the production costs of the ECCs. At present, most ECCs are made of micro-silica sand from river sand, which is a nonrenewable resource. Because of the surge in demand for building raw materials, China uses approximately 20 billion tons of sand and stone every year, which account for half of the world's consumption. The phenomenon of indiscriminate excavation and mining frequently occurs, which has caused great damage to the environment and has led to a sharp increase in the price of materials, such as construction sand, and a decrease in sand reserves [33]. Today, ECCs cannot be widely used in practical projects due to their high cost, and the existing research on reducing ECC costs has focused on the optimization and selection of fibers. However, the micro-silica sand used in the ECCs is also one of the important reasons for the high cost. Thus, finding new and alternative sand sources is important and urgent.

Desert sand is a very rich natural resource, which is widely distributed all over the world. The total desert area in China is approximately 700,000 km^2, which accounts for 7% of its total land area. China has eight deserts with huge reserves of desert sand [34]. Desert sand is ultrafine sand with an average particle size generally below 0.2 mm. Currently, it is used in some concrete materials [35–38]; however, the use of desert sand in ECCs is rarely reported. If desert sand can be reasonably used in ECC materials, it will not only reduce engineering costs, but also protect environmental resources and help achieve the sustainable development of concrete.

Based on the above research contents, firstly, the chemical composition of the desert sand is tested to determine whether it contains harmful elements that can affect ECCs. Secondly, ECCs are prepared with desert sand, which are then compared with the ECCs of ordinary sand with different particle sizes. Through a uniaxial tensile test, three-point bending test and single-seam tensile test on the ECC specimens, the influences of desert sand and ordinary sand with different grain sizes on the ECC tensile strength, deformation capacity, initial crack strength, cement-matrix-fracture toughness, multiple cracking characteristics and strain-hardening properties were studied. The schematic flow diagram of this study is shown in Figure 1.

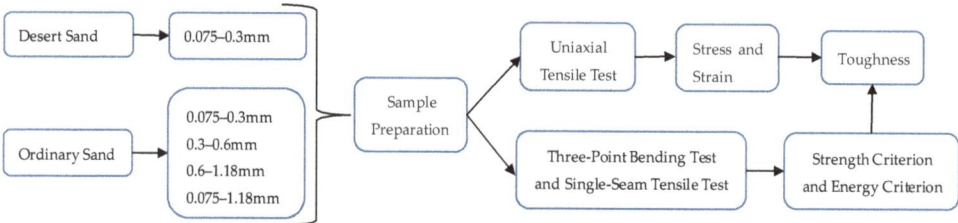

Figure 1. Schematic flow diagram of this study.

2. Materials and Methods

2.1. Materials and Preparation of ECCs

The cement used in this study was P·II 42.5 ordinary Portland cement produced by Xinjiang Tianshan Cement Co., Ltd. (Urumqi, China), and the mineral powder was high-quality first-grade S95. The fine aggregates were desert sand from Wuqia County, Kezhou and ordinary river sand from Kashi. The fiber used in this study was polyethylene (PE) fiber. The polycarboxylic acid superplasticizer was produced by the Kashi Water Reducing Agent Factory, as was the 200,000-viscosity hydroxypropyl methyl cellulose. Group A was composed of desert sand with a particle size of 0.075–0.3 mm. The ordinary river sand was sieved into four particle size grades with particle size ranges from 0.075 to 0.3 mm for Group B, from 0.3 to 0.6 mm for Group C, from 0.6 to 1.18 mm for Group D and from 0.075 to 1.18 mm for Group E. The five groups of sands with different types and particle sizes were taken as the research objects, as shown in Figure 2. The particle size of the desert sand was very close to that of the ordinary sand, with a particle size from 0.075 to 0.3 mm. The chemical composition of the desert sand is listed in Table 1. The sulfide and sulfate contents were relatively low (calculated according to the mass of SO_3) and the chloride content was very low (calculated using the mass of chloride ion), which satisfy the limit for harmful substances in the project. Table 2 lists the physical and mechanical properties of the PE fiber.

Figure 2. Desert sand and ordinary sand: (**a**) Group A, 0.075–0.3 mm; (**b**) Group B, 0.075–0.3 mm; (**c**) Group C, 0.3–0.6 mm; (**d**) Group D, 0.6–1.18 mm; (**e**) Group E, 0.075–1.18 mm.

Table 1. Chemical composition of desert sand.

Component	SiO_2	Al_2O_3	Fe_2O_3	CaO	MgO	TiO_2	MnO	SO_3
Proportion (%)	74.48	9.53	2.78	4.34	1.76	0.34	0.001	0.08
Component	Cl	MnO	P_2O_5	Cl^{-1}	Na^{+1}	K^{+1}	Alkali content	Loss on ignition
Proportion (%)	0.013	0.001	0.09	0.005	0.007	0.003	2.67	3.34

Table 2. Physical and mechanical properties of PE fiber.

Diameter (μm)	Length (mm)	Tensile Strength (MPa)	Elastic Modulus (GPa)	Elongation (%)	Density (g/cm^3)
24	12	3000	120	5	0.97

The five groups of ECCs were configured according to the five groups of sand and adopted the same mix proportion, as shown in Table 3. The success of the ECC preparation is closely related to the manufacturing process [39–41], especially regarding the mixing sequence and mixing time of materials. Cement, slag, sand and other materials were added to a 5 L planetary mixer according to the proportions listed in Table 3. After the materials were slowly mixed for 2 min, one-third of the water was added and stirred for 1 min. Then, the water reducer and one-third of the water were uniformly mixed and poured into the mixer. Subsequently, the remaining one-third of the water was poured into the mixed materials for 2 min of mixing. At this time, the fluidity of the cement matrix was very good. Then, the thickener was added for quick mixing for 1.5 min. Here, the cement matrix was relatively thick and dough-like. Finally, the PE fiber was uniformly dispersed into the cement matrix within 6 min. No obvious agglomeration occurred when the mixture was manually kneaded, which is a key indicator of the success of the ECC preparation. The final mixing time depends on the even dispersion of the fiber in the cement matrix without agglomeration. The prepared ECC material was put into a specially made transparent acrylic template and was subjected to full vibrations until large air holes no longer appeared, which was observed from the side and bottom of the template. The templates were removed after the prepared samples were kept for 24 h. The specimens were placed in a constant-temperature and humidity-curing room at a temperature of 20 ± 1 °C and a relative humidity of 95% for curing.

Table 3. ECC mixing proportions (kg/m^3).

Cement	Slag	Sand	Water	Fiber	Water Reducer	Thickener
617.2	411.4	308.6	250	12	6	0.5

2.2. Test-Scheme Design

2.2.1. Uniaxial Tensile Test

The main methods for testing the tensile properties of materials include uniaxial tensile, splitting tensile and bending tests. Among them, the most direct test method that can best reflect the tensile properties of the materials is the uniaxial tensile test. Although researchers in various countries have been trying to agree on the standardization of the uniaxial tensile test, it has not yet been standardized. At present, dumbbell-shaped specimens and plates are widely used in uniaxial tensile tests, and dumbbell-shaped specimens were used in this study [42]. Both ends of the dumbbell-shaped test piece consist of clamping ends. The gauge length of the test piece is 100 mm, and the size of the test section is 100 mm × 30 mm × 15 mm, as shown in Figure 3. A set of displacement sensors were installed at both ends of the test piece with a measurement range from 0 to 20 mm, and the final deformation was measured according to the average value of two sets of displacement sensors, as shown in Figure 4. An electronic tensile-testing machine with a measurement range of 5 kN, which was manufactured by Jinan Chuanbai Instrument and Equipment Co., Ltd. (Jinan, China), was used as the loading equipment, and displacement-control loading was adopted at a loading rate of 0.5 mm/min. The test pieces were divided into five groups according to the aggregate size and sand type. The curing ages were 7 and 28 days. Three test blocks were made at each curing age.

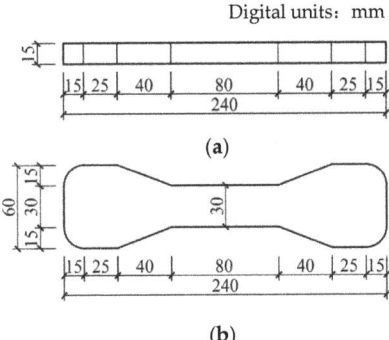

Figure 3. Schematic diagram of uniaxial tensile specimen: (**a**) profile; (**b**) plan.

Figure 4. Uniaxial tensile test device.

2.2.2. Three-Point Bending Test

To quantitatively analyze the strain–strengthening and toughness properties of the desert sand and ordinary sand with different grain sizes, the fracture energy of the cement matrix needs to be determined according to the strength criteria and energy criteria [3]. When the three-point bending beam is a standard specimen (the span–height ratio of the specimen is four), the fracture energy of the ECC cement matrix can be tested according to the formula recommended by Tada [42]. A cement matrix specimen without PE fiber was made according to the mixing proportions listed in Table 3. The specimen size was 350 mm × 75 mm × 40 mm. Each group contained three test pieces, for a total of five groups with fifteen test pieces. After the matrix was cured under standard conditions for 28 days, the three-point bending test was performed at a loading rate of 0.5 mm/min. The test-piece span was 300 mm, and an incision that was 30 mm deep and less than 1 mm wide was made at the bottom of the test piece. The sizes of the test piece and test device are shown in Figures 5 and 6, respectively.

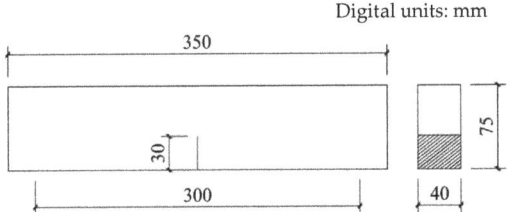

Figure 5. Dimensions of three-point bending test piece.

Figure 6. Three-point bending test loading device.

2.2.3. Single-Seam Tensile Test

To determine the parameters of the strength criteria and energy criteria, the maximum peak stress and maximum complementary energy should be determined based on the relationship between the bridging-fiber stress and crack width, in addition to the fracture energy of the cement matrix that is determined with the three-point bending test. To determine the relationship between the ECC stress and crack opening width, the multiple cracking needed to be artificially limited to ensure that single-seam cracking occurred in the case of failure. The size of the test piece was the same as that of the dumbbell-type test piece mentioned earlier. Each group contained three test pieces, for a total of five groups of fifteen test pieces. After 28 days of curing under standard conditions, an annular notch with a width of less than 1 mm was cut into the middle of the test piece. The depth of the notch is shown in Figure 7. The notch is cut using diamond-cutting pieces. During the notch-making process, care is taken to avoid damage to the other parts of the test piece to ensure that the ECC only experiences single-seam cracking when under uniaxial tension. The tensile test device is shown in Figure 8.

Digital units: mm

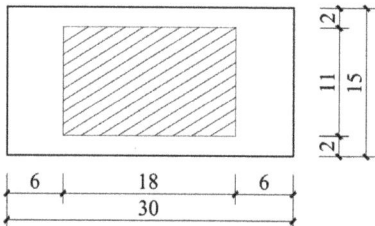

Figure 7. Cutting depth of the single-seam cracking specimen.

Figure 8. Single-seam cracking tensile test device.

2.3. Evaluation Method of ECC Toughness

On the basis of the design theory of ECC micromechanics [3], the ECCs must satisfy both the strength criteria and energy criteria to achieve the characteristics of multiple cracking and strain hardening; otherwise, stress softening occurs during the tensile process.

(1) Strength Criteria

The strength criteria set the boundary condition of tensile stress for cracks starting from the initial defect, as they control the cracking process. Continuous multiple cracking must satisfy Equation (1).

$$\sigma_c < \min\{\sigma_0\} \tag{1}$$

where σ_c is the initial crack strength of the matrix and σ_0 is the peak stress.

(2) Energy Criteria

The crack-distribution phenomenon of the test conforms to the flat crack-propagation mode and satisfies the energy criteria. The stability of the crack width under a constant external load must satisfy Equation (2).

$$\sigma_0 \delta_0 - \int_0^{\delta_0} \sigma(\delta) d\delta = J'_b \geq J_{\text{tip}} \tag{2}$$

where δ_0 is the displacement corresponding to the bridging-fiber peak stress, $\sigma(\delta)$ denotes the relationship between the bridging-fiber stress and crack opening width, J'_b is the maximum complementary energy and J_{tip} is the fracture energy of the matrix material.

Theoretically, $\sigma_0/\sigma_c \geq 1$ and $J'_b/J_{\text{tip}} \geq 1$ can achieve stable tensile strain-hardening characteristics. However, in practice, because of the existence of uncertainty factors such as material fluctuation, an uneven manufacturing process, and test errors, satisfying the requirements for strain-hardening characteristics is difficult. Research shows that only PCM = $\sigma_0/\sigma_c \geq 1.3$ and PSH = $J'_b/J_{\text{tip}} \geq 2.7$ can meet the characteristics of multiple cracking and strain hardening [43], where MCP is the multiple cracking performance and PSH is the pseudostrain hardening.

Based on the calculations of Equations (3)–(6) recommended by Tada [41], which are combined with the three-point bending test, the fracture energy of the cement matrix can be calculated.

$$J_{\text{tip}} = \frac{K_m^2}{E_m} \tag{3}$$

$$K_m = \frac{3(F + 10^{-3} mg/2) 10^{-3} L \sqrt{a}}{2bh^2} f(\alpha) \tag{4}$$

$$f(\alpha) = \frac{1.99 - \alpha(1-\alpha)(2.15 - 3.93\alpha + 2.7\alpha^2)}{(1+2\alpha)(1-\alpha)^{3/2}} \tag{5}$$

$$\alpha = \frac{a}{h} \tag{6}$$

where J_{tip} (J/m^2) is the fracture energy, K_m (MPa·m$^{1/2}$) is the fracture toughness, E_m (GPa) is the tensile modulus of elasticity, F (kN) is the three-point bending peak load, m (kg) is the test-piece quality, g (m/s^2) is the gravitational acceleration, L (m) is the span of the three-point bending test piece, a (m) is the notch depth, b (m) is the width of the test piece, h (m) is the height of the test piece and $f(\alpha)$ is the test-piece shape parameters.

According to the single-seam tensile test, the relationship between the stress and crack opening width can be obtained, and the maximum complementary energy can be obtained by integrating it with the axis where the stress is located, as shown in Figure 9.

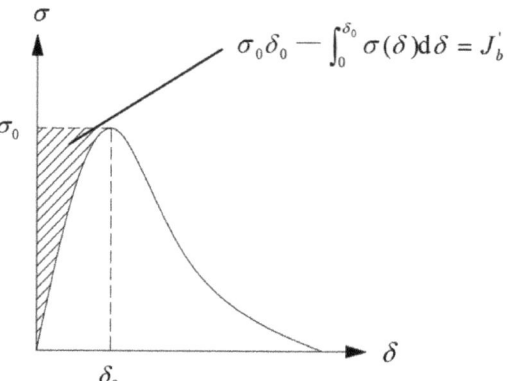

Figure 9. Relationship between the crack opening width and bridging-fiber stress.

3. Results and Discussion
3.1. Uniaxial Tensile Results and Stress–Strain Analysis

Figure 10 shows the failure mode of each group of specimens under uniaxial tension, and Figure 11 shows the stress–strain curve under uniaxial tension. The test phenomenon and stress–strain curve show that in the initial stage of loading, the ECC specimen is in the elastic stage and the tensile force is mainly borne by the cement matrix. When the cracking strength of the cement matrix is reached, the specimen exhibits the first crack, and the stress–strain curve suddenly drops. Cracks continue to appear as the tensile force increases and the stress–strain curve repeatedly fluctuates, exhibiting strain-hardening characteristics. During the failure process of the test piece, the continuous sound of fibers being pulled out or broken can be heard. The reason for this is that the PE fiber plays a bridging role after the cement matrix breaks, bears the load from the matrix and makes the surrounding matrix continuously generate new cracks until the PE fiber fails in its bridging role, resulting in multiple cracks and a large deformation.

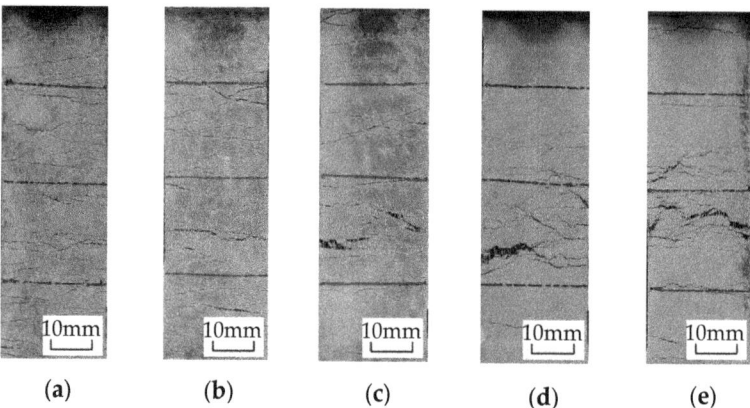

Figure 10. Failure mode of uniaxial tensile specimen: (**a**) Group A; (**b**) Group B; (**c**) Group C; (**d**) Group D; (**e**) Group E.

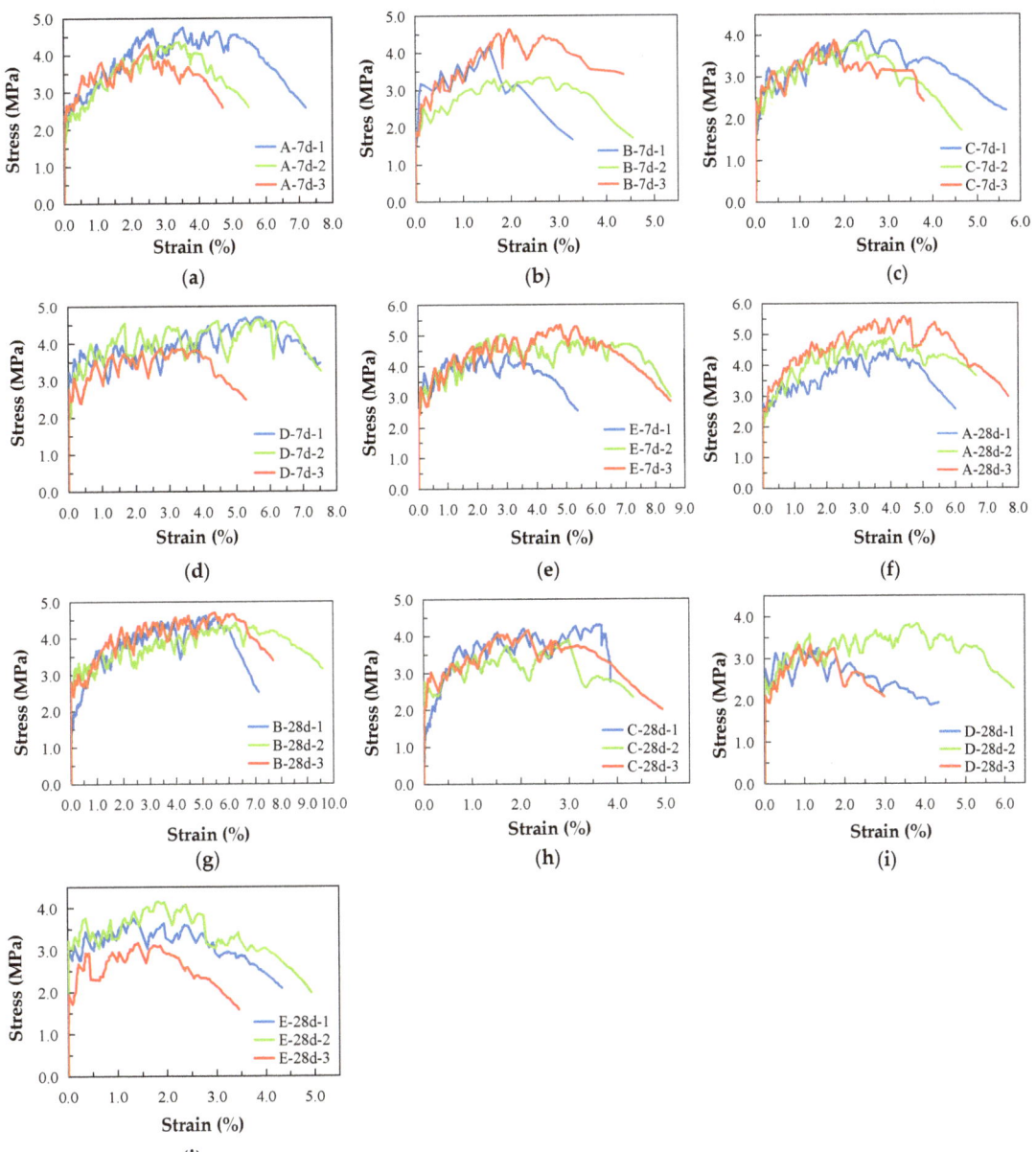

Figure 11. Uniaxial tensile stress–strain curve of ECCs with different sand sizes and ages: (**a**) Group A-7d; (**b**) Group B-7d; (**c**) Group C-7d; (**d**) Group D-7d; (**e**) Group E-7d; (**f**) Group A-28d; (**g**) Group B-28d; (**h**) Group C-28d; (**i**) Group D-28d; (**j**) Group E-28d.

The shape and distribution of the two groups of fractures are basically consistent based on the comparative analysis of Group A with desert sand and Group B with ordinary sand as both have a spacing of approximately 2–3 mm and a fracture width of less than 100 μm. The number of cracks is 40–50, which demonstrates the characteristics of the dense, saturated multiple cracks. The deformation of the two groups is mainly generated by the cumulative width of these multiple cracks. For ordinary sand, Group B exhibits dense, saturated multiple cracking characteristics. Although Groups C, D and E also exhibit multiple cracking characteristics, they exhibit unsaturated multiple cracking. The cracks are unevenly distributed, and the maximum spacing of the cracks can reach 15 mm, which indicates that the reinforcement effect of the fibers has not been fully utilized. The deformation of materials is mainly obtained through the final main-crack cracking, which limits the final deformation ability of the materials. In total, the number of cracks becomes increasingly fewer as the particle size increases. A coarse particle size is not conducive to the realization of multiple cracking in the ECCs, especially for particles larger than 0.6 mm. Desert sand and ordinary sand with a 0.075–0.3 mm particle size show excellent multiple cracking characteristics and the ability to control the crack width.

The uniaxial tensile properties of the ECCs with different particle sizes and ages are shown in Figure 12. From the perspective of the initial crack strength, the 7-day initial crack strengths of Group A with desert sand and Group B with ordinary sand are 2.09 MPa and 2.06 MPa, respectively, and the 28-day initial crack strengths are 2.51 MPa and 2.64 MPa, respectively, under the same change trend. For ordinary sand, the initial crack strength of the ECCs tends to increase along with the increase in the sand particle size, but the initial crack strength of Groups C–E at 28 days does not significantly increase, which may be caused by the random difference in the cement matrix's defect size.

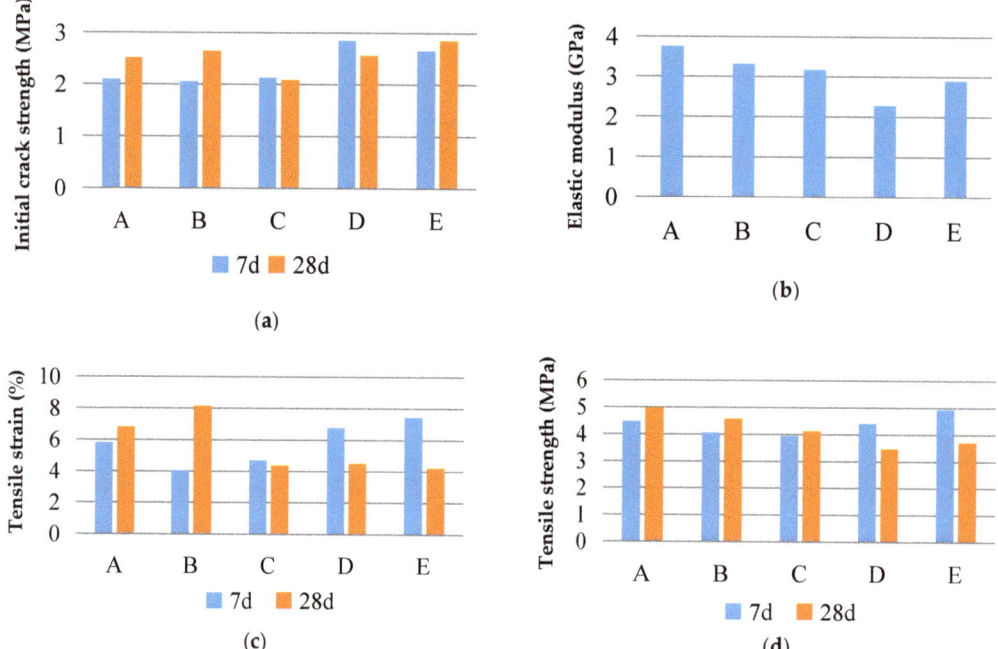

Figure 12. Uniaxial tensile properties of specimens with different particle sizes and ages: (**a**) initial crack strength; (**b**) elastic modulus; (**c**) tensile strain; (**d**) tensile strength.

According to the 28-day tensile elastic modulus, for Groups A and B, the tensile elastic moduli are 3.76 GPa and 3.31 GPa, respectively. The tensile elastic modulus of the ECCs with desert sand is higher, but the difference is not significant. For ordinary sand, the tensile elastic modulus of Group B with fine particles is higher than the tensile elastic modulus of Groups C–E. The tensile elastic modulus decreases as the particle size increases.

Comparison and analysis of the desert and ordinary sand reveal that the tensile strengths of Group A at 7 and 28 days are 4.47 MPa and 4.97 MPa, respectively, and those of Group B are 4.06 MPa and 4.57 MPa, respectively. The tensile strength of the desert sand is slightly higher than that of Group B with ordinary sand. The tensile strains of Group A at 7 and 28 days are 5.80% and 6.78%, respectively, whereas those of Group B are 4.06% and 8.13%, respectively. The tensile strain of ordinary sand at 28 days is slightly higher than that of the desert sand, and the tensile deformation of both is much higher, being 2% more than the minimum limit value of the ECCs defined by the general rules, which indicates that the deformation capacity of the ECC materials prepared from desert sand is excellent and can completely replace ordinary sand.

The tensile strength and tensile deformation of ordinary sand with a 0.075–0.3 mm particle size show an increasing trend alongside the increase in age, whereas when the particle size exceeds 0.6 mm, a decreasing trend occurs. The reason for this is that the strain-hardening property of the coarse-grain-sized sand decreases as the grain size increases, which leads to the premature occurrence of the main cracks in the test piece and leads to a decrease in the tensile deformation and tensile strength. The tensile strain of the four groups of the ECC specimens with different particle sizes exceeds 2%, which is consistent with the identification of the ECCs in this study. Sand with a grain size of less than 0.3 mm can effectively improve the deformation capacity and tensile strength of the ECCs. The 0.3–0.6 mm grain size increases with age, and the change range of the tensile strength and tensile deformation is not obvious. A coarse-grain size of more than 0.6 mm can reduce the deformation capacity and tensile strength, which is not conducive to achieving strain strengthening in the ECC materials.

3.2. Three-Point Bending Test Results and Fracture Energy of the Cement Matrix

Figure 13 shows the failure mode of the three-point bending test piece. Because PE fiber was not added to the cement matrix, each group of the test pieces was brittle when they were damaged, and they all broke from the notch. Equations (3)–(6), which are recommended by Tada [42], show that, based on the peak load of the specimen failure, the mass of the specimen and the elastic modulus measured under uniaxial tension, the fracture energy of the five groups of matrixes can be obtained, as listed in Table 4 and Figure 14.

For Group A and Group B, the fracture energies are 72.5 J/m^2 and 67 J/m^2, respectively, which are small and conducive to achieving a high toughness. For ordinary sand, the matrix fracture energies of Groups B to E are 67.0 J/m^2, 90.6 J/m^2, 109.6 J/m^2 and 96.5 J/m^2, respectively. The matrix fracture energy of Group B was the smallest, and that of Group C, Group D and Group E was 35.2%, 63.6% and 44.0% higher than that of Group B. It can be seen that the larger the particle size is, the higher the fracture energy is.

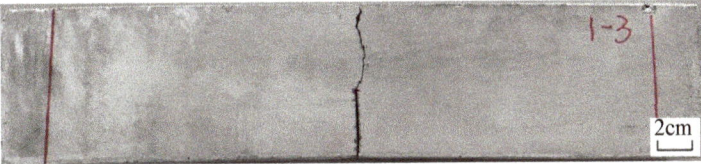

Figure 13. Failure mode of the three-point bending test specimen.

Table 4. Fracture energy of the ECC matrix with different sand sizes.

Group	Peak Load F (kN)	Fracture Toughness K_m (MPa·m$^{1/2}$)	Fracture Energy J_{tip} (J/m^2)
A	0.71	0.522	72.5
B	0.64	0.471	67.0
C	0.73	0.536	90.6
D	0.68	0.500	109.6
E	0.72	0.529	96.5

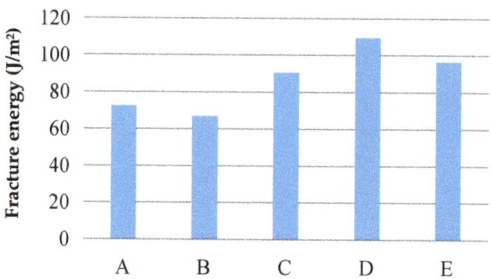

Figure 14. Fracture energy of the cement matrix with different sand particle sizes.

3.3. Single-Seam Tensile Test Results and Complementary Energy

Figure 15 shows the failure mode of the single-seam tensile specimen. The failure occurs at the notch, and no cracks appear in other places. The stress–displacement curve is shown in Figure 16. The peak stress and the opening width at the peak stress can be obtained. The complementary energy J'_b can be obtained by integrating the axis of the stress, as listed in Table 5.

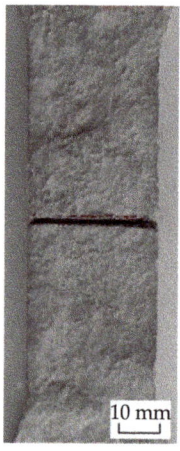

Figure 15. Failure mode of the single-seam tensile specimen.

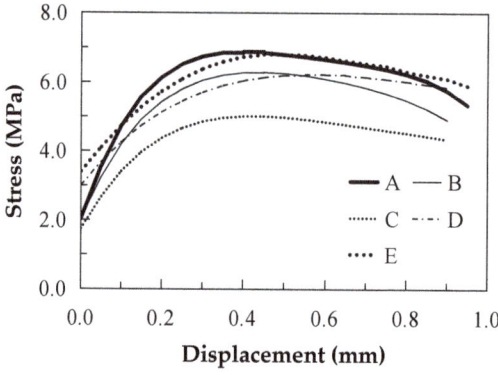

Figure 16. Tensile stress–displacement curve of the single seam with different sand particle sizes.

Table 5. Complementary energy of the ECC single-seam cracking test.

Group	Peak Stress σ_0 (MPa)	Opening Width Corresponding to Peak Stress δ_0 (mm)	Complementary Energy J'_b (J/m^2)
A	7.24	0.40	635
B	6.14	0.48	547.5
C	5.07	0.45	397.7
D	6.27	0.63	553.9
E	6.83	0.52	578.0

3.4. Analysis and Discussion of Fracture Toughness

The multiple cracking performance (MCP) and the pseudostrain hardening (PSH) could be obtained, respectively, according to the matrix's fracture energy J_{tip}, the complementary energy J'_b and the initial crack strength, which was measured by the three-point bending test, the single-seam tensile test and the uniaxial tensile test, respectively, as shown in Table 6 and Figure 17.

Table 6. MCP and PSH values of the desert sand and ordinary sand with different particle sizes.

Group	MCP	PSH
A	2.88	8.76
B	2.33	8.17
C	2.44	5.39
D	2.45	5.05
E	2.40	6.00

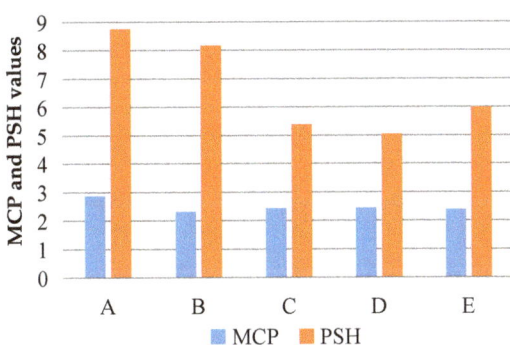

Figure 17. Comparative analysis of MCP and PSH values for the desert sand and ordinary sand.

The MCP of Group A with desert sand and Group B with ordinary sand is 2.88 and 2.33, respectively. The desert sand group's MCP is slightly larger than that of the ordinary sand group, and both are greater than 1.3, which meets the requirements of the strength criteria in Equation (1). Both groups of specimens have obvious characteristics of multiple cracking, which have been verified via the uniaxial tensile stress–strain curve and the specimen failure phenomenon. The PSH of the two groups of specimens is 8.76 and 8.17, respectively, which are both much larger than 2.7. The desert sand group's PSH is slightly larger than that of the ordinary sand group, and both meet the requirements of the energy criteria in Equation (2). The higher PSH of the desert sand group is conducive to achieving a high toughness.

For ordinary sand, the MCP of Groups B to E is 2.33, 2.44, 2.45 and 2.40, respectively, which are all greater than 1.3 and meet the requirements of the strength criteria in Equation (1). The four groups of specimens have the characteristic of multiple cracking.

The PSH of the four groups of ordinary sand specimens are 8.17, 5.39, 5.05 and 6.00, which are all greater than 2.7 and meet the requirements of the energy criteria in Equation (2). It can be seen that the smaller the grain size of the ordinary sand is, the easier it is to achieve stable multiple cracking and strain hardening. Similar conclusions were reflected in M. Sahmaran's research [33]. In that research, dolomite limestone sand and gravel sand with a maximum particle size of 1.19 mm and 2.38 mm were used to replace micro-silica sand with maximum particle sizes of 0.2 mm when preparing the ECCs, and the production cost of the ECCs was reduced. The tensile strength and deformation of the ECC materials prepared by using larger grains of sand were reduced to varying degrees.

4. Conclusions

(1) The results show that the uniaxial tensile strength of the ECCs with desert sand is 4.97 MPa, whereas the maximum tensile strain is 6.17%. The high toughness of the ECCs with desert sand is shown to be outstanding according to the results of the stress–strain curves, the strength criteria and the energy criteria. It is verified that desert sand is feasible to be used as a substitute to ordinary sand for the ECC preparation.

(2) The performance of the ECCs with ordinary sand is closely related to the particle size of sand. The tensile strength, tensile deformation and toughness of the ECC materials are decreased when the particle size is increased. The maximum tensile strength and tensile deformation were obtained for the particle size from 0.075 to 0.3 mm, which are 4.57 MPa and 8.13%, respectively. In engineering projects, the ECCs should be prepared with a particle size below 0.3 mm, and the maximum particle size should not exceed 0.6 mm.

(3) The compression and bending properties of the ECCs with desert sand, as well as the interfacial connection between the fiber and matrix at the microscale, should be further studied.

Author Contributions: Conceptualization, Z.L. and Y.H.; methodology, Z.L. and Y.H.; formal analysis, Z.L., Y.H., G.H. and X.G.; investigation, Z.L., X.G., G.H. and H.W.; data curation, Z.L. and Y.H.; resources, X.G. and H.W.; visualization, Z.L. and X.G.; writing—original draft preparation, Z.L.; writing—review and editing, Z.L., Y.H. and G.H.; funding acquisition, Z.L. All authors have read and agreed to the published version of the manuscript.

Funding: This research was funded by the Natural Science Foundation of Xinjiang Uygur Autonomous Region (grant number: 2022D01B03), the Kashgar Science and Technology Plan Project (grant number: KS2021030) and the Horizontal Project of Kashi University (grant number: 022022184).

Institutional Review Board Statement: Not applicable.

Informed Consent Statement: Not applicable.

Data Availability Statement: The data used to support the findings of this study are included within the article.

Conflicts of Interest: The authors declare no conflict of interest.

References

1. Kulkarni, S.B.; Clinton, P. Durability of concrete materials for components of high rise structures. *J. Res. Eng. Appl. Sci.* **2019**, *4*, 145–149.
2. Guo, J.; Wang, K.; Qi, C. Determining the mineral admixture and fiber on mechanics and fracture properties of concrete under sulfate attack. *J. Mar. Sci. Eng.* **2021**, *9*, 251. [CrossRef]
3. Li, V.C.; Leung, C.K. Steady-state and multiple cracking of short randon fiber composites. *J. Eng. Mech.* **1992**, *118*, 2246–2264.
4. Li, V.C. From micromechanics to structural engineering—The design of cementitous composites for civil engineering applications. *JSCE J. Struct. Mech. Earthq. Eng.* **1993**, *10*, 37–48.
5. Zhong, L.; Li, V.C. Crank bridging in fiber reinforced cementitious composites with slip-hardening interfaces. *J. Mech. Phys. Solids* **1997**, *45*, 763–787.
6. Li, V.C. Tailoring ECC for special attributes: A review. *Int. J. Concr. Struct. Mater.* **2012**, *6*, 135–144. [CrossRef]
7. Hung, C.C.; Chen, Y.S. Innovative ECC jacketing for retrofitting shear-deficient RC members. *Constr. Build. Mater.* **2016**, *111*, 408–418. [CrossRef]

8. Zhou, Y.; Xi, B.; Sui, L.; Zheng, S.; Xing, F.; Li, L. Development of high strain-hardening light weight engineered cementitious composites: Design and performance. *Cem. Concr. Compos.* **2019**, *104*, 103370. [CrossRef]
9. Zhang, J.; Li, V.C. Monotonic and fatigue performance in bending of fiber-reinforced engineered cementitious composites in overlay system. *Cem. Concr. Res.* **2002**, *32*, 415–423. [CrossRef]
10. Li, V.C. On Engineered Cementitious Composites (ECC). *J. Adv. Concr. Technol.* **2003**, *1*, 215–230. [CrossRef]
11. Mechtcherine, V.; Millon, O.; Butler, M.; Thoma, K. Mechanical behaviour of strain hardening cement-based composites under impact loading. *Cem. Concr. Compos.* **2011**, *33*, 2930–2937. [CrossRef]
12. Li, V.C.; Wang, S.; Wu, S. Tensile strain-hardening behavior of polyvinyl alcohol engineered cementitious composite (PVA-ECC). *ACI Mater. J.* **2001**, *98*, 483–492.
13. Wang, Q.; Zhang, G.; Tong, Y.; Gu, C. Prediction on permeability of engineered cementitious composites. *Crystals* **2021**, *11*, 526. [CrossRef]
14. Djerbi, A.; Bonnet, S.; Khelidj, A.; Baroghel-bouny, V. Influence of traversing crack on chloride diffusion into concrete. *Cem. Concr. Res.* **2008**, *38*, 877–883. [CrossRef]
15. Sahmaran, M.; Li, M.; Li, V.C. Transport properties of engineered cementitious composites under chloride exposure. *ACI Mater. J.* **2007**, *104*, 604–611.
16. Chen, G.; Tang, W.; Chen, S.; Wang, S.; Cui, H. Prediction of self-healing of engineered cementitious composite using machine learning approaches. *Appl. Sci.* **2022**, *12*, 3605. [CrossRef]
17. Ahn, T.-H.; Kishi, T. Crack self-healing behavior of cementitious composites incorporating various mineral admixtures. *J. Adv. Concr. Technol.* **2010**, *8*, 171–186. [CrossRef]
18. Liu, H.; Zhang, Q.; Gu, C.; Su, H.; Li, V.C. Self-healing of microcracks in engineered cementitious composites under sulfate and chloride environment. *Constr. Build. Mater.* **2017**, *153*, 948–956. [CrossRef]
19. Yu, J.; Lin, J.; Zhang, Z.; Li, V.C. Mechanical performance of ECC with high-volume fly ash after sub-elevated temperatures. *Constr. Build. Mater.* **2015**, *99*, 82–89. [CrossRef]
20. Turk, K.; Kina, C.; Nehdi, M.L. Durability of engineered cementitious composites incorporating high-volume fly ash and limestone powder. *Sustainability* **2022**, *14*, 10388. [CrossRef]
21. Zhang, H.; Shao, Y.; Zhang, N. Carbonation behavior of engineered cementitious composites under coupled sustained flexural load and accelerated carbonation. *Materials* **2022**, *15*, 6192. [CrossRef]
22. Bu, L.; Qiao, L.; Sun, R.; Lu, W. Time and crack width dependent model of chloride transportation in engineered cementitious composites (ECC). *Materials* **2022**, *15*, 5611. [CrossRef]
23. Chung, K.L.; Luo, J.; Yuan, L.; Zhang, C.; Qu, C. Strength correlation and prediction of engineered cementitious composites with microwave properties. *Appl. Sci.* **2017**, *7*, 35. [CrossRef]
24. Atmajayanti, A.T.; Hung, C.-C.; Yuen, T.Y.P.; Shih, R.-C. Influences of sodium lignosulfonate and high-volume fly ash on setting time and hardened state properties of engineered cementitious composites. *Materials* **2021**, *14*, 4779. [CrossRef]
25. Huang, X.; Ranade, R.; Wen, N.; Li, V.C. On the use of recycled tire rubber to develop low modulus ECC for durable concrete repairs. *Constr. Build. Mater.* **2013**, *46*, 134–141. [CrossRef]
26. Pachideh, G.; Toufigh, V. Strength of SCLC recycled springs and fibers concrete subject to high temperatures. *Struct. Concr.* **2022**, *23*, 285–299. [CrossRef]
27. Gong, G.; Guo, M.; Zhou, Y.; Zheng, S.; Hu, B. Multiscale investigation on the performance of engineered cementitious composites incorporating PE fiber and limstone calcined clay cement (LC3). *Polymers* **2022**, *14*, 1291. [CrossRef] [PubMed]
28. Amin, M.N.; Ashraf, M.; Kumar, R.; Khan, K.; Saqib, D.; Ali, S.S.; Khan, S. Role of sugarcane bagasse ash in developing sustainable engineered cementitious composites. *Front. Mater.* **2020**, *7*, 65. [CrossRef]
29. Xu, L.; Huang, B.; Lao, J.; Dai, J. Tailoring strain-hardening behavior of high-strength engineered cementitious composites (ECC) using hybrid silica sand and artificial geopolymer aggregates. *Mater. Des.* **2022**, *220*, 110876. [CrossRef]
30. Toufigh, V.; Pachideh, G. Cementitious mortars containing pozzolana under elevated temperatures. *Struct. Concr.* **2022**, *23*, 3294–3312. [CrossRef]
31. Pachideh, G.; Gholhaki, M. Assessment of post-heat behavior of cement mortar incorporating silica fume and granulated blast-furnace slag. *J. Struct. Fire Eng.* **2020**, *11*, 221–246. [CrossRef]
32. Andrew, R.M. Global CO_2 emissions from cement production. *Earth Syst. Sci. Data* **2018**, *10*, 195–217. [CrossRef]
33. Şahmaran, M.; Lachemi, M.; Hossain, K.M.A.; Ranade, R.; Li, V.C. Influence of aggregate type and size on ductility and mechanical properties of engineered cementitious composites. *ACI Mater. J.* **2009**, *106*, 308–316.
34. Yuan, X.; Ren, Y.; Xi, L. Analysis on the difference of sea sand, river sand, desert sand and machine-made sand. *Technol. Wind.* **2020**, *11*, 110–111.
35. Ji, Y.; Liu, C.; Ding, Z. Research and analysis on present situation of desert sand concrete; E3S Web of Conferences. In Proceedings of the 2021 2nd International Academic Conference on Energy Conservation, Environmental Protection and Energy Science (ICEPE 2021), Dali, China, 21–23 May 2021.
36. Shi, F.; Li, T.; Wang, W.; Liu, R. Research on the effect of desert sand on pore structure of fiber reinforced mortar based on X-CT technology. *Materials* **2021**, *14*, 5572. [CrossRef] [PubMed]
37. Liu, Y.; Yang, W.; Chen, X.; Liu, H.; Yan, N. Effect of desert sand on the mechanical properties of desert sand concrete (DSC) after elevated temperature. *Adv. Civ. Eng.* **2021**, *2021*, 3617552. [CrossRef]

38. Chen, Q.; Liu, H.; Han, L.; Wang, Y. Comparative studies of dynamic mechanical properties of desert sand concrete and ordinary concrete. *Shock Vib.* **2022**, *2022*, 8680750. [CrossRef]
39. Zhou, J.; Qian, S.; Ye, G.; Copuroglu, O.; van Breugel, K.; Li, V.C. Improved fiber distribution and mechanical properties of engineered cementitious composites by adjusting mixing sequence. *Cem. Concr. Compos.* **2012**, *34*, 342–348. [CrossRef]
40. Li, M.; Li, V.C. Rheology, fiber dispersion, and robust properties of engineered cementitious composites. *Mater. Struct.* **2013**, *46*, 405–420. [CrossRef]
41. Li, V.C. *Engineered Cementitious Composites (ECC): Bendable Concrete for Sustainable and Resilient Infrastructure*, 1st ed.; Springer: Berlin/Heidelberg, Germany, 2018; pp. 73–99.
42. Tada, H.; Paris, P.C.; Irwin, G.R. *The Stress Analysis of Cracks Handbook*, 3rd ed.; ASME: New York, NY, USA, 2000; p. 58.
43. Kanda, T.; Li, V.C. Practical design criteria for saturated pseudo strain hardening behavior in ECC. *J. Adv. Concr. Technol.* **2006**, *4*, 59–72. [CrossRef]

Disclaimer/Publisher's Note: The statements, opinions and data contained in all publications are solely those of the individual author(s) and contributor(s) and not of MDPI and/or the editor(s). MDPI and/or the editor(s) disclaim responsibility for any injury to people or property resulting from any ideas, methods, instructions or products referred to in the content.

Article

Mechanical Properties of Fly Ash-Based Geopolymer Concrete Incorporation Nylon66 Fiber

Muhd Hafizuddin Yazid [1,2], Meor Ahmad Faris [1,3], Mohd Mustafa Al Bakri Abdullah [1,2,*], Muhammad Shazril I. Ibrahim [3], Rafiza Abdul Razak [1], Dumitru Doru Burduhos Nergis [4,*], Diana Petronela Burduhos Nergis [4], Omrane Benjeddou [5] and Khanh-Son Nguyen [6]

1. Center of Excellence Geopolymer & Green Technology (CEGeoGTech), Universiti Malaysia Perlis (UniMAP), Kangar 01000, Malaysia
2. Faculty of Chemical Engineering & Technology, Universiti Malaysia Perlis (UniMAP), Kangar 01000, Malaysia
3. Department of Civil Engineering, Faculty of Engineering, Universiti Malaya, Kuala Lumpur 50603, Malaysia
4. Faculty of Materials Science and Engineering, Gheorghe Asachi Technical University of Iasi, 700050 Iasi, Romania
5. Department of Civil Engineering, College of Engineering, Prince Sattam Bin Abdulaziz University, Alkharj 16273, Saudi Arabia
6. Faculty of Materials Technology, Ho Chi Minh City University of Technology—HCMUT, Ho Chi Minh City 70000, Vietnam
* Correspondence: mustafa_albakri@unimap.edu.my (M.M.A.B.A.); doru.burduhos@tuiasi.ro (D.D.B.N.)

Abstract: This study was carried out to investigate the effect of the diamond-shaped Interlocking Chain Plastic Bead (ICPB) on fiber-reinforced fly ash-based geopolymer concrete. In this study, geopolymer concrete was produced using fly ash, NaOH, silicate, aggregate, and nylon66 fibers. Characterization of fly ash-based geopolymers (FGP) and fly ash-based geopolymer concrete (FRGPC) included chemical composition via XRF, functional group analysis via FTIR, compressive strength determination, flexural strength, density, slump test, and water absorption. The percentage of fiber volume added to FRGPC and FGP varied from 0% to 0.5%, and 1.5% to 2.0%. From the results obtained, it was found that ICBP fiber led to a negative result for FGP at 28 days but showed a better performance in FRGPC reinforced fiber at 28 and 90 days compared to plain geopolymer concrete. Meanwhile, NFRPGC showed that the optimum result was obtained with 0.5% of fiber addition due to the compressive strength performance at 28 days and 90 days, which were 67.7 MPa and 970.13 MPa, respectively. Similar results were observed for flexural strength, where 0.5% fiber addition resulted in the highest strength at 28 and 90 days (4.43 MPa and 4.99 MPa, respectively), and the strength performance began to decline after 0.5% fiber addition. According to the results of the slump test, an increase in fiber addition decreases the workability of geopolymer concrete. Density and water absorption, however, increase proportionally with the amount of fiber added. Therefore, diamond-shaped ICPB fiber in geopolymer concrete exhibits superior compressive and flexural strength.

Keywords: geopolymers; geopolymer concrete; polymer fiber reinforced geopolymers; interfacial bonding

1. Introduction

Traditional Portland cement (OPC) is regarded as the most widely used construction material in the world for the production of mortars. Large amounts of energy derived from the combustion of fossil fuels are used in the production of OPC, resulting in the emission of greenhouse gases, such as carbon dioxide (CO_2). According to earlier research, 1.5 metric tons of raw materials are required to produce one metric ton of cement, resulting in the emission of 0.8 metric tons of CO_2 into the environment [1].

Many studies have been conducted in an effort to reduce the OPC contents in concrete mixtures by partially or entirely substituting the OPC with a mineral addition or industrial by-product such as fly ash, slag, or silica fume in order to reduce CO_2 emissions [2]. Due to the substitution of aluminosilicate materials, geopolymers have been introduced as alternatives to OPC in the construction field. Geopolymers can be made from any raw materials that have a high silica (SiO_2) and alumina (Al_2O_3) composition as their main constituents, can react with a concentrated alkaline solution, and have thermal energy for curing to speed up the reactions [3].

One of the commonly used aluminosilicate materials is fly ash. Fly ash is a by-product that is produced from burning anthracite or bituminous coal. Fly ash is widely available over the world, possesses pozzolanic properties, and is high in alumina and silicate, but its application has been limited so far. Despite the fact that coal-burning power plants are environmentally harmful, the amount of energy produced by them is increasing due to the huge global supply of high-quality coal and the low cost of energy produced from these sources [4–6].

Due to the material's high compressive strength and low tensile strength, geopolymer has been shown to possess mechanical properties similar to those of hardened cement (brittle). Fiber reinforced concrete (FRC), also known as conventional concrete, is made by randomly adding tiny fibers to the concrete mixture to increase its brittleness. If a crack develops in plain concrete while it is being loaded, it spreads quickly and results in a loss of load carrying capacity. In contrast, with FRC, the break is intercepted by the fibers scattered throughout the matrix, causing it to slow down and even come to a stop. This process, known as the "crack bridging effect", increases the toughness of the concrete and maintains its capability of supporting a load even after the first crack appears [2].

The type of fiber, fiber content, bonding strength between the fiber and matrix, and mechanical properties of the fiber are significant in improving the mechanical qualities of geopolymer concrete (GPC). Steel fiber (SF) and polypropylene fiber (PF) are the two most common forms of fiber [7]. Metallic fibers frequently enhance flexural strength due to their higher stiffness, whereas non-metallic fibers regulate plastic matrix shrinkage due to their larger aspect ratio and surface contact area [8]. PF is believed to improve the performance of concrete owing to its high impact resistance, greater strain to failure, fine crack-free finish, increased water permeability resistance, and subsequently improved durability.

On the other hand, fibers from a variety of materials, including metal-based fibers such as steel and stainless-steel alloys; carbon-based fibers such as PAN rayon and mesophase pitch; synthetic fibers such as polyvinyl alcohol, polypropylene, and polyethylene; natural fibers such as jute, sisal, bamboo, and coconut; and inorganic fibers such as silica and basalt, are frequently used in composite materials [9].

The high strength, high modulus fibers, such as steel, glass, asbestos, carbon, and etc., are primarily used to acquire superior strain hardening after peak load, fracture toughness, and resistance to fatigue/thermal shocks, whereas the low modulus, high elongation fibers, such as nylon, polypropylene, PET (Polyethylene terephthalate), polyester, and shredded tire wastes, are potentially used in, but not limited to, enhancement of energy absorbed [10].

The most common steel and polypropylene fibers employed in nylon fiber research were quite few. A synthetic substance is nylon. Nylon is a smooth, thermoplastic substance that may be melted and processed into a variety of "films, fibers, or forms". Nylon fiber was chosen because it has excellent hardness, resilience, and durability qualities. It is also easily available in a wide range of colors, can be dyed, is resistant to soil and filth, has high abrasion capabilities, and can be cut into various cross sections. Nylon is resistant to a range of materials, hydrophilic, heat stable, and generally inactive. After the first fracture, nylon is most effective in increasing concrete's load-bearing capacity, flexural toughness, and impact resistance [11].

The prospect of adding fibers as reinforcement to a geopolymer matrix is therefore the subject of investigation. These materials' flexural strength and fracture toughness should both be strengthened by the addition of fibers, as well as the energy that the geopolymer can take before suffering damage. Strengthening geopolymers with short fibers is particularly effective because of how easily they can be dispersed. A fracture becomes more ductile and less brittle as fibers are added. The material's cracks are less numerous, and they are smaller in size, with a maximum crack width. This is especially true of microcracks, which are less likely to spread [12].

The main objective of this study is to investigate the strength of the fly ash geopolymer concrete reinforced fibers at two different curing times, which are 28 days and 90 days. The nylon66 fiber has been introduced in this study due to its good properties. This study also focuses on the effect of interlocking plastic bead (diamond end shape) toward the geopolymer concrete strength properties.

2. Materials and Methods

Preparation of NFRGC

In this study, Nylon66 Fiber Reinforcement Geopolymer Concrete (NFRGC) was used in the formation of geopolymers alongside other materials, including fly ash, alkali activator, and aggregates. Sodium silicate (Na_2SiO_3) and sodium hydroxide (NaOH) were combined to create an alkali activator with a ratio of 2.5. 12M of NaOH concentration was used in this research, which was achieved by diluting the NaOH pellet in distilled water at the desired concentration. Meanwhile, the ratio of fly ash to an alkali activator was fixed at 2.0. All of the selected ratios for the formation of fly ash geopolymers in Tables 1 and 2 were based on the previous findings [13].

Table 1. Mix design of nylon66 fiber reinforced geopolymer concrete for compressive.

Plastic Fiber Addition (Kg/m^3)	Fly Ash (kg/m^3)	Coarse Aggregate (kg/m^3)	Fine Aggregate (kg/m^3)	Sodium Silicate (kg/m^3)	Sodium Hydroxide (kg/m^3)
0	640.00	864.00	576.00	229.00	91.00
0.012	630.44	851.10	567.40	225.58	89.42
0.024	620.99	838.34	558.89	222.20	88.30
0.036	611.63	825.70	550.47	218.85	86.97
0.048	602.36	813.19	542.13	215.53	85.65

Table 2. Mix design of nylon66 fiber reinforced geopolymer concrete for flexural.

Plastic Fiber Addition (kg/m^3)	Fly Ash (kg/m^3)	Coarse Aggregate (kg/m^3)	Fine Aggregate (kg/m^3)	Sodium Silicate (kg/m^3)	Sodium Hydroxide (kg/m^3)
0	3200.00	4320.00	2880.00	1145.00	455.00
0.06	3152.20	4255.50	2837.00	1127.90	447.10
0.12	3104.95	4191.70	2794.45	1101.00	441.50
0.18	3058.18	4128.50	2752.35	1094.25	438.85
0.24	3011.80	4065.96	2710.65	1077.65	428.25

The fly ash class C used in this study was taken from the plant Manjung, Perak, Malaysia. There are two types of aggregates used in this study: fine and coarse. River sand was used as fine aggregate, and granite was used as coarse aggregate, with sizes of 4.7 mm and 20 mm, respectively. The combination ratio for both aggregates is 60% coarse and 40% fine by weight. Meanwhile, the ratio between geopolymers and aggregate is 40% geopolymers and 60% aggregate.

Fly ash and alkali activators are mixed at a ratio of 2.0 to create the geopolymer paste. Nylon66 fibers of diamond form were used in this experiment, which involved Interlocking Chain Plastic Beads (ICPB). The volume fraction of samples with compressive and flexural strengths of 0%, 0.5%, 1.0%, 1.5%, and 2.0% was used to determine the amount of nylon66 fibers to add to the geopolymer concrete mixture. Addition of 0%, 0.5%, 1.0%, 1.5%, and 2.0% by volume were tested for compressive strength testing. Additional information on Nylon fiber specification used in the production of NFRGC is summarized in Figure 1.

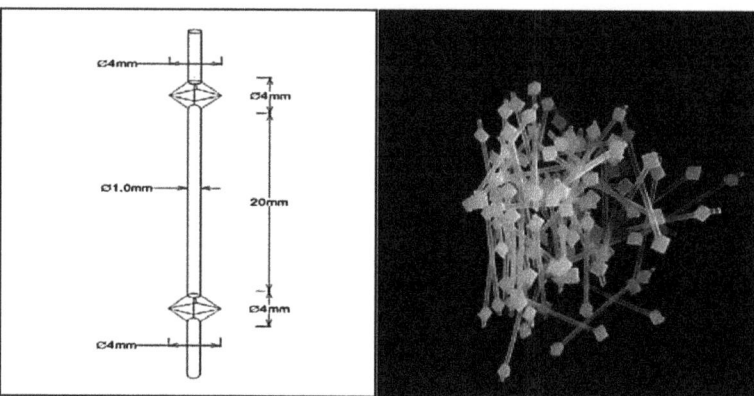

Figure 1. The plastic fiber and specification.

To create the required and precise shape and dimensions of the plastic bead, a unique mold was created. The beads' shape was created using the plastic injection molding process. With controlled speed and pressure, a molten nylon66 resin mixture colored with white was injected into the mold. The substance was then freed from the mold after cooling and taking on the appropriate form. The procedure was repeated in order to obtain more units. Six (6) different beads per set make up the linked plastic beads. They had two and three-bead systems cut off. This study does not include a fiber-type processing step. The fiber type was applied in this study to investigate the effect of ICPB in geopolymer concrete, since studies on the diamond shape of ICPB are still limited. Nylon66 is noted as a polymer, and thus has poor bonding between matrix and fiber compared to metallic fibers, but excellent corrosion resistance. In this study, new virgin material was used instead of recycled material to reduce impurities.

There is no standard shape for aggregates because they all have unique forms; however, spherical and angular aggregates are the most popular and function well. Additionally, as aggregate shapes are inherently uneven after crushing and display similarities in terms of shape, size, and surface roughness, it is impossible to design or manufacture something that is precisely like an aggregate. Based on the situation, diamond-shaped beads were chosen as the form for the beads. The diamond shape is both round and slightly angular.

The NFRGC samples were cast in (100 mm × 100 mm × 100 mm) and (500 mm × 100 mm × 100 mm) molds for physical (workability, density, and water absorption) and mechanical (compressive and flexural) testing. Following a 24 h curing period, samples were removed from the mold and allowed to cure for 28 and 90 days at room temperature.

A slump test was used to evaluate the NFRGC's workability. The ASTM C143 guidelines were followed for performing the slump test. After mixing, three layers of newly created geopolymer concrete were poured into a slump cone. Twenty-five tamping rod strokes were used to compress each layer. From the cone's top, fresher NFRGC was scraped off. The freshly constructed NFRGC was then immediately raised vertically to eliminate the concrete cone's workability. The slump was calculated by determining the separation between the top of the slump cone and the original center, which had been shifted, of the top surface of the new NFRGC.

A density test was conducted on the 28-day sample. A sample was submerged in water at room temperature for 24 h. In a water tank, the NFRGC sample was positioned apart from one another without touching. The top of the sample surface was no more than 150 mm relative to the still water line. To guarantee there was a 3 mm space between the sample and the bottom of the water container, the immersed sample was set on a wire mesh.

The 24 h immersed sample of NFRGC was weighed and recorded (Wi). The sample was then taken out of the water tank and left to dry for one minute. A moist towel was used to remove any apparent water from the sample's surface. Afterwards, the sample was weighed and recorded as being saturated (Ws). After that, the sample was dried for 24 h at 110 °C in an oven. Following that, the dried sample was weighed and given a dried weight label (Wd).

A Universal Testing Machine (UTM) Automatic Max was used to determine the compressive strength of sample NFRGC in accordance with standard BS 1881-116. (Instron, 5569, Norwood, MA, USA). This testing was done on samples that were cured for 28 and 90 days at room temperature. A load speed adjustment of 0.1 kN/s was made.

The flexural test was carried out to gauge the sample's flexural strength. Using the UTM model Automatic Max, the sample was put through a 4-point bending test (Instron, 5569, Norwood, MA, USA). The testing was carried out according to ASTM C1018. This study used a constant deflection rate that ranged from 0.05 to 0.10 mm/min. The lower and top supports were 300 mm and 100 mm in height, respectively. After being cured at room temperature for 28 and 90 days, the sample was examined.

3. Results and Discussion

3.1. Chemical Composition

Table 3 provides a summary of the chemical composition of the fly ash used to make geopolymer concrete with both types of fibers. There are five major elements that contribute to the properties of geopolymers, comprising SiO_2, CaO, Fe_2O_3, Al_2O_3 and MgO. It is worth noting that fly ash is composed of silicon oxide, aluminum oxide, iron oxides, and other minor oxides. Major components including SiO_2 and Al_2O_3 contribute almost 90% of the total weight of fly ash. Meanwhile, Fe_2O_3 content is less than 5% of the total weight of fly ash. From Table 3, the total composition of SiO_2 and Al_2O_3 are 43.9%, followed by CaO with 22.30%, Fe_2O_3 with 22.99%, and MgO with less than 1%. According to the chemical composition obtained, the fly ash used in this study was classified as Class C fly ash [14]. In addition, the fly ash used meets the basic requirements for a source of material to be used as a precursor due to its high Si and Al content, which is significant for creating geopolymer bonds.

Table 3. Chemical composition of fly ash.

Composition	Weight%
SiO_2	30.80
CaO	22.30
Fe_2O_3	22.99
Al_2O_3	13.10
MgO	4.00
K_2O	1.60
TiO_2	0.89
SO_3	2.67
MnO	0.21
Others	1.44

Si-O-Al appeared as one of the most significant linkages that affected the strength of the geopolymer, and the combination of Si and Al maps demonstrated how it formed. The geopolymer typically has a favorable setting time due to being high in calcium (Ca) content. Although there was a significant difference in the amount of Ca in the two geopolymers, it was discovered that the strength growth was gradual. Meanwhile, increasing curing temperature and time resulted in increased strength. The presence of Si and Al components in geopolymer composites influences strength development because more geopolymer chains are formed, which strengthens the geopolymer composite materials. The majority of the geopolymer's basic structure is made up of Si-O-Al, demonstrating the importance of Si and Al components in producing strong development. The presence of Mg, however, slowed the geopolymer's strength growth. This has disrupted the Ca-Si-O-backbone Al structure, reducing the geopolymer's ability to produce strength. In addition, due to the development of hydrotalcite group phases and a decrease in the amount of readily available Al element, appropriate Ca enables the formation of low Al C-(A)-S-H.

3.2. FTIR Spectera

The infrared analysis spectra of the applied fly ash are shown in Figure 2. The figure shows several peaks at 428 cm^{-1}, 532.64 cm^{-1}, 733.22 cm^{-1}, 1253.53 cm^{-1}, 1663.95 cm^{-1}, 3222.98 cm^{-1}, and 3782 cm^{-1}. Absorption bands at 733.22 cm^{-1}, 532.64 cm^{-1}, and 428.59 cm^{-1} were labelled as O Si O links in quartz, and Si O and Al O bonds in zeolite frameworks, and the band surrounding 1000−960 cm−1 represents bonds of Si–O–T (T is tetrahedral Si or Al) of the geopolymer gel. Absorption bands at regions at 450 cm^{-1} can represent Si–O–Al linkage; Si O bond characterizes to bending vibration at 400−500 cm^{-1}, and the stretching vibration at 800−1000 cm^{-1}. Although, absorption bands in regions at 980 cm^{-1} can be related to O–Si–O bond bending vibration, or symmetric stretching vibrations of the Si–O–Si (Al) bridge [15,16].

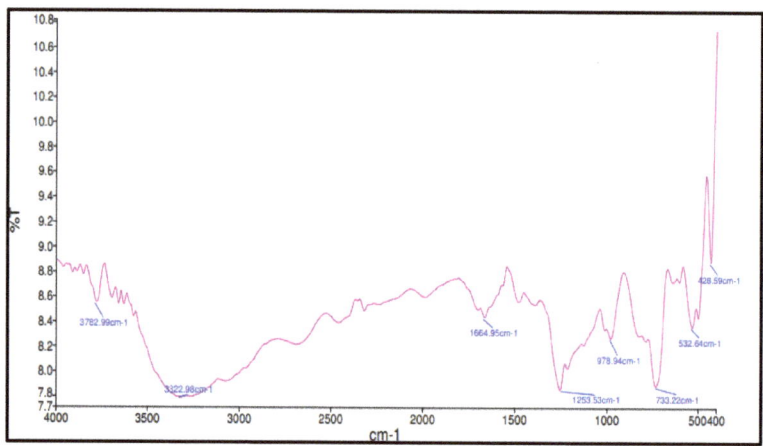

Figure 2. FTIR spectrum of fly ash.

The asymmetric stretching vibrations of the silicon tetrahedral (SiO4-4) found in the chain structure of the Si-O terminal bonds can also be attributed to several additional bands found in regions around 1253.53 cm^{-1} [17]. Meanwhile, the stretching vibration of O-H and H-O-H due to water and silanol group occurs within a range of 3222.98 cm^{-1} to 3782 cm^{-1}. This indicates a stretching vibration of O-H and H-O-H from 82 water molecules which are weakly bonded that appear at the surface, or are trapped in a large cavity inside the geopolymer sample. In addition, a wavenumber of 1664.95 cm^{-1} represents bending vibration of H-O-H.

Meanwhile, infrared analysis spectra for the geopolymer concrete are illustrated in Figure 3. The result shows observation at peak 3775.37 cm^{-1}, 3454.07 cm^{-1}, 1638.68 cm^{-1}, 1544.29 cm^{-1}, 1411.56 cm^{-1}, 1062.90 cm^{-1}, and 671.18 cm^{-1}.

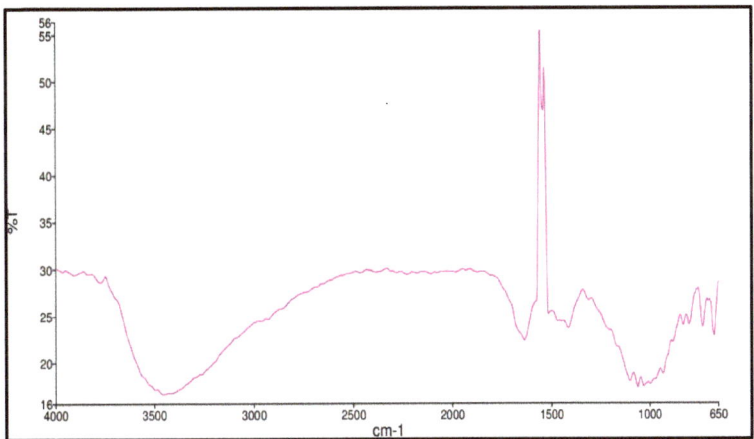

Figure 3. FTIR spectrum of geopolymers.

The intensity of absorption bands at 671.18 cm^{-1} is connected to the stretching vibration of the Si-O-Si symmetry and the bending vibration of the O-Si-O bonds (Al) [17]. The size of these bands is due to the material being amorphous. There is also a vibration band for the stretching of Si-O-A at 1062.90 cm^{-1}. The Si-O-Al was determined by the peaks found between 700 and 1100 cm^{-1} [18]. In the peak from 1400 to 1450 cm^{-1} it was noticed the temperature rose to 1000 °C. As the peak shifted to 1411.56 cm^{-1}, the strength of the composite decreased. The band at 1411.56 cm^{-1} displays the feature of the asymmetric O-C-O stretching mode, which shows the existence of sodium carbonate due to the interaction between too much sodium and ambient carbon dioxide [19].

Three bands located at 1638.68 cm^{-1}, 3375.37 cm^{-1}, and 3454.07 cm^{-1} were associated with the water molecules. As a result of the inclusion of nanoparticles, the overall spectra also demonstrated an increase in the intensity of the Si-O-Al band, suggesting a rise in the quantity of N-A-S-H gel [20]. Simultaneously, the frequency moved to a higher wavenumber at 1544.29 cm^{-1} as rising solid/liquid ratios, which suggested calcite vibration. Calcite and amorphous silica were produced when tobermorite decalcified, which caused the wavenumber to change [21]. Peak calcium-based component intensity demonstrated the dominance of high strength geopolymer structure.

3.3. Compressive Strength

Fiber reinforced geopolymers at 28 days, as well as Geopolymer concrete with nylon66 fiber (NF) reinforcement's compressive strength for both samples at 28 and 90 days. The compressive strength of geopolymer concrete appears to increase with plastic fiber addition, up to a maximum value at 0.50% of fiber addition. This is because nylon66 fibers, which restrict cracks from spreading during compression loads, and linked interlocking plastic beads act as reinforcing agents by interlocking with each other in the aggregate skeleton. The main strategy used in this study is to fill the spaces between the fine and coarse aggregate with beads to give them an interlocking strength using a linked plastic system, as illustrated in the schematic picture in Figure 4. The weak interfacial connection between the matrix and the fiber caused by hydrophobic surface characteristics was significantly improved by the linked interlocking plastic beads. As a result of the nylon66 fibers' contribution, the geopolymer binder slid out of the nylon66 fibers' diamond-shaped ends with greater resistance than the straight fiber without anchorage.

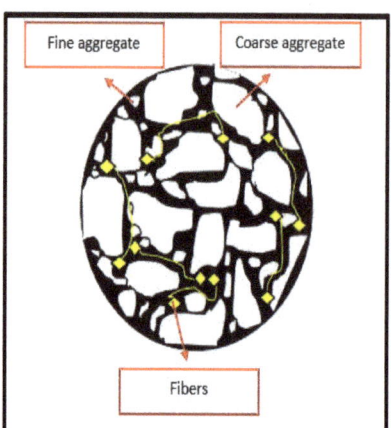

Figure 4. The chained interlocking plastic beads schematic diagram illustrated.

The compressive strength of geopolymers with fiber addition is depicted in Figure 5. The results demonstrate that fly ash geopolymers without nylon66 fiber addition have higher strengths, and the strength starts to decrease with the inclusion of nylon66 fiber. Even though the reduction in strength is about 35% at 0.50v% fiber addition, the strength obtained is still notably higher (35.59 MPa). According to the results, adding nylon66 fiber or linked interlocking plastic beads did not improve the compressive strength properties of geopolymers. The interfacial connection between the matrix and fiber is believed to be weak due to the smooth or hydrophobic surfaces of the polymers, and the fiber cannot inhibit the spread of cracks in geopolymers. However, some regions in the geopolymer matrix that include nylon66 fibers are believed to fill the voids between fly ash particles with beads to give them an interlocking strength and contribute to good strength. Insertion of fiber greater than 0.50% disturbs the CASH bonding in the geopolymer matrix and diminishes its compressive strength. According Patrycya et al. [22], the optimum result obtained on geopolymers reinforced with hooked-end steel fiber and melamine fiber was circa 0.5% amount of fiber by weight. The result shows that plain GPC is 40 MPa; steel fiber 0.5% is 40 MPa, and 1.0% is 39 MPa; and melamine fiber 0.5% is 50 MPa, and 1.0% is 45 MPa. Melamine fiber has better resistance to force. Based on the research, fiber shape gave an effect to the compressive strength on geopolymers; hooked-end type steel fiber held the matrix with greater force during crack propagation [23]. The schematic function of fiber that was used in this study for the geopolymer concrete was illustrated previously in Figure 4. The addition of nylon66 fibers' (ICPB) diamond shape on reinforced geopolymers was intended to investigate the effect on the compressive strength between GP and GPC.

Figure 6a depicts the compressive strength of Nylon66 Fiber Reinforced Geopolymer Concrete (NFRGC) after 28 days of room temperature curing. It was discovered that adding 0.5% of fiber resulted in a higher compressive strength with a value of 67.6 MPa, which then decreased to 53.3 MPa with the addition of fiber at 2.0%. Geopolymer concrete has a high compressive strength, and suitable fiber addition as well as fiber type were discovered to increase properties depending on the application. Chained interlocking plastic-bead fibers increase the strength of NFRGC as compared to geopolymers. This is due to the capacity of Nylon66 fibers to delay the spread of cracks during compression loads. This can be attributed to RTS fiber's high stiffness and hydrophilicity, which allow it to absorb more energy and form a strong fiber-matrix interaction [23,24].

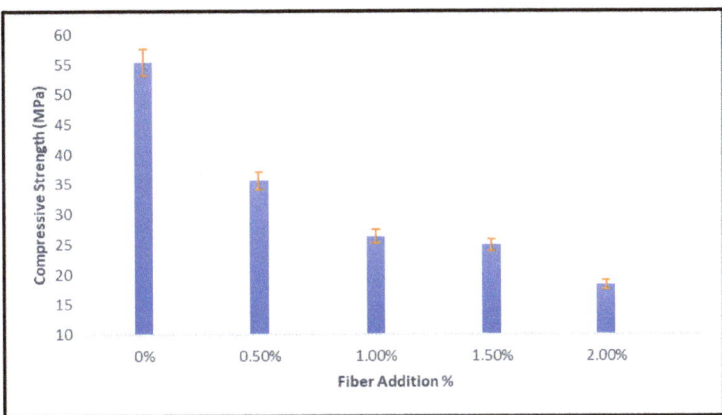

Figure 5. Compressive strength of geopolymers.

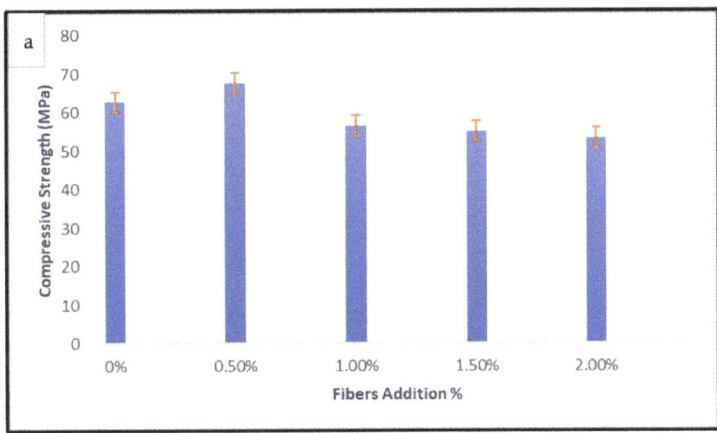

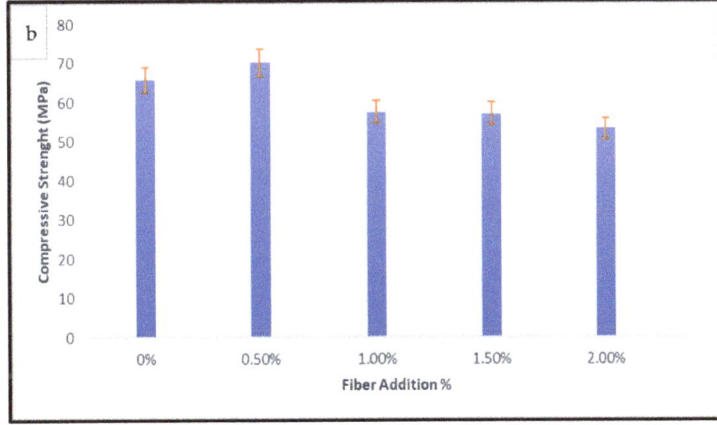

Figure 6. (**a**) Compressive strength of NFRGC for 30 days and (**b**) Compressive strength of NFRGC for 90 days.

Since Nylon66 fiber has a lower young modulus than steel fiber, increasing fiber addition in geopolymers results in a negative trend in concrete. As additional fiber is added, the compressive strength decreases while the toughness increases due to the higher elasticity of nylon66 fiber over geopolymer concrete. Fiber length influences compressive strength or toughness strength, and research has shown that short fiber is ideal for these qualities as well as to avoid microcrack propagation. Composites with an irregular internal structure resulted in reduced compressive strengths and variable compressive behaviour. As a result, the incorporation of metallic fibers could improve the mechanical properties of GPC, whilst the high fraction of nylon66 fibers in GPC could lower its mechanical performance. The substantial standard deviations of the findings of the hybrid replacement series with more fiber made this apparent. Meanwhile, Figure 6b illustrates the NFRGC's compressive strength after 90 days.

The compressive strength of NFRGC is affected by the curing period. After 90 days, the compressive strength of the NFRGC has increased in comparison to 28 days. After 90 days, the compressive strength increased to 70.13 MPa, from 67.6 MPa at 28 days, with the addition of 0.5% fiber. However, once the fiber inclusion exceeds 0.5%, the compressive strength of geopolymers decreases. When the NF volume exceeds 0.5%, the decrease in compressive strength is primarily due to the difficulty of fiber distribution, especially in large volume fractions, which is caused by poor workability and inadequate compaction. In contrast, NFRGC with the lowest nylon66 fiber concentration achieved the highest compressive strength, owing to the mechanical properties of nylon66 fiber. The fibers in concrete contributed to energy dissipation via the bridging effect of their shape and mechanical properties. The frictional bonding that develops as a result of the resistance to pulling out the nylon66 fibers, caused by friction between the fibers and the geopolymer matrix, contribute to the NFRGC's high strength [25].

In addition, as the fiber content increased to 1%, compressive strength decreased substantially from 70.13 to 57.5 MPa. This is believed to be attributed to the material's poor compaction and significant voids. Due to the material's high degree of flexibility, high volume fractions of Nylon66 fiber make compaction difficult, resulting in a loose and porous geopolymer matrix. The relative density of fiber-reinforced geopolymer composites, on the other hand, was decreased by adding more fibers. This is due to the fact that the air bubbles caused by imperfect vibration in the composite products caused the relative density to increase. This condition hinders the consolidation of the fresh mixture, and even the long exterior vibrations are ineffective at compacting the concrete. As observed, an increase in fiber content above 2% has a negative impact on composite density.

The compressive strength of fiber reinforced concrete increased initially and subsequently declined as the nylon66 fiber content grew from 0% to 2%, with a 0.5% optimal point where the internal structure of geopolymer concrete was considerably enhanced. The main factor causing the decline in compressive strength when the NF percentages are more than 0.5% is the difficulty in dispersing fiber, especially in large volume fractions, which contributes to poor workability and insufficient compaction [26].

Judging by previous work, there are no studies focusing on Nylon as fiber in concrete. However, other types of fiber such as PP and PF were the guidance in this research. According to Wang et al. [27] the compressive strength of polypropylene (PP) fiber reinforced geopolymer concrete with fiber length of 12 mm was observed to be slightly higher than that of 3 mm. The fiber type was straight fibers. Longer fibers performed better in terms of bridge effects because of the increased contact area between them and the geopolymer concrete, which led to a greater frictional force. In comparison to shorter strands, longer fiber could connect more air spaces. This research shows that effect of length contributes to the contact area and helps improve the bonding of polymer fibers and geopolymers. Compared to our study using long fiber, short fiber can also be improved by size and shape.

Piti et al.'s [2] study used the PF crimped type in fiber reinforced geopolymers, based on the compressive strength result that 0.5% fiber content is the optimum result. The plain geopolymer's strength was 40.08 MPa, and the strength improved to 47.0 MPa after being reinforced by fibers at 0.5%. This illustrated that the trend of compressive strength was decreased to 34.49 MPa at 1.0% and so on.

According to Ranjbar et al.'s [15] investigation on the mechanisms of interfacial bond in steel and polypropylene reinforced geopolymer composite, after curing PF reinforced geopolymers for 56 days, the compressive strength of the plain geopolymers was the highest compared to others with fiber added. They illustrated that 0.5% content was the best, with 45 MPa compressive strength, compared with 1%, 2%, 3%, and 4% contents.

The compression results between 28 and 90 days followed the same pattern as the optimal result, indicating that a fiber content of 0.5% is the best result. The 90-day outcome was somewhat enhanced due to the geopolymers themselves. Curing time and temperature are significant factors in the hydration process of geopolymerization, with higher temperatures accelerating the hydration process and contributing to the geopolymer's high strength. An extended curing period, however, influences the performance of geopolymers. In this instance, curing time enhances the compressive strength of geopolymers. The addition of NF reinforced in fly ash-based geopolymers progressively causes a geopolymer bonding reaction in the NFRGC, and the interlocking between the fiber, aggregate, and geopolymer matrix gets better with time.

The failure mode of the NFRGC cube when compressed is shown in Figure 7. All NFRGC specimens maintained their forms with little debris even after compression-induced failure, which is often characterized by evident large fissures.

Figure 7a shows the geopolymer concrete breaks into parts due to the brittle properties of geopolymer concrete. The fibers provided greater energy for resisting tensile tension in the cube, which prevented tensile fracture growth. Without fiber, geopolymer concrete can withstand the high load of energy. Addition of nylon66 fibers make the major crack propagation directly occur without displaying signs of crack growth prior to breakage.

In comparison to geopolymer concrete without fiber addition, 0.5% has the highest compressive strength of all results, despite having significant fracture propagation. The addition of fiber improves geopolymers' ability to absorb energy, and the interlocking plastic beads aid in limiting crack propagation, resulting in the major crack spreading from the minor crack after 0.5% fiber was added. The inclusion of more fibers reduces compressive strength; however, crack propagation was decreased from major to minor due to the energy supplied into the fibers during the compression test to slow or stop the crack growth. The tensile strength was only slightly different from the value reported in the work of Arsalan et al. [25], which included NF as a fiber addition to the concrete mix. In addition to selecting the proper fiber fraction, geopolymer concrete must also have equally distributed fibers in order to achieve the desired amount of strength.

During the compressive test, the greater fiber volume controlled the development of cracks. Geopolymer density decreases as fiber volume increases, whereas compressive strength and toughness increase. With the interlocking chain fiber, it is feasible to reduce energy transmission from the geopolymer concrete itself. It exchanges energy with the fiber to slow the spread of cracks. Nylon66 fiber can restrict the spread of cracks in geopolymer concrete, as shown in Figure 7e. It exhibits the symptoms of material breakdown as the crack spreads.

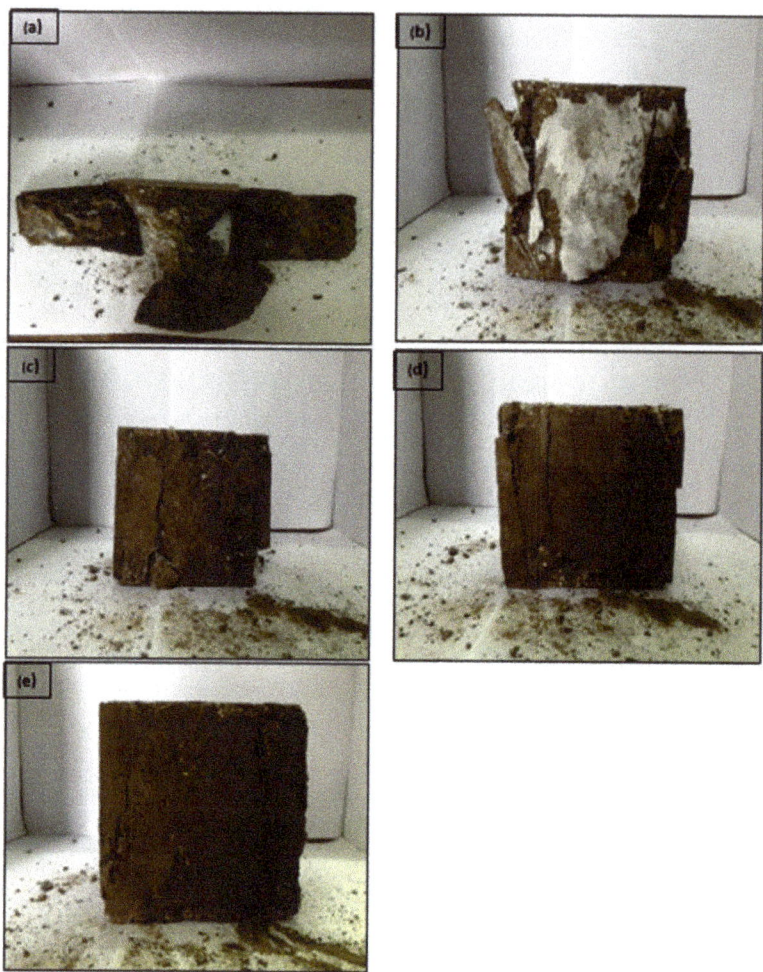

Figure 7. Crack pattern of geopolymer concrete (**a**) without fiber, (**b**) 0.5%, (**c**) 1.0%, (**d**) 1.5%, and 2.0% (**e**) with a schematic of failure mode.

3.4. Flexural Strength

Figures below illustrate the flexural strength of geopolymer concrete and NFRGC with fiber addition after 28 and 90 days of curing, respectively. The flexural strength of GPC was observed to be enhanced with fiber added compared to plain GPC, and this improvement increased as the volume percent of fiber in the GPC increased.

From Figure 8, it was found that the inclusion of geopolymers led to an increment of the flexural strength of a concrete to a maximum of 0.5%, or 4.43 MPa. Meanwhile, normal geopolymer concrete without fiber addition exhibits a flexural strength of 4.39 MPa. Furthermore, when the nylon66 content exceeded 0.5%, the flexural strength began to decline. This was due to the samples' poor workability when nylon66 fibers were added in large quantities. It is believed that this poor workability has an impact on the distribution of nylon66 fibers. As a result, when loading was applied, the absorption capacity inside the sample was unbalanced, thus causing crack formation.

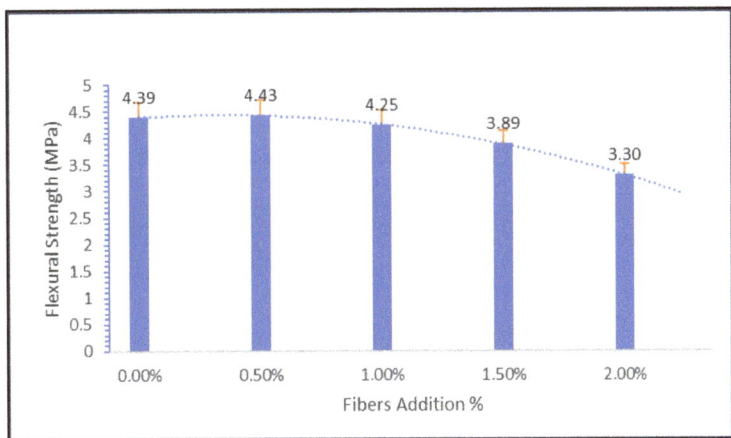

Figure 8. Flexural strength for 28 days.

Due to the limited availability of fiber, none of the three varieties have the same forms, surface smoothness, or aspect ratio, making direct comparisons using normalized or deleted measurements of the same level difficult. The tensile strength of the fiber has the greatest influence on post-crack behavior, and these factors only have an impact on the first fracture load. Furthermore, for each type of fiber used, various fiber volume fractions are generated, taking into account cost, density, and fiber dispersion in the concrete mix.

As shown in Figure 9, the nylon66 fiber played a role in enhancing flexural strength by inhibiting crack propagation during flexural testing by bridging at the crack region. With the addition of Nylon66 fibers, the sample was able to sustain a larger flexural force prior to complete failure. Photographic observation of the crack and final fracture in NFRGC and geopolymer concrete with various fiber additions is shown in Figure 9.

Figure 9. Shows the sample of bending test (**a**) geopolymer concrete without fiber addition and (**b**) with fiber addition.

The fiber failure mode demonstrates which type of feature dominates the flexural performance of geopolymer concrete. The majority of fibers do not draw out. All fibers are not extracted, particularly in the NFRGC. In this case, the binding property between the fiber and the concrete significantly influences how effectively the structure bends. The majority of the fibers rip apart at the fracture surface, indicating that fiber tensile behavior has a major influence on reinforced concrete flexural performance.

With an increase in the percentage of fiber volume, the number of fibers spread over the fracture surface increases, and the post-cracking performance is also enhanced. Fracture toughness, also known as post-crack performance, is expressed by the energy absorbed by a sample during deformation and failure.

Figure 10 shows that until the addition of 0.5%, the flexural strength of the geopolymer concrete increases to 4.99 MPa. Meanwhile, the plain geopolymer concrete without fiber addition has a flexural strength of 4.87 MPa. Flexural performance decreases when nylon66 fiber addition exceeds 0.5%. This is due to the samples' poor workability when substantial amounts of Nylon66 fibers are introduced. It is believed that the low workability affects the distribution of nylon66 fibers. As a result, when loading is applied, the absorption capacity within the sample is unbalanced. The sample frequently cracks due to the fewer fibers available. The results of the comparison between 28 and 90 days show that the curing period is the most important factor in the development of the flexural strength.

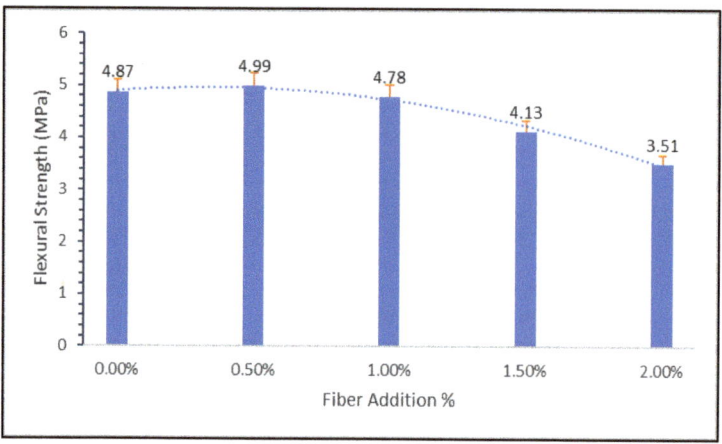

Figure 10. Flexural strength for 90 days.

Results for 90 days are better than those for 30 days since geopolymers are still hydrating slowly at that point. The pattern matches the compressive strength result exactly. In comparison to 30 days, the geopolymerization at 90 days enhanced the geopolymers' characteristics, leading to better microstructure properties. According to Figure 10, it was found that Nylon66 fiber improved the flexural strength of geopolymer concrete. Regardless of the fiber type, the improvement in first-crack strength was expected due to the increase in fiber volume fraction.

Moreover, the nylon66 fiber reinforced geopolymer concrete was noted to be primarily responsible for the fiber bridging effect. Therefore, the characteristics of strain hardening and the flexural strength may be adversely affected by an increase in fiber content due to the uneven geometry of fiber from the recycling process.

When the specimen was subjected to the bending load, the area between the two loading pins, where the flexural stress was at its highest, began to deform and split. Once the matrix's bending strength was exceeded, the first crack in composite materials began to form. After that, the crack continued to spread until it reached a nylon66 fiber with a low rigidity. The fracture attempted to penetrate through the fiber at this point due to the flexural tension being applied, which caused it to elongate, rupture, or pull out.

According to Wang et al. [27], the fiber addition was found to significantly improve the flexural strength of PPRGPC. The percentage of addition was varied at 0.1%, 0.15%, and 2.0%, respectively. This study used the PP fiber straight type. According to the result obtained, plain GPC obtained a flexural strength of 4 MPa and the strength was increased to 4.3 MPa with 0.1% fiber addition. Meanwhile, the flexural strength started to decrease for fiber addition at 0.15% (4.1 MPa) and 2.0% (4 MPa).

The geopolymerization product that had adhered to the nylon66 fibers' surface suggested that the binding strength might be high enough to activate this mechanism. The interfacial bond strength, in contrast, is weaker than the applied stress. By redistributing the localized stress, fiber bridging caused the specimens to develop many microcracks. Increases in ductility and post-cracking toughness were brought on by the ongoing process of microcrack development.

Flexural strength at 28 and 90 days yielded results with several decimals since there was little significant variation in strength. The result is 5 MPa to 3 MPa between 28 and 90 days. Flexural strength displays the improvement and gap strength for each fiber added for 0.5%, 1%, 1.5%, and 2.0% with numerous decimals. The fiber distribution inside the geopolymers cannot be controlled, which is a problem.

3.5. Water Absorption

The results of the water absorption test are illustrated in Figure 11. With more fiber additions, the water absorption of geopolymer concrete increased. For geopolymer concrete, the nylon66 fibers with a 2.0% concentration have the maximum water absorption (0.057). This is because the workability reduced with the addition of nylon66 fibers as discovered in this study, which might cause an increase in the creation of pores.

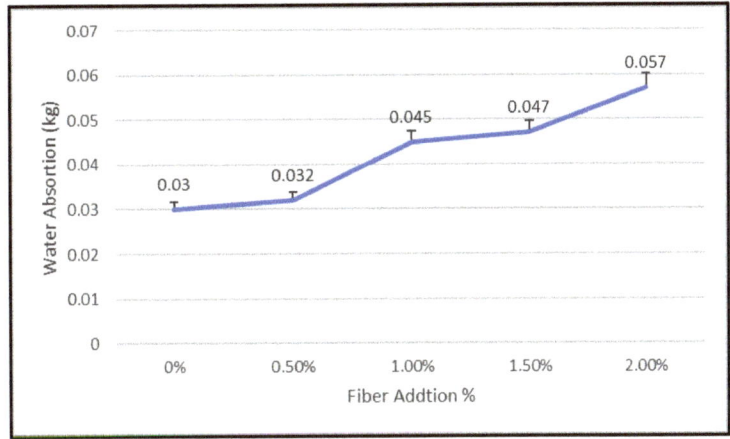

Figure 11. Water absorption of geopolymer concrete with addition of plastic fibers.

Permeability is a measure of how efficiently water, air, and other chemicals, such as chloride ions, can be absorbed by geopolymer concrete. Similar to OPC concrete, geopolymer concrete also contains pores that enable the absorption of particular compounds. Higher porosity leads to higher water absorption, which lowers the density of the concrete. Meanwhile, less porosity leads to a higher density of geopolymers, decreasing water absorption. Figure 11 shows the significant relation between water absorption and density, as well as how fiber addition appears to improve water absorption due to higher density. The weak interfacial interaction of Nylon66 fiber with the matrix could lead to the formation of a void, which would increase water absorption as fiber addition increased. Furthermore, as the amount of nylon66 fiber in geopolymer concrete increases, the degree of compaction in the mix decreases, encouraging the volume of air voids in the geopolymer concrete.

According to Jawad et al. [11], when using nylon fibers, water absorption is increased by 3–6%. As compared to samples of ordinary concrete, samples of concrete reinforced with nylon fibers absorb slightly more water. Improved connection between microchannels in the concrete's outer surface and binder matrix may be to blame for this. Additionally, studies show that the addition of fibers improves concrete captivity and water absorption due to the lengthening of the microchannels in the microstructure.

This degradation would affect the performance of the geopolymer concrete's fiber reinforcement, including its compressive strength, flexural properties, fiber matrix interfacial bonding, and durability against blasting. It is essential that geopolymer concrete has minimal water absorption for better performance. The geopolymer concrete samples obtained from this study have a high potential for corrosion resistance due to low water absorption and the use of fiber material. Water absorption was investigated by weighing the sample after it was removed from the water, since in NRGPC, it alters the qualities of fibers made with poor resistance to corrosion. Because nylon66 fiber has a low water absorption rate and is unaffected by corrosion, the amount of fiber used in this investigation was not measured.

3.6. Slump Test

Using a standard slump cone, the slump was measured. In this study, the geopolymer concrete's consistency and workability were assessed using the slump test. Figure 12 shows the decrease trend of workability for geopolymer concrete with addition of nylon66 fiber.

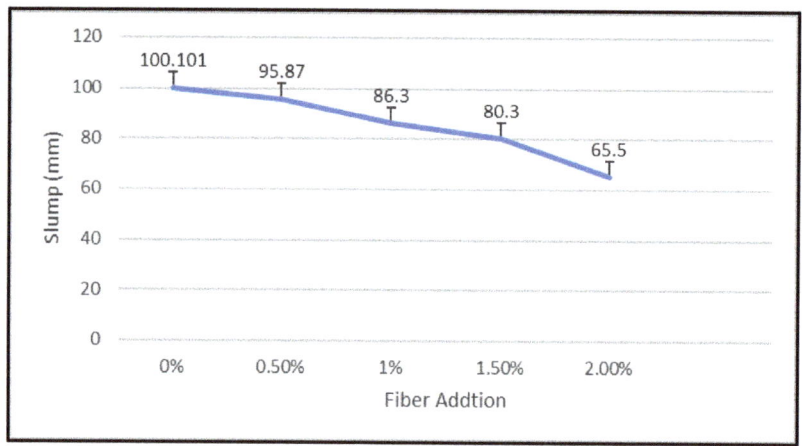

Figure 12. Slump of geopolymer concrete with addition of plastic fibers.

Figure 12 shows that the slump test result for geopolymer concrete without the inclusion of nylon66 fibers was 100.101 mm. The slump of geopolymer concrete with nylon66 fiber addition reduced with increasing additions of nylon66 fibers from 0% to 2%, which are 95.87 mm (0.5%), 86.3 mm (1%), 80.3 mm (1.5%), and 65.5 mm (2.0%). This has demonstrated that the presence of nylon66 fibers makes geopolymer concrete less workable. This finding suggests that the 65.5 to 100 mm range has low and medium workability.

This outcome also proved that the presence and addition of fibers significantly negatively impacted the workability of geopolymer concrete. This is due to increased friction between the geopolymer concrete matrix and fibers. The addition of more fibers causes the viscosity of new geopolymer concrete to increase because more binder is absorbed by the fibers' higher surface area, resulting in low slump.

In addition, the fiber and coarse aggregate particles were noted to have compatible dimensions, which contribute to resisting the relative mobility of the latter. The flow of fresh geopolymer concrete was resisted in this condition, making it more difficult for coarse particles to move. This interlocking of fiber and aggregate is depicted in Figure 4. As a result, the difficulty of the relative movement between the coarse aggregates and the movement of the mixture increases with the number of fibers added. The mixture flows much more slowly and becomes less workable. The slump test was carried out using a slump cone and a mixture of fresh NFRGC, measuring the distance between the surface of the latter and the top of the slump to gauge the combination's workability.

3.7. Density

Measured as mass per unit volume, density is the quantity of a substance. All samples were weighed after curing for 28 days at room temperature, and their masses were split by the mold's 100 mm × 100 mm × 100 mm dimensions. The impact of adding fiber to geopolymer concrete's density is seen in Figure 13.

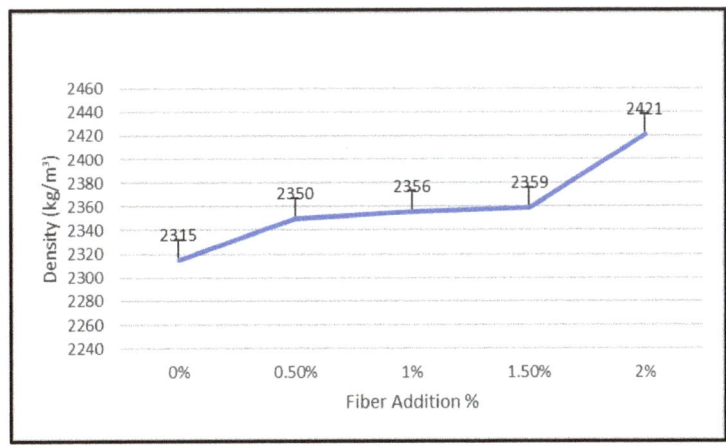

Figure 13. Density of plastic fibers against fiber addition.

According to the findings, 2421 kg/m^3 was the density of the addition of 2% plastic fiber. In direct proportion to the addition of more nylon66 fibers, the geopolymer's density rose. This result is illustrated by the range 2315–2421 kg/m^3 after fiber was added. The results show that the density of GPC and NFRGC increase 4% at 2% fiber added.

The change in density value with the addition of fiber in geopolymer concrete does not reveal any discernible trend in which the density rises and then somehow falls with the addition of a particular fiber. If the substitute fibers have nearly equal specific gravities, the density of any fiber-reinforced concrete often does not change considerably.

This tendency according to past investigations is the reverse of what we discovered. Fiber has increased based on weight rather than volume in this case. The GPC is influenced by the weight of nylon66 fibers itself. More fiber is added, which boosts the NFRGC's performance. We employed the dry test, with a total density of GPC of 2400 kg/m^3, in this experiment. The sample illustrates that the increase in the fiber addition reduces the shrinkage that can cause the weight loss.

4. Conclusions

The purpose of this study was to determine the effect of varying the percentage of nylon66 fiber in fly ash geopolymer concrete on strength performance. Furthermore, chemical, physical, and mechanical testing were performed for evaluating fiber properties, raw material characteristics, and geopolymers. Based on the analysis and experimental data results, the following conclusion can be drawn:

1. The majority of the geopolymers' basic structure is made up of Si-O-Al, indicating the significance of the Si and Al components in creating strong strength development. The presence of Mg in the geopolymer, on the other hand, hindered the geopolymers' ability to gain strength. This has disrupted the Ca-Si-O-backbone Al structure, reducing the geopolymers' ability to produce strength.
2. Geopolymers reinforced by nylon66 fiber exhibit negative data. The interfacial connection between the matrix and fiber is weak due to the plastic's smooth or hydrophobic surfaces, and the fiber cannot stop the spread of cracks in geopolymers. However, some geopolymer matrix spaces containing plastic fibers fill the spaces between fly ash

particles with beads to provide interlocking strength and contribute to good strength. More than 0.50% fiber insertion disrupts the CASH bonding in the geopolymer matrix and reduces its compressive strength.
3. Solid to liquid ratio 2.0, alkali activator 2.5, and 12 M NaOH alongside the aggregate ratio were found to be an optimized combination for the mixture process and molding.
4. NFRGC results show that 0.5% fiber addition yields the best results for 28 days (67.7 MPa) and 90 days (70.13 MPa). Due to the development of the geopolymer itself, 90 days NFRGC shows better data than 28 days. In addition, the properties of geopolymers are affected by their curing time. Figure 7 also shows evidence that increasing the volume of fiber increases energy absorption, which aids in the reduction of crack propagation and fracture before they fracture. The ICPB diamond-shaped nylon66 fibers helps to control crack propagation and reduce crack or fracture on the NFRGC by holding the aggregate and matrix together.
5. For the NFRGC, 0.5% was concluded to be the optimum addition due to the flexural strengths obtained for 28 days and 90 days (4.43 MPa and 4.99 MPa, respectively). Addition of plastic fiber at excess of 0.5% reduces the flexural strength. Short fiber showed a small contribution to the compressive and the young modulus of fibers but improve the energy absorbed, based on Figure 9, the comparison between GPC and NFRGC during the flexural test. Due to the higher volume of fiber friction, GPC fractures at the first crack and NFRGC fractures at the final crack. The dominant mode of fracture for nylon66 fibers is no pull out.
6. Based on Figure 9, the contribution of additional fiber improves the crack propagation and slows the fracture process by changing the process from major to minor crack propagation.
7. The water absorption of geopolymer concrete increased as fiber additions increased. The highest water absorption was obtained for geopolymer concrete with the addition of plastic fibers at 2.0% (0.057), and the lowest was obtained at 0.5% with a value of 0.032. This is due to a decrease in workability caused by the addition of nylon66 fibers, which resulted in an increase in pore formation. Water absorption is low in comparison to 0.5% and plain GPC. When comparing water absorption between 0.5% addition and plain GPC, a small range was found, but a larger range was observed when comparing to other volume fiber ratios. This is due to the variety of fiber shapes available, including cylindrical and diamond, as well as the uncontrollable fiber arrangement inside the geopolymers, which resulted in variations in water absorption. Furthermore, since nylon66 fiber is resistant to corrosion, it did not significantly affect the NFRGC.
8. The slump of geopolymer concrete with nylon66 fiber addition decreased as the plastic fiber additions increased. Increasing the fiber content increases the difficulty of relative movement between the coarse aggregates and motion of the mixture, resulting in less workability and flow. The main point is that ICPB diamond-shaped fiber contributes to low workability and a higher viscosity of the NRGPC, and holds the aggregates, giving high resistance in moving the mixture.
9. The change in density value with the addition of fiber in geopolymer concrete does not reveal any discernible trend in which the density rises, because the fiber also has its own density, which may lead to an increase of density of NFRGC.
10. There needs to be more study in these fibers with different materials and dimensions.

Author Contributions: M.H.Y.—conceptualization, writing, original draft preparation; M.A.F.—methodology, writing, original draft preparation; M.M.A.B.A.—formal analysis, project administration, methodology; M.S.I.I.—methodology, data curation; R.A.R.—data curation, formal analysis; D.D.B.N.—writing, review and editing, funding acquisition; D.P.B.N.—writing, review and editing, funding acquisition; O.B.—investigation, methodology; K.-S.N.—validation, investigation. All authors have read and agreed to the published version of the manuscript.

Funding: Ministry of Science, Technology and Innovation, Malaysia through MOSTI-Ted01 Grant TDF05211386. This work was also supported by Gheorghe Asachi Technical University of Iași—TUIASI- Romania, Scientific Research Funds, FCSU-2022.

Institutional Review Board Statement: Not applicable.

Informed Consent Statement: Not applicable.

Data Availability Statement: Not applicable.

Acknowledgments: The authors gratefully acknowledge the support of the Ministry of Science, Technology and Innovation, Malaysia through MOSTI-Ted01 Grant TDF05211386.This study was supported by the Center of Excellence Geopolymer and Green Technology (CEGeoGTech), Universiti Malaysia Perlis (UniMAP). This paper was also supported by "Gheorghe Asachi" Technical University from Iași (TUIASI), through the project "Performance and excellence in postdoctoral research 2022".

Conflicts of Interest: The authors declare no conflict of interest.

References

1. Baykara, H.; Cornejo, M.H.; García, E.; Ulloa, N. Heliyon preparation, characterization, and evaluation of compressive strength of polypropylene fi ber reinforced geopolymer mortars. *Heliyon* **2020**, *6*, e03755. [CrossRef] [PubMed]
2. Sukontasukkul, P.; Pongsopha, P.; Chindaprasirt, P.; Songpiriyakij, S. Flexural performance and toughness of hybrid steel and polypropylene fibre reinforced geopolymer. *Constr. Build. Mater.* **2018**, *161*, 37–44. [CrossRef]
3. Mermerdaş, K.; Algın, Z.; Oleiwi, S.M.; Nassani, D.E. Optimization of lightweight GGBFS and FA geopolymer mortars by response surface method. *Constr. Build. Mater.* **2017**, *139*, 159–171. [CrossRef]
4. Aliabdo, A.A.; Abd Elmoaty, A.E.M.; Salem, H.A. Effect of cement addition, solution resting time and curing characteristics on fly ash based geopolymer concrete performance. *Constr. Build. Mater.* **2016**, *123*, 581–593. [CrossRef]
5. Nergis, D.D.B.; Vizureanu, P.; Sandu, A.V.; Nergis, D.P.B.; Bejinariu, C. XRD and TG-DTA study of new phosphate-based geopolymers with coal ash or metakaolin as aluminosilicate source and mine tailings addition. *Materials* **2022**, *15*, 202. [CrossRef]
6. Abdila, S.R.; Abdullah, M.M.A.B.; Ahmad, R.; Nergis, D.D.B.; Rahim, S.Z.A.; Omar, M.F.; Sandu, A.V.; Vizureanu, P. Syafwandi potential of soil stabilization using ground granulated blast furnace slag (GGBFS) and fly ash via geopolymerization method: A review. *Materials* **2022**, *15*, 375. [CrossRef]
7. Wang, Y.; Aslani, F.; Liu, Y. The effect of tensile and bond characteristics of NiTi shape memory alloy, steel and polypropylene fibres on FRSCC beams under three-point flexural test. *Constr. Build. Mater.* **2020**, *233*, 117333. [CrossRef]
8. Abu Aisheh, Y.I.; Atrushi, D.S.; Akeed, M.H.; Qaidi, S.; Tayeh, B.A. Influence of polypropylene and steel fibers on the mechanical properties of ultra-high-performance fiber-reinforced geopolymer concrete. *Case Stud. Const. Mater.* **2022**, *17*, e01234. [CrossRef]
9. Pakravan, H.; Latifi, M.; Jamshidi, M. Hybrid short fiber reinforcement system in concrete: A review. *Constr. Build. Mater.* **2017**, *142*, 280–294. [CrossRef]
10. Farhan, K.Z.; Johari, M.A.M.; Demirboğa, R. Impact of fiber reinforcements on properties of geopolymer composites: A review. *J. Build. Eng.* **2021**, *44*, 102628. [CrossRef]
11. Ahmad, J.; Zaid, O.; Pérez, C.L.-C.; Martínez-García, R.; López-Gayarre, F. Experimental research on mechanical and permeability properties of nylon fiber reinforced recycled aggregate concrete with mineral admixture. *Appl. Sci.* **2022**, *12*, 554. [CrossRef]
12. Korniejenko, K.; Lin, W.-T.; Šimonová, H. Mechanical properties of short polymer fiber-reinforced geopolymer composites. *J. Compos. Sci.* **2020**, *4*, 128. [CrossRef]
13. Faris, M.A.; Abdullah, M.M.; Muniandy, R.; Abu Hashim, M.F.; Błoch, K.; Jeż, B.; Garus, S.; Palutkiewicz, P.; Mohd Mortar, N.A.; Ghazali, M.F. Comparison of hook and straight steel fibers addition on density, water absorption and mechanical properties. *Mater. Aeticle* **2021**, *14*, 1310. [CrossRef] [PubMed]
14. Risdanareni, P.; Puspitasari, P.; Jaya, E.J. Chemical and physical characterization of fly ash as geopolymer material. *MATEC Web Conf.* **2017**, *97*, 01031. [CrossRef]
15. Ranjbar, N.; Talebian, S.; Mehrali, M.; Kuenzel, C.; Metselaar, H.S.C.; Jumaat, M.Z. Mechanisms of interfacial bond in steel and polypropylene fiber reinforced geopolymer composites. *Compos. Sci. Technol.* **2016**, *122*, 73–81. [CrossRef]
16. He, S.; Sun, H.; Tan, D.G.; Peng, T. Recovery of titanium compounds from Ti-enriched product of alkali melting Ti-bearing blast furnace slag by dilute sulfuric acid leaching. *Procedia Environ. Sci.* **2016**, *31*, 977–984. [CrossRef]
17. Ramos, F.J.T.V.; Vieira, M.D.F.M.; Tienne, L.G.P.; de Oliveira Aguiar, V. Evaluation and characterization of geopolymer foams synthesized from blast furnace with sodium metasilicate. *J. Mater. Res. Technol.* **2020**, *9*, 12019–12029. [CrossRef]
18. El Didamony, H.; Assal, H.; El Sokkary, T.; Gawwad, H.A. Kinetics and physico-chemical properties of alkali activated blast-furnace slag/basalt pastes. *HBRC J.* **2012**, *8*, 170–176. [CrossRef]
19. Assaedi, H.; Shaikh, F.; Low, I. Effect of nano-clay on mechanical and thermal properties of geopolymer. *J. Asian Ceram. Soc.* **2016**, *4*, 19–28. [CrossRef]
20. Zaharaki, D.; Komnitsas, K.; Perdikatsis, V. Use of analytical techniques for identification of inorganic polymer gel composition. *J. Mater. Sci.* **2010**, *45*, 2715–2724. [CrossRef]

21. Lo, K.W.; Lin, K.L.; Cheng, T.W.; Chang, Y.M.; Lan, J.Y. Effect of nano-SiO_2 on the alkali-activated characteristics of spent catalyst metakaolin-based geopolymers. *Constr. Build. Mater.* **2017**, *143*, 455–463. [CrossRef]
22. Bazan, P.; Kozub, B.; Łach, M.; Korniejenko, K. Evaluation of hybrid melamine and steel fiber reinforced geopolymers composites. *Materials* **2020**, *13*, 5548. [CrossRef] [PubMed]
23. Ranjbar, N.; Zhang, M. Fiber-reinforced geopolymer coposites: A. review. *Cem. Concr. Compos.* **2019**, *107*, 103498. [CrossRef]
24. Palacios, M.; Puertas, F. Effect of carbonation on alkali-activated slag paste. *J. Am. Ceram. Soc.* **2006**, *89*, 3211–3221. [CrossRef]
25. Guo, L. Sulfate resistance of hybrid fiber reinforced metakaolin geopolymer composites. *Compos. Part B* **2020**, *183*, 107689. [CrossRef]
26. Arsalan, H.H.; Nyazi, R.M.; Yassin, A.I. Effect of polypropylene fiber content on strength and workability properties of concrete. *Polytechnic J.* **2019**, *9*, 7–12.
27. Wang, Y.; Zheng, T.; Zheng, X.; Liu, Y.; Darkwa, J.; Zhou, G. Thermo-mechanical and moisture absorption properties of fly ash-based lightweight geopolymer concrete reinforced by polypropylene fibers. *Constr. Build. Mater.* **2020**, *251*, 118960. [CrossRef]

Article

Mechanical Behavior of Crushed Waste Glass as Replacement of Aggregates

Ali İhsan Çelik [1], Yasin Onuralp Özkılıç [2,*], Özer Zeybek [3], Memduh Karalar [4], Shaker Qaidi [5,6], Jawad Ahmad [7], Dumitru Doru Burduhos-Nergis [8,*] and Costica Bejinariu [8]

1. Tomarza Mustafa Akincioglu Vocational School, Department of Construction, Kayseri University, Kayseri 38940, Turkey
2. Department of Civil Engineering, Faculty of Engineering, Necmettin Erbakan University, Konya 42000, Turkey
3. Department of Civil Engineering, Faculty of Engineering, Mugla Sitki Kocman University, Mugla 48000, Turkey
4. Faculty of Engineering, Department of Civil Engineering, Zonguldak Bulent Ecevit University, Zonguldak 67100, Turkey
5. Department of Civil Engineering, College of Engineering, University of Duhok, Duhok 42001, Iraq
6. Department of Civil Engineering, College of Engineering, Nawroz University, Duhok 42001, Iraq
7. Department of Civil Engineering, Military College of Engineering (Nust), Risalpur 24080, Pakistan
8. Faculty of Materials Science and Engineering, Gheorghe Asachi Technical University of Iasi, 700050 Iasi, Romania

* Correspondence: yozkilic@erbakan.edu.tr (Y.O.Ö.); doru.burduhos@tuiasi.ro (D.D.B.-N.)

Citation: Çelik, A.İ.; Özkılıç, Y.O.; Zeybek, Ö.; Karalar, M.; Qaidi, S.; Ahmad, J.; Burduhos-Nergis, D.D.; Bejinariu, C. Mechanical Behavior of Crushed Waste Glass as Replacement of Aggregates. *Materials* 2022, *15*, 8093. https://doi.org/10.3390/ma15228093

Academic Editor: Alessandro P. Fantilli

Received: 13 October 2022
Accepted: 9 November 2022
Published: 15 November 2022

Publisher's Note: MDPI stays neutral with regard to jurisdictional claims in published maps and institutional affiliations.

Copyright: © 2022 by the authors. Licensee MDPI, Basel, Switzerland. This article is an open access article distributed under the terms and conditions of the Creative Commons Attribution (CC BY) license (https://creativecommons.org/licenses/by/4.0/).

Abstract: In this study, ground glass powder and crushed waste glass were used to replace coarse and fine aggregates. Within the scope of the study, fine aggregate (FA) and coarse aggregate (CA) were changed separately with proportions of 10%, 20%, 40%, and 50%. According to the mechanical test, including compression, splitting tensile, and flexural tests, the waste glass powder creates a better pozzolanic effect and increases the strength, while the glass particles tend to decrease the strength when they are swapped with aggregates. As observed in the splitting tensile test, noteworthy progress in the tensile strength of the concrete was achieved by 14%, while the waste glass used as a fractional replacement for the fine aggregate. In samples where glass particles were swapped with CA, the tensile strength tended to decrease. It was noticed that with the adding of waste glass at 10%, 20%, 40%, and 50% of FA swapped, the increase in flexural strength was 3.2%, 6.3%, 11.1%, and 4.8%, respectively, in amount to the reference one (6.3 MPa). Scanning electron microscope (SEM) analysis consequences also confirm the strength consequences obtained from the experimental study. While it is seen that glass powder provides better bonding with cement with its pozzolanic effect and this has a positive effect on strength consequences, it is seen that voids are formed in the samples where large glass pieces are swapped with aggregate and this affects the strength negatively. Furthermore, simple equations using existing data in the literature and the consequences obtained from the current study were also developed to predict mechanical properties of the concrete with recycled glass for practical applications. Based on findings obtained from our study, 20% replacement for FA and CA with waste glass is recommended.

Keywords: eco-friendly concrete; waste glass; crushed; powder; hardened; fresh; slump; compressive; splitting; flexural; equation

1. Introduction

In proportion to the increasing urbanization in the world, raw material resources are decreasing and industrial wastes are increasing [1–11]. This situation causes serious environmental problems. Scientists who are aware of the problem offer various construction element suggestions in order to consume fewer raw materials and to permanently evaluate the wastes by the construction industry. Recycling waste materials to utilize in the construction projects reduces the use and costs of raw materials, even when more than one ton of glass is recycled, 1.2 tons of raw materials can be saved [12]. In general, Europe's glass recycling rate is 71.48%, while Slovenia and Belgium have a high rate of 98%. In Turkey, around 9% of glass waste can be recycled. While 52.9% of glass containers produced in the United States are disposed of in landfills, 26.63% of glass can be recycled [13]. Recycling glass as glass again takes time and is costly. Because it is subjected to processes at 1200–1400 °C, in order to get rid of wastes such as dirt and rust from the final product, it needs to be melted again and a product can be obtained [14]. Reusing waste glass in the production of glass provides 5–10% economic benefits [15]. However, adding it as aggregate to structural products, such as concrete, may provide more economic benefits [16–18]. Many recycled materials are used in concrete by replacing them with cement or aggregates [19–21]. Recycled crushed glass is used in numerous studies around the world as a substitute for CA and FA for the development of sustainable concrete [22–24]. One of the most important of these waste materials is crushed waste glass (CWG) in aggregate size. The use of broken glass particles by substituting aggregates in concrete production is an important and interesting issue. Since glass does not change its structure when reprocessed, its chemical properties do not deteriorate and it is very easy to reprocess. Some glass products have a limited lifespan depending on their intended use. This increases the potential of broken waste glass particles. This recycling potential needs to be processed quickly [25]. In addition, the degradation rate of CWG is lower than industrial products, such as plastic, paper, and rubber. Therefore, the usage of broken glass fragments as a substitute for aggregates in concrete production will yield impressive consequences [26].

In recent years, much research was conducted on the mechanical properties of CWG added concretes. The CWG can be substituted for both fine and CAs [27]. The nature of glass reactivity causes significant effects when added to concrete. For example, some may cause excessive expansion in concrete when used as 100% of the total aggregate [28]. Studies show that with CWG aggregates, as the rate of change increases, the strength decreases faster. For example, in a study that used 15–60% substitution, it decreased the CS by 8% with the addition of 15% CWG, while it decreased the CS by 15% with the addition of 30% CWG. When this proportion is increased to 60%, the CS decreases by 49% [29]. Singh and Siddique, in their study, by replacing 10–50% CWG with FA, found that the strength decreased as the CWG proportion increased. However, when CWG was used with methacholine, they discovered that the strength improved as the methacholine addition increased [26]. This result shows that additional reinforcing materials can be added to increase the mechanical properties of the concrete whose environmental impact is increased with CWG. Harrison et al. used fine glass particles by substituting cement and FA in certain proportions in hopes of benefiting the mechanical properties of concrete. According to the consequences obtained, the cement preserves its mechanical properties when CWG particles are added up to 20%, but when the additions of 30% and above cause insufficient $CaCO_3$ in the cement, the mechanical properties are adversely affected. It was also found that with the change in FA conservative, its mechanical properties are up to 20% substitution [27]. It was confirmed in some studies that the CS of concrete increases as the size of CWG particles is substituted with aggregate decreases. In their study, Shi et al. observed that finer glass particles improved the 28-day CS of concrete more than larger ones [30]. Chen et al., on the other hand, found that finer-ground glass particles performed better with similar behavior in 7-day and 28-day concrete CS [31]. Khmiri's et al. found that with up to 20% cement replacement, CWG ground up to 20 μm improved the late mechanical properties of concrete, while 40 μm CWG reduced the CS [32]. Letelier and Al-Hashmi found that 38 μm CWG had optimal mechanical properties at 20% cement substitute (by weight). In another study, it was revealed that 38 μm CWG had optimal mechanical properties at 20% cement replacement (by weight). It was determined that 10% and 30% substitutions gave consequences close to the reference sample [33]. Mostofinejad et al., examining 30% of cement by weight for ground CWG, found that the replacement had a strength reduction of 40% and 42%, respectively [34]. Tamanna et al. used 3000 μm CWG by replacing 20%, 40%, and 60% coarse sand. As a result of the 7, 28, and 56-day compressive strength (CS) and flexural strength (FS) tests, they discovered that the 20% substitution gave good consequences, while the 40% and 60% substitution had negative effects on the CS and FS [35]. Penacho et al. investigated the effects of replacing 20%, 50%, and 100% CWG with FA on the mechanical properties of concrete. As a result of the analysis, he found that samples with higher sand substitution for fine CWG gave a greater strength increase. The increase in strength is related to the positive effect of pozzolanic reaction with the fine glass particles [36]. Lee et al. investigated the size effect of glass waste on the mechanical properties of concrete. As a result of the research, they found that the CS was better in samples consisting of particles smaller than 600 μm [37]. Corinaldesi et al. performed tests by replacing glasses with glass particles smaller than 36 μm, 36–50 μm, and 50–100 μm with aggregate. They performed compressive and FS tests at 180 days to study the influence of particle size. As a result of the observation, they found that the CS decreased at 30% FA substitution. However, at the 70% substitution proportion, they found that 50–100 μm samples showed milder increases over the reference sample [38]. Tejaswi et al. In a study they conducted, they found that replacing 20% by weight of CWG with FA gave a result close to the control sample, but replacing 10%, 30%, 40%, and 50% lowered the CS [39]. Batayneh et al. reported that replacing the CWG with the addition of fly ash increased the CS; however, they revealed that it did not change the splitting strength [40]. With a similar statement, Gerges et al. found that the rate of substitution of fine glass with sand had little effect on the compressive, splitting stress, or flexural strengths of concrete [41]. As a result of their study, Mohammed and Hama determined that when glass is added to concrete alone, it improves properties, such as CS, FS, and splitting tensile strength (STS). Compared to the reference sample, an increase of 14.12%, 1.7%, 6.01%, 52.63%, and 57.32% was observed in elastic modulus, energy capacity, and bond strength, respectively [42]. In a study by, Asa et al., it was found that the concrete to which 5% and 20% glass fragments were added led to a decrease of 3.8–10.6 percent and 3.9–16.4 percent, respectively, in the CS and tensile strength at the end of the 21st day, but the use of mineral additives changed the properties of the mixtures. They found that it improved after 7, 14, and 21 days of testing [25]. Walczak et al. used cathode ray tubes glass instead of sand, and they found that the compression strength increased about 16% and the bending strength increased about 14% [43].

Though several investigations were performed on this topic, as presented in above, there were changes in the consequences gained from the literature. Therefore, there is still a necessity to examine the mechanical productivity of concrete with the fractional replacement of CWG, and the perfect amount of it. For this purpose, an investigational study was performed on some investigation samples. The effects of different production-based features of concrete with different amounts of crushed CWG as replacement of aggregates were studied. More importantly is that empirical equations are developed to predict the capacity of concrete with CWG, considering both the literature and the data obtained from the experimental study.

2. Experimental Program

To amount the mechanical belongings of the concrete with recycled CWG, different mixtures were cast. Eight different mixtures were chosen. Four of them were considered for FAs and rest of them were considered for CAs. Four different proportions of 10%, 20%, 40%, and 50% were selected (Table 1). CAs with a size of 5–13 mm were utilized while FAs with sizes of 0–4 mm were utilized. FA represents FA replacement and CA represents CA replacement. CWG was collected from Akcihan Glass, Istanbul, Turkey, involving a waste window glass (soda lime glass). The size of fine glass is a combination of glass with an equal amount of 1.7–4 mm and 100–200 micron, while the size of coarse glass is a combination of glass with an equal amount of 9–12 mm and 5–8 mm.

Table 1. Different mixtures of the concrete with recycled glass.

REF. SAMPLE	Without Glass
CA10%	10% CA was swapped with waste coarse waste glass
CA20%	20% CA was swapped with waste coarse waste glass
CA40%	40% CA was swapped with waste coarse waste glass
CA50%	50% CA was swapped with waste coarse waste glass
FA10%	10% FA was swapped with waste fine waste glass
FA20%	20% FA was swapped with waste fine waste glass
FA40%	40% FA was swapped with waste fine waste glass
FA50%	50% FA was swapped with waste fine waste glass

In the mixture, CEM 32.5 Portland cement was utilized. Water-to-cement proportion was selected as 0.5. The proportion of fine to CAs was selected as unity. The proportion of cement to total fine and CAs was selected as 0.2. Figure 1 shows slump test consequences of each mixture. The highest slump value was observed in the reference mixture. Adding recycled glass in concrete reduced slump value. Replacing aggregate resulted in a significant decrease in workability, especially after 20%. This is because glass particles have sharper and irregular geometric forms than sand particles, which can cause high friction, resulting in less fluidity [44]. Moreover, the binder effect of glass powder in FA replacement reduced workability even more than replacement of CA.

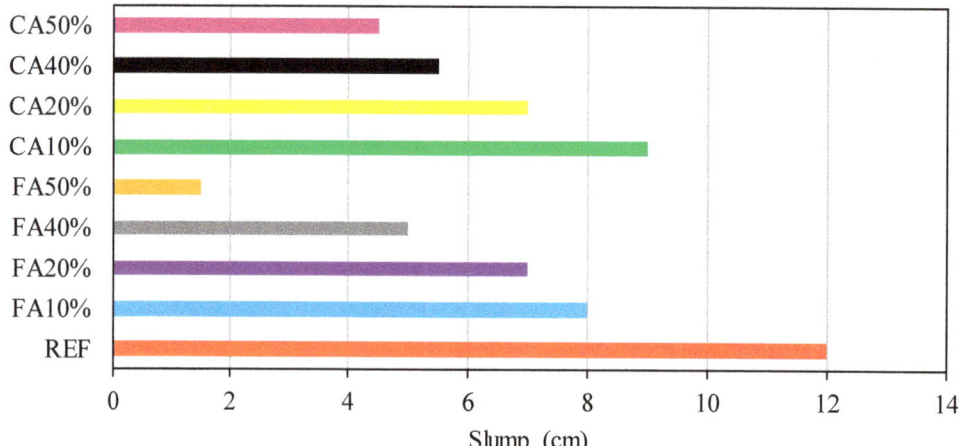

Figure 1. Slump test.

3. Experimental Consequences and Discussions

3.1. Compressive Strength (CS)

Figure 2 demonstrates the CS test consequences at a 28-day period formed with 100% reference and numerous proportions of CWG of fine and CA exchanging reference aggregates. As recognized in Figure 2, in the midst of the low CWG proportion, the CSs of concrete mixtures with CWG consumption as a replacement for FA were greater than those of the consistent concrete mixes lacking CWG. As presented in Figure 2, fine and CA for 10%, 20%, 40%, and 50% (FA10%, FA20%, FA40%, FA50%, and CA10%, CA20%, CA40%, and CA50%) were exchanged with CWG; this inclination was reversed while fractional replacement of CA with CWG decreased the CS of concrete. Statistical examination of specimen consequences indicated the important proceeds of CWG (as FA was swapped with CWG) to the CS of concrete. As the evaluation of the FA is swapped with CWG powder, the CS of FA10% (10% FA were swapped with CWG) is 1% greater than that of reference concrete. Similarly, the CS of FA20% (20% FA were swapped with CWG) is 9%

greater than that of FA10%. At the evaluation of CS of FA40% (40% FA were swapped with CWG) is 6.2% larger than that of FA20%. Lastly, at the evaluation of CS of FA50% (50% FA were swapped with CWG) is 5.2% greater than that of FA40%. While compared with reference concrete, the CS of FA50% (50% FA were swapped with CWG) is comparable with that of reference concrete. The evaluation of CS of FA50% (50% FA were swapped with CWG) is 12% greater than that of reference concrete. The increase in the capacity with FA can be explained with the use of glass powder since the glass powder has a binder effect. Alternatively, at the evaluation of CS of CA10%, CA20%, CA40%, and CA50% (CAs were swapped with CWG), the CS test consequences for the great CWG proportion downgraded trends comparable to those for the small CWG proportion. In this case too, the CS of CA10% is 27% larger than the corresponding mix (CA50%). While CA was swapped with CWG, a remarkable downgrade in strength was detected when compared with the reference concrete and FA50%. This decrease in CS can be explained by two reasons. One of the reasons is that glass waste has a smooth surface compared to normal aggregates, which causes a decrease in the bond among the particles and the cement matrix, and the decrease in CS increases as the glass percentage increases. The second reason is that the absorption of CWG is less than normal aggregates, and this causes higher slump values. For this reason, there is an extra amount of free water that evaporates, leaving a little more space, which causes a decrease in CS [45].

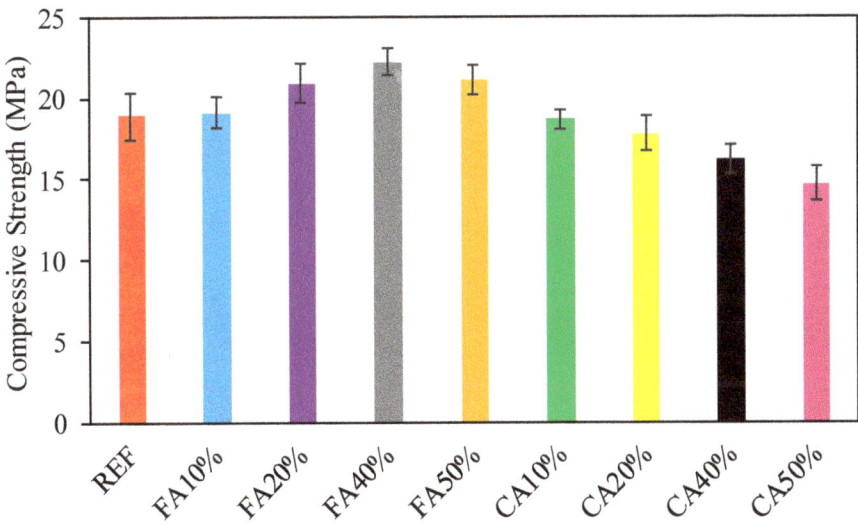

Figure 2. Consequences of CS.

3.2. Splitting Tensile Strength (STS)

An evaluation of the relation STS of the numerous concrete proportion mixtures made with the reference, and several proportions of recycled aggregate replacing FA with and without CWG and CA, which were swapped with CWG, is presented in Figure 3. As presented in Figure 3, the consequences of STS generally follow a similar trend as the CS. As detected from Figure 3, the significant progress in STS of concrete is up to 13%, with CWG used as fractional replacement for FA. The reason for this can be shown as the progress of hydration, decreased permeability of glass-mixed concrete, good bond strength among the glass aggregate, and the surrounding cement paste due to the irregular geometry of the glass. The pozzolanic reaction may also offset this trend at a later stage of hardening and help improve the STS at 28 days [44]. Upon changing the amount of CWG, outcomes of tensile strength are affected correspondingly to the CA. This decrease in strength can be explained due to the surface structure of CWG. Similar results were also observed by others [45–47]. As presented in Figure 3, tensile strength optimum values are established at 40% CWG, requiring maximum distributed tensile. The correlation of CS in competition with STS is provided in Figure 4. As presented in Figure 4, the regression model among the CS and STS is performed to be flat. Additionally, as detected in Figure 4, the regression stroke characterizes a strong relationship among CS, compared with STS needing an R^2 worth of more than 92%.

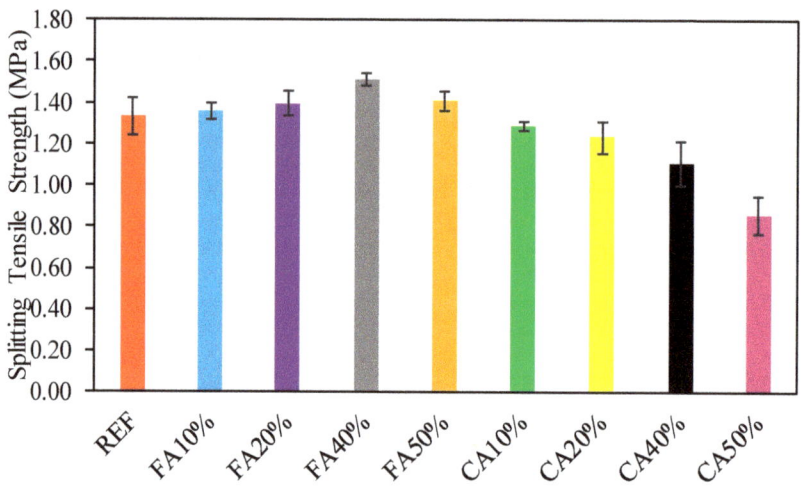

Figure 3. Consequences of STS.

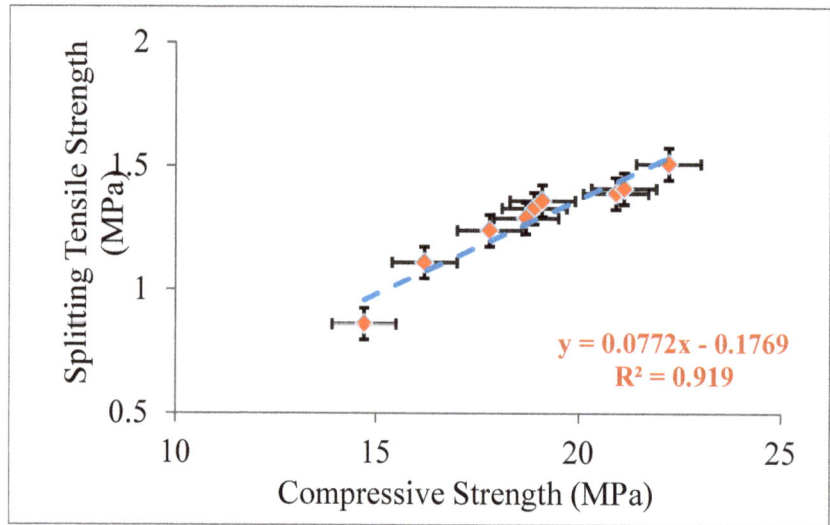

Figure 4. Splitting tensile strength variation versus compressive strength.

3.3. Flexural Strength (FS)

FS of investigational specimens were recognized on specimens after CST. The assimilated types of strength alternated among 5.2 MPa and 7.0 MPa. The result of the FS of the specimens is offered in Figure 5. In Figure 5, FS with different proportions of CWG is shown. It was noticed that with the addition of CWG at 10%, 20%, 40%, and 50% of FA swapped, the increase in FS was 3.2%, 6.3%, and 11.1% and 4.8%, respectively, in percentage to the reference sample (6.3 MPa). This could be dedicated to pozzolanic reactions, which accelerate with time, and offset the hardening process and aid the increase in FS [44]. A related behavior was also described by Shehata et al. [48]. It must be noted that higher CWG replacement may have an opposing effect on the FS [48]. Figure 6 shows a rectilinear relationship among the tensile FS of the example and the substance of the CWG addition. Alternatively, if FAs were swapped with CWG, it was observed that the increase in FS was 11.1%, respectively, in amount to the reference sample (6.3 MPa). However, if the amount of CWG is employed in relation to the CA contented, it may be noticed that the FS and CWG contented are associated (Figure 6)). This can be attributed to the change in the interfacial transition zone properties of the glass-containing mixtures [49]. Additionally, as detected in Figure 6, the regression shows a good relationship among FS contrasted with CWG having an R^2 worth of more than 95%.

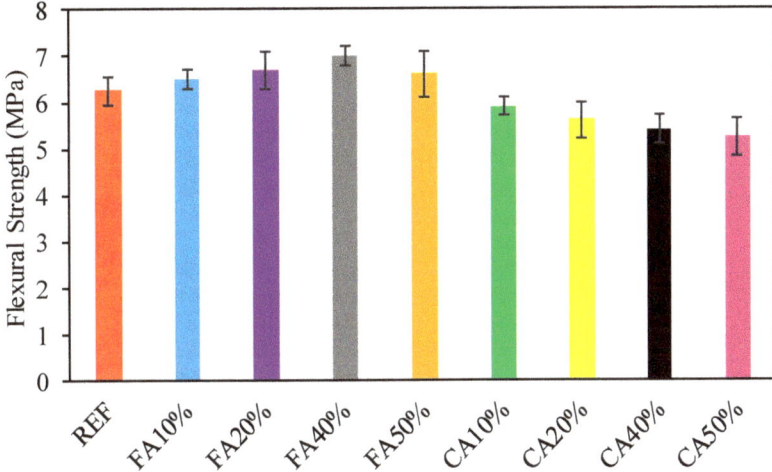

Figure 5. Consequences of FS.

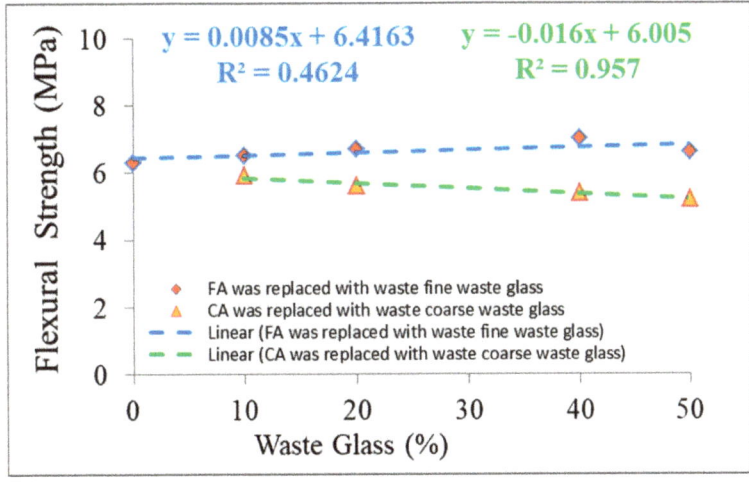

Figure 6. Consequences of tensile FS tests of concrete examples with changed CWG substances.

4. Scanning Electron Microscope (SEM) Analysis

SEM investigation was achieved with samples taken from some samples of this experimental study. The images are magnified 500 times to show the interaction among binder materials and aggregates. In SEM analysis, there are images of two different mixtures. The first four are images of concrete samples containing only CGW, and the last four are images of samples containing both CGW and glass powder. Figure 7a shows the situation where CGW is swapped by aggregate. It is seen that the glass pieces, in the range of 4–9 mm, are homogeneously dispersed in the homogeneous concrete by providing good compatibility with the cement binder. Figure 7b shows the slightly larger and closer state glass particles dispersed in the concrete and their harmonious positions. In Figure 7c, it can be said that the interface among the glass piece and the binding cement remains in a discrete form. Figure 7d shows that gaps and holes are formed in some regions. This problem can be eliminated with better compression and vibration. Glass powder makes a good pozzolanic effect because it has a higher specific surface area than cement [30,50]. It can be seen in Figure 7e that the binding property of the CWG powder cement and the pozzolanic property of the glass powder results in good bonding. Etringite formation is an important issue in Portland cement concretes due to early phase hydration [51,52]. Although the hydration of glass powder and cement show similarities, some ettringite formation is seen in Figure 7e. While the samples were being prepared, a good mixing in the mixer ensured the homogeneous distribution of the additional powder particles. Although a very small part was examined in the SEM analysis, the compressive and FS consequences show a homogeneous distribution among homogeneous binders with an increasing productivity in the range of 10–40%. Figure 7f shows the standing of small and large CGWs in concrete. While there is a line among the large particle and the cement, it is seen that the smaller piece of glass has a better bonding with the cement. It is stated that the use of glass particles as aggregate can increase cracks and voids, and this will cause a decrease in strength [53]. Figure 7g shows the distribution of glass powders and small glass particles in the concrete. It can be said that the use of both CWG together for concrete is a good match. In Figure 7h, it is seen that there are gaps and holes in places. These can be considered as mini problems. It is stated in some studies that glass powder with finer particles has a high pozzolanic effect and provides better strength in concrete. The consequences obtained confirm the statements in the literature [53–56].

Figure 7. *Cont.*

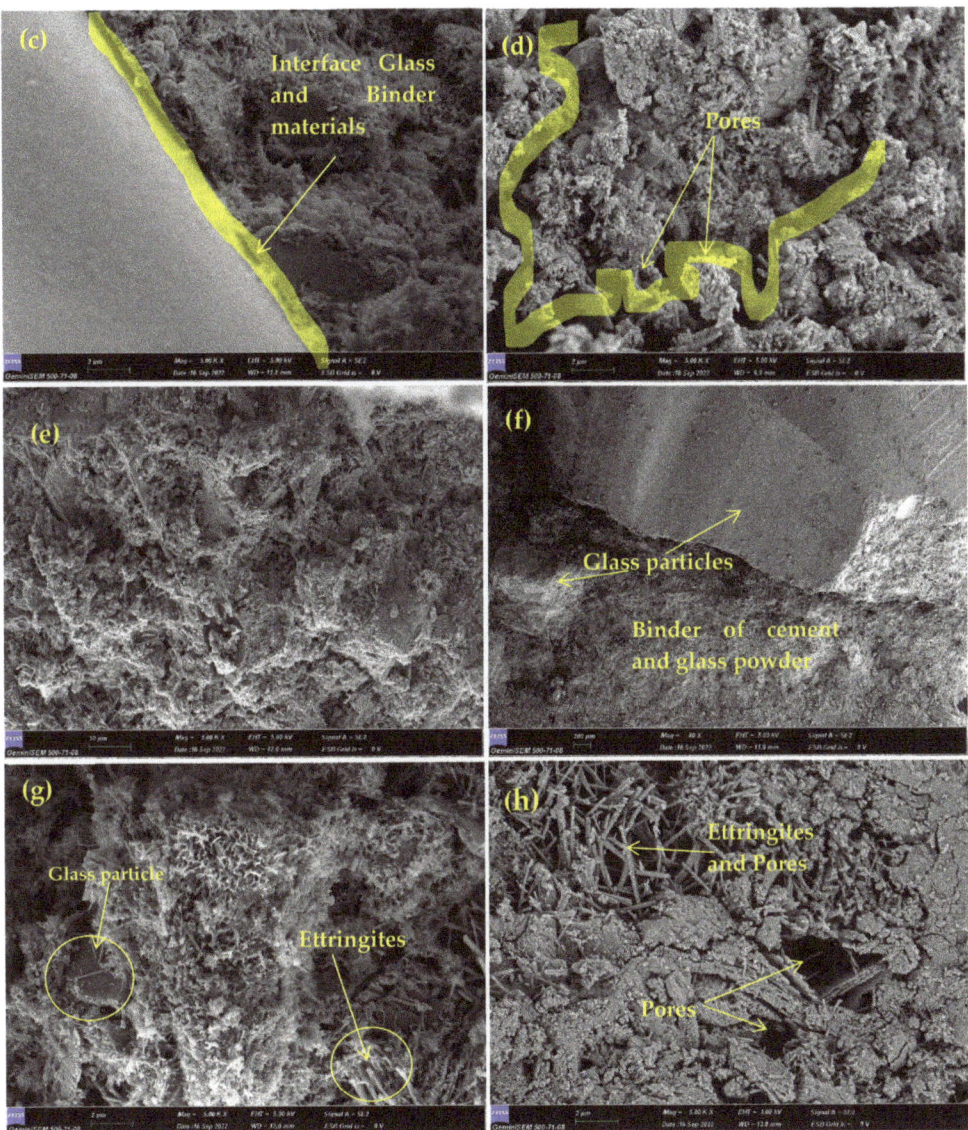

Figure 7. Consequences of SEM analysis.

5. Evaluation of Current Findings with Previous Studies

Since it has the potential to be preferred as aggregate in concrete mixture, the effect of CWG on engineering properties was examined by many investigators. CWG were used as either FA or CA replacement in concrete. To develop empirical equations, strength values (CS, STS, and FS) for plain concrete and concrete formed from CWG were collected from previous experimental studies [28,29,33,35,39,40,44,46,47,57–80]. The obtained strength values of concrete formed with CWG (f) were primary regularized by the strength value of concrete without glass (f'). These normalized strength values (f/f') were then shown as a function of different replacement proportions. The changes in the normalized strength values were depicted in Figures 8–13.

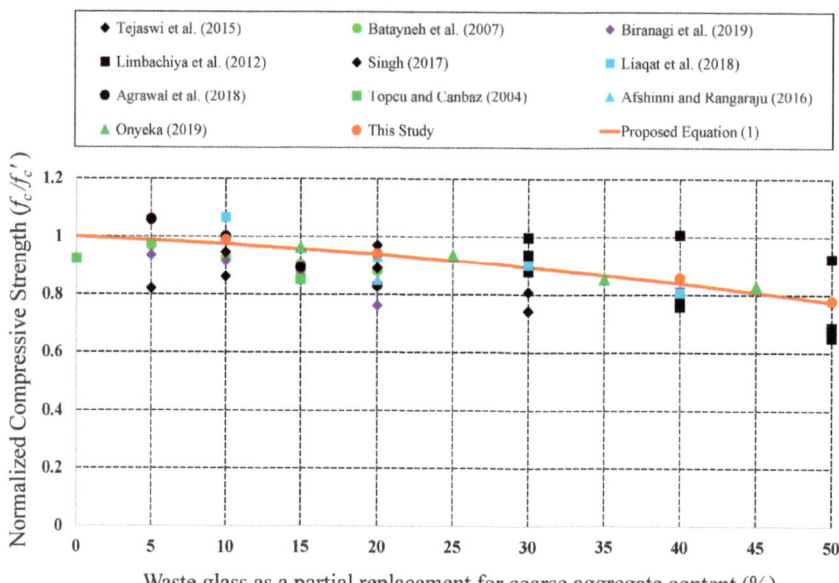

Figure 8. Discrepancy of the normalized CS of the concrete formed with CWG as a partial replacement for CA [39,40,47,58–62,78,79].

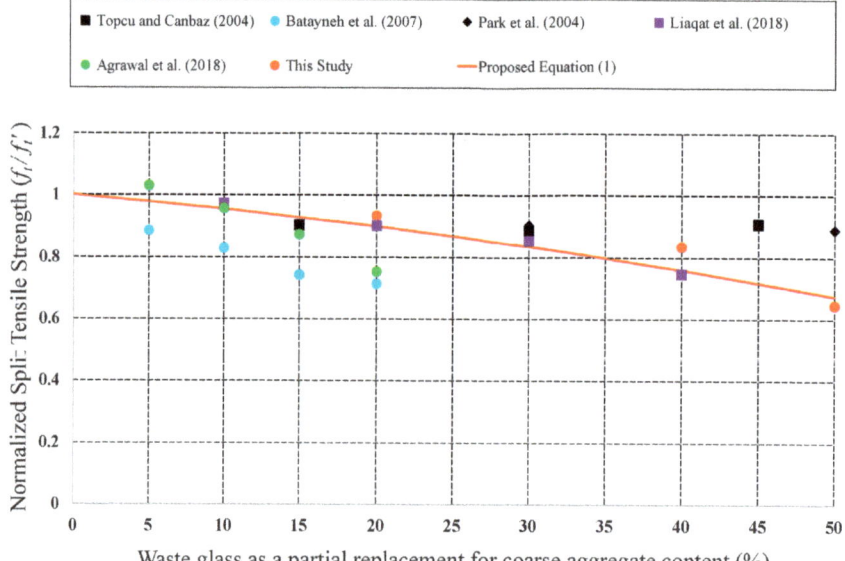

Figure 9. Discrepancy of the regulated STS of the concrete formed with CWG as a partial replacement for CA [40,46,47,60,79].

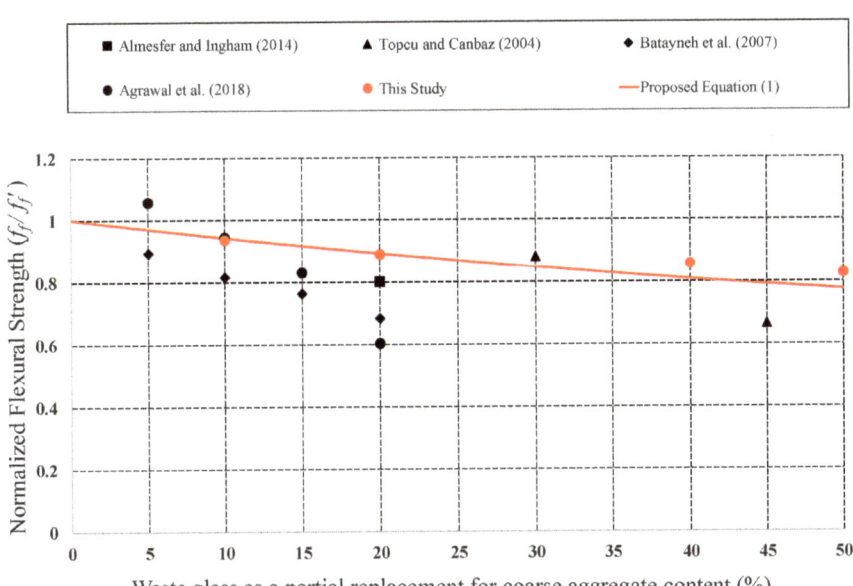

Figure 10. Discrepancy of the normalized FS of the concrete produced with CWG as a partial replacement for CA [40,47,57,79].

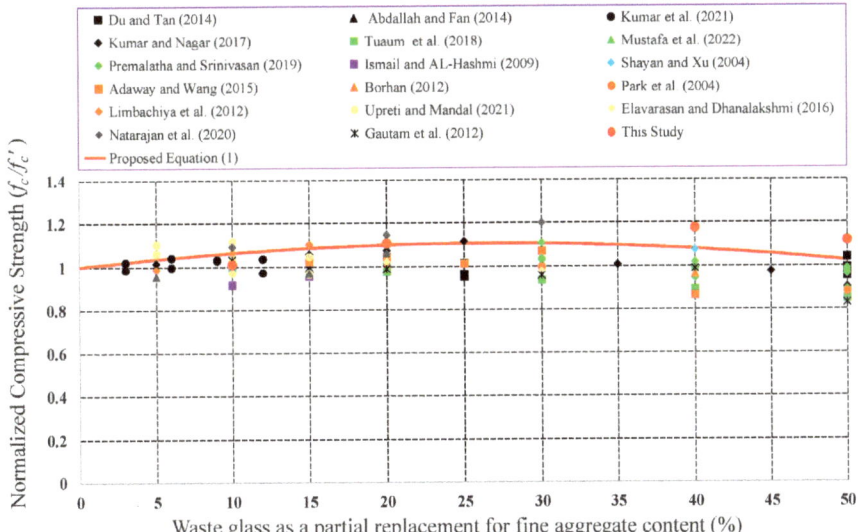

Figure 11. Discrepancy of the normalized CS of the concrete formed with CWG as a partial replacement for FA [28,33,44,46,58,63–67,69–74,77].

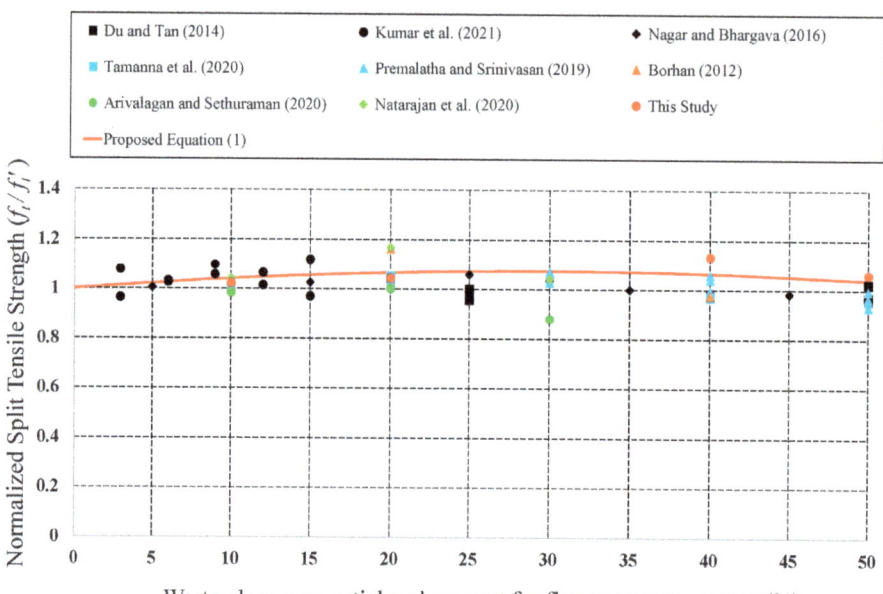

Figure 12. Discrepancy of the regulated Splitting tensile strength of the concrete produced with CWG as a partial replacement for FA [29,35,63,64,69,70,73,80].

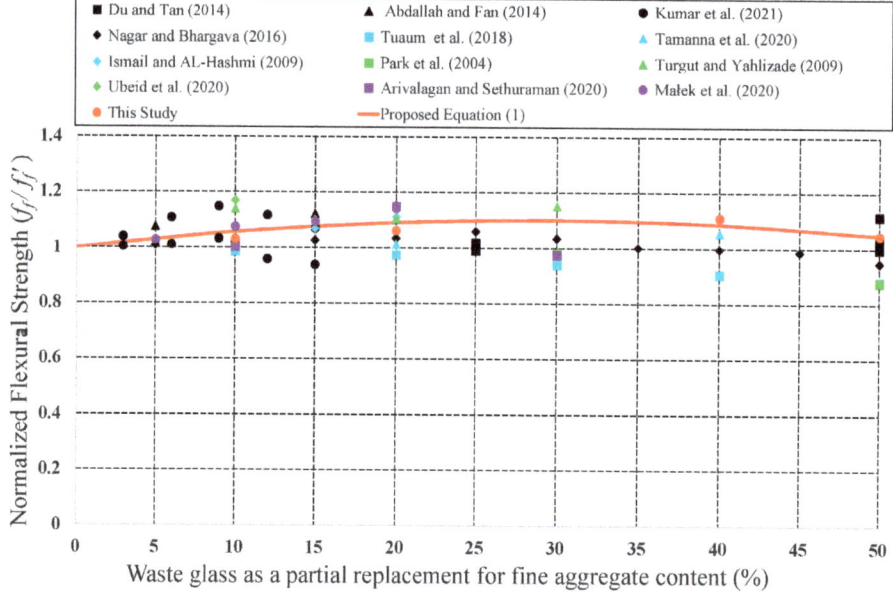

Figure 13. Discrepancy of the normalized FS of the concrete produced with CWG as a partial replacement for FA [29,33,35,44,46,63,64,66,68,75,76,80].

Considering both our findings and previous studies, an empirical equation was developed as follows to predict the CS, STS, and FS, respectively:

$$f = \left[1 + c_1 \times (WGR) + c_2 \times (WGR)^2\right] \times f' \quad (1)$$

where f = strength values to be calculated (f_c = CS; f_s = STS; f_f = FS); c_1 and c_2 = the coefficients given in Table 2; WGR: CWG proportion (0 < WGR < 50); and f' = strength values of the plain concrete.

Table 2. The parameters used in Equation (1).

Strength Values (f)		c_1	c_2
Concrete produced with CWG as a partial replacement for CA	f_c	−0.002	−0.00005
	f_s	−0.004	−0.00005
	f_f	−0.006	0.00003
Concrete produced with CWG as a partial replacement for FA	f_c	0.008	−0.00015
	f_s	0.005	−0.00009
	f_f	0.007	−0.00012

As presented in Equation (1), engineering properties of concrete formed with CWG were identified as a function of the quantity of the CWG. The developed expressions for compressive, flexural, and splitting tensile strength of concrete produced with CWG can be employed in project phases.

6. Conclusions and Summary

In this study, the effects of different productions based features of concrete with different amounts of CWG as replacements of aggregates were investigated. For this purpose, fine and CAs were altered for 10%, 20%, 40%, and 50% (FA10%, FA20%, FA40%, FA50%, and CA10%, CA20%, CA40%, and CA50%). Penetrability and slump properties of produced concrete samples were also examined. Then, CS, STS, and FS of the produced test examples were investigated. Furthermore, SEM analyses were also performed to compare the strength consequences obtained from the experimental study. These specifications were then compared with those of reference concrete. Lastly, practical equations were derived to easily estimate the CS, STS, and FS of produced concrete samples. Considering our findings, the following consequences can be obtained from this study:

According to the slump test consequences, while the slump value reduces, the quantity of CWG rises. In other words, the workability of concrete reduces when the rate of glass powder as replacement for aggregates increases.

In this study, fine and CA for 10%, 20%, 40%, and 50% (FA10%, FA20%, FA40%, FA50%, and CA10%, CA20%, CA40%, and CA50%) were exchanged with CWG, and this tendency was inverted while fractional replacement of CA with CWG decrease the CS of concrete. In other words, while CA was swapped with CWG, a remarkable downgrade in strength was detected when compared with the reference concrete.

As mentioned above, the consequences of STS generally follow a similar trend as the CS. The significant progress in STS of the concrete is up to 13% with CWG used as fractional replacement for FA. Upon changing the quantity of CWG, consequences of STS are affected correspondingly to the CA.

While aggregates were swapped with CWG, there was an increase in the FS values up to a certain value of the quantity of the CWG. It was observed that with the addition of CWG at 10%, 20%, 40%, and 50% of FA swapped, the increase in FS was 3.2%, 6.3%, and 11.1% and 4.8%, respectively, compared to reference concrete sample.

From the SEM analyses, it can be observed that the use of both CWG together for concrete is a good match. Furthermore, similar to the studies in the literature, it is also shown in this study that glass powder with finer particles has a high pozzolanic effect and provides better strength in concrete.

The developed empirical equations for the CS, STS, and FS are quite general and they have the potential to be implemented into design guidelines of the concretes with CWG.

The use of 20% replacement for fine and CA with CWG is recommended, considering both workability and strength.

Author Contributions: Conceptualization, Y.O.Ö. and D.D.B.-N.; methodology, Y.O.Ö. and D.D.B.-N.; data curation, Ö.Z., M.K., D.D.B.-N. and Y.O.Ö.; investigation, A.İ.Ç., Y.O.Ö., Ö.Z. and M.K.; writing—original draft preparation, A.İ.Ç., Y.O.Ö., Ö.Z., S.Q. and J.A.; writing—review and editing, S.Q., J.A., D.D.B.-N. and C.B.; funding acquisition, D.D.B.-N. and C.B. All authors have read and agreed to the published version of the manuscript.

Funding: This work was supported by Gheorghe Asachi Technical University of Iași—TUIASI-Romania, Scientific Research Funds, FCSU-2022.

Institutional Review Board Statement: Not applicable.

Informed Consent Statement: Not applicable.

Data Availability Statement: Not applicable.

Acknowledgments: This paper was financially supported by the Project "Network of excellence in applied research and innovation for doctoral and postdoctoral programs"/InoHubDoc, project co-funded by the European Social Fund financing agreement no. POCU/993/6/13/153437. This paper was also supported by "Gheorghe Asachi" Technical University from Iași (TUIASI), through the Project "Performance and excellence in postdoctoral research 2022".

Conflicts of Interest: The authors declare no conflict of interest.

References

1. de Azevedo, A.R.G.; Amin, M.; Hadzima-Nyarko, M.; Saad Agwa, I.; Zeyad, A.M.; Tayeh, B.A.; Adesina, A. Possibilities for the application of agro-industrial wastes in cementitious materials: A brief review of the Brazilian perspective. *Clean. Mater.* **2022**, *3*, 100040. [CrossRef]
2. Martínez-García, R.; Jagadesh, P.; Zaid, O.; Șerbănoiu, A.A.; Fraile-Fernández, F.J.; de Prado-Gil, J.; Qaidi, S.; Grădinaru, C.M. The Present State of the Use of Waste Wood Ash as an Eco-Efficient Construction Material: A Review. *Materials* **2022**, *15*, 5349. [CrossRef] [PubMed]
3. Karalar, M.; Özkılıç, Y.O.; Deifalla, A.F.; Aksoylu, C.; Arslan, M.H.; Ahmad, M.; Sabri, M.M.S. Improvement in Bending Performance of Reinforced Concrete Beams Produced with Waste Lathe Scraps. *Sustainability* **2022**, *14*, 12660. [CrossRef]
4. Karalar, M.; Bilir, T.; Çavuşlu, M.; Özkılıç, Y.O.; Sabri, M.M.S. Use of Recycled Coal Bottom Ash in Reinforced Concrete Beams as Replacement for Aggregate. *Front. Mater.* **2022**, *675*, 1064604.
5. Farhan Mushtaq, S.; Ali, A.; Khushnood, R.A.; Tufail, R.F.; Majdi, A.; Nawaz, A.; Durdyev, S.; Burduhos Nergis, D.D.; Ahmad, J. Effect of Bentonite as Partial Replacement of Cement on Residual Properties of Concrete Exposed to Elevated Temperatures. *Sustainability* **2022**, *14*, 11580. [CrossRef]
6. El-Mandouh, M.A.; Hu, J.-W.; Mohamed, A.S.; Abd El-Maula, A.S. Assessment of Waste Marble Powder on the Mechanical Properties of High-Strength Concrete and Evaluation of Its Shear Strength. *Materials* **2022**, *15*, 7125. [CrossRef]
7. de Azevedo, A.R.G.; Marvila, M.T.; de Oliveira, M.A.B.; Umbuzeiro, C.E.M.; Huaman, N.R.C.; Monteiro, S.N. Perspectives for the application of bauxite wastes in the development of alternative building materials. *J. Mater. Res. Technol.* **2022**, *20*, 3114–3125. [CrossRef]
8. Çelik, A.İ.; Özkılıç, Y.O.; Zeybek, Ö.; Özdöner, N.; Tayeh, B.A. Performance Assessment of Fiber-Reinforced Concrete Produced with Waste Lathe Fibers. *Sustainability* **2022**, *14*, 11817. [CrossRef]
9. Basaran, B.; Kalkan, I.; Aksoylu, C.; Özkılıç, Y.O.; Sabri, M.M.S. Effects of Waste Powder, Fine and Coarse Marble Aggregates on Concrete Compressive Strength. *Sustainability* **2022**, *14*, 14388. [CrossRef]
10. Alani, A.A.; Lesovik, R.; Lesovik, V.; Fediuk, R.; Klyuev, S.; Amran, M.; Ali, M.; de Azevedo, A.R.G.; Vatin, N.I. Demolition Waste Potential for Completely Cement-Free Binders. *Materials* **2022**, *15*, 6018. [CrossRef]
11. Aksoylu, C.; Özkılıç, Y.O.; Hadzima-Nyarko, M.; Işık, E.; Arslan, M.H. Investigation on Improvement in Shear Performance of Reinforced-Concrete Beams Produced with Recycled Steel Wires from Waste Tires. *Sustainability* **2022**, *14*, 13360. [CrossRef]
12. Parfitt, J. *Analysis of Household Waste Composition and Factors Driving Waste Increases*; WRAP for the Strategy Unit, Government Cabinet Office: London, UK, 2002.
13. Epa, U. *Advancing Sustainable Materials Management: 2014 Fact Sheet*; United States Environmental Protection Agency, Office of Land and Emergency: Washington, DC, USA, 2015.

14. Hagger, S.E. Sustainable industrial design and waste management. In *Cradle-to-Cradle for Sustainable Development*; Academic Press: Cambridge, MA, USA, 2007.
15. Energy, P.R. *Sustainability Report*; Pacific Rubiales Energy: Toronto, ON, Canada, 2019.
16. Qaidi, S.; Najm, H.M.; Abed, S.M.; Özkılıç, Y.O.; Al Dughaishi, H.; Alosta, M.; Sabri, M.M.S.; Alkhatib, F.; Milad, A. Concrete Containing Waste Glass as an Environmentally Friendly Aggregate: A Review on Fresh and Mechanical Characteristics. *Materials* **2022**, *15*, 6222. [CrossRef] [PubMed]
17. Ahmad, J.; Martinez-Garcia, R.; Algarni, S.; de-Prado-Gil, J.; Alqahtani, T.; Irshad, K. Characteristics of Sustainable Concrete with Partial Substitutions of Glass Waste as a Binder Material. *Int. J. Concr. Struct. Mater.* **2022**, *16*, 21. [CrossRef]
18. Aslam, F.; Zaid, O.; Althoey, F.; Alyami, S.H.; Qaidi, S.M.A.; de Prado Gil, J.; Martínez-García, R. Evaluating the influence of fly ash and waste glass on the characteristics of coconut fibers reinforced concrete. *Struct. Concr.* **2022**; *early view*. [CrossRef]
19. Zeybek, Ö.; Özkılıç, Y.O.; Çelik, A.İ.; Deifalla, A.F.; Mahmood, A.; Sabri, M.M.S. Performance evaluation of fiber-reinforced concretes produced with steel fibers extracted from waste tire. *Front. Mater.* **2022**, 692.
20. Karalar, M.; Özkılıç, Y.O.; Aksoylu, C.; Sabri, M.M.S.; Alexey, N.B.; Sergey, A.S.; Evgenii, M.S. Flexural Behavior of Reinforced Concrete Beams using Waste Marble Powder towards Application of Sustainable Concrete. *Front. Mater.* **2022**, 701.
21. Qaidi, S.; Al-Kamaki, Y.S.S.; Al-Mahaidi, R.; Mohammed, A.S.; Ahmed, H.U.; Zaid, O.; Althoey, F.; Ahmad, J.; Isleem, H.F.; Bennetts, I. Investigation of the effectiveness of CFRP strengthening of concrete made with recycled waste PET fine plastic aggregate. *PLoS ONE* **2022**, *17*, e0269664. [CrossRef]
22. Ekop, I.E.; Okeke, C.J.; Inyang, E.V. Comparative study on recycled iron filings and glass particles as a potential fine aggregate in concrete. *Resour. Conserv. Recycl. Adv.* **2022**, *15*, 200093. [CrossRef]
23. Li, S.; Zhang, J.; Du, G.; Mao, Z.; Ma, Q.; Luo, Z.; Miao, Y.; Duan, Y. Properties of concrete with waste glass after exposure to elevated temperatures. *J. Build. Eng.* **2022**, *57*, 104822. [CrossRef]
24. Song, W.; Zou, D.; Liu, T.; Teng, J.; Li, L. Effects of recycled CRT glass fine aggregate size and content on mechanical and damping properties of concrete. *Constr. Build. Mater.* **2019**, *202*, 332–340. [CrossRef]
25. Asa, E.; Anna, A.S.; Baffoe-Twum, E. An investigation of mechanical behavior of concrete containing crushed waste glass. *J. Eng. Des. Technol.* **2019**, *17*, 1285–1303. [CrossRef]
26. Singh, H.; Siddique, R. Utilization of crushed recycled glass and metakaolin for development of self-compacting concrete. *Constr. Build. Mater.* **2022**, *348*, 128659. [CrossRef]
27. Harrison, E.; Berenjian, A.; Seifan, M. Recycling of waste glass as aggregate in cement-based materials. *Environ. Sci. Ecotechnology* **2020**, *4*, 100064. [CrossRef] [PubMed]
28. Shayan, A.; Xu, A. Value-added utilisation of waste glass in concrete. *Cem. Concr. Res.* **2004**, *34*, 81–89. [CrossRef]
29. Arivalagan, S.; Sethuraman, V.S. Experimental study on the mechanical properties of concrete by partial replacement of glass powder as fine aggregate: An environmental friendly approach. *Mater. Today Proc.* **2021**, *45*, 6035–6041. [CrossRef]
30. Shi, C.; Wu, Y.; Riefler, C.; Wang, H. Characteristics and pozzolanic reactivity of glass powders. *Cem. Concr. Res.* **2005**, *35*, 987–993. [CrossRef]
31. Chen, Z.; Wang, Y.; Liao, S.; Huang, Y. Grinding kinetics of waste glass powder and its composite effect as pozzolanic admixture in cement concrete. *Constr. Build. Mater.* **2020**, *239*, 117876. [CrossRef]
32. Khmiri, A.; Samet, B.; Chaabouni, M. A cross mixture design to optimise the formulation of a ground waste glass blended cement. *Constr. Build. Mater.* **2012**, *28*, 680–686. [CrossRef]
33. Ismail, Z.Z.; Al-Hashmi, E.A. Recycling of waste glass as a partial replacement for fine aggregate in concrete. *Waste Manag.* **2009**, *29*, 655–659. [CrossRef]
34. Mostofinejad, D.; Hosseini, S.M.; Nosouhian, F.; Ozbakkaloglu, T.; Tehrani, B.N. Durability of concrete containing recycled concrete coarse and fine aggregates and milled waste glass in magnesium sulfate environment. *J. Build. Eng.* **2020**, *29*, 101182. [CrossRef]
35. Tamanna, N.; Tuladhar, R.; Sivakugan, N. Performance of recycled waste glass sand as partial replacement of sand in concrete. *Constr. Build. Mater.* **2020**, *239*, 117804. [CrossRef]
36. Penacho, P.; de Brito, J.; Veiga, M.R. Physico-mechanical and performance characterization of mortars incorporating fine glass waste aggregate. *Cem. Concr. Compos.* **2014**, *50*, 47–59. [CrossRef]
37. Lee, G.; Poon, C.S.; Wong, Y.L.; Ling, T.C. Effects of recycled fine glass aggregates on the properties of dry–mixed concrete blocks. *Constr. Build. Mater.* **2013**, *38*, 638–643. [CrossRef]
38. Corinaldesi, V.; Gnappi, G.; Moriconi, G.; Montenero, A. Reuse of ground waste glass as aggregate for mortars. *Waste Manag.* **2005**, *25*, 197–201. [CrossRef] [PubMed]
39. Tejaswi, S.S.; Rao, R.C.; Vidya, B.; Renuka, J. Experimental investigation of waste glass powder as partial replacement of cement and sand in concrete. *IUP J. Struct. Eng.* **2015**, *8*, 14.
40. Batayneh, M.; Marie, I.; Asi, I. Use of selected waste materials in concrete mixes. *Waste Manag.* **2007**, *27*, 1870–1876. [CrossRef] [PubMed]
41. Gerges, N.N.; Issa, C.A.; Fawaz, S.A.; Jabbour, J.; Jreige, J.; Yacoub, A. Recycled glass concrete: Coarse and fine aggregates. *Eur. J. Eng. Technol. Res.* **2018**, *3*, 1–9.
42. Mohammed, T.K.; Hama, S.M. Mechanical properties, impact resistance and bond strength of green concrete incorporating waste glass powder and waste fine plastic aggregate. *Innov. Infrastruct. Solut.* **2022**, *7*, 1–12. [CrossRef]

43. Walczak, P.; Małolepszy, J.; Reben, M.; Rzepa, K. Mechanical properties of concrete mortar based on mixture of CRT glass cullet and fluidized fly ash. *Procedia Eng.* **2015**, *108*, 453–458. [CrossRef]
44. Abdallah, S.; Fan, M. Characteristics of concrete with waste glass as fine aggregate replacement. *Int. J. Eng. Tech. Res. (IJETR)* **2014**, *2*, 11–17.
45. Oan, A.F.; Alrefaei, A. Using Glass Wastes as Partial Replacement of Coarse Aggregates in Concrete. *Key Eng. Mater.* **2022**, *921*, 207–215. [CrossRef]
46. Park, S.B.; Lee, B.C.; Kim, J.H. Studies on mechanical properties of concrete containing waste glass aggregate. *Cem. Concr. Res.* **2004**, *34*, 2181–2189. [CrossRef]
47. Topcu, I.B.; Canbaz, M. Properties of concrete containing waste glass. *Cem. Concr. Res.* **2004**, *34*, 267–274. [CrossRef]
48. Shehata, I.; Varzavand, S.; Elsawy, A.; Fahmy, M. The use of solid waste materials as fine aggregate substitutes in cementitious concrete composites. In Proceedings of the 1996 Semisesquicentennial Transportation Conference Proceedings, Ames, IA, USA, 13–14 May 1996.
49. Degirmenci, N.; Yilmaz, A.; Cakir, O.A. *Utilization of Waste Glass as Sand Replacement in Cement Mortar*; NISCAIR-CSIR: Delhi, India, 2011.
50. Deng, Q.; Lai, Z.; Xiao, R.; Wu, J.; Liu, M.; Lu, Z.; Lv, S. Effect of Waste Glass on the Properties and Microstructure of Magnesium Potassium Phosphate Cement. *Materials* **2021**, *14*, 2073. [CrossRef] [PubMed]
51. Lubej, S.; Anžel, I.; Jelušič, P.; Kosec, L.; Ivanič, A. The effect of delayed ettringite formation on fine grained aerated concrete mechanical properties. *Sci. Eng. Compos. Mater.* **2016**, *23*, 325–334. [CrossRef]
52. Talero, R. Performance of Metakaolin and Portland Cements in Ettringite Formation as Determined by Le Chatelier-Ansttet Test: Kinetic and Morphological Differences and New Specification. *Silic. Ind.* **2007**, *11–12*, 191–204.
53. Sun, J.; Wang, Y.; Liu, S.; Dehghani, A.; Xiang, X.; Wei, J.; Wang, X. Mechanical, chemical and hydrothermal activation for waste glass reinforced cement. *Constr. Build. Mater.* **2021**, *301*, 124361. [CrossRef]
54. Liu, M.; Xue, Z.; Zhang, H.; Li, Y. Dual-channel membrane capacitive deionization based on asymmetric ion adsorption for continuous water desalination. *Electrochem. Commun.* **2021**, *125*, 106974. [CrossRef]
55. Qu, S.; Xu, W.; Zhao, J.; Zhang, H. Design and implementation of a fast sliding-mode speed controller with disturbance compensation for SPMSM system. *IEEE Trans. Transp. Electrif.* **2021**, *7*, 2611–2622. [CrossRef]
56. Xu, D.-S.; Huang, M.; Zhou, Y. One-dimensional compression behavior of calcareous sand and marine clay mixtures. *Int. J. Geomech.* **2020**, *20*, 04020137. [CrossRef]
57. Almesfer, N.; Ingham, J. Effect of Waste Glass on the Properties of Concrete. *J. Mater. Civ. Eng.* **2014**, *26*, 06014022. [CrossRef]
58. Limbachiya, M.; Meddah, M.S.; Fotiadou, S. Performance of granulated foam glass concrete. *Constr. Build. Mater.* **2012**, *28*, 759–768. [CrossRef]
59. Singh, S. Partial replacement of coarse aggregate with waste glass in concrete. *Int. J. Innov. Res. Sci. Eng. Technol.* **2017**, *6*, 6723–6728.
60. Liaqat, M.; Shah, M.L.; Baig, M.A. Effect of waste glass as partial replacement for coarse aggregate in concrete. *Int. J. Tech. Innov. Mod. Eng. Sci.* **2018**, *4*, 2455–2585.
61. Onyeka, F. Effect of partial replacement of coarse aggregate by crushed broken glass on properties of concrete. *Int. J. Civ. Eng. Technol. (IJCIET)* **2019**, *10*, 356–367.
62. Afshinnia, K.; Rangaraju, P.R.J.C.; Materials, B. Impact of combined use of ground glass powder and crushed glass aggregate on selected properties of Portland cement concrete. *Constr. Build. Mater.* **2016**, *117*, 263–272. [CrossRef]
63. Du, H.; Tan, K.H. Concrete with recycled glass as fine aggregates. *ACI Mater. J.* **2014**, *111*, 47–58.
64. Kumar, M.H.; Mohanta, N.R.; Samantaray, S.; Kumar, N.M. Combined effect of waste glass powder and recycled steel fibers on mechanical behavior of concrete. *SN Appl. Sci.* **2021**, *3*, 1–18.
65. Kumar, S.; Nagar, B. Effects of waste glass powder on compressive strength of concrete. *Int. J. Trend Sci. Res. Dev.* **2017**, *1*, 289–298. [CrossRef]
66. Tuaum, A.; Shitote, S.; Oyawa, W. Experimental study of self-compacting mortar incorporating recycled glass aggregate. *Buildings* **2018**, *8*, 15. [CrossRef]
67. Mustafa, T.; Beshlawy, S.E.; Nassem, A.; Materials, B. Experimental study on the behavior of RC beams containing recycled glass. *Constr. Build. Mater.* **2022**, *344*, 128250. [CrossRef]
68. Małek, M.; Łasica, W.; Jackowski, M.; Kadela, M.J.M. Effect of waste glass addition as a replacement for fine aggregate on properties of mortar. *Materials* **2020**, *13*, 3189. [CrossRef] [PubMed]
69. Premalatha, J.; Srinivasan, R. Properties of Concrete with Waste Glass Powder (GP) as Fine Aggregate Replacement. *Int. J. Recent Technol. Eng. (IJRTE)* **2019**, *8*, 2277–3878.
70. Borhan, T. Properties of glass concrete reinforced with short basalt fibre. *Mater. Des.* **2012**, *42*, 265–271. [CrossRef]
71. Adaway, M.; Wang, Y. Recycled glass as a partial replacement for fine aggregate in structural concrete–Effects on compressive strength. *Electron. J. Struct. Eng.* **2015**, *14*, 116–122. [CrossRef]
72. Elavarasan, D.; Dhanalakshmi, D.G. Experimental study on waste glass as a partial replacing material in concrete for fine aggregate'. *Int. J. Adv. Res. Biol. Eng. Sci. Technol. (IJARBEST)* **2016**, *2*, 116–120.

73. Natarajan, S.; Ravichandran, N.; Basha, N.A.S.; Ponnusamy, N. Partial Replacement of Fine Aggregate by Using Glass Powder. In Proceedings of the IOP Conference Series: Materials Science and Engineering, Ulaanbaatar, Mongolia, 10–13 September 2020; p. 012050.
74. Gautam, S.; Srivastava, V.; Agarwal, V. Use of glass wastes as fine aggregate in Concrete. *J. Acad. Indus. Res.* **2012**, *1*, 320–322.
75. Ubeid, H.S.; Hama, S.M.; Mahmoud, A.S. Mechanical Properties, Energy Impact Capacity and Bond Resistance of concrete incorporating waste glass powder. In Proceedings of the IOP Conference Series: Materials Science and Engineering, Pahang, Malaysia, 30 March 2020; p. 012111.
76. Turgut, P.; Yahlizade, E. Engineering, E. Research into concrete blocks with waste glass. *Int. J. Civ. Environ. Eng.* **2009**, *3*, 186–192.
77. Upreti, S.; Mandal, B. Waste-glass as a partial replacement for fine aggregate in concrete. *Int. Res. J. Mod. Eng. Technol. Sci.* **2021**, *3*, 489–495. [CrossRef]
78. Biranagi, P.; Urs, S.R.; Shridhar, K.; Sunitha, H.K.; Vinoda, M.N. Study on effect of waste glass as partial replacement for coarse aggregate in concrete. *Int. Res. J. Eng. Technol. (IRJET)* **2019**, *6*, 1052–1063.
79. Agrawal, V.; Singh, J.P.; Sharma, C.R. Studies of Characteristics Strength and Durability of concrete using waste glass as partial replacement of coarse aggregate. *Int. J. Res. Eng. Appl. Manag. (IJREAM)* **2018**, *4*, 483–488.
80. Nagar, B.; Bhargava, V.P. Effect of glass powder on various properties of concrete. *Int. J. Sci. Eng. Technol.* **2016**, *4*, 567–573.

Article

Strength and Durability of Sustainable Self-Consolidating Concrete with High Levels of Supplementary Cementitious Materials

Moslih Amer Salih [1,*], Shamil Kamil Ahmed [2], Shaymaa Alsafi [3], Mohd Mustafa Al Bakri Abullah [4,5,*], Ramadhansyah Putra Jaya [6], Shayfull Zamree Abd Rahim [5], Ikmal Hakem Aziz [5] and I Nyoman Arya Thanaya [7]

[1] Department of Surveying Techniques, Technical Institute of Babylon, Al-Furat Al-Awsat Technical University (ATU), Najaf 54003, Iraq
[2] Tech Remix LLC, CGRC+Q8C, Jiddah St, Al Jerf Industrial 1, Ajman P.O. Box 4778, United Arab Emirates
[3] Department of Water Resources, Faculty of Engineering, Al-Mustansiriyah University, Baghdad 10052, Iraq
[4] Faculty of Chemical Engineering & Technology, Universiti Malaysia Perlis (UniMAP), Arau 02600, Malaysia
[5] Centre of Excellence Geopolymer & Green Technology (CEGeoGTech), Universiti Malaysia Perlis (UniMAP), Perlis 01000, Malaysia
[6] Faculty of Civil Engineering Technology, Universiti Malaysia Pahang, Lebuhraya Tun Razak, Kuantan 26300, Malaysia
[7] Department of Civil Engineering, Udayana University, Bali 80361, Indonesia
* Correspondence: moslih.a.salih@atu.edu.iq (M.A.S.); mustafa_albakri@unimap.edu.my (M.M.A.B.A.)

Abstract: Self-consolidating concrete (SCC) has been used extensively in the construction industry because of its advanced characteristics of a highly flowable mixture and the ability to be consolidated under its own weight. One of the main challenges is the high content of OPC used in the production process. This research focuses on developing sustainable, high-strength self-consolidating concrete (SCC) by incorporating high levels of supplementary cementitious materials. The overarching purpose of this study is to replace OPC partially by up to 71% by using fly ash, GGBS, and microsilica to produce high-strength and durable SCC. Two groups of mixtures were designed to replace OPC. The first group contained 14%, 23.4%, and 32.77% fly ash and 6.4% microsilica. The second group contained 32.77%, 46.81%, and 65.5% GGBS and 6.4% microsilica. The fresh properties were investigated using the slump, V-funnel, L-box, and J-ring tests. The hardened properties were assessed using a compressive strength test, while water permeability, water absorption, and rapid chloride penetration tests were used to evaluate the durability. The innovation of this experimental work was introducing SCC with an unconventional mixture that can achieve highly durable and high-strength concrete. The results showed the feasibility of SCC by incorporating high volumes of fly ash and GGBS without compromising compressive strength and durability.

Keywords: self-consolidating concrete; SCC; fly ash; GGBS; microsilica; sustainable concrete; high strength; durability

1. Introduction

Self-consolidating concrete (SCC) was invented in 1980 as a promising solution to cast concrete for structures with dense reinforced formwork sections [1,2]. Technically, SCC can be placed and consolidated in congested reinforced sections under its weight and flow around reinforcement by improving filling capacity. The cohesiveness of the concrete obtained from optimized mixture design and proper handling of concrete during pouring facilitates casting concrete without segregation and bleeding [1,3–7]. In addition, the characteristics of SCC provide more technical solutions by eliminating vibrating equipment, reducing noise pollution, and lowering labor costs and construction time. Applying SCC in the construction industry showed more positive aspects, such as reduced labor associated with lower human risk in construction sites. From the microstructural point of view, the proper mixture design of the SCC improves the interfacial transition zone (ITZ) between aggregate, reinforcement, and bulk cement paste, which enhances durability by decreasing permeability [2,7–9]. Given all the advantages of SCC in practice, emphasis has been placed on optimizing its constituent composition by incorporating supplementary cementitious materials (SCMs). Moreover, emphasis has been placed on investigating the effect of the water-to-binder ratio in the mixture design with the use of chemical admixtures such as superplasticizers and viscosity-modifying admixtures. SCMs and mixture design with the type of applications collectively affect SCC's fresh and hardened properties [10–12]. However, with the increasing global trend towards sustainable development in construction, more research is indispensable to reduce the high content of Portland cement used in SCC production. SCC cost was considered one of the major drawbacks in the production process due to the high content of the OPC utilized in the mixture [13]. It has been estimated that cement production was 4100 million tons [14]. The production process of Portland cement releases at least 930 kg/ton of carbon dioxide into the atmosphere [15], which is considered one of the main challenges that many countries have targeted by adopting long-term measures to minimize CO_2 emissions [16]. Furthermore, consuming natural resources for the constituent components of concrete exerts a considerable impact annually that jeopardizes sustainability. Due to the increasing population worldwide and rapid urbanization, there is global demand for Portland cement, which augments a massive demand in the construction industry for infrastructural development [17]. Therefore, inspecting more sustainable and environmentally friendly construction materials is crucial in developing advanced concrete-tech binders. Developing green concrete by incorporating SCMs such as fly ash, GGBS, and microsilica is a promising solution for producing environmentally friendly concrete by reducing mixtures' OPC quantity and lowering CO_2 emissions [18–22]. The SCMs have long-lasting effects on the environment because of their nature as non-biodegradable waste materials. Incorporating pozzolanic materials can improve concrete durability and increase the life span of the structures by reducing the required maintenance and repair in addition to cement reduction [23,24]. The most commonly used SCMs for replacing OPC in SCC are ground granulated blast furnace slag (GGBS), fly ash (FA), silica fume (SF), and Microsilica (MS) [25–27]. Fly ash, GGBS, and microsilica are by-products generated from different manufacturing processes and are not produced intentionally. The SCMs have been used as essential constituents to enhance concrete performance and durability when exposed to different aggressive environments [28,29]. Previously, fly ash, GGBS, and microsilica have been applied to replace OPC partially in SCC to enhance fresh and hardened properties and reduce its carbon footprint owing to the high content of binder used in its mixture design [30–33]. This technique was intended to lower CO_2 emissions associated with OPC production and improve concrete durability [34–36]. Moreover, it is intended as a method to enhance the environment by applying a green combined binder with sustainability in addition to the durability factor [37]. However, increasing the cement replacement level while maintaining the engineering properties and the durability of SCC is still challenging [38]. Previously, it has been found that 10% silica fume and 10% GGBS gave the best results for the durability and mechanical properties in SCC; however, the recommendations were that 6% silica fume and 8% GGBS should be incorporated as a partial replacement for the OPC separately for better performance [39]. A silica fume to OPC ratio was used in three different percentages (4.85%, 10.5%, and 14%) to produce SCC; however, a better mechanical performance was exhibited in comparison to the normal vibrated concrete [40]. Zhao et al. [41] incorporated 20–40% FA as a partial replacement for OPC to investigate its performance in SCC and concluded that a decrease in mechanical properties was registered at 7 and 90 days in both mechanical properties. In another study, Liu [42] investigated the substitution of FA as a partial replacement for cement to study its effect on SCC. The research showed a decrease in compressive strength as FA increased from 20% to 80%. The results showed that 40% replacement with FA revealed insignificant compressive strength loss. Siddique's results [43] also showed that up to 35% replacement with FA in SCC resulted in compressive strength reduction as well as split tensile strength.

Replacing cement with FA was investigated by Uysal and Sumer [44]; they concluded that FA up to 25% may result in more developed compressive strength compared to the 100% OPC SCC. Previous studies investigated the influence of silica fume and fly ash on the performance of the SCC. Cement was replaced with 10% SF and 30% FA, and the results showed an improvement in the compressive strength [45]. The impact of GGBS and SF on SCC compressive strength was determined after replacing the cement with 30%, 50%, 65%, and 80% GGBS. SF was incorporated with 50% GGBS in three percentages, 5%, 10%, and 15%. The study showed that SF has a recognized impact on compressive strength when used with 50% GGBS [46]. In another study, SF was incorporated as a partial replacement for OPC up to 25%. The results showed an enhancement in tensile strength, whereas a decrease in compressive strength was registered; however, researchers concluded that no more than 5% SF may be used as an enhancement factor in SCC [47]. Micro- and nanosilica were investigated as replacements for OPC in high-performance SCC, and the results showed the dominancy of nanosilica in its effect on the strength properties due to its high reactivity. It is concluded that particle size distribution with a wider range may create low porosity and low water demand and enhance packing density [48]. Higher resistance of sorptivity characteristics of SCC was registered when combined with FA and SF; however, partial replacement of OPC with only 20% FA showed a reduction in sorptivity [49].

In general, fewer and limited studies have been performed regarding the durability of the SCC with the maximum amounts of binary blended replacement of OPC by SF and GGBS; moreover, fewer studies have been conducted to investigate the durability of SCC using SF and FA. In the present experimental investigation, this study investigates up to 70% cement replacement with binary mixtures of microsilica, fly ash, and GGBS yet aims to maintain the engineering properties of SCC for infrastructure applications. Essentially, the novelty in this work is the sustainable mixture design associated with high-strength and durable SCC with high content of SCMs as a partial replacement for cement. This research will achieve two significant goals: the first one is the sustainability of SCC as a high-strength building material, and the second one is the advanced durability which will provide protection against an aggressive environment. Two groups of SCCs were designed to study fresh properties such as flowability and viscosity, in addition to compressive strength as the hardened property. In order to evaluate SCC durability and service life [50], a water absorption test, water permeability test, and chloride ion penetration test were applied. The first group of mixtures contained a binary system with up to 38.74% low calcium fly ash having 0.12 CaO/SiO_2 in addition to the microsilica. The second group contained GGBS at up to 71.16% having 1.33 CaO/SiO_2 in addition to the microsilica. Microsilica was incorporated in a constant quantity of 30 kg/m^3 in all mixtures, which is equal to 6.4%. The main objective of the current work is to confirm the possibility of producing high-strength and durable SCC by incorporating a high percentage of SCMs as a partial replacement for OPC.

2. Materials and Methods

This study used ordinary Portland cement with 42.5 N grade in compliance with BS EN 196 [51] and standard BS EN 197–1:2000 CEM I [52]. The chemical and physical properties of OPC are shown in Tables 1 and 2. The supplementary cementitious materials were ground granulated blast furnace slag GGBS complies with the BS EN 15167–1:2006 [53]S, Indian low calcium fly ash (FA), and microsilica (MS). Microsilica (MS) was used in a constant quantity (30 kg/m^3) in all mixtures. Tables 3 and 4 show the chemical analysis for all SCMs and the residue on 45 micron sieve, respectively. Polycarboxylate high-range superplasticizer (HRSP) type F and G [54,55] compatible with the ASTM C494 and BSEN [55,56] was used to produce SCC. It is a high-performance concrete superplasticizer based on modified polycarboxylate ether, and it has a unique carboxylic ether polymer with long lateral chains. The superplasticizer, an effective cement dispersant and high-range water reducer, was used to fix constant water content and control flow in all mixtures. In addition to that, it can produce high-flowing concrete without segregation, high early strength, and high workability with lower water content and lower permeability. It is a high-range superplasticizer that can be used for ready-mix concrete, self-consolidating concrete, precast concrete, and underwater concreting. Moreover, it is used for concrete containing microsilica, GGBS, and fly ash with extremely low w/c [57]. Polycarboxylate high-range superplasticizer (HRSP) was used for the admixture in this research.

Table 1. OPC physical properties.

Physical Properties	Specification	Results
Fineness (Air Permeability)	-	3280 cm^2/gm
Initial setting time	$\geq$60	205 Minutes
Final setting time		265 Minutes
Compressive Strength—2 Days	$\geq$10	25.05 MPa
Compressive Strength—7 Days	-	40.22 MPa
Compressive Strength—28 Days	$\geq$42.50 and $\leq$62.50	53.77 MPa

Table 2. OPC chemical analysis.

Parameter	C_3S	C_2S	C_3A	C_4AF
Results	57.37%	13.83%	6.82%	11.32%

Table 3. Chemical composition for cementitious materials.

Chemical Composition %	SiO_2	Al_2O_3	Fe_2O_3	CaO	MgO	TiO_2	SO_3	Cl	Na_2O	K_2O	L.O.I
GGBS	31.27	13.34	0.64	41.55	6.90	0.98	0.11	0.01	-	-	-
FA	47.78	29.74	5.2	5.57	3.20	1.99	0.63	-	0.97	0.96	2.42
MS	92.38	-	-	-	-	-	-	-	0.46	-	5.01

Table 4. Residue on 45 micron sieve for cementitious materials.

GGBS	1.78%
FA [58]	13.30%
MS	2%

In order to overcome the problem of natural fine sand shortage, a mixture of fine washed sand and dune sand was used as part of the concrete ingredients in all mixtures. According to the sieve analysis, the dune sand particle size is 50% passing sieve size with 0.150 mm and 1% passing sieve size with 0.075 mm. Coarse aggregate was used in two sizes, 20 mm and 10 mm.

2.1. Experimental Program

The experimental program was designed to produce SCC with a high replacement level of OPC content by incorporating accurate amounts of several combinations of FA with MS and GGBS with MS. Table 5 shows the mixture proportions used in this experimental work. Seven SCC mixtures were prepared with a constant water-to-binder ratio (w/b) of 0.33. Different characteristics of SCC were investigated according to the ASTM [59] and European guidelines [60].

Table 5. SCC mixture proportions.

Mixture Code	Mixture without MS (%)	Mixture with MS (%)	OPC kg/m³	SCM kg/m³			Aggregate		Sand		HRSP kg/m³	Water kg/m³
				FA	GGBS	MS	20 mm	10 mm	Washed Sand	Dune Sand		
OPC	100% OPC	100% OPC	470	0	0	0	331	395	726	386	6.8	155
20.43FAMS	14.0% FA	20.43% (FA + MS)	374	66	0	30	338	401	711	374	6.5	155
29.5FAMS	23.4% FA	29.5% (FA + MS)	330	110	0	30	336	399	707	372	6.7	155
38.74FAMS	32.77% FA	38.74% (FA + MS)	286	154	0	30	335	397	704	370	6.8	155
38.74GGBSMS	32.77% GGBS	38.74% (GGBS + MS)	286	0	154	30	339	402	713	375	6.5	155
52.6GGBSMS	46.81% GGBS	52.6% (GGBS + MS)	220	0	220	30	338	401	711	374	6.8	155
71.16GGBSMS	65.5% GGBS	71.16% (GGBS + MS)	132	0	308	30	337	399	708	373	7.0	155

2.2. Testing Procedures

Fresh properties of SCC were determined by using the slump-flow test to determine the concrete flowability [59,61] (Figure 1) and V-funnel [62], L-box [63], and J-ring [64] tests, as shown in Figures 2–4, respectively. SCC viscosity was assessed by measuring the flow rate using the V-funnel test. The L-box test was used to measure the passing ability of SCC [2], and the flow spread with passing ability was measured by using the J-ring test. The durability of SCC was measured by applying different tests that have been used regularly for standard concrete [20–22]. Water absorption (Figure 5) was determined according to BS 1881: 122 [65]. Water permeability (Figure 6) was determined according to BS EN 12390 [66]. The rapid chloride penetration test (RCPT) (Figure 7) was conducted for all concrete mixtures to measure the electrical conductance and ability to resist chloride ion penetration. The RCP test was conducted according to ASTM C 1202 [67]. Figure 8 shows sample preparation.

Figure 1. Slump-flow test.

Figure 2. V-funnel test.

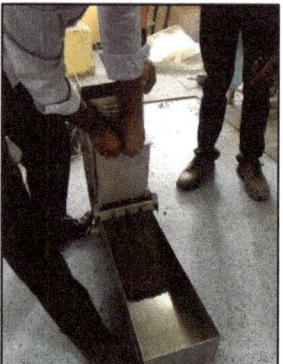

Figure 3. L-box test.

Figure 4. J-ring flow test.

Figure 5. Water absorption test device.

Figure 6. Water permeability test device.

Figure 7. RCPT device.

Figure 8. Sample preparation.

3. Results and Discussion
3.1. Workability
3.1.1. Slump Flow

Figure 9 shows the initial slump flow and the slump retention results for the self-consolidating concrete mixtures. As can be seen, two periods were chosen to measure a range of flow of SCC under its weight, initially before casting and 60 min after casting.

In general, the initial slump and 60 min slump results of SCC mixtures with SCMs were increased in comparison to the reference SCC mixture that was produced with 100% OPC. The increase in flow is related to the high replacement levels of FA, GGBS, and MS. The enhancing effect of supplementary cementitious materials on the flowability of concrete was reported in previous studies [20,68]. This behavior is attributed to the positive effect of SCM particles because of their high surface area on the packing density of the mixtures and the lower reactivity of the SCMs compared to the OPC.

FA presented different effects on SCC initial flow compared to the effect of GGBS. As shown in Figure 9, FA with MS showed gentle concave initial flow and a decrease in the measurements. Replacement of OPC with 20.43%, 29.5%, and 38.74% FA and MS showed 740 mm, 730 mm, and 720 mm initial flow, respectively. On the other hand, the replacement of OPC with GGBS showed a sudden increase in the initial flow. As can be seen, 38.74%, 52.6%, and 71.16% GGBS and MS replacement showed 720, 750, and 750 mm initial flow, respectively. The effect of particle size and the large surface area that was added to the mixture effectively changed the behavior of the mixtures and the initial slump.

Two periods were applied to measure the slump in this investigation for different reasons: The first reason was the SCC workability and high-range superplasticizer (HRSP) dosage compatibility with ingredients having different particle sizes; moreover, the time tolerance for SCC to be handled and cast was considered. On the other hand, HRSP admixture was added to the mixture in order to keep the w/b ratio fixed at 0.33. The figure shows that the dosage was gradually increased with the increase in the replacement ratio of the cement [69].

Figure 10 shows the gradual increase in HRSP dosage with fly ash and GGBS mixtures. As can be seen, the admixture dosage was increased gradually, which may be attributed to the higher specific area of the cementitious materials [70]. Slump and workability showed that incorporating FA and GGBS in addition to a constant quantity of MS results in almost converging quantities of HRSP admixture needed to keep a constant water-to-binder ratio of 0.33. Moreover, it has been reported that MS may increase the water demand in the concrete mixture due to its very fine smooth spherical glassy particles that provide a high surface area compared to FA, GGBS, and OPC [71].

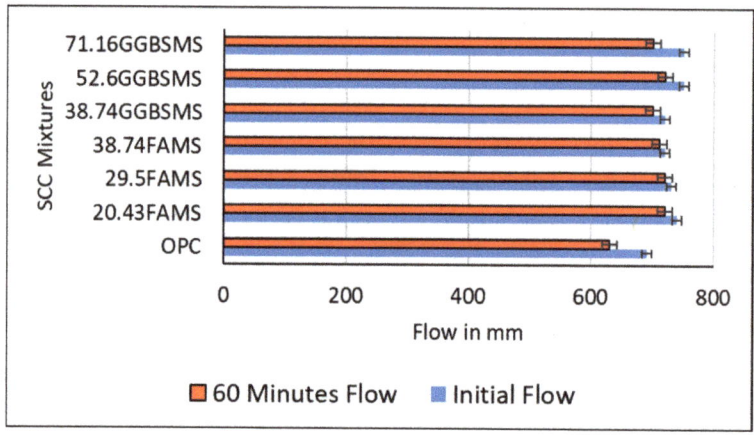

Figure 9. Slump-flow results for SCC.

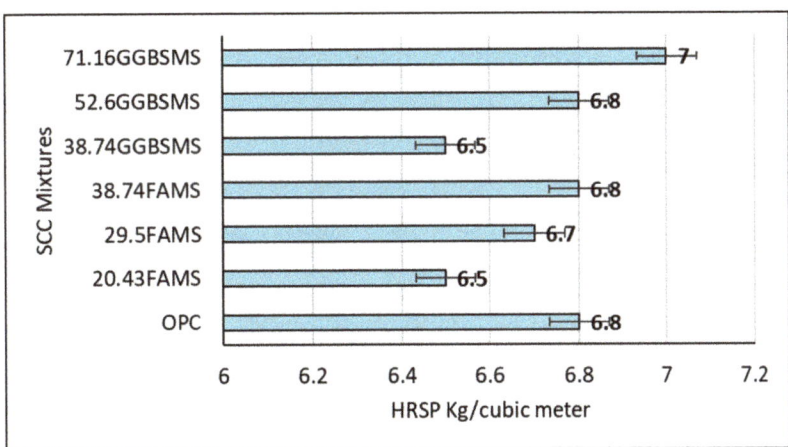

Figure 10. HRSP quantity used in SCC.

Generally, SCC produced with different amounts of supplementary cementitious materials has shown an acceptable range of slump and workability. Such flow ability may provide appropriate time for handling and casting the mixture for different applications and environments. According to the European guidelines for self-compacting concrete, the slump flow between 660 and 750 mm for SCC mixtures is suitable for many normal applications such as walls and columns [2,72]. It has been reported that FA and GGBS show a slow hydration reaction; however, providing sufficient moisture content will allow the reaction to be continued over a longer period of time. This mechanism will affect the concrete ability to flow and setting time; moreover, it will affect the strength development [73].

3.1.2. V-Funnel Test

Figure 11 shows the V-funnel test results for the SCC mixtures produced with FA and GGBS. As can be seen, the incorporation of FA and GGBS at different replacement levels registered different rates of flow in the V-funnel test. Mixtures produced with FA and MS had a lower rate of flow which increased as the percentage of replacement increased in comparison to the reference mixture. The registered rate of flow was 8, 5, and 4 s for SCC having 20.43%, 29.5%, and 38.74% FA and MS, respectively. The incorporation of GGBS showed a different effect in comparison to the reference mixture. As can be seen, there was an increase in the rate of flow with the increase in GGBS percentage. The SCC mixture with 38.74%, 52.6%, and 71.16% GGBS and MS showed an increase in the rate of flow, which was 8, 10, and 12 s, respectively. Overall, the V-funnel test can provide an indication of SCC viscosity by measuring the time required for the mixture to pass the V-funnel. The concrete viscosity increases with the increase in flow time. The results showed that FA decreased the concrete viscosity while GGBS increased the SCC viscosity.

According to the European guidelines, SCC with low viscosity will present a very quick initial flow that will then stop, whereas SCC with high viscosity may continue to flow over an extended time (creep over) [60]. The results may reflect the ability of the produced mixtures to show adequate filling capability even with congested reinforcement and the capability for the mixture to be self-leveled with the best surface finish; however, it has been reported previously that SCC may suffer from bleeding and segregation [2,8].

In this investigation, based on visual observation during the V-funnel test, mixtures showed no bleeding and no segregation, which reflects an advanced design and performance. The rate of flow showed better times in all mixtures which are lower than 100% OPC-SCC. It is practical to mention that viscosity is also a critical parameter and is required to be measured for SCC where a good surface finish is in demand when reinforcement is very dense [8].

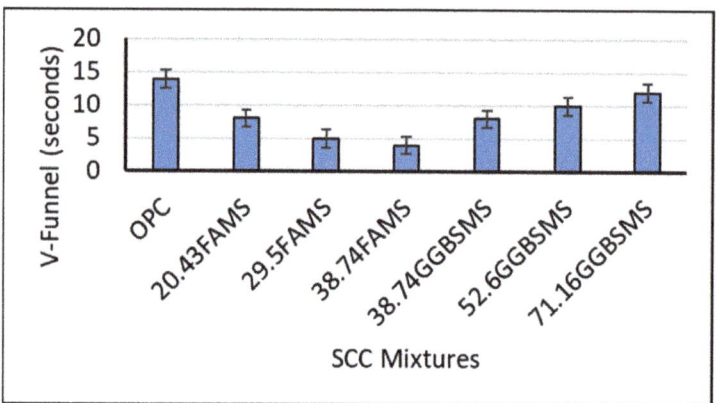

Figure 11. V-funnel results of SCC mixtures.

3.1.3. L-Box Test

Figure 12 shows the passing ability (PA) ratio of all mixtures. For the OPC-SCC control mixture, the PA ratio was 0.85, which meets the requirements mentioned in the standards [60]. SCC produced by replacing OPC with FA or GGBS with the constant amount of MS showed a higher passing ratio. The SCC mixtures showed the ability to reach equal depths for vertical section height and a horizontal section height of the L-box container. According to the European guidelines for self-compacting concrete [60], the conformity criteria for L-box are classified into two classes based on the number of steel bars installed in the L-box. The first class is PA1 (two bars with 59 mm gap), and the second class is PA2 (three bars with 41 mm gap). This classification is related to the number of smooth steel bars (12 + 0.2 mm) installed at the gate of the filling hopper of the L-box used in this investigation. This test represents the ability of concrete to flow in spaces and pass through steel reinforcing bars or tight openings without aggregate segregation or blocking; moreover, it represents the ability of concrete to flow without leaving voids at the time of casting. European guidelines showed that the passing ability ratio should be ≥ 0.75, whereas British Standards (BSI) showed that the passing ability (PA) ratio for SCC must be ≥ 0.8 and should not exceed 1.0. Figure 12 shows that all the SCC mixtures had a passing ability equal to 1.0 except for SCC with 35% GGBS (PA = 0.9); however, the PA ratio of SCC with 35% GGBS was higher than the OPC SCC passing ability ratio. In general, the results in this investigation showed that the SCC mixtures designed with replacement levels from 20.43% to 71.16% GGBS/FA have the ability to be self-leveled horizontally when placed in formwork. Moreover, cementitious materials have participated successfully in producing SCC with an appropriate passing ability, and no segregation or blocking was observed for the mixtures [2,63].

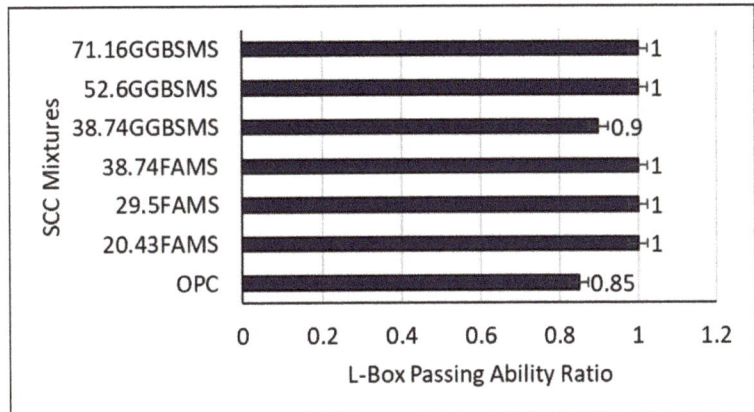

Figure 12. L-box test results (passing ability ratio).

3.1.4. J-Ring Test

Figure 13 shows the J-Ring test results expressed to the nearest 10 mm for all mixtures. In this experimental work, the J-ring test was used to assess SCC passing ability [60,74]. This is crucial to be sure that SCC can flow through spaces between and around congested reinforcement, through tight openings, and around any other obstructions that might prevent SCC from flowing during the casting process without segregation, blocking, or leaving voids. As can be seen from the figure, the spread flow ability of OPC-SCC is 600 mm, whereas all developed sustainable high-strength SCCs showed a higher ability to flow and spread within the casting process.

The results showed that there was an increase in the flow spread of concrete produced with binder containing FA and MS. The increase was 15%, 17%, and 17% for SCC produced by replacing OPC with 20.43%, 29.5%, and 38.74% of the FA + MS system, respectively. Replacing OPC with high levels of GGBS and MS also showed an increase in the spread flow ability. The results registered 12%, 15%, and 10% increases in the flow spread for mixtures produced with 38.74%, 52.6%, and 71.16% GGBS + MS, respectively. Generally, the test depicted the capability of SCC produced with high levels of replacement to fill the formwork without segregation or blocking even with congested reinforcement and the possibility of full compaction based on its weight.

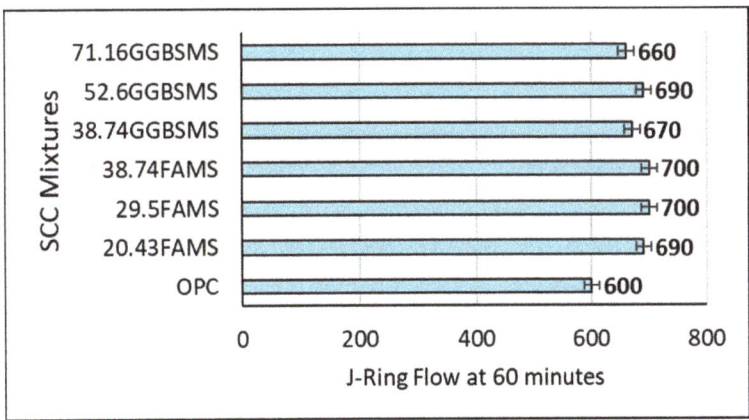

Figure 13. J-ring flow mm of SCC mixture.

3.2. Mechanical Performance

Compressive Strength

Figure 14 shows the compressive strength test results for SCC produced with 100% OPC and the binary binder system of FA with MS and GGBS with MS as a partial replacement for OPC. The test was conducted at 3, 7, and 28 days of curing, and the average of three specimens for each test was recorded. As can be seen from Figure 14, the OPC-SCC mixture had 62.5, 68.1, and 72.6 MPa at the ages of 3, 7, and 28 days, respectively. The results showed the ability of the mixture design to produce high-strength self-consolidated concrete at the early and late ages of 3 and 28 days, respectively. Most of the results showed an increase in strength with the incorporation of FA and GGBS. The increase in strength at the age of 28 days was 18%, 15%, and 10% for mixtures with 20.43%, 29.5%, and 38.74% FA and MS, respectively.

Furthermore, the increase in strength at the age of 28 days was 22%, 24%, and 13.4% for mixtures with 38.74%, 52.6%, and 71.16% GGBS with MS, respectively. It is observed that there was a slight reduction in strength for all mixtures with the increase in replacement levels at all ages; however, all mixtures showed high strength results in comparison to the reference SCC-OPC mixture. In general, the mixture design used in this experimental work showed the ability to produce high-strength concrete at early ages, meeting advanced concrete requirements. This may enable the de-molding of work forms and increase the constructability during the production cycle while maintaining a high sustainability index due to the high amount of replacement levels. The overall effect of cementitious materials was clear in increasing the strength property by replacing OPC in self-consolidating concrete [17,75]. It has been reported previously that a reduction in compressive strength property was registered for binary and ternary mixtures, and that was attributed to the low content of CaO which may cause a delay in hydraulic reaction [76], whereas, in this investigation, high-strength SCC was achieved.

The mixture proportions of the binder in this work may be a combination of synergistic ingredients that can chemically react well, producing a higher concentration of hydration products. On the other hand, the effect of the dune sand commingled with fine sand may have filled different size voids in the structure of the paste and aggregate, producing well-compacted concrete [21]. The results showed that GGBS was able to be used as an effective replacement material with good homogeneity and high synergy with MS to produce high-strength self-consolidating eco-friendly concrete despite the high level of replacement. The incorporated MS was effectively active during the chemical reactions with the presence of FA and GGBS, producing high-early-strength self-consolidating concrete. MS works as a booster to continue chemical reactions in the system, generating high-strength concrete at the age of 28 days. It is like a reactor that works to activate the potential chemical power in SCMs and react with $Ca(OH)_2$ to form greater quantities of calcium silicate hydrate (C-S-H). This mechanism may work as a densification factor to fill different voids between paste ingredients and also fill small spaces between fine particles, thus enhancing the structure of the SCC by increasing the packing density and producing a denser microstructure. The synergy between FA, GGBS, and MS previously was reported in [77]. Moreover, the polycarboxylate high-range water superplasticizer used in this mix design was capable of accelerating and boosting the chemical reactions, increasing the hydration products.

The homogeneity of the mixed materials for binder production, crushed fine sand, dune sand, and coarse aggregate was also a vital factor in producing concrete with a density between 2460 and 2485 kg/m^3, as presented in Table 6.

Table 6. Fresh density for SCC mixtures.

Mixture	OPC	20.43% FAMS	29.5% FAMS	38.74% FAMS	38.74% GGBSMS	50.6% GGBSMS	71.16% GGBSMS
kg/m^3	2490	2475	2480	2460	2485	2470	2470

The ITZ in normal concrete, which is the space between the binder paste and the coarse aggregate particles, exhibits lower strength than bulk cement paste, which is attributed to the gathering of more voids. This weakness is due to the accumulation of bleed water underneath aggregate particles, resulting in difficulty in packing solid particles near the surface. This behavior leads to more calcium hydroxide (CH) forming and concentrating in this region than elsewhere.

In this investigation, 6.4% MS played a sophisticated role at an early age. MS increased the bond strength between the paste and aggregate particles. According to the ACI Committee 234R-06, MS will react with calcium hydroxide (CH), producing more calcium silicate hydrate (C-S-H), and it is expected that all CH will be consumed in the early ages, producing a well-crystallized form of CSH-I. Without the pozzolanic reaction of the added MS, CH crystals will grow large and tend to be strongly oriented parallel to the surface of the aggregate particles. CH is weaker than C-S-H, and when the crystals are large and strongly oriented parallel to the aggregate surface, they are easily cleaved. A weak transition zone results from the combination of high void content and large, strongly oriented CH crystals. Microsilica produces a denser structure in the transition zone with a consequent increase in microhardness and fracture toughness. The presence of MS as part of the binder in fresh concrete also may reduce bleeding and greater cohesiveness.

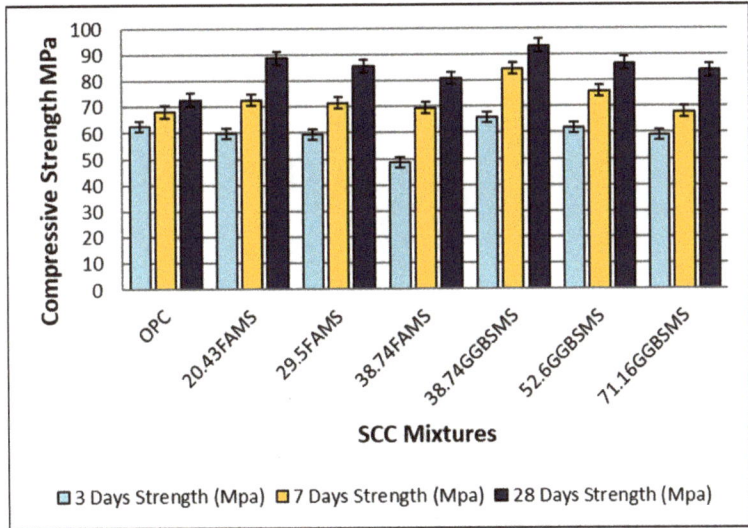

Figure 14. Compressive strength test results.

Moreover, different fine particles such as dune sand with MS may increase the packing of the solid materials as mentioned above. This behavior is related to the interlocking mechanism of the microparticles increasing the packing of solid materials by filling the spaces between cement and coarse aggregate grains [78]. It has been reported previously that GGBS can be used at an optimum level of up to 55% [70], whereas in this investigation, a high compressive strength was able to be produced with a higher replacement level of up to 71.16%.

3.3. Durability

3.3.1. Water Permeability

Figure 15 shows the water permeability results for all samples exposed to a water pressure of 500 ± 50 kPa for a period of time extended up to 72 ± 2 h [66]. As can be seen from the figure, 100% OPC-SCC showed a 3 mm water penetration depth, whereas improved results showed water penetration resistance by FA and GGBS SCC. The SCC with 20.43% FA and MS replacement showed the same permeability as the reference concrete; however, there was a 67% reduction in water permeability when the replacement level increased to 29.5% FA and MS. The SCCs with 38.74% FA and MS and 38.74%, 52.6%, and 71.16% GGBS and MS showed zero water permeability. The results with zero water permeability can be related to the development of the hydration products with the homogeneous combination of the binder ingredients. The hydration products for FA with MS and GGBS with MS may be increased due to the pozzolanic reactivity acceleration. The permeability of concrete is a congregation of the size, shape, distribution, tortuosity, and continuity of the pores; overall, it is not a simple function. It has been reported previously that there is a good relation between concrete durability and maximum continuous pore radius [79]. In this investigation, it is suggested that different particle sizes for the fine sand with dune sand and large surface area for the particles of the cementitious materials in addition to the compactness of the hydration produced led to reduced pore size and cut off pore continuity. The mixture design mechanism priority targeted a great increase in the packing density by filling the micro- and nanopaste void–pore systems, which reduced the coefficient of permeability [80–83]. Moreover, the synergistic interaction of FA with OPC and MS or GGBS with OPC and MS may have refined the pore system generated in the cement gel that created and developed a very dense and complex structure inhibiting the penetration of water within the investigated duration of 72 h [84]. Improvement in concrete permeability may also be related to the superplasticizer effect, which is designed to lower concrete permeability, in addition to its different advantages with SCC.

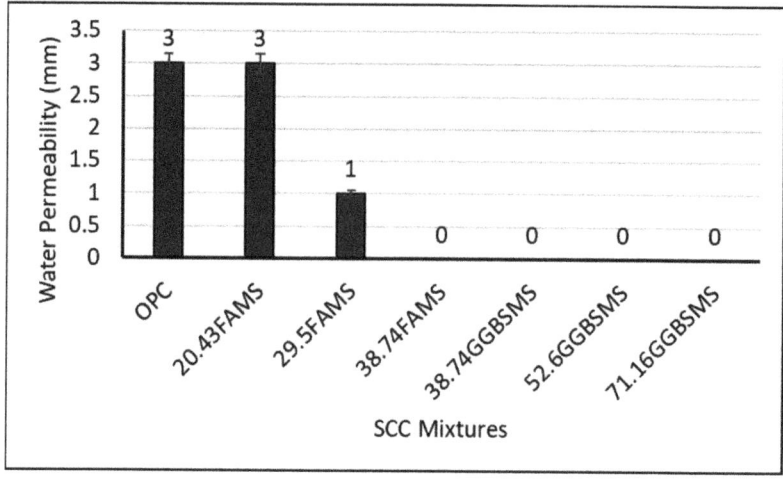

Figure 15. Water penetration depth of SCC mixtures at 28 days.

3.3.2. Water Absorption

Figure 16 shows the water absorption of SCC mixtures in percentage at 28 days. As can be seen, the water absorption results for all SCCs produced with SCMs showed lower results in comparison to the reference OPC-SCC. The water absorption test was performed according to BS 1881: Part 122 [65] and involved immersing specimens in water for 30 min after drying according to a certain procedure. It included calculating the increase in sample mass resulting from full water immersion and expressed as a percentage of the dried specimen. Replacing OPC with FA or GGBS and 30% MS showed a significant effect on enhancing the ability of SCC against absorbing water in a sophisticated way. The reduction in water absorption in mixtures with 20.43%, 29.5%, and 38.74% FA and MS was 33.3%, 40%, and 40%, respectively; mixtures with 38.74%, 52.6%, and 71.16% GGBS and MS showed 33.3%, 47%, and 53.3% reductions, respectively, as compared to the reference mixture. It has been reported previously that water may ingress to the surface of unsaturated concrete by capillary suction based on the initial water content [85–87]; moreover, capillary adsorption may be strongly connected to the size distribution of the pores in addition to pore volume and pore radius. Based on the work of Powers [88], two sizes of pores were identified; the smaller pores are the gel pores less than 10 nm in diameter working as part of the hydration products, and the larger pores are the capillary pores that occur due to excess water. In this investigation, the reduction in water absorption may be related to the synergistic interaction between supplementary cementitious materials. The hydration products of FA, GGBS, and water with different quantities of OPC in the presence of MS were developed and allowed the microsilica to react as any finely divided amorphous silica-rich constituent in the presence of CH. Calcium ions combined with the microsilica to form extra C-S-H through the pozzolanic reaction mechanism to produce a well-crystallized form of C-S-H type I which is formed during early age of curing [71,78]. It has been reported previously that fly ash and silica fume showed considerable a reduction in volume of large pores generated in concrete [89]. The same conclusion was reported when mixing silica fume and GGBS, which showed high early strength and later age strength development that may be related to the increase in the hydration products that reduced the pore volume size and structure in the mixture, resulting in reduced water absorption [90].

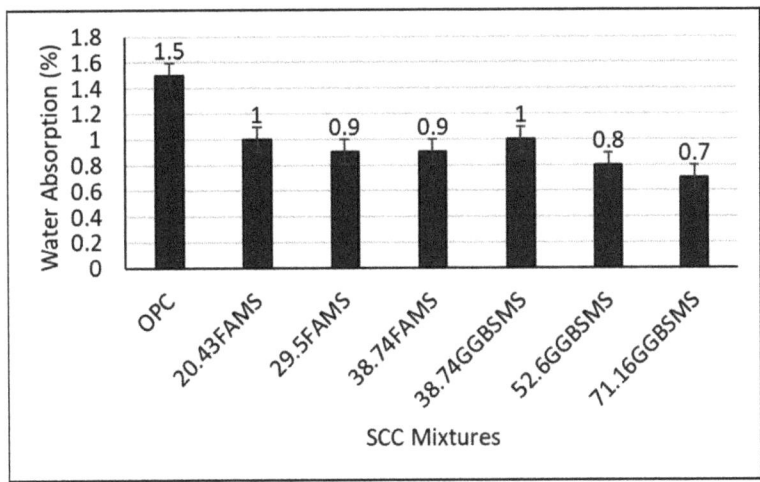

Figure 16. Water absorption of SCC mixtures at 28 days.

3.3.3. Rapid Chloride Penetration Test (RCPT)

The RCPT was applied as quality control and to evaluate SCC chloride penetration. The evaluation included electrical conductance to provide a rapid induction of the chloride ion penetration resistance into the SCC. In this test method, according to ASTM standards, the amount of electrical current passed through 51 mm thick slices of 102 mm nominal diameter cores of cylinders for 6 h is monitored. Numerical results for the RCPT represent the total electric charge that can pass through the concrete [67]. It is important to mention that many factors affect chloride ion penetration, such as type of curing, w/b, the presence of polymeric admixtures, air-void system, aggregate type, degree of consolidation, and age of the sample when the test is applied.

As can be seen from Figure 17, the total charge passed through SCC produced with 100% ordinary Portland cement was 2700 coulombs, and this sample is classified as concrete with moderate chloride ion penetrability as reported previously [67]. In this investigation, SCC produced with FA and MS as well as GGBS and MS showed an advanced ability to reduce chloride ion penetrability effectively. Adding supplementary cementitious materials as a partial replacement for cement was extremely effective in producing SCC with very low chloride ion penetrability. All the charges passed through concrete samples had results between 170 and 340 coulombs, which are lower in an effective level than the result for the reference OPC-SCC having the same w/b.

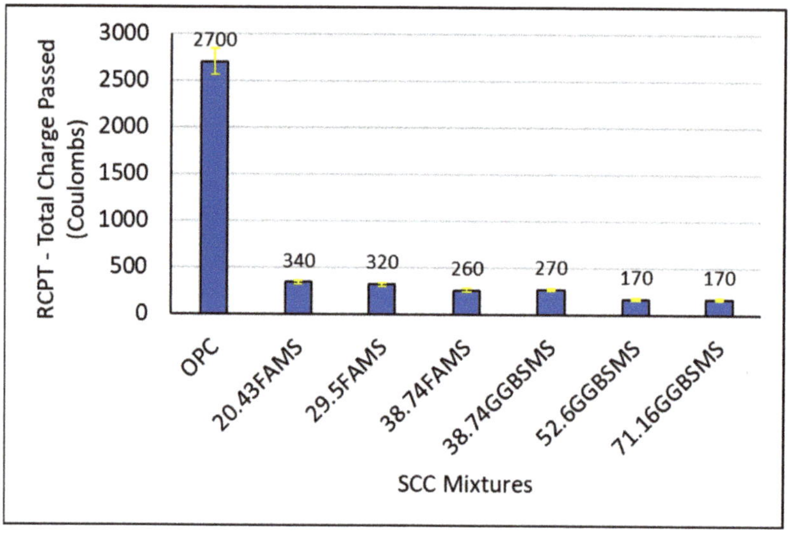

Figure 17. Total charge passed (RCP) of SCC mixtures at 28 days.

Figure 17 shows that the reduction in the charges passed was 87%, 88%, and 90% compared to the reference OPC-SSC for SCC mixtures with 20.43%, 29.5%, and 38.74% FA and MS, respectively. The reduction in charge passed was 90%, 94%, and 94% compared to the reference OPC-SCC for SCC mixtures with 38.74%, 52.6%, and 71.16% GGBS and MS, respectively. Because of the diversity and non-homogeneous mixture of materials, the chloride ion penetration in concrete is a complex process of diffusion; moreover, other environmental factors are involved in the measurement (e.g., chloride ion concentration in seawater or structure location). Previously, it has been reported that the penetration process for chloride ions may be related to the pore system in the body of the concrete. The ions start the intrusion process into the pore system because of the diffusion process which will start due to the capillary suction [91]. The addition of SCMs has reduced the penetration of chloride ions efficiently and lowered ion diffusion ability to a very low level. This behavior may be attributed to the pozzolanic reaction resulting from the addition of SCMs which causes pore refinement. This process eventually reduced the concrete permeability, as shown in Figures 13 and 14, which is also in agreement with the results of [82]. Combining or incorporating MS in the SCC mixture design as an activation factor was crucial to accelerate, enhance, and activate the chemical reactions with the presence of fly ash and GGBS. Microsilica worked as a reactor to activate the potential chemical power in SCMs and was an effective addition in increasing packing density, producing a denser microstructure [77].

4. Conclusions

In this experimental study, self-compacting concrete was produced with high replacement levels of OPC by SCMs. The results showed the ability to produce high-strength, highly durable concrete with a high quality of sustainability by reducing the OPC used in the SCC. Fly ash and GGBS were used as partial replacement materials with a constant quantity of microsilica. The following conclusions can be drawn based on the results registered from the experiments:

- In this investigation, GGBS and MS were able to be used at levels up to 71%. A sustainable and durable SCC was successfully produced.

- The high surface area of the SCM particles has increased both the initial slump and 60 min slump for the SCC in comparison to the 100% OPC-SCC. An appropriate time for handling and casting was registered. Moreover, SCC with high contents of SCMs showed the ability to be self-leveled with passing ability through reinforcement without segregation.
- Viscosity for SCC was decreased in mixtures containing fly ash, while with GGBS the viscosity was increased. The results showed adequate FA-SCC filling ability and low rate of flow with congested reinforcement, whereas GGBS increased the viscosity and consequently increased the SCC rate of flow. There was no bleeding and no segregation, which reflects an advanced mixture design. Flow spreadability was increased for SCCs with high levels of SCMs, which reflects the ability to flow and fill congested reinforcement formwork without segregation or blocking. SCC showed a higher passing ratio based on L-box test results.
- The synergy of high-content MS with FA or MS with GGBS was a clear factor in producing high-strength SCC. MS worked as a reactor to activate the potential chemical power in SCMs and react with $Ca(OH)_2$ to form more calcium silicate hydrate (C-S-H).
- The combination of very fine SCMs in SCC showed an advanced interaction producing very dense cement gel with good compactness for the paste. The result showed sophisticated water permeability which is related to the effect of the increase in the hydration products and good compactness of different aggregate sizes and fine sand particles. Water permeability in SCC concrete was reduced to zero due to the effective changes in the gel pore system. Water penetration ability was reduced due to the final hydration products of high replacement levels of SCMs which reduced pore volume and changed pore structure.

Author Contributions: M.A.S., S.K.A., S.A., M.M.A.B.A., writing—original draft preparation; M.A.S., R.P.J., S.Z.A.R., I.H.A., I.N.A.T., writing—review and editing; M.A.S., supervision; M.M.A.B.A., S.Z.A.R., R.P.J., and I.N.A.T., funding acquisition. All authors have read and agreed to the published version of the manuscript.

Funding: The support provided by Universiti Malaysia Pahang (research grant number PDU213219) is highly appreciated.

Institutional Review Board Statement: Not applicable.

Informed Consent Statement: Not applicable.

Data Availability Statement: Data are available in a publicly accessible repository.

Acknowledgments: The authors acknowledge the laboratory workers in TECH REMIX LLC Ajman, UAE; the Al-Furat Al-Awsat Technical University (ATU), Iraq; and Faculty of Chemical Engineering & Technology and Center of Excellence Geopolymer & Green Technology (CEGeoGTech) Universiti Malaysia Perlis for the support and give special thanks to those who contributed to this project directly or indirectly.

Conflicts of Interest: The authors declare no conflict of interest.

References

1. Okamura, H.; Ouchi, M. Self-compacting concrete. *J. Adv. Concr. Technol.* **2003**, *1*, 5–15. [CrossRef]
2. Concrete, S.-C. The European Guidelines for Self-Compacting Concrete. *BIBM* **2005**, *22*, 563.
3. Esfandiari, J.; Loghmani, P. Effect of perlite powder and silica fume on the compressive strength and microstructural characterization of self-compacting concrete with lime-cement binder. *Measurement* **2019**, *147*, 106846. [CrossRef]
4. Shi, C.; Wu, Z.; Lv, K.; Wu, L. A review on mixture design methods for self-compacting concrete. *Constr. Build. Mater.* **2015**, *84*, 387–398. [CrossRef]
5. Mohamed, H.A. Effect of fly ash and silica fume on compressive strength of self-compacting concrete under different curing conditions. *Ain Shams Eng. J.* **2011**, *2*, 79–86. [CrossRef]
6. ACI Concrete Terminology; American Concrete Institute. In *Ferrocement*. 2013. Available online: https://www.concrete.org/topicsinconcrete/topicdetail/ferrocement (accessed on 11 May 2019).

7. Khayat, K.H.; De Schutter, G. *Mechanical Properties of Self-Compacting Concrete: State-of-the-Art Report of the RILEM Technical Committee 228-MPS on Mechanical Properties of Self-Compacting Concrete*; Springer Science & Business Media: Berlin/Heidelberg, Germany, 2014; Volume 14.
8. BS EN 206:2013; Concrete-Specification, Performance, Production and Conformity. The British Standards Institution, BSI Standards Limited: London, UK, 2014.
9. Shi, C.; Wu, Y. Mixture proportioning and properties of self-consolidating lightweight concrete containing glass powder. *ACI Mater. J.* **2005**, *102*, 355.
10. Wang, X.; Wang, K.; Taylor, P.; Morcous, G. Assessing particle packing based self-consolidating concrete mix design method. *Constr. Build. Mater.* **2014**, *70*, 439–452. [CrossRef]
11. Siddique, R.; Aggarwal, P.; Aggarwal, Y. Mechanical and durability properties of self-compacting concrete containing fly ash and bottom ash. *J. Sustain. Cem.-Based Mater.* **2012**, *1*, 67–82. [CrossRef]
12. Esmaeilkhanian, B.; Khayat, K.H.; Yahia, A.; Feys, D. Effects of mix design parameters and rheological properties on dynamic stability of self-consolidating concrete. *Cem. Concr. Compos.* **2014**, *54*, 21–28. [CrossRef]
13. Owsiak, Z.; Grzmil, W. The evaluation of the influence of mineral additives on the durability of self-compacting concretes. *KSCE J. Civ. Eng.* **2015**, *19*, 1002–1008. [CrossRef]
14. Ashish, D.K. Feasibility of waste marble powder in concrete as partial substitution of cement and sand amalgam for sustainable growth. *J. Build. Eng.* **2018**, *15*, 236–242. [CrossRef]
15. Technical Report No. 74, Cementitious Materials, The Effect of Ggbs, Fly Ash, Silica Fume and Limestone Fines on the Properties of Concrete. 2011. Available online: www.concrete.org.uk (accessed on 11 May 2019).
16. Cement Association of Canada. Sustainable Cement Manufacturing. Available online: https://www.cement.ca/sustainability/# (accessed on 11 May 2019).
17. Rajhans, P.; Panda, S.K.; Nayak, S. Sustainable self compacting concrete from C&D waste by improving the microstructures of concrete ITZ. *Constr. Build. Mater.* **2018**, *163*, 557–570.
18. Revilla-Cuesta, V.; Skaf, M.; Faleschini, F.; Manso, J.M.; Ortega-López, V. Self-compacting concrete manufactured with recycled concrete aggregate: An overview. *J. Clean. Prod.* **2020**, *262*, 121362. [CrossRef]
19. Rojo-López, G.; Nunes, S.; González-Fonteboa, B.; Martínez-Abella, F. Quaternary blends of portland cement, metakaolin, biomass ash and granite powder for production of self-compacting concrete. *J. Clean. Prod.* **2020**, *266*, 121666. [CrossRef]
20. Salih, M.A. Strength And Durability of High Performance Concrete Containing Fly Ash and Micro Silica. *Intern. J. Civ. Eng. Technol. (IJCIET)* **2018**, *9*, 104–114.
21. Salih, M.; Ahmed, S. Mix Design for Sustainable High Strength Concrete by Using GGBS and Micro Silica as Supplementary Cementitious Materials. *Int. Rev. Civ. Eng. (IRECE)* **2020**, *11*, 45. [CrossRef]
22. Salih, M. Durability of Concrete Containing Different Levels of Supplementary Cementitious Materials. *Int. Rev. Civ. Eng. (IRECE)* **2018**, *9*, 241.
23. Le, H.T.; Ludwig, H.-M. Effect of rice husk ash and other mineral admixtures on properties of self-compacting high performance concrete. *Mater. Des.* **2016**, *89*, 156–166. [CrossRef]
24. Łaźniewska-Piekarczyk, B. The influence of chemical admixtures on cement hydration and mixture properties of very high performance self-compacting concrete. *Constr. Build. Mater.* **2013**, *49*, 643–662. [CrossRef]
25. Kavitha, S.; Kala, T.F. Evaluation of strength behavior of self-compacting concrete using alccofine and GGBS as partial replacement of cement. *Indian J. Sci. Technol.* **2016**, *9*, 1–5. [CrossRef]
26. Sharbaf, M.; Najimi, M.; Ghafoori, N. A comparative study of natural pozzolan and fly ash: Investigation on abrasion resistance and transport properties of self-consolidating concrete. *Constr. Build. Mater.* **2022**, *346*, 128330. [CrossRef]
27. Tadi, C.; Rao, T.C. Investigating the performance of self-compacting concrete pavement containing GGBS. *Mater. Today: Proc.* **2022**, *49*, 2013–2018. [CrossRef]
28. Sun, Z.; Young, C. Bleeding of SCC pastes with fly ash and GGBFS replacement. *J. Sustain. Cem. Based Mater.* **2014**, *3*, 220–229. [CrossRef]
29. Memon, S.A.; Shaikh, M.A.; Akbar, H. Utilization of Rice Husk Ash as viscosity modifying agent in Self Compacting Concrete. *Constr. Build. Mater.* **2011**, *25*, 1044–1048. [CrossRef]
30. Abd El Aziz, M.; Abd El Aleem, S.; El Heikal, M.; Didamony, H. Hydration and durability of sulphate-resisting and slag cement blends in Caron's Lake water. *Cem. Concr. Res.* **2005**, *35*, 1592–1600. [CrossRef]
31. Khan, M.I.; Siddique, R. Utilization of silica fume in concrete: Review of durability properties. *Resour. Conserv. Recycl.* **2011**, *57*, 30–35. [CrossRef]
32. Hanif, A.; Lu, Z.; Li, Z. Utilization of fly ash cenosphere as lightweight filler in cement-based composites—A review. *Constr. Build. Mater.* **2017**, *144*, 373–384. [CrossRef]
33. Elemam, W.E.; Abdelraheem, A.H.; Mahdy, M.G.; Tahwia, A.M. Prediction and Optimization of Self-Consolidating Concrete Properties. *ACI Mater. J.* **2022**, *119*, 91–104.
34. Li, G.; Zhao, X. Properties of concrete incorporating fly ash and ground granulated blast-furnace slag. *Cem. Concr. Compos.* **2003**, *25*, 293–299. [CrossRef]
35. Berndt, M.L. Properties of sustainable concrete containing fly ash, slag and recycled concrete aggregate. *Constr. Build. Mater.* **2009**, *23*, 2606–2613. [CrossRef]

36. Liu, Y.; Zhou, X.; Lv, C.; Yang, Y.; Liu, T. Use of Silica Fume and GGBS to Improve Frost Resistance of ECC with High-Volume Fly Ash. *Adv. Civ. Eng.* **2018**, *2018*, 11. [CrossRef]
37. Schneider, M. Process technology for efficient and sustainable cement production. *Cem. Concr. Res.* **2015**, *78*, 14–23. [CrossRef]
38. Choudhary, R.; Gupta, R.; Nagar, R.; Jain, A. Mechanical and abrasion resistance performance of silica fume, marble slurry powder, and fly ash amalgamated high strength self-consolidating concrete. *Constr. Build. Mater.* **2021**, *269*, 121282. [CrossRef]
39. Mohan, A.; Mini, K. Strength and durability studies of SCC incorporating silica fume and ultra fine GGBS. *Constr. Build. Mater.* **2018**, *171*, 919–928. [CrossRef]
40. Kanellopoulos, A.; Savva, P.; Petrou, M.F.; Ioannou, I.; Pantazopoulou, S. Assessing the quality of concrete–reinforcement interface in Self Compacting Concrete. *Constr. Build. Mater.* **2020**, *240*, 117933. [CrossRef]
41. Zhao, H.; Sun, W.; Wu, X.; Gao, B. The properties of the self-compacting concrete with fly ash and ground granulated blast furnace slag mineral admixtures. *J. Clean. Prod.* **2015**, *95*, 66–74. [CrossRef]
42. Liu, M. Self-compacting concrete with different levels of pulverized fuel ash. *Constr. Build. Mater.* **2010**, *24*, 1245–1252. [CrossRef]
43. Siddique, R. Properties of self-compacting concrete containing class F fly ash. *Mater. Des.* **2011**, *32*, 1501–1507. [CrossRef]
44. Uysal, M.; Sumer, M. Performance of self-compacting concrete containing different mineral admixtures. *Constr. Build. Mater.* **2011**, *25*, 4112–4120. [CrossRef]
45. Sharma, R.; Khan, R.A. Sustainable use of copper slag in self compacting concrete containing supplementary cementitious materials. *J. Clean. Prod.* **2017**, *151*, 179–192. [CrossRef]
46. Tavasoli, S.; Nili, M.; Serpoush, B. Effect of GGBS on the frost resistance of self-consolidating concrete. *Constr. Build. Mater.* **2018**, *165*, 717–722. [CrossRef]
47. Vivek, S.S.; Dhinakaran, G. Fresh and hardened properties of binary blend high strength self compacting concrete. *Eng. Sci. Technol. Int. J.* **2017**, *20*, 1173–1179. [CrossRef]
48. Massana, J.; Reyes, E.; Bernal, J.; León, N.; Sánchez-Espinosa, E. Influence of nano- and micro-silica additions on the durability of a high-performance self-compacting concrete. *Constr. Build. Mater.* **2018**, *165*, 93–103. [CrossRef]
49. Leung, H.Y.; Kim, J.; Nadeem, A.; Jaganathan, J.; Anwar, M.P. Sorptivity of self-compacting concrete containing fly ash and silica fume. *Constr. Build. Mater.* **2016**, *113*, 369–375. [CrossRef]
50. Castro, J.; Bentz, D.; Weiss, J. Effect of sample conditioning on the water absorption of concrete. *Cem. Concr. Compos.* **2011**, *33*, 805–813. [CrossRef]
51. BS 196-6:1992; Methods of Testing Cement. In Part 6: Determination of Fineness. BSI: London, UK, 1992.
52. BS EN 197-1:2000; In Part 1: Composition, Specifications and Conformity Criteria for Common Cements. BSI: London, UK, 2000.
53. BS EN 15167-1:2006; Ground Granulated Blast Furnace Slag for Use in Concrete, Mortar and Grout. BSI: London, UK, 2006.
54. Neville, A.M. *Properties of Concrete*; Longman: London, UK, 2011; Volume 4.
55. ASTM C 494/C 494M-19; Standard Specification for Chemical Admixtures for Concrete. ASTM International: West Conshohocken, PA, USA, 2019.
56. BS EN 934-2; Admixtures for Concrete, Mortar and Grout. BSI: London, UK, 1998.
57. CONMIX LTD. MegaFlow 2000, Polycarboxylated High Range Superplasticiser, 2020. Available online: www.conmix.com (accessed on 11 May 2019).
58. C311, A; Standard Test Methods for Sampling and Testing Fly Ash or Natural Pozzolans for Use as a Mineral Admixture in Portland-Cement Concrete. ASTM International: West Conshohocken, PA, USA, 2000.
59. ASTM C1611/C1611M-05; Standard Test Method for Slump Flow of Self-Consolidating Concrete. ASTM International: West Conshohocken, PA, USA, 2009.
60. EFNARC. The European Guidelines for Self-Compacting Concrete. In *Specification, Production and Use*; EFNARC: London, UK, 2005.
61. EN 12350-8; Testing Fresh Concrete—Part 8: Self-Compacting Concrete—Slump-Flow Test. BSI: London, UK, 2010.
62. EN 12350-9; Testing Fresh Concrete—Part 9: Self-Compacting Concrete—V Funnel Test. BSI: London, UK, 2010.
63. EN 12350-10; Testing Fresh Concrete—Part 10: Self-Compacting Concrete—L Box Test. BSI: London, UK, 2010.
64. EN 12350-12; Testing Fresh Concrete—Part 12: Self-Compacting Concrete—J Ring Test. BSI: London, UK, 2010.
65. BS 1881—122:2011 + A1; Testing Concrte Method for Determination of Water Absorption. BSI: London, UK, 2020.
66. EN 12390-8; Depth of Penetration of Water under Pressure. BSI: London, UK, 2000.
67. C1202-17a; Standard Test Method for Electrical Indication of Concrete's Ability to Resist Chloride Ion Penetration. ASTM: West Conshohocken, PA, USA, 2017.
68. Kumar, M.P.; Mini, K.M.; Rangarajan, M. Ultrafine GGBS and calcium nitrate as concrete admixtures for improved mechanical properties and corrosion resistance. *Constr. Build. Mater.* **2018**, *182*, 249–257. [CrossRef]
69. Li, L.G.; Zheng, J.Y.; Zhu, J.; Kwan, A.K.H. Combined usage of micro-silica and nano-silica in concrete: SP demand, cementing efficiencies and synergistic effect. *Constr. Build. Mater.* **2018**, *168*, 622–632. [CrossRef]
70. Oner, A.; Akyuz, S. An experimental study on optimum usage of GGBS for the compressive strength of concrete. *Cem. Concr. Compos.* **2007**, *29*, 505–514. [CrossRef]
71. Aldred, J.M.; Holland, T.C.; Morgan, D.R.; Roy, D.M.; Bury, M.A.; Hooton, R.D.; Olek, J.; Scali, M.J.; Detwiler, R.J.; Jaber, T.M. *ACI 234R Guide for the Use of Silica Fume in Concrete*; ACI–American Concrete Institute–Committee: Farmington Hills, MI, USA, 2006; Volume 234.

72. EFNARC. *Guidelines for Self-Compacting Concrete*; Association House: London, UK, 2002; Volume 32, p. 34.
73. Concrete Society. *Technical Report No. 40, The Use of GGBS and PFA in Concrete*; The Concrete Society: Slough, UK, 2008.
74. *C1621/C1621M*; Standard Test Method for Passing Ability of Self-Consolidating Concrete by J-Ring. ASTM: West Conshohocken, PA, USA, 2017.
75. Rajhans, P.; Panda, S.K.; Nayak, S. Sustainability on durability of self compacting concrete from C&D waste by improving porosity and hydrated compounds: A microstructural investigation. *Constr. Build. Mater.* **2018**, *174*, 559–575.
76. Guo, Z.; Jiang, T.; Zhang, J.; Kong, X.; Chen, C.; Lehman, D.E. Mechanical and durability properties of sustainable self-compacting concrete with recycled concrete aggregate and fly ash, slag and silica fume. *Constr. Build. Mater.* **2020**, *231*, 117115. [CrossRef]
77. Celik, K.; Meral, C.; Petek Gursel, A.; Mehta, P.K.; Horvath, A.; Monteiro, P.J.M. Mechanical properties, durability, and life-cycle assessment of self-consolidating concrete mixtures made with blended portland cements containing fly ash and limestone powder. *Cem. Concr. Compos.* **2015**, *56*, 59–72. [CrossRef]
78. ACI Committee. *234R-06: Guide for the Use of Silica Fume in Concrete*; ACI Committee: Farmington Hills, MI, USA, 2012.
79. Midgley, H.G.; Illston, J.M. The penetration of chlorides into hardened cement pastes. *Cem. Concr. Res.* **1984**, *14*, 546–558. [CrossRef]
80. Ng, P.-L.; Kwan, A.K.-H.; Li, L.G. Packing and film thickness theories for the mix design of high-performance concrete. *J. Zhejiang Univ. Sci. A* **2016**, *17*, 759–781. [CrossRef]
81. Kwan, A.K.H.; Wong, H.H.C. Packing density of cementitious materials: Part 2—packing and flow of OPC + PFA + CSF. *Mater. Struct.* **2007**, *41*, 773. [CrossRef]
82. Mehta, P.K.; Monteiro, P.J. *Concrete: Microstructure, Properties, and Materials*; McGraw-Hill: New York, NY, USA, 2006; Volume 3.
83. Dordi, C.M.; Vyasa Rao, A.N.; Santhanam, M. Microfine Gound Granulated Blast Furnace Slag For High Performance Concrete. In Proceedings of the Third International Conference on Sustainable Construction Materials and Technologies, Kyoto, Japan, 18–21 August 2013.
84. Neville, A.M. *Properties of Concrete*; Longman: London, UK, 1995; Volume 4.
85. Lamond, J.F.; Pielert, J.H. *Significance of Tests and Properties of Concrete and Concrete-Making Materials*; ASTM: West Conshohocken, PA, USA, 2006.
86. Hall, C. Water sorptivity of mortars and concretes: A review. *Mag. Concr. Res.* **1989**, *41*, 51–61. [CrossRef]
87. Henkensiefken, R.; Castro, J.; Bentz, D.; Nantung, T.; Weiss, J. Water absorption in internally cured mortar made with water-filled lightweight aggregate. *Cem. Concr. Res.* **2009**, *39*, 883–892. [CrossRef]
88. Powers, T.C.; Brownyard, T.L. Studies of the Physical Properties of Hardened Portland Cement Paste. *J. Proc.* **1946**, *43*, 101–132.
89. Mehta, P.; Gjørv, O. Properties of portland cement concrete containing fly ash and condensed silica-fume. *Cem. Concr. Res.* **1982**, *12*, 587–595. [CrossRef]
90. Bickley, J.; Ryell, J.; Rogers, C.; Hooton, R. Some characteristics of high-strength structural concrete. *Can. J. Civ. Eng.* **1991**, *18*, 885–889. [CrossRef]
91. Yazıcı, H. The effect of silica fume and high-volume Class C fly ash on mechanical properties, chloride penetration and freeze–thaw resistance of self-compacting concrete. *Constr. Build. Mater.* **2008**, *22*, 456–462. [CrossRef]

Review

Geopolymer Ceramic Application: A Review on Mix Design, Properties and Reinforcement Enhancement

Nurul Aida Mohd Mortar [1,2,*], Mohd Mustafa Al Bakri Abdullah [1,2,*], Rafiza Abdul Razak [2,3], Shayfull Zamree Abd Rahim [2,4], Ikmal Hakem Aziz [2,*], Marcin Nabiałek [5], Ramadhansyah Putra Jaya [6], Augustin Semenescu [7], Rosnita Mohamed [2] and Mohd Fathullah Ghazali [2,4]

1. Faculty of Chemical Engineering and Technology, Universiti Malaysia Perlis, Arau 02600, Malaysia
2. Center of Excellence Geopolymer and Green Technology (CEGeoGTech), Universiti Malaysia Perlis, Kangar 01000, Malaysia
3. Faculty of Civil Engineering and Technology, Universiti Malaysia Perlis, Kangar 01000, Malaysia
4. Faculty of Mechanical Engineering and Technology, Universiti Malaysia Perlis, Kangar 01000, Malaysia
5. Department of Physics, Częstochowa University of Technology, 42-201 Częstochowa, Poland
6. Faculty of Civil Engineering Technology, Universiti Malaysia Pahang, Kuantan 26300, Malaysia
7. Faculty of Materials Science and Engineering, University POLITEHNICA Bucharest, 313 Splaiul Independentei, 060042 Bucharest, Romania
* Correspondence: nurulaida@unimap.edu.my (N.A.M.M.); mustafa_albakri@unimap.edu.my (M.M.A.B.A.); ikmalhakem@unimap.edu.my (I.H.A.)

Abstract: Geopolymers have been intensively explored over the past several decades and considered as green materials and may be synthesised from natural sources and wastes. Global attention has been generated by the use of kaolin and calcined kaolin in the production of ceramics, green cement, and concrete for the construction industry and composite materials. The previous findings on ceramic geopolymer mix design and factors affecting their suitability as green ceramics are reviewed. It has been found that kaolin offers significant benefit for ceramic geopolymer applications, including excellent chemical resistance, good mechanical properties, and good thermal properties that allow it to sinter at a low temperature, 200 °C. The review showed that ceramic geopolymers can be made from kaolin with a low calcination temperature that have similar properties to those made from high calcined temperature. However, the choice of alkali activator and chemical composition should be carefully investigated, especially under normal curing conditions, 27 °C. A comprehensive review of the properties of kaolin ceramic geopolymers is also presented, including compressive strength, chemical composition, morphological, and phase analysis. This review also highlights recent findings on the range of sintering temperature in the ceramic geopolymer field which should be performed between 600 °C and 1200 °C. A brief understanding of kaolin geopolymers with a few types of reinforcement towards property enhancement were covered. To improve toughness, the role of zirconia was highlighted. The addition of zirconia between 10% and 40% in geopolymer materials promises better properties and the mechanism reaction is presented. Findings from the review should be used to identify potential strategies that could develop the performance of the kaolin ceramic geopolymers industry in the electronics industry, cement, and biomedical materials.

Keywords: geopolymer; kaolin; ceramic; zirconia; reinforcement

1. Introduction

Geopolymer manufactured raw materials are extremely rich in silica and alumina, which is an advantage given that over 65% of the Earth's crust is composed of alumina and silica minerals [1,2]. Geopolymer consists of a three-dimensional network of aluminosilicate tetrahedral atoms that are covalently bound to one another [3–6]. Geopolymers are a relatively recent type of construction material created from industrial by-products and cementitious materials with high alumina and silica content [1,6–8]. Due to climate change strategic initiatives, the market for geopolymer products has expanded dramatically in recent years. In addition, green systems are a key problem in the building industry, and the use of geopolymers through the geopolymerisation process has piqued the interest of scientists worldwide [5,6,9,10]. It is a "new" category of materials that has attracted a great deal of interest and risen gradually in research article investigations over the past decade.

Geopolymers have traditionally been considered an alternative to Portland cement-based materials with significant environmental and durability benefits. These advantages need to be compensated for by many brief mix designs and technologies when compared to conventional Portland cement [7,11,12]. Blended cements use a wide variety of non-conventional ingredients, such as geopolymer binders and pozzolan-based compounds. However, geopolymers have many additional potential uses; for example, that are advantageous due to thermal stability in the fabrication of thermally resistant structural elements [5,13–15], as adhesives [16,17], for the solidification of hazardous wastes [18–21], or as catalytic support [7,8,22].

Blended cements are stronger and less likely to crack than conventional cement, not to mention being eco-friendlier. Inorganic geopolymers are synthesised in an alkaline environment from silica–alumina gels [6,7,9,13,23]. When viewed with scanning electron microscopy (SEM), the structure is composed of interconnected chains or networks of inorganic molecules that are held together by covalent bonds. One atom of silicon or aluminium is connected to four atoms of oxygen to form a tetrahedron. These tetrahedrons form a three-dimensional network with one oxygen atom in common between each of the tetrahedrons [17,18,23–25]. The most used raw materials are natural minerals, such as kaolin [9,24,26,27] and calcined clays [28–32], and industrial wastes, such as fly ash [4,13,33–36], slag [35–37], red mud [28,38,39], and waste glass [40–42].

Kaolin converts to a pozzolan material named metakaolin (MK) after high temperature of thermal treatment. Regarding the issues of sustainability, kaolin as a geopolymer material can satisfy the world demand for ceramic industries. This review also discovered findings on the potential use of kaolin as a raw material with and without thermal treatment. However, there have only been a few research studies conducted on the use of kaolin as a raw material in a ceramic geopolymer application. This article also discussed a comprehensive review of the characterization of kaolin, addition of kaolin geopolymer, and the potential of zirconia reinforcement in ceramics. Furthermore, the experimental results by the researchers regarding the percentage ratio of zirconia addition to improve the properties of the ceramic geopolymers are also presented. At the conclusion of this review, the feasibility of future research into the low-cost manufacture of ceramics from geopolymer derived from kaolin is evaluated. Therefore, it is necessary to undertake a thorough literature analysis on current understandings regarding the functionality of geopolymer regarding its application on ceramic.

2. Mix Design and Manufacture Method for Kaolin Geopolymer in Ceramics

A new type of building materials with improved strength, durability, and other qualities entered the market in the nineteenth and twentieth centuries [43,44]. Ceramic components can be made from a wide variety of metallic and non-metallic atom combinations, and each atom combination typically lends itself to a number of structural configurations [30,40,45,46]. To address the rising needs and requirements in a wide range of application fields, scientists were compelled to develop numerous novel ceramic materials.

Inorganic solid powders with carefully controlled purity, particle size, and particle dispersion are used to create ceramic geopolymers [9,17,22]. To create a ceramic with specific material properties, various precursors are mixed in the process. This powdered mixture is mixed with a binder so that it can be machined in a "raw" state, moulded to exact specifications, and then sintered in a controlled furnace [40,41,47]. The raw ceramic must be heated to a temperature below its melting point to be sintered. By removing the moisture and binder, fine ceramic products with high hardness and density are created by condensing the microscopic gaps between the particles and fusing them together [30,46,48]. The formulation of geopolymer materials for ceramic applications is shown in Table 1.

Table 1. Mix design of geopolymer materials in ceramic application.

Authors	Raw Materials	Curing Method	Activator Molarity	Formulation
Jamil et al. 2020 [22]	• Kaolin (Associated Kaolin Industries Sdn. Bhd., Malaysia)	• Curing in 50 × 50 mm mould	• NaOH molarity (8 M)	• Solid liquid ratio (1, 1.5, 2) • Sodium silicate (Na_2SiO_3) to sodium hydroxide (NaOH) ratio (4:1) • Particle size (kaolin: ~13.3 μm, GGBS: 41.4 μm) • Two step sintering temperature (1st: 500 °C, 2nd: 900 °C)
Ma et al. 2021 [46]	• Kaolin (95%, Fengxian Reagent Factory, China) • SCWS (99.9%, Beijing, Forsman Technology Co., Ltd., China, d = 500 nm, l = 13 μm) • Silica sol (40%, Jiangsu Xiagang, Indus, China)	• Cast into polystyrene containers and cured at 60 °C the incubator (7 day)	• $H_2O/K_2O = 11$ (mole ratio)	• Calcining kaolin at 800 °C • Stirring 24 h on the rotating ball mill at 60 °C • Treated in a tube furnace (RHTH120/600/18, Nabertherm, Germany) at (1100–1200 °C) • $SiO_2/Al_2O_3 = 4$, $SiO_2/K_2O = 4$ • SCWS contents (0.5 wt%, 1 wt%, 2 wt%, 3 wt%, and 4 wt%)
Yun Ming et al. 2017 [48]	• Metakaolin	• Pre curing (80 °C, 4 h) • Curing (RT, 40, 60, 80, 100 °C at 6, 12, 24, 48, 72 h) • Curing day (7, 28 day)	• NaOH molarity (8 M)	• Kaolin (sintered at 800 °C for 2 h) • Na_2SiO_3 to NaOH ratio (0.8 and 0.2)

According to Jamil et al. [22], the phase transition of the sintered kaolin-ground granulated blast surface slag (GGBS) geopolymer was aided by the addition of GGBS to kaolin, which accelerated the geopolymer's setting time. Kaolin's structural alterations were influenced by the high alkalinity of NaOH (8 M), which made it capable of reacting with GGBS. The sintered kaolin-slag geopolymer's characteristics alter as the solid to liquid (SL) ratio rises. Akermanite and albite are two new phases that are formed when the solid content is at its highest (SL:2). The morphology of the sintered kaolin-GGBS geopolymer indicates enhanced densification and pore creation with increasing solid-to-liquid ratios. Additionally, two steps of the sintering profile, as shown in Figure 1, mitigated the beginning of fractures as the dihydroxylation mechanism is retarded. In this research, the use of kaolin as a raw material without calcination gives good feedback on energy consumption and green method by skipping the sintering stage. The effect of kaolin geopolymer at post-sintering temperatures, however, is not explored further in the thermal gravimetric and thermal analysis.

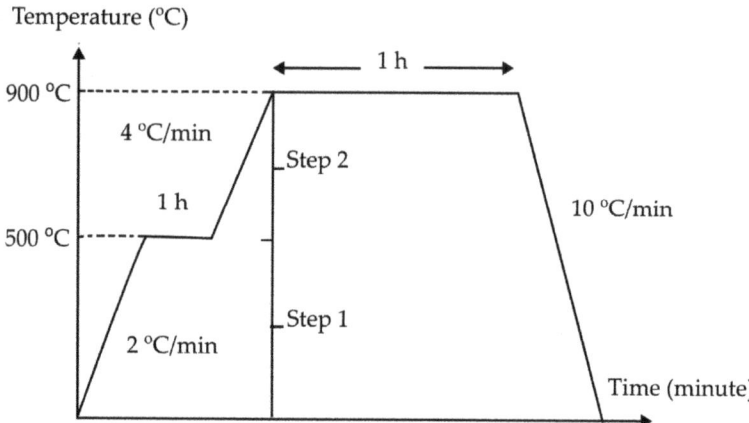

Figure 1. Two-steps sintering profile of kaolin-GGBS geopolymer [22].

Ma et al. [46] revealed that the flexural strength SiC whiskers (SCWS) reinforced geopolymer composites (SCWS/KGP) composites could be improved with the presence of SiC whisker and reached the peak value when the SCWS content was 2 wt%. The production process for the composite of SiC whiskers (SCWS) and KGP (Kaolin Geopolymer) is shown in Figure 2. The improvement in the KGP composites' flexural strength is mostly attributable to the strong interface bonding between the SiC whiskers and the geopolymer matrix. When its content reached 4 wt%, whiskers aggregation was observed, which negatively impacted the mechanical performance of SCWS/KGP composites. Additionally, geopolymer evolved into high density, twin-structure leucite ceramics after being heated to 1100 °C and 1200 °C. While this was going on, there was no interfacial reaction between the leucite matrix and the SiC whisker, which preserved its chemical stability. Due to leucite formation and a strong interfacial contact between the whisker and matrix, the composite treated at 1200 °C with 2 wt% SiC whisker demonstrated a 124.8% higher flexural strength than the composites before high temperature treatment. Nevertheless, this research does not compute the compressive strength, which is the interfacial zone between the whisker and the matrix, because shrinkage can be determined by the whisker that is subjected to compressive stresses.

Yun Ming et al. [48] confirmed the existence of zeolite Y in metakaolin-based geopolymer powder-based geopolymers with one-part mixing. Figure 3 depicts the production procedures for geopolymer powder, one-component geopolymer, and ceramic geopolymers. The one-part mixing geopolymers attained a maximum compressive strength of 10 MPa after 28 days. The sintering of the compressed geopolymer powder changed the amorphous phases into nepheline phases without passing through intermediate phases. At 1200 °C, the greatest flexural strength of ceramic geopolymers was 90 MPa. This method reduced the probability of cracking in geopolymers that had already been cured. However, it was recommended to reduce the sintering temperature to produce nepheline ceramic geopolymers, as the sintering temperature indicated in this study was too high.

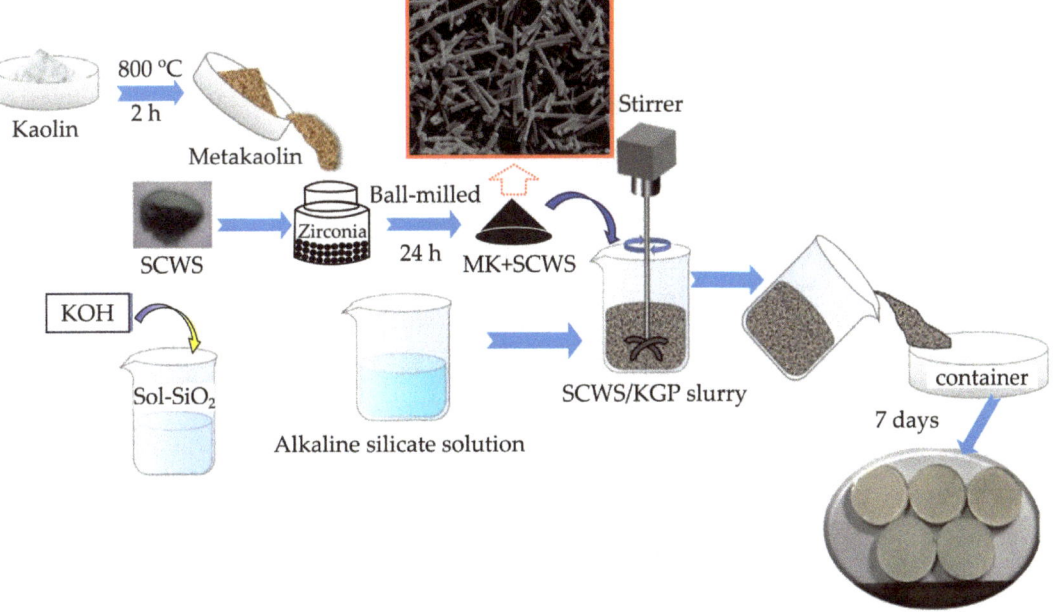

Figure 2. Preparation procedure for SCWS (SiC whiskers)/KGP (Kaolin Geopolymer) composite [46].

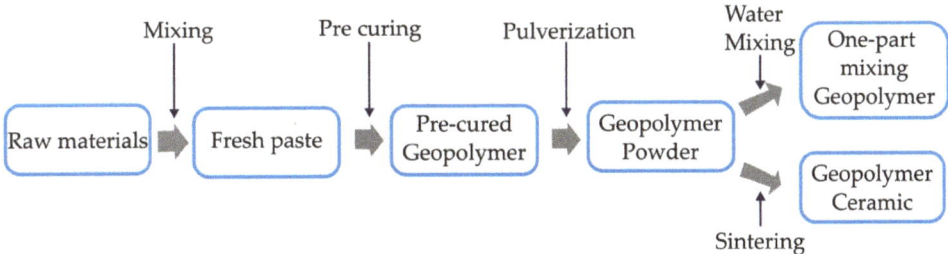

Figure 3. Steps to produce geopolymer powder, one-part-mixing geopolymer and ceramic geopolymer [48].

3. Factors Affecting the Suitability of Kaolin Geopolymers in Ceramics

Alumina is now used in the production of ceramic membranes. Kaolin, which is extensively used as a substitute for alumina due to its unique chemical and physical properties, provides a membrane with low plasticity and high refractoriness [22,30,49]. Additionally, kaolin shows hydrophilic properties. It possesses good chemical and fire resistance in addition to a comparatively high mechanical strength [50–52]. Therefore, geopolymers have the potential to be employed as construction and building materials that are environmentally beneficial. Under thermal activation, kaolin geopolymers become more stable, and kaolin clay transforms into the reactive phase of metakaolin. When metakaolin was employed as an aluminosilicate precursor [35,48], its characterization was simplified.

Ceramics cannot transfer high internal loads via plastic deformation due to their brittleness properties [53–55]. Despite all its benefits, ceramics as a building material also has several drawbacks. The basic structure of kaolin is a highly disturbed phyllosilicate network consisting primarily silicon and aluminium, which confer the major advantage of having a particle size distribution that is relatively homogeneous [22,50,56,57].

3.1. Curing Process

In general, efforts are undertaken to develop ceramics with superior mechanical qualities by incorporating amorphous phases, whiskers, fibres, particles, and even metallic phase and pores. Another method for improving the ballistic impact on a ceramic surface is to promote finer particle size, which prevents the initiation and spread of failures such as pores, flaws, fractures, and cracks [58,59]. Table 2 provides a review of prior research on geopolymer materials for diverse ceramic applications.

Table 2. Kaolin geopolymer on various ceramic applications.

Authors	Raw Materials	Curing Process	Application
Kovarik et al. 2017 [9]	Kaolin	Expose to 1000 °C for 30 min, then let it cool at room temperature.	Ceramic grog
Keppert et al. 2020 [60]	Red kaolin	Thermal curing at 60 °C	Cementitious materials
Mohamad Zaimi et al. 2020 [61]	Kaolin	24 h curing at a temperature of 80 °C	Electronic packaging industries
Kovarik et al. 2021 [62]	Calcined kaolin and blast furnace slag	• Sintering temperature 1300 °C for 3 h. • Curing temperature 70 °C	Ceramic foam
Cheng et al. 2021 [15]	Coal-series kaolin	• At a rate of 5 °C/min, the maximum temperature was kept at 600 °C, 650 °C, and 700 °C for 2 h. • The crucibles were taken out of the furnace and the calcined kaolin was quickly cooled to room temperature.	Geopolymeric cement materials
Aziz et al. 2021 [63]	Natural perlite and kaolinic clay	Placed until the test age in a curing chamber with a relative humidity of > 90%	Ceramic insulator
Sarde et al. 2022 [64]	Kaolin	Calcined at 600 °C for 2 h	Electoceramic (Dielectric character)
Marsh et al. 2019 [65]	Kaolin	Pre-dried and allowed to cool in a 105 °C oven.	Soil construction
Wang et al. 2022 [66]	Nano-ZnO/melamine polyphosphate (MPP) and silica fume clay	Thermal acceleration rate of 10 °C·min^{-1} from 40 to 1000 °C under a pure N_2 atmosphere.	Ceramic coating

Numerous studies of kaolin as a ceramic material show it is widely used for high performance ceramic materials, which are divided according to the end use application into electronic packaging industries [61], electroceramics (dielectric [64], insulator [63]), ceramic foam [62], ceramic grog [9], cementitious materials [15,60], soil construction [65], and ceramic coatings [66].

Kaolin has narrower interlayer spacing and less cation exchange capacity than other clay mineral materials [17,65,67]. Kaolin is the principal clay formed by chemical weathering; it is coarse in particle size and inflexible compared to other clays. It is the most researched clay mineral in this field, and its extensive use is attributed to its capacity to change into the metastable and more reactive phase of metakaolin following dehydroxylation at temperatures between 550 °C and 800 °C [52,62,68]. Similarly, kaolin's basic structure consists of a highly disturbed phyllosilicate matrix comprising primarily silicon and aluminium, with little variation in particle size [69,70].

3.2. Si and Al Composition

Geopolymer materials composed of Si and Al have more potential as ceramics. As one of the most important clay minerals, kaolin has a high porosity, strong mechanical stability, and low thermal conductivity in geopolymers [22,50,71]. During the firing of kaolin, the type and quantity of secondary phases can have a significant impact on the thermal properties of the raw materials [48,68]. Iron oxide is very significant [49,72]; Fe_2O_3 can exist in raw materials as either mineral complexes or silicate structures. The addition of Fe_2O_3 in kaolin not only increases the quantity of mullite phase at lower temperatures (1050 °C), but also improves the crystallisation of mullite at higher temperatures [19,73,74]. Table 3 lists the Si and Al content of numerous kaolin types successfully employed in geopolymer production.

Table 3. Si and Al content of kaolin for geopolymer synthesis.

Authors	Content (wt%)		Particle Size, D_{50} (μm)	Surface Area, (m^2/g)
	SiO_2	Al_2O_3		
Kovarik et al. 2017 [9]	52.1	41.9	4.0	13.0
Borges et al. 2017 [75]	54.5	44.2	4.5	N/A
Belmokhtar et al. 2017 [76]	53.6	42.2	4.8	6.2
Lahoti et al. 2017 [77]	53.0	43.8	1.3	N/A
Belmokhtar et al. 2018 [51]	47.2	37.12	6.20	4.72
Kwasny et al. 2018 [78]	32.04	24.99	N/A	1.57
Marsh et al. 2019 [2]	57.76	22.85	2.0	17.6
	60.73	24.05	2.0	33.7
	60.20	11.60	2.0	36.9
Jamil et al. 2020 [22]	54.0	31.7	13.3	N/A
Matalkah et al. 2020 [79]	52.1	26.2	Less than 100	2.67
Nnaemeka et al. 2020 [80]	45.3	38.38	N/A	N/A
Tiffo et al. 2020 [50]	38.00	40.10	90	N/A
Rania and Samir 2021 [81]	48.21	39.85	1–80	4.78
Aziz et al. 2021 [63]	55.14	28.52	63	N/A
Mehmet et al. 2022 [49]	70.32	18.87	N/A	N/A
Alexandre and Lima 2022 [26]	36.3	34.9	2	N/A
	47.08	39.19	2	

The percentage content range of Si and Al was from 32.04% to 70.32% and 11.60% to 44.2%, respectively, while the lowest particle size was 1.3 μm and lowest surface area is 1.57 m^2/g. According to previous research, mechanical activation altered the particle size and specific surface as well as the kaolin's reactivity with respect to the geopolymerization reaction, hence increasing the compressive strength of the geopolymers [69,81–83]. This increase was attributed to the smaller particle size and altered shape, which allowed for a faster dissolution of the particles in the activating solution [82,84]. The initial crystalline structure of clay minerals that already exist is broken during dehydroxylation, making the substance reactive; obviously, the higher the level of dehydroxylation, such as amorphousness, the more reactive the material [22,49,50,76]. Kaolin as a geopolymer material which has high Si Al content is highly suitable in ceramic application

There is a critical alkaline concentration that can achieve the maximum compressive strength, and a higher concentration does not favour the formation of geopolymers, according to prior research that used kaolin as a single source material to synthesise geopolymer products and investigate the effect of different alkaline activator concentrations on the compressive strength [22,56,85,86]. Important criteria that influence the degree of the geopolymerization process and the reaction duration with the alkali activator, respectively, are the Si-to-Al ratio and the ageing time [26,63,87]. Table 4 summarises the utilisation of alkali activator in past studies on kaolin.

Table 4. Previous research on kaolin and alkali activator.

Authors	Activator			Molarity	Raw Material
Prasanphan et al. [86]	NaOH	Na_2SiO_3	•	10 M NaOH	
Jamil et al. 2020 [22]	NaOH	Na_2SiO_3	•	6 to 8 M NaOH	
Aziz et al. [63]	NaOH	Na_2SiO_3	•	8 M NaOH	Kaolin
Alexandre et al. 2022 [26]	KOH		•	16 M of KOH	

The reaction of alkaline activators has been observed; NaOH and Na_2SiO_3 are commonly used at a range of 6 to 10 molar NaOH concentration. In the synthesis of kaolin geopolymers, the higher molar concentration of KOH, 16 Molar, is employed. In addition, high concentrations of KOH are utilised to enhance the solubility of Al^{3+} and Si^{4+} ions [26,88].

In addition to the selection of raw materials and manufacturing conditions, geopolymers can exhibit a vast array of qualities and characteristics. In general, the properties of geopolymers are highly reliant on the composition of the reactants, particularly the Si/Al ratio, process, and method, mixing design, and alkali activator type. In addition, the most important component in determining the application sectors of geopolymers is the sintering process, which is the most crucial step in the fabrication of ceramic geopolymers.

3.3. Sintering Temperature

Sintering is the process of producing a solid mass of material under pressure and heat without fully melting it. In this process, atoms in raw materials diffuse across particle boundaries and fuse to form a single solid object. The sintering process flow is shown in Figure 4.

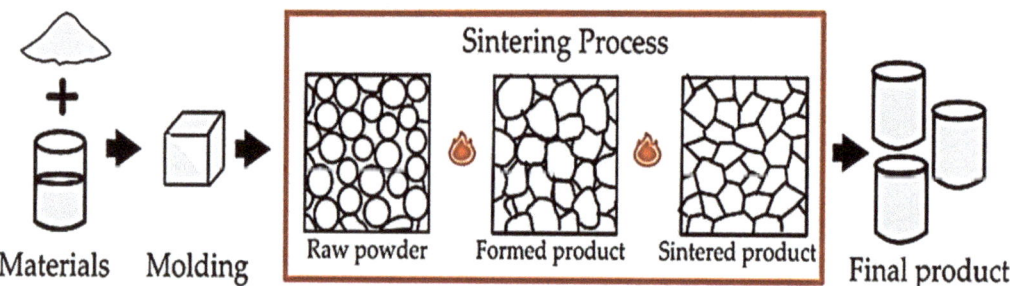

Figure 4. Sintering process flow.

Zhang et al. [57] revealed, possibly for the first time in China, the alkali activation reactivity of calcined kaolin from Guangxi province. The thermal treatment of kaolin is typically necessary to obtain more reactive precursors that result in geopolymers with high strength was discovered. Significant attempts have been made to identify the best heating temperature for the increase of the kaolin geopolymer's strength. Past research findings regarding the sintering temperature are presented in Table 5.

Table 5. Sintering temperature from past research.

Authors	Raw Material	Sintering Temperature Range (°C)	Optimum Sintering Temperature Range (°C)	Phase Formation
Naghsh and Shams 2017 [89]	Kaolin	400–800	600	• Major crystalline phase is kaolinite. • Amorphous phase in which the kaolin (peak at 20–40°)
Sornlar et al. 2021 [90]		600	600	• Amorphous phase increase with certain crystalline phases still present (illite and quartz)
Alexandre and Lima 2022 [26]		750	750	• Neoformation of hematite, from the dehydroxylation of goethite
Majdoubi et al. 2021 [91]		300–1100	800	• Undergo a significant transformation from amorphous to entirely crystalline
Merabtene et al. 2019 [92]		800	800	• Muscovite is transformed into Anorthite through its reaction with CaO, which also results in the development of other minerals like Quartz and Leucite.
Villaquirán and Mejia 2018 [93]		300–1500	900	• Below 900 °C—no obvious structure change • Exceed 900 °C—formation of macroporous mullite ceramic
Jamil et al. 2020 [22]		200–1200	900	• Kaolinite phase at 2θ of ~13°, ~25°, and ~26° • Disappearance of the gehlinite phase • Phases of akermanite and albite in SL ratio 2
Liew et al. 2017 [48]		900–1300	1200	• Major crystalline phase is kaolinite. • Formation of crystalline nepheline

Naghsh and Shams [89] demonstrated that as the calcination temperature increased from 400 °C to 800 °C, the MK dissolution extent in NaOH solution increased continuously. Sornlar et al. [90] discovered that dehydroxylation of kaolin to metakaolin occurred upon calcination at 600 °C, resulting in a large increase of amorphous phase with some crystalline phases (illite and quartz) remaining in the resulting metakaolin powder. According to reports, the amount of amorphous phase in the metakaolin had a significant effect on the curing and strength development of the geopolymer. Due to the relatively wide specific surface area, which may necessitate a high water-to-binder ratio to obtain satisfactory workability, MK is not utilised in the majority of construction situations.

Alexandre and Lima [26] conducted additional research using KOH as an alkaline activator. After calcination, minerals such as anatase and quartz in samples sintered at 750 °C remain stable or metastable. In addition, the de-hydroxylation of goethite has led to the neoformation of hematite in these samples. However, current knowledge suggests that the use of KOH as an alkali activator does not optimise geopolymerization, as the K-O link is weaker than the Na-O bond.

Majdoubi et al. [91] showed that the crystalline phase of kaolinite disappears between 300 °C and 1300 °C of sintering temperature. The constant emergence of several characteristic peaks of kaolinite suggests that calcination at temperatures between 700 °C and 800 °C was not performed perfectly. At 800 °C, it is readily apparent that the full absence of the halo characterises the amorphization of our material, indicating that the three-dimensional network of geopolymers is no longer in place and has experienced a significant transformation from amorphous to completely crystalline. When the temperature approaches 1100 °C, the aluminium phosphate phosphocristobalite phase and SiO_2 cristobalite dominate the X-ray powder diffraction (XRD) graph. This phase is the most resistant at high temperatures, which explains why the geopolymer's resistance is very weak. The slight variation between the three halos is a result of the increase in calcination temperature: the higher the temperature, the greater the intensity. After heat treatment, it was also noticed that the typical peaks of quartz and muscovite become more distinct and intense.

Merabtene et al. [92] discovered that calcined kaolin at 800 °C for 24 h, followed by quick air cooling and the selection of 800 °C possessed an excellent precursor for geopolymer synthesis. The heating scheme of 800 °C confirms the existence of carbonates such as calcite ($CaCO_3$), as indicated by XRD and Fourier transform infrared (FTIR) investigations. Nevertheless, the sintering temperature must be increased to 900 °C due to the increasing kaolin reactivity between 800 °C and 900 °C. It appears to be connected to the altered oxygen atom environment during dehydration.

Villaquirán and Meja [93] determined the sintering temperature to be between 300 °C and 1500 °C. In the absence of dehydration, the compressive strength of the geopolymer reached its maximum value as the calcination temperature rose to 900 °C and decreased drastically at 1000 °C. However, the 300 °C sintering temperature range is not really significant because past research has shown that kaolinite gradually loses OH cation between 700 °C and 900 °C during the sintering temperature.

Jamil et al. [22] emphasised from the outset that 6 M to 8 M of NaOH is sufficient to achieve alkalinity. The partial conversion of Al from its original 4-coordinated state to its 6-coordinated state is known as the phase transformation from monoclinic to tetragonal. As a result, the sintering temperature ranges between 200 °C and 1200 °C. The kaolin geopolymer after pre-sintering, which can be explained by the beginning of crystallisation of the amorphous geopolymer network, the total disappearance of the gehlinite phase, and the beginning of the appearance of the akermanite and albite phase, is due to the phase change from the more stable hexagonal aluminium phosphate. This phase is responsible for the observed colour change; the geopolymer has turned white due to the presence of crystalline Al, which indicates the presence of the akermanite and albite phases. Previous geopolymer research [20] had already revealed this modification.

Liew et al. [48] evaluated the reactivity of kaolin calcined at 900 °C, 1000 °C, 1100 °C and 1200 °C in a furnace at a heating rate of 5 °C/min and soaking duration of 3 h and discovered that the ceramic geopolymers had a maximum flexural strength of 90 MPa at 1200 °C. This study discovered the highest sintering temperature among previous studies. There is, in reality, no universally ideal temperature for MK production. Nevertheless, it is acceptable to select various calcination schemes (temperature and duration) because the mineral composition and particle size of kaolin, as well as the heating procedure, all have a role (stable or fluidized).

The porosity and apparent density variation with kaolin content have different aspects according to the sintering temperature. The sintering reactions between kaolin and alkali activator absolutely influences the chemical and mechanical properties of kaolin ceramic geopolymers, respectively. Although calcination temperature had a positive impact on aluminium alloys, the high calcination temperature had a significant negative impact on the sustainability of the environment. As a comparison, the calcination of kaolin to obtain metakaolin takes place from 600 °C to 800 °C. Concerning the environmental impact and sustainability of geopolymeric cements produced from natural kaolin, the use of non-calcined kaolin aids in the reduction of manufacturing costs and environmental implications, resulting in a green ceramic [2,22,94].

4. Properties of the Kaolin Ceramic Geopolymers

Figure 5 shows that the chemical structure of kaolin in the 3D network structure of geopolymers consists of a Si_2O_2Al framework spatially connected chains of $[SiO_4]$ and $[AlO_4]$ tetrahedral. The Si and Al share oxygen corners for each other and produce the charge-balancing metal cations.

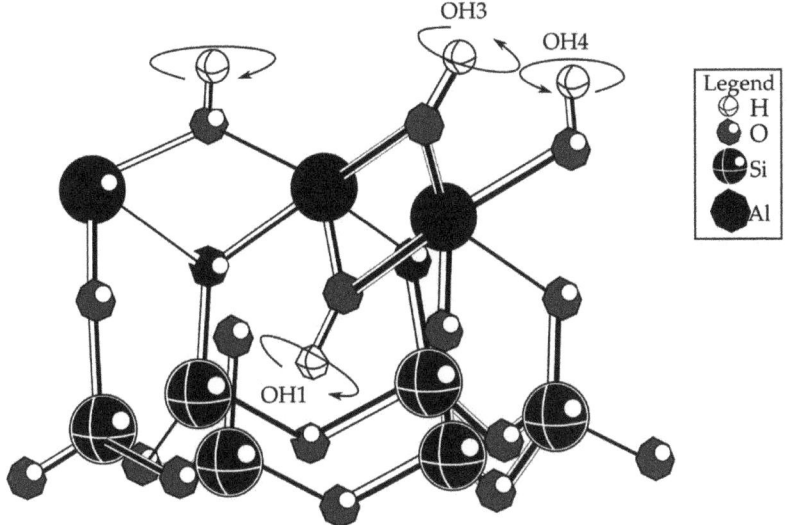

Figure 5. Chemical structure of kaolin [95].

Kaolin is a common mineral found in soils and sediments, and it has a wide range of applications. This clay mineral is a 1:1 layer aluminosilicate in which an alumina octahedral sheet and a silica tetrahedral sheet are fused to produce a layer held together by hydrogen bonding [73,93,96]. This clay possesses no exchangeable cations [89,97,98], because isomorphic substitution and cationic vacancies are near to zero.

Theoretically, any pozzolanic compound with a high alumina and silica concentration is acceptable for geopolymer synthesis under highly alkaline circumstances [18,49,56,76]. However, several considerations must be made for the geopolymerization reaction following the addition of an alkaline activator. The measurement of the physical properties and chemical composition of the raw material is one of the most essential variables in this category, as it determines the alkalinity level of the activator [19,22]. It is essential to completely analyse the samples and, based on this, to optimise the composition and amount of the activating solution and curing conditions due to the diversity of the raw materials, which may vary from batch to batch, whether mineral or waste products, for example [17,18,99,100].

Clays are hydrous aluminium silicates with a composition of approximately Al_2O_3–$2SiO_2$–$2H_2O$ [67,101]. In order to lower costs, contemporary research on the manufacture of ceramic support has centred on the use of less expensive raw materials, such as apatite powder, fly ash, natural raw clay, dolomite, and kaolin. Among these ceramic materials, kaolin has emerged as a potential raw material that is frequently employed for separation applications at a lower cost [57,73]. Moreover, kaolin is one of the least expensive and most abundant support raw materials, and it is readily accessible [22,33,102].

Geopolymers derived from kaolin has demonstrated great promise in the construction and building industries as well as engineering applications. Previous research indicates that changes in the reactivity of source materials employed in the synthesis of waste-based geopolymers have a substantial impact on the final characteristics of the ceramic geopolymer. At the raw materials selection stage, the attributes of kaolin correspond to its mineralogical compositions and thermal treatment histories. Consequently, it merits additional research into its compressive strength, chemical and mineralogical composition, morphological development, and phase analytic features.

4.1. Compressive Strength

The compressive strength of a geopolymer is contingent on several factors. These factors include the strength of the gel phase, the ratio of gel phase to undissolved Al-Si particles, the distribution and hardness of the undissolved Al-Si particle size, the amorphous nature of the geopolymer or the degree of crystallinity, and surface reactions between the gel phase and undissolved Al-Si particles [76,103–105]. For instance, curing at elevated temperatures for more than two hours appears to promote the development of compressive strength. Nevertheless, curing at 70 °C appears to increase compressive strength significantly more than curing at 30 °C during the same period. Table 6 displays the compressive strength and influencing parameters from prior studies.

Table 6. Compressive strength and factors affecting.

Authors	Raw Material	Compressive Strength	Significant Design Parameter	Factors Affecting Compressive Stress
Hajkova, 2018 [19]	Calcined kaolinite claystone	68 MPa at 28 days curing	• Constant weight ratio water glass: kaolinite: calcium hydroxide • Water glass density (1.2, 1.3, 1.4, 1.5, and 1.6 g·cm^{-3})	• Lower total pore volume and increasing the compressive strength (at higher density of water glass-1.5 g·cm^{-3})
Matalkah et al., 2020 [79]	Kaolin	48 MPa	• Added calcium oxide and sodium hydroxide (0, 5, 10, 15, and 20% by weight of kaolin) • 5 wt.% sodium hydroxide and 10 wt.% calcium oxide	• Ca makes C-S-H phases more likely, which could make the geopolymer paste denser. • NaOH could make it easier for Si and Al to leach out of the kaolin particles and into the solutions, which led to more geopolymerization. • Formation of N-A-S-H gel
Tiffo et al., 2020 [50]	Kaolin	29.1 MPa heated at 1100 °C	• Used aluminium hydroxide and oxyhydroxide to replace kaolin (30% by mass)	• Formation of stable crystalline phases • Nepheline and carnegieite are partially dissolved, which lead to closed pores and a drop in compressive strength.
Ababneh et al., 2020 [106]	Kaolin	7-day 19.83 MPa (heat cure) 16.47 MPa (room-cure)	• 62.5 wt.% kaolin, 30 wt.% calcium oxide, 5 wt.% sodium carbonate and 2.5 wt.% sodium silicate	• Due to the loss of water, a high curing temperature could also make the resulting geopolymer matrix more porous. • Presence of calcium oxide in the aluminosilicate generate C-S-H gel

Figure 6 shows the compressive strength using kaolin from prior studies summarized from Table 1. The best compressive strength by using kaolin as raw materials, 68 MPa was obtained at 28 days of curing by using high density water glass about 1.5 g·cm^{-3}. Choosing a suitable water glass density and type of alkali activator helps to ensure that the geopolymer mortar used as ceramic has good compressive strength. Generally, the ratio of each binder either solid or liquid will result in different compressive strength due to phase and bond formation in geopolymer system.

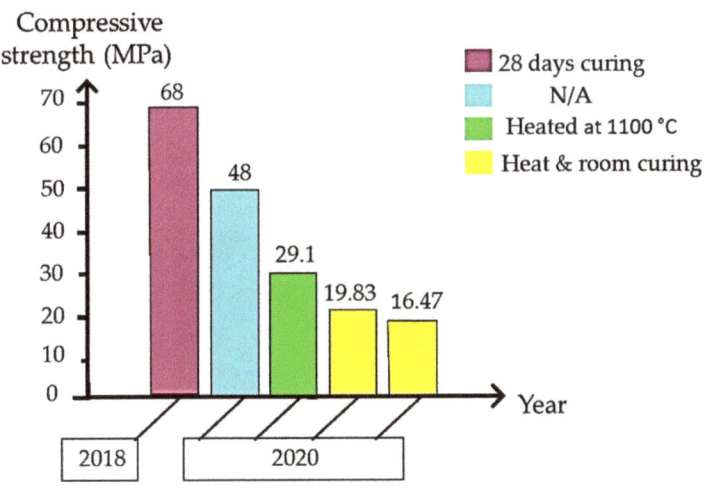

Figure 6. Reported compressive strength using kaolin.

4.2. Chemical and Mineralogical Composition

The chemical composition of raw materials as determined by X-ray fluorescence impacts the development of geopolymerization and the kind and quantity of zeolite [12,22,63,107]. The primary chemical component of kaolin, kaolinite, is dehydroxylated at temperatures reaching 550 °C, hence converting its long-range organised microstructure to an amorphous state. Consequently, kaolinite has been converted to metakaolin [48,50,74]. In geopolymer synthesis, the kind and temperature of thermal treatment affect the reactivity of metakaolin. Due to its relatively well-defined chemical structure, chemical composition, and properties, which increase its reactivity [50,57], kaolin is also considered one of the most essential precursors for geopolymer synthesis [22,49,50,56,68,108]. According to several studies, the use of kaolin as a raw material for the synthesis of geopolymer is environmentally friendly because it generates less carbon dioxide than the production of Portland cement. Table 7 displays the chemical composition of kaolin geopolymer.

Table 7. Chemical composition of kaolin geopolymer.

Authors	Al_2O_3	SiO_2	MgO	Fe_2O_3	SO_3	K_2O	Na_2O	CaO	TiO_2	P_2O_5	LOI
Heah et al. 2012 [109]	31.7	54.0	0.11	4.89	0	6.05	0	0	1.41	0	1.74
Hajkova 2018 [19]	41.45	52.03	0.13	1.05	0.20	0.79	0	0.15	1.62	0.06	2.52
Belmokhtar et al. 2018 [51]	37.12	47.2	0.39	0.83	0	2.2	0.05	0.03	0.04	0	0
Mehmet et al. 2022 [49]	18.87	70.32	0.32	0.58	1.33	0.87	0.04	1.44	-	0.1	6.03

Alumina (Al_2O_3) and silica (SiO_2) content, in general, have the greatest impact on the geopolymerization. Other mineral compositions also play a part, including magnesium oxide, MgO (speeds up the hydration reaction and may cause low porosity and high bulk density due to the large volume of the hydrate) [49], iron(III) oxide, Fe_2O_3 (able to exhibit adsorptive, ion-exchange, and catalytic properties similar to those of zeolitic aluminosilicate molecular sieves) [110], and calcium oxide, CaO (acts as to harden at room temperature without affecting the mechanical properties of the final product) [22].

4.3. Morphological and Phase Analysis

Scanning electron micrographs of kaolin's microstructure were examined. These show that morphology changes as the calcining temperature rises, and it also gradually affects the strength, hardness, apparent density, and volume shrinkage ratio. Figure 7 shows plate like kaolin SEM micrographs. It is evident that the kaolin morphology consisted of crystals with sharp edges, hexagonal shapes, rods, plates with corrosion, and irregular shapes [22,49,111]. The geopolymer made from kaolin has the benefit of being reliably created, with known properties during both preparation and development. However, the rheological issues caused by its plate-shaped particles make the system more complex to process and require more water [51,111].

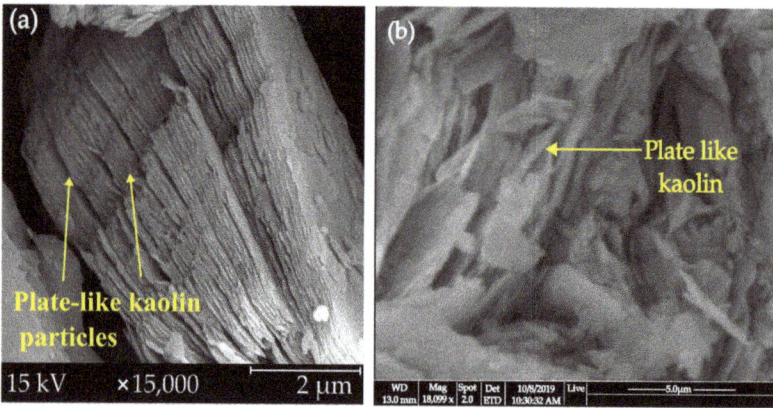

Figure 7. SEM images of the plate like kaolin (**a**) [111] (**b**) [79].

Figure 8 shows the SEM images for the room-temperature cured binders (a), paste specimen after immersed in water (b), room-temperature cured binders with 10% NaOH (c), and for the room temperature cured binders with 10% CaO (d). When the specimen was cured at room temperature, several microcracks with discontinuous gel developed, however a rather big crack was identified in the paste specimen following immersion in water. As illustrated in the figure, this produced a dense and strong alkali aluminosilicate matrix (c). The formation of such cracks could explain why the compressive strength of geopolymer specimens decreased following immersion in water. A solid gel with a well-packed structure was observed in the CaO paste, with visible amounts of crystalline or weakly crystalline C-S-H phase. The inclusion of C-S-H phases may provide stiffness to the geopolymer paste, improving the mechanical properties of a kaolin-based geopolymer [79,99,112]. The addition NaOH to the system could enhance the leaching of Si and Al from the kaolin particles to the solutions and resulted in increased geopolymerization and formation of sodium silicate hydrate gel [23,36,113,114].

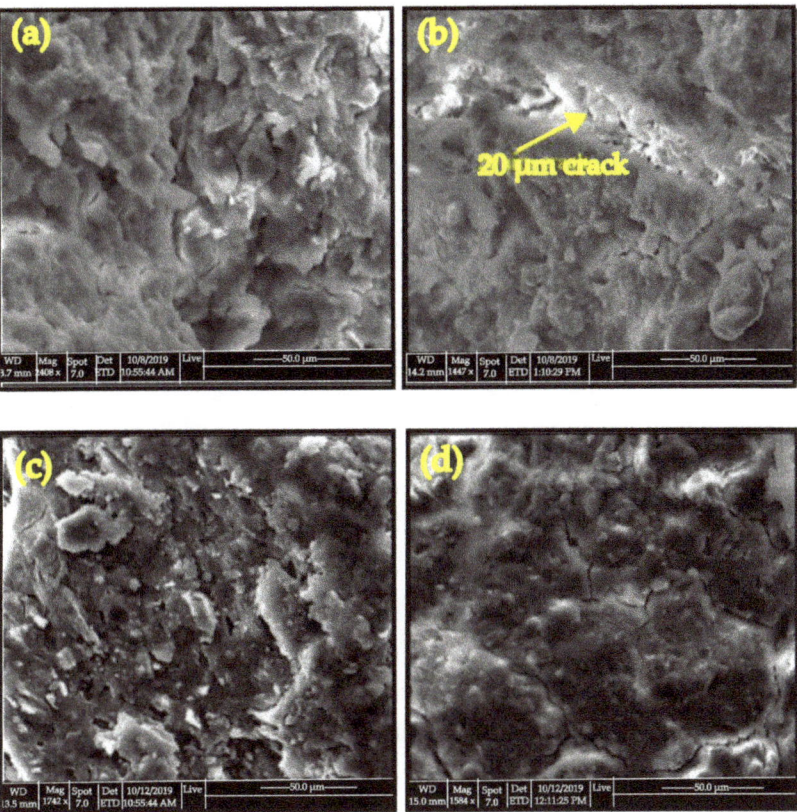

Figure 8. SEM images for (**a**) a room-temperature cured paste, (**b**) paste specimen after immersed in water, (**c**) paste with 10% NaOH and (**d**) paste with 10% CaO [79].

The Si/Al ratio has a strong influence on the microstructure of geopolymers, and the other three parameters (Al/Na, water/solids, and H_2O/Na_2O ratios) have less of an impact. This is shown by the fact that geopolymers with the same Si/Al ratio have similar microstructures, but there are big differences when the Si/Al ratio changes [25,115,116]. The geopolymer system is a two-phase gel made of water and an aluminosilicate binder. The water acts as a reaction intermediate and is released when the gel solidifies to create pores and a two-phase structure [15,105,117]. In contrast, water plays an active role in cement hydration and ultimately affects paste porosity in the OPC system [49,56,118]. Porosity in geopolymers is determined by solution chemistry during geopolymerization, which is primarily a function of Si/Al ratio and alkali metal cation type. Absolute pore volume is governed by nominal water content [50,64,119].

Only three of the kaolin's classic phases, quartz, muscovite, and kaolinite, can be seen in the XRD diffraction pattern depicted as in Figure 9 [22,51,108]. One of the crystalline phases found in kaolin is kaolinite, which makes up 35% of the crystalline phases and is converted to metakaolin through calcination [35,117,120]. The decomposition of the mineral calcite into CaO and CO_2, which is evidenced by the mass loss at around 677 °C in the thermogravimetric analysis (TGA), is what causes the calcite reflections in the XRD pattern of the calcined ceramic industrial sludge to vanish [51].

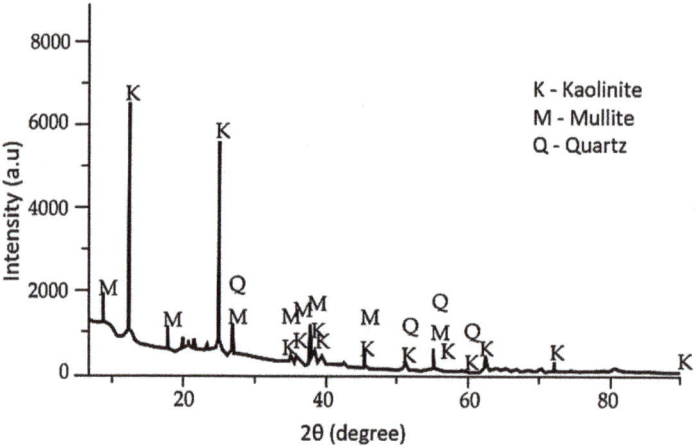

Figure 9. XRD patterns of kaolin (K) [51].

The XRD patterns for calcined kaolin and the comparable hydrates with and without additions are shown in Figure 10, where, relative to calcined kaolin, the kaolin-based geopolymer paste has less crystalline peaks. Several crystalline hydrated phases, including quartz (Q), muscovite (M), and gypsum, were discovered (G). The addition of NaOH or CaO has only a minimal impact on the crystalline phases of the resulting hydrates.

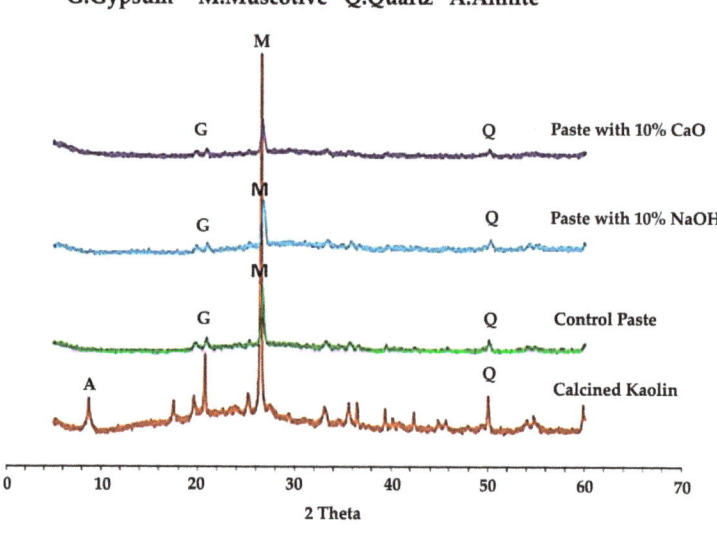

Figure 10. X-ray Diffraction Analysis for kaolin-based geopolymer [79].

Kaolin is highly recommended for use in ceramic geopolymers based on its excellent properties as a ceramic. In addition, the properties of the ceramic geopolymer can be enhanced by addition of reinforcement. There is a critical need for ceramic reinforcement to enhance its physical and mechanical properties. Geopolymers have several positive properties, including high strength, high density, few pores, an elastic modulus, and little shrinkage; yet, brittle and can easily break. Reinforcement or addition in kaolin geopolymers may solve this problem.

5. Reinforcement in Kaolin Ceramic Geopolymers

In comparison to OPC-based materials, geopolymers exhibited improved mechanical characteristics and resilience to fire, sulphates, and acids. When used as OPC products, geopolymers, however, exhibit brittle failure due to low tensile strength, which may place several restrictions on its potential structural uses. Usually the material properties of kaolin ceramic geopolymers have a greater impact on the performance of reinforced geopolymer composite than binders do. Table 8 summarises the various types of additions, the percentages of additions, and the results of studies that examined the enhancement of properties in kaolin geopolymer ceramic.

Table 8. Addition as a reinforcement in ceramic composite.

Authors	Raw Materials	Addition	Percentage/Ratio Addition	Finding Descriptions
Wu and Tian, 2013 [98]	Kaolin	Rubber composites(NR, SBR, BR, NBR, EPDM, MVQ, and CR)	40 parts per hundred rubber (phr) and 50 parts per hundred rubbers (phr)	• Superior tensile strength and weaker elongation at break • Plate-like structure of kaolin helps rubber release heat and makes rubber composites more stable at high temperatures.
Selmani et al., 2017 [121]	Kaolin	Commercial Metakaolin	0%, 16%, 33% and 50%	• Increasing the metakaolin percentage confirms the existence of different networks. • More reinforcements (illite and mica) caused more networks to form and more impurities (illite and calcite) to be coated by the excess alkaline solution.
Jamil et al., 2020 [22]	Kaolin	Ground granulated blast furnace slag (GGBS)	Kaolin:GGBS (4:1 wt.%)	• Accelerate the geopolymer's setting time and contributed to the phase transformation of sintered kaolin-GGBS geopolymer.
Tiffo et al. 2020 [50]	Kaolin	Amorphous aluminium hydroxide and aluminium oxyhydroxide	0%, 10%, 20% and 30% by mass	• Enhances compressive strength and thermal stability • 30% by mass of aluminium oxyhydroxide heated at 1100 °C, compressive strength of 29.1 MPa • 10% by mass of amorphous aluminium hydroxide gave 60.2 MPa at 1150 °C. • Formation of stable crystalline phases
Coudert et al., 2021 [122]	Kaolin	Fly ash	10%, 20% and 40% of fly ash with reference to dry mass of solids (fly ash + kaolin).	• Porosity is less because small kaolinite platelets fill the pores. • Decrease in the soil's ability to be compressed and an increase in the yield stress
Perumal et al., 2021 [67]	Kaolin	Surfactants (Hydrogen peroxide, chemical that lowers surface tension)	5, 10, 15 M NaOH Water binder ratio (0.55 and 0.65)	• Strength improvement mainly by bubble stabilization avoiding the bubble coalescence and, by reducing the pore size • Surfactants and H_2O_2 bringing down the viscosity values by 20–60% and the effect is higher at higher H_2O_2 dosage
Kaya et al., 2022 [49]	Kaolin	Zeolite	Replacing 10%, 20%, and 30% kaolin with zeolite	• 4%, 5%, and 6% increase in unit weight • Increased compressive strength and flexural strength • 3%, 7%, and 12% increase in ultrasonic pulse velocity (UPV) of the geopolymer specimens due to the formation of dense structure owing to lower porosity of kaolin than zeolite

Wu and Tian [98] reported on rubber addition, which significantly improved deformation and yield strength, as is the case for matrices with higher SiO_2/Al_2O_3 molar ratio and 40 and 50 parts per hundred rubber (phr) rubber. The tensile strength, elongation at break, and hardness of nitrile butadiene rubber (NBR) filled with 40 phr kaolin were satisfactory, and the minimal wear indicated the optimum wear resistance. Additionally, 50 phr kaolin-filled EPDM and CR have adequate elongation at break, hardness, and wear but not tensile strength, which is lower than that of 40 phr kaolin-filled EPDM. However, the durability of rubber is not particularly long. Additionally, rubber cannot handle high temperatures because it may cause a tendency for the material to rupture and degrade, which reduces the composite's ability to endure tensile strain. In order to create new geopolymer materials from a blend of commercial metakaolins and calcined clays, Selmani et al. [121] focused on valorizing naturally existing clays. Metakaolin MK1 was replaced by metakaolin MK2, which produced different compositions with the following codes: G1, G2 (16%), G2 (33%) and G2 (50%). However, because of impurities, adding natural clay reduced the compressive strength of the geopolymer composites (illite, calcite, iron).

According to Jamil et al. [22], the sluggish rate of the Al content's dissolution makes it necessary to spend more time to produce kaolin with a strong chemical interaction. A kaolin-GGBS ceramic geopolymer was created by adding ground granulated blast furnace slag (GGBS) to reduce the rate of dissolution. However, research in measuring compressive strength to gauge the brittleness of kaolin geopolymer composites is still limited. In addition, Tiffo et al. [50] reported that, to give the kaolin geopolymer its physical and mechanical properties, researchers substituted amorphous aluminium hydroxide and aluminium oxyhydroxide in varying amounts. As a result, the replacement successfully contributed to the development of heated kaolin-based geopolymers that are thermally stable and have a high compressive strength. The result is not visible, though, until 28 days into the curing process. This is crucial to demonstrate that the geopolymer system has no additional reaction mechanisms because of the removal of kaolin.

Coudert et al.'s [122] study was primarily concerned with the application of an alkali activated fly ash-based binder to improve the engineering properties of soft clay-rich soils and to replace conventional stabilisers (lime or cement). By using optical microscopy, microstructural measurements of the alkali activated fly ash binder treated soil over time were made. In a way like the alkali-activated fly ash binder, after 24 h of curing there are scattered dark patches that look like calcium-rich nodules all over the sample. After 28 days, these nodules are surrounded by larger black zones that are made up of newly formed compounds. At 28 days, hollowed grey nodular structures can be identified as the binder and linked to the breakdown of calcium-rich particles. However, combining micro-indentation with scanning electron imaging would also make it possible to measure regional variations in hardness. Therefore, the degree of calcium particle reactivity can be used to understand how important local microstructural differences are for macroscopic mechanical performances.

In the fresh-state, alkali-activated slurry, Perumal et al. [67] examined how surfactants function as stabilisers for the gas-liquid interface, enhancing the establishment of interconnected porosity utilising impure kaolin. Depending on the paste characteristics, surfactant type, and content, the pore structure produced by direct foaming can have a wide size distribution. Lower strength is generally the result of a less homogenous pore structure. The effect of three different molarities of alkali activator (5 M, 10 M, or 15 M NaOH) and water binder ratio (0.55 and 0.65) on the mechanical property of kaolin-based geopolymer has been described, however the research on the effects of Si/Al ratio and ageing duration has not been covered.

The effects of Micro additions of Fe_2O_3 and MgO on the mechanical and physical characteristics of the geopolymer binder were investigated by Kaya et al. [49]. The binder was developed by substituting zeolite for kaolin at percentages of 10%, 20%, and 30%. Additionally, by replacing 4%, 6%, and 8% of the Fe_2O_3 and MgO in the binder with zeolite, the quantities of Fe_2O_3 and MgO were enhanced. The binder was activated using NaOH that contained 15% Na by weight (Na/binder). Because kaolin has a denser structure due to its lower porosity than zeolite, replacing zeolite with it causes an increase in the unit weight, compressive strength, flexural strength, and UPV of the geopolymer specimens. However, to correspond with the development of hematite (Fe_2O_3) and periclase, the formation of sodium aluminosilicate and calcium silicate hydrates as hydration products was not further discussed (MgO). Therefore, these authors were the first to realize the potential of nanoparticles to impart toughness and strength of geopolymer structure.

Addition of Zirconia in Ceramic Geopolymers

To increase the strength and toughness of ceramics, for instance, zirconium dioxide (ZrO_2) may be added. This would be done by taking advantage of the tetragonal to monoclinic phase transformation that is brought on by the presence of a stress field before a break. On the other hand, zirconia brings improvement in compressive strength, fracture toughness, crack deflection, crack bridging, and micro-cracks.

Due to the better chemical and thermal properties to standard additives, nanosized particles are one possibility to increase the mechanical performance of such geopolymers [121,123,124]. Additionally, nanoparticles can function as a filler to lower the nanoporosity at the level of the interfacial transition zone between aggregated particles as well as a catalyst to speed up the geopolymerization reaction [55,124,125]. Table 9 shows the properties of zirconia, including melting point, boiling point, density, and molar mass.

Table 9. Zirconia properties [126].

Properties	Value
Melting Temperature (°C)	2715
Boiling temperature (°C)	4300
Density (g/m^2)	5.68
Molar mass (g/mol)	123.2

Temperature and time during the sintering process should be studied because they directly affect the grain size, yttrium segregation, and amount of cubic phase in zirconia, which in turn affects its physical, mechanical, and optical properties [14,45,125]. Increasing the sintering temperature increases the grain size of zirconia, which may improve its physical qualities but makes it more susceptible to low-temperature irradiation (LTD). Although Al_2O_3 has good hardness, abrasion resistance, and chemical inertness at elevated temperatures, it has relatively low toughness [56,121,123], which leads to early failure. To boost its fracture toughness, Yttria-Stabilized Zirconia (YSZ) is used as a strengthening agent. The product of this mixture is Zirconia-reinforced Alumina (ZTA). It undergoes a phase transformation from tetragonal to monoclinic that results in a transformation strengthening process [45,127].

Greater ZrO_2 content in a kaolin-based mullite ZrO_2 composite yielded greater density and flexural strength. Due to the decreased viscosity of the produced glassy phases in the sintered samples [56,128], the presence of ZrO_2 may have increased the thermal shock resilience of the sample. The proposed method involves adding zirconia to kaolin or metakaolin (calcined kaolinite) in order to produce at high temperatures mullite and zircon ($ZrSiO_4$)-based ceramics, according to the following Equation (1) [56]:

$$3(Al_2O_3 \cdot 2SiO_2) + 4ZrO_2 \rightarrow 3Al_2O_3 \cdot 2SiO_2 + 4ZrSiO_4 \tag{1}$$

Zircon, which does not suffer any structural change until its dissociation at around 1500 °C, possesses a number of desirable qualities, including a high resistance to alkali corrosion and an extremely low thermal expansion coefficient (4.1×10^{-6} °C^{-1}) between room temperature and 1400 °C, and a low heat conductivity [56,121,124]. Table 10 shows the impacts of the addition of zirconia in past research.

Table 10. Impact of the addition of zirconia in past research.

Authors	Percentage Addition (%)	Raw Materials	Properties Improvement and Mechanism Reaction
Phair et al., 2000 [129]	0, 1%, 3%, 5%, 7% by mass of FA	Fly ash	• Increase compressive strength • Chemically, non-aluminosilicate materials are thought to be based on zirconia's ability to make insoluble sodium polysialate, which then forms a 3D polysialate grid structure. • Zeolite production could not occur using zirconia as a nucleation germ or template.
Mecif et al., 2010 [56]	0%, 10%, 20%, 30% and 40% wt	Metakaolin	• High thermal stability, low thermal expansion, and conductivity • High creep resistance associated with strong strength, and fracture toughness • The amount of flux in the mixture cannot get any denser because the clay content is decreased, and silica is being used up when $ZrSiO_4$ is made. • Disappearance of cristobalite occurs during zircon formation
Kenawy et al., 2016 [130]	0, 5, 10, 15 and 20 wt%	Calcined kaolin at 1000 °C for 2 h	• High thermal shock resistance and flexural strength • Reduction in the viscosity of the glassy phases that develop in sintered samples • Glassy phase, which might help serious faults repair or make materials appear more durable throughout the sintering process • Continuous solid solution at the grain boundary between ZrO_2 and mullite strengthens the grain-boundary mechanism
Zawrah et al., 2018 [124]	0, 10%, 15% by weight of metakaolin	Metakaolin	• Compressive strength was increased (10% gives 74 MPa) • No new phases were produced because zircon did not take part in the geopolymerization process. Instead, it filled the spaces between the polysialate networks.

Phair et al. [129] demonstrated that the incorporation of just 3% mass of zirconia to a geopolymeric matrix significantly increased the compressive strength by 30%. Incorporating 5% or more zirconia, caused considerable brittleness due to the adverse bulk physical effects of extra filler on the 3D polysialate network. However, no clear evidence exists to establish that the absence of zeolite crystallisation is primarily attributable to the high CaO level. Furthermore, Mecif et al. [56] discovered that $ZrSiO_4$ production, which occurs at temperatures above 1150 °C, is promoted by the presence of fusing impurities such as K, Fe, Ca, and Mn in clays, as well as a reduction in zirconia particle size. It was also discovered that the rise in the porosity ratio of the final products for zirconia levels more than 20 wt percent was dictated by a decrease in the flux amount due to the reduced clay content. Sintering a mixture of 38 wt% of fine zirconia powder and 62 wt% of the more reactive clay at 1400 °C for 2 h produced ceramics that are mostly composed of zircon and mullite.

Kenawy et al. [130] hypothesized that the comparatively lower density with greater ZrO_2 contents could be the result of thermal expansion mismatches between ZrO_2 and the mullite matrix. This may cause interior fissures and a weakened matrix, resulting in a reduced density. Moreover, the higher the ZrO_2 content, the greater the viscosity of the produced glassy phases and, consequently, the lower the particle diffusion and rearrangement. Regardless, this researcher did not explore the effect on compressive strength. Moreover, the previous researchers theorised that zirconia promotes a 3D polysialate grid structure through the creation of insoluble sodium polysialate, based on research by Zawrah et al. [124] to determine the chemical foundation for the increase in compressive strength. This 3D polysialate minimises the mobility of sodium while maintaining the matrix's structural integrity and charge balance. To clarify the grain/particle sizes, phases, chemical species, and yttrium distribution, as zirconia materials behave differently at different sintering temperatures, additional research is required.

Geopolymers, which combine some characteristics of organic polymers, cements, and ceramics due to the unusual polycondensed network structure, have attracted a great deal of interest from researchers as a green cementitious material due to the advantageous and distinctive properties. Additional research is necessary to comprehend the properties of kaolin ceramic geopolymers reinforced with zirconia for use in ceramic technology.

6. Conclusions and Future Works

Geopolymer material with a low calcium content holds great potential for future ceramic applications because geopolymers have superior characteristics to OPC counterparts and have various performance-related qualities. Even though the characteristics of geopolymer have been thoroughly explored in the literature, there are still certain elements that require additional investigation. Furthermore, it is evident from this brief analysis that addition of zirconia into kaolin geopolymer applications still has a great deal of room for research and improvement. Zirconia-reinforced geopolymer matrices produce materials with enhanced compressive, fracture toughness, crack resistance, and thermal stability, relative to the unreinforced matrix. In conclusion, the essential formulation parameters for geopolymers were assembled and compared based on chemical composition, thermal processing, and consequent mechanical properties. Based on the facts reviewed, it is obvious that additional research must be conducted to optimise formulas for the development of zirconia-reinforced geopolymers with improved characteristics. Several conclusions can be drawn from a review of the existing literature on the properties of geopolymer materials in ceramic industry:

i. Geopolymer materials mentioned above have the potential to be converted into environmentally friendly ceramics because the silico-aluminate used to make them is derived from natural sources and industrial by-products. However, utilizing geopolymer materials derived from kaolin without calcination that have comparable properties to metakaolin-based materials is a significant contribution to sustainability as it reduces energy consumption during the sintering process.
ii. Many factors, such as selection of alkali activator, chemical composition of raw materials and sintering temperature, could have great possibility to produce ceramic geopolymers. Therefore, more data is needed to finally establish a clear relationship between characterization of raw materials and the thermal process.
iii. Typically, geopolymers are weak under tension and fail brittlely. Numerous studies have focused on the inclusion of different types of reinforcement into geopolymers to acquire appropriate mechanical and thermal properties for each application, particularly ceramic applications, to overcome this weakness. The method describing suitable non-destructive testing for evaluating geopolymers alongside the destructive/strength test results is recommended for reviewing in the future.
iv. To impart the strength beyond the standard properties, addition materials are needed. Generally, incorporation of zirconia in kaolin, using a proper alkali activator ratio, chemical composition of raw materials and optimum sintering temperature could increase the compressive strength, and usually result into toughness properties.
v. Although adding inorganic polymer or natural fibres incurs low-cost and is usually flexible and can be used at high content as reinforcement in geopolymers, it is not possible due to its toughness property.
vi. It should be noted that a geopolymer material has demonstrated significant feasibility and application prospects to be used as an environmentally friendly ceramic material, which may be a suitable replacement for the conventional ceramic materials and process in the future.

The significant of this review to the ceramic industry is to produce ceramic geopolymers with high compressive strength at optimal sintering temperature of kaolin. The potential of addition of zirconia may enhance the properties of kaolin geopolymers to reduce energy consumption and increase residual compressive strength without affected the growth of ceramics grains during sintering.

Author Contributions: Conceptualization, N.A.M.M., M.M.A.B.A., R.A.R., M.N. and R.P.J.; data curation, N.A.M.M., M.M.A.B.A., S.Z.A.R., R.A.R. and A.S.; formal analysis N.A.M.M., R.A.R., R.P.J., A.S. and R.M.; investigation, N.A.M.M., M.M.A.B.A., M.F.G., M.N. and A.S.; methodology, M.M.A.B.A., M.F.G. and R.A.R.; project administration; N.A.M.M., I.H.A. and R.M., validation, M.M.A.B.A., R.A.R., I.H.A. and S.Z.A.R.; writing—review and editing, N.A.M.M., M.M.A.B.A., R.A.R., S.Z.A.R. and R.M. All authors have read and agreed to the published version of the manuscript.

Funding: This study was supported by the Center of Excellence Geopolymer and Green Technology (CEGeoGTECH) UniMAP and Faculty of Technology Chemical Engineering, UniMAP. The authors appreciated the support from Research Materials Funding (RESMATE-9001-00629) and Universiti Malaysia Pahang under research grant number PDU213219.

Institutional Review Board Statement: Not applicable.

Informed Consent Statement: Not applicable.

Data Availability Statement: Not applicable.

Acknowledgments: The authors acknowledge the Center of Excellence Geopolymer and Green Technology, University Malaysia Perlis for the contribution. Special thanks to those who contributed to this project directly or indirectly.

Conflicts of Interest: The authors declare no conflict of interest.

References

1. Silva, G.; Kim, S.; Aguilar, R.; Nakamatsu, J. Natural Fibers as Reinforcement Additives for Geopolymers—A Review of Potential Eco-Friendly Applications to the Construction Industry. *Sustain. Mater. Technol.* **2020**, *23*, e00132. [CrossRef]
2. Marsh, A.; Heath, A.; Patureau, P.; Evernden, P.; Walker, P. Influence of Clay Minerals and Associated Minerals in Alkali Activation of Soils. *Constr. Build. Mater.* **2019**, *229*, 116816. [CrossRef]
3. Keppert, M.; Vejmelková, E.; Bezdička, P.; Doleželová, M.; Čáchová, M.; Scheinherrová, L.; Pokorný, J.; Vyšvařil, M.; Rovnaníková, P.; Černý, R. Red-Clay Ceramic Powders as Geopolymer Precursors: Consideration of Amorphous Portion and CaO Content. *Appl. Clay Sci.* **2018**, *161*, 82–89. [CrossRef]
4. Król, M.; Rożek, P.; Chlebda, D.; Mozgawa, W. Influence of Alkali Metal Cations/Type of Activator on the Structure of Alkali-Activated Fly Ash—ATR-FTIR Studies. *Spectrochim. Acta Part A Mol. Biomol. Spectrosc.* **2018**, *198*, 33–37. [CrossRef] [PubMed]
5. Favier, A.; Hot, J.; Habert, G.; Roussel, N.; D'Espinose De Lacaillerie, J.-B. Flow Properties of MK-Based Geopolymer Pastes. A Comparative Study with Standard Portland Cement Pastes. *Soft Matter* **2014**, *10*, 1134–1141. [CrossRef]
6. Topçu, I.B.; Toprak, M.U.; Uygunoğlu, T. Durability and Microstructure Characteristics of Alkali Activated Coal Bottom Ash Geopolymer Cement. *J. Clean. Prod.* **2014**, *81*, 211–217. [CrossRef]
7. Kashani, A.; Ngo, T.D.; Mendis, P. The Effects of Precursors on Rheology and Self-Compactness of Geopolymer Concrete. *Mag. Concr. Res.* **2019**, *71*, 557–566. [CrossRef]
8. Rahman, M.R.; Paswan, R.; Singh, S.K. Geopolymeric Materials in Infrastructure Development: A Way Forward. *Indian Concr. J.* **2020**, *94*, 30–39.
9. Kovářík, T.; Rieger, D.; Kadlec, J.; Křenek, T.; Kullová, L.; Pola, M.; Bělský, P.; France, P.; Říha, J. Thermomechanical Properties of Particle-Reinforced Geopolymer Composite with Various Aggregate Gradation of Fine Ceramic Filler. *Constr. Build. Mater.* **2017**, *143*, 599–606. [CrossRef]
10. Khater, H.M. Effect of Nano-Silica on Microstructure Formation of Low-Cost Geopolymer Binder. *Nanocomposites* **2016**, *2*, 84–97. [CrossRef]
11. Nikolov, A.; Nugteren, H.; Rostovsky, I. Optimization of Geopolymers Based on Natural Zeolite Clinoptilolite by Calcination and Use of Aluminate Activators. *Constr. Build. Mater.* **2020**, *243*, 118257. [CrossRef]
12. Villa, C.; Pecina, E.T.; Torres, R.; Gómez, L. Geopolymer Synthesis Using Alkaline Activation of Natural Zeolite. *Constr. Build. Mater.* **2010**, *24*, 2084–2090. [CrossRef]
13. He, R.; Dai, N.; Wang, Z. Thermal and Mechanical Properties of Geopolymers Exposed to High Temperature: A Literature Review. *Adv. Civ. Eng.* **2020**, *2020*, 7532703. [CrossRef]
14. Rendtorff, N.M.; Garrido, L.B.; Aglietti, E.F. Thermal Behavior of Mullite-Zirconia-Zircon Composites. Influence of Zirconia Phase Transformation. *J. Therm. Anal. Calorim.* **2011**, *104*, 569–576. [CrossRef]

15. Cheng, S.; Ge, K.; Sun, T.; Shui, Z.; Chen, X.; Lu, J.X. Pozzolanic Activity of Mechanochemically and Thermally Activated Coal-Series Kaolin in Cement-Based Materials. *Constr. Build. Mater.* **2021**, *299*, 123972. [CrossRef]
16. Davidovits, J. Properties of Geopolymer Cements. In *First International Conference on Alkaline Cements and Concretes*; Kiev State Technical University: Kiev, Ukraine, 1994; pp. 131–149.
17. Ma, J.; Ye, F.; Zhang, B.; Jin, Y.; Yang, C.; Ding, J.; Zhang, H.; Liu, Q. Low-Temperature Synthesis of Highly Porous Whisker-Structured Mullite Ceramic from Kaolin. *Ceram. Int.* **2018**, *44*, 13320–13327. [CrossRef]
18. Luo, Y.; Meng, J.; Wang, D.; Jiao, L.; Xue, G. Experimental Study on Mechanical Properties and Microstructure of Metakaolin Based Geopolymer Stabilized Silty Clay. *Constr. Build. Mater.* **2022**, *316*, 125662. [CrossRef]
19. Hajkova, P. Kaolinite Claystone-Based Geopolymer Materials: Effect of Chemical Composition and Curing Conditions. *Minerals* **2018**, *8*, 444. [CrossRef]
20. Kovářík, T.; Křenek, T.; Rieger, D.; Pola, M.; Říha, J.; Svoboda, M.; Beneš, J.; Šutta, P.; Bělský, P.; Kadlec, J. Synthesis of Open-Cell Ceramic Foam Derived from Geopolymer Precursor via Replica Technique. *Mater. Lett.* **2017**, *209*, 497–500. [CrossRef]
21. van Deventer, J.S.J.; Provis, J.L.; Duxson, P.; Lukey, G.C. Reaction Mechanisms in the Geopolymeric Conversion of Inorganic Waste to Useful Products. *J. Hazard. Mater.* **2007**, *139*, 506–513. [CrossRef]
22. Jamil, N.H.; Al Bakri Abdullah, M.M.; Pa, F.C.; Mohamad, H.; Ibrahim, W.M.A.W.; Chaiprapa, J. Influences of SiO_2, Al_2O_3, CaO and MgO in Phase Transformation of Sintered Kaolin-Ground Granulated Blast Furnace Slag Geopolymer. *J. Mater. Res. Technol.* **2020**, *9*, 14922–14932. [CrossRef]
23. Amran, Y.H.M.; Alyousef, R.; Alabduljabbar, H.; El-Zeadani, M. Clean Production and Properties of Geopolymer Concrete; A Review. *J. Clean. Prod.* **2020**, *251*, 119679. [CrossRef]
24. Malkawi, A.B.; Nuruddin, M.F.; Fauzi, A.; Almattarneh, H.; Mohammed, B.S. Effects of Alkaline Solution on Properties of the HCFA Geopolymer Mortars. *Procedia Eng.* **2016**, *148*, 710–717. [CrossRef]
25. Ababneh, A.; Matalkah, F.; Matalkeh, B. Effects of Kaolin Characteristics on the Mechanical Properties of Alkali-Activated Binders. *Constr. Build. Mater.* **2022**, *318*, 126020. [CrossRef]
26. Alexandre, I.; Barreto, R.; Lima, M. Synthesis of Geopolymer with KOH by Two Kaolinitic Clays from the Amazon: Influence of Different Synthesis Parameters on the Compressive Strength. *Mater. Chem. Phys.* **2022**, *287*, 126330. [CrossRef]
27. Gasparini, E.; Tarantino, S.C.; Conti, M.; Biesuz, R.; Ghigna, P.; Auricchio, F.; Riccardi, M.P.; Zema, M. Geopolymers from Low-T Activated Kaolin: Implications for the Use of Alunite-Bearing Raw Materials. *Appl. Clay Sci.* **2015**, *114*, 530–539. [CrossRef]
28. Bonet-Martínez, E.; Pérez-Villarejo, L.; Eliche-Quesada, D.; Carrasco-Hurtado, B.; Bueno-Rodríguez, S.; Castro-Galiano, E. Inorganic Polymers Synthesized Using Biomass Ashes-Red Mud as Precursors Based on Clay-Kaolinite System. *Mater. Lett.* **2018**, *225*, 161–166. [CrossRef]
29. Tahmasebi Yamchelou, M.; Law, D.; Brkljača, R.; Gunasekara, C.; Li, J.; Patnaikuni, I. Geopolymer Synthesis Using Low-Grade Clays. *Constr. Build. Mater.* **2020**, *268*, 121066. [CrossRef]
30. Wang, H.; Li, H.; Wang, Y.; Yan, F. Preparation of Macroporous Ceramic from Metakaolinite-Based Geopolymer by Calcination. *Ceram. Int.* **2015**, *41*, 11177–11183. [CrossRef]
31. Ondova, M.; Stevulova, N.; Meciarova, L.; Ondova, M.; Stevulova, N.; Meciarova, L. X-ray Fluorescence Method and Thermal Analysis of Concrete Prepared Based on a Share. *Tech. Univ. Kosice* **2013**, *756*, 193–199.
32. Shaikh, F.U.A.; Hosan, A. Mechanical Properties of Steel Fibre Reinforced Geopolymer Concretes at Elevated Temperatures. *Constr. Build. Mater.* **2016**, *114*, 15–28. [CrossRef]
33. Dinesh, H.T.; Shivakumar, M.; Dharmaprakash, M.S.; Ranganath, R.V. Influence of Reactive SiO_2 and Al_2O_3 on Mechanical and Durability Properties of Geopolymers. *Asian J. Civ. Eng.* **2019**, *20*, 1203–1215. [CrossRef]
34. Al-Majidi, M.H.; Lampropoulos, A.; Cundy, A.B. Steel Fibre Reinforced Geopolymer Concrete (SFRGC) with Improved Microstructure and Enhanced Fibre-Matrix Interfacial Properties. *Constr. Build. Mater.* **2017**, *139*, 286–307. [CrossRef]
35. Samson, G.; Cyr, M.; Gao, X.X. Formulation and Characterization of Blended Alkali-Activated Materials Based on Flash-Calcined Metakaolin, Fly Ash and GGBS. *Constr. Build. Mater.* **2017**, *144*, 50–64. [CrossRef]
36. Moukannaa, S.; Bagheri, A.; Benzaazoua, M.; Sanjayan, J.G.; Pownceby, M.I.; Hakkou, R. Elaboration of Alkali Activated Materials Using a Non-Calcined Red Clay from Phosphate Mines Amended with Fly Ash or Slag: A Structural Study. *Mater. Chem. Phys.* **2020**, *256*, 123678. [CrossRef]
37. Al Bakri, A.M.M.; Kamarudin, H.; Bnhussain, M.; Nizar, I.K.; Rafiza, A.R.; Izzat, A.M. Chemical Reactions in the Geopolymerisation Process Using Fly Ash-Based Geopolymer: A Review. *J. Appl. Sci. Res.* **2011**, *7*, 1199–1203.
38. Yang, Z.; Mocadlo, R.; Zhao, M.; Sisson, R.D.; Tao, M.; Liang, J. Preparation of a Geopolymer from Red Mud Slurry and Class F Fly Ash and Its Behavior at Elevated Temperatures. *Constr. Build. Mater.* **2019**, *221*, 308–317. [CrossRef]
39. Geng, J.; Zhou, M.; Li, Y.; Chen, Y.; Han, Y.; Wan, S.; Zhou, X.; Hou, H. Comparison of Red Mud and Coal Gangue Blended Geopolymers Synthesized through Thermal Activation and Mechanical Grinding Preactivation. *Constr. Build. Mater.* **2017**, *153*, 185–192. [CrossRef]
40. Dadsetan, S.; Siad, H.; Lachemi, M.; Mahmoodi, O.; Sahmaran, M. Optimization and Characterization of Geopolymer Binders from Ceramic Waste, Glass Waste and Sodium Glass Liquid. *J. Clean. Prod.* **2022**, *342*, 130931. [CrossRef]
41. Parthasarathy, T.A.; Jefferson, G.J.; Kerans, R.J. Analytical Evaluation of Hybrid Ceramic Design Concepts for Optimized Structural Performance. *Mater. Sci. Eng. A* **2007**, *459*, 60–68. [CrossRef]

42. Guazzato, M.; Albakry, M.; Quach, L.; Swain, M.V. Influence of Surface and Heat Treatments on the Flexural Strength of a Glass-Infiltrated Alumina/Zirconia-Reinforced Dental Ceramic. *Dent. Mater.* **2005**, *21*, 454–463. [CrossRef] [PubMed]
43. Liang, Y.; Dutta, S.P. Application Trend in Advanced Ceramic Technologies. *Technovation* **2001**, *21*, 61–65. [CrossRef]
44. Cao, Q.; Wang, Z.; He, W.; Guan, Y. Fabrication of Super Hydrophilic Surface on Alumina Ceramic by Ultrafast Laser Microprocessing. *Appl. Surf. Sci.* **2021**, *557*, 149842. [CrossRef]
45. Saxena, A.; Singh, N.; Kumar, D.; Gupta, P. Effect of Ceramic Reinforcement on the Properties of Metal Matrix Nanocomposites. *Mater. Today Proc.* **2017**, *4*, 5561–5570. [CrossRef]
46. Ma, S.; He, P.; Zhao, S.; Yang, H.; Wang, Q. Formation of SiC Whiskers/Leucite-Based Ceramic Composites from Low Temperature Hardening Geopolymer. *Ceram. Int.* **2021**, *47*, 17930–17938. [CrossRef]
47. Yucel, M.T.; Aykent, F.; Akman, S.; Yondem, I. Effect of Surface Treatment Methods on the Shear Bond Strength between Resin Cement and All-Ceramic Core Materials. *J. Non-Cryst. Solids* **2012**, *358*, 925–930. [CrossRef]
48. Yun-ming, L.; Cheng-yong, H.; Li, L.; Ain, N.; Mustafa, M.; Bakri, A.; Soo, T.; Hussin, K. Formation of One-Part-Mixing Geopolymers and Geopolymer Ceramics from Geopolymer Powder. *Constr. Build. Mater.* **2017**, *156*, 9–18. [CrossRef]
49. Kaya, M.; Koksal, F.; Gencel, O.; Junaid, M.; Minhaj, S.; Kazmi, S. Influence of Micro Fe_2O_3 and MgO on the Physical and Mechanical Properties of the Zeolite and Kaolin Based Geopolymer Mortar. *J. Build. Eng.* **2022**, *52*, 104443. [CrossRef]
50. Tiffo, E.; Bike Mbah, J.B.; Belibi Belibi, P.D.; Yankwa Djobo, J.N.; Elimbi, A. Physical and Mechanical Properties of Unheated and Heated Kaolin Based-Geopolymers with Partial Replacement of Aluminium Hydroxide. *Mater. Chem. Phys.* **2020**, *239*, 122103. [CrossRef]
51. Belmokhtar, N.; El Ayadi, H.; Ammari, M.; Ben Allal, L. Effect of Structural and Textural Properties of a Ceramic Industrial Sludge and Kaolin on the Hardened Geopolymer Properties. *Appl. Clay Sci.* **2018**, *162*, 1–9. [CrossRef]
52. Wang, M.R.; Jia, D.C.; He, P.G.; Zhou, Y. Influence of Calcination Temperature of Kaolin on the Structure and Properties of Final Geopolymer. *Mater. Lett.* **2010**, *64*, 2551–2554. [CrossRef]
53. Huo, W.; Zhang, X.; Chen, Y.; Lu, Y.; Liu, J.; Yan, S.; Wu, J.M.; Yang, J. Novel Mullite Ceramic Foams with High Porosity and Strength Using Only Fly Ash Hollow Spheres as Raw Material. *J. Eur. Ceram. Soc.* **2018**, *38*, 2035–2042. [CrossRef]
54. Guazzato, M.; Albakry, M.; Quach, L.; Swain, M.V. Influence of Grinding, Sandblasting, Polishing and Heat Treatment on the Flexural Strength of a Glass-Infiltrated Alumina-Reinforced Dental Ceramic. *Biomaterials* **2004**, *25*, 2153–2160. [CrossRef]
55. Wildan, M.; Edrees, H.J.; Hendry, A. Ceramic Matrix Composites of Zirconia Reinforced with Metal Particles. *Mater. Chem. Phys.* **2002**, *75*, 276–283. [CrossRef]
56. Mecif, A.; Soro, J.; Harabi, A.; Bonnet, J.P. Preparation of Mullite- and Zircon-Based Ceramics Using Kaolinite and Zirconium Oxide: A Sintering Study. *J. Am. Ceram. Soc.* **2010**, *93*, 1306–1312. [CrossRef]
57. Zhang, Z.H.; Zhu, H.J.; Zhou, C.H.; Wang, H. Geopolymer from Kaolin in China: An Overview. *Appl. Clay Sci.* **2016**, *119*, 31–41. [CrossRef]
58. Shikano, H. *Ceramic Filter for Trapping Diesel Particles*; Elsevier B.V.: Amsterdam, The Netherlands, 2018; ISBN 9780444641106.
59. Luhar, S.; Cheng, T.W.; Nicolaides, D.; Luhar, I.; Panias, D.; Sakkas, K. Valorisation of Glass Waste for Development of Geopolymer Composites—Mechanical Properties and Rheological Characteristics: A Review. *Constr. Build. Mater.* **2019**, *220*, 547–564. [CrossRef]
60. Keppert, M.; Scheinherrová, L.; Doleželová, M.; Vejmelková, E.; Černý, R. Phase Composition of Ceramic-Based Alkali-Activated Polymers: Combination of X-Ray Diffraction and Thermal Analysis. *J. Therm. Anal. Calorim.* **2020**, *142*, 157–166. [CrossRef]
61. Zaimi, N.S.M.; Salleh, M.A.A.M.; Abdullah, M.M.A.B.; Ahmad, R.; Mostapha, M.; Yoriya, S.; Chaiprapa, J.; Zhang, G.; Harvey, D.M. Effect of Kaolin Geopolymer Ceramic Addition on the Properties of Sn-3.0Ag-0.5Cu Solder Joint. *Mater. Today Commun.* **2020**, *25*, 101469. [CrossRef]
62. Kovářík, T.; Hájek, J.; Pola, M.; Rieger, D.; Svoboda, M.; Beneš, J.; Šutta, P.; Deshmukh, K.; Jandová, V. Cellular Ceramic Foam Derived from Potassium-Based Geopolymer Composite: Thermal, Mechanical and Structural Properties. *Mater. Des.* **2021**, *198*, 109355. [CrossRef]
63. Aziz, A.; Bellil, A.; El Amrani, I.; Hassani, E.; Fekhaoui, M.; Achab, M.; Dahrouch, A.; Benzaouak, A. Geopolymers Based on Natural Perlite and Kaolinic Clay from Morocco: Synthesis, Characterization, Properties, and Applications. *Ceram. Int.* **2021**, *47*, 24683–24692. [CrossRef]
64. Sarde, B.; Patil, Y.; Dholakiya, B.; Pawar, V. Effect of Calcined Kaolin Clay on Mechanical and Durability Properties of Pet Waste-Based Polymer Mortar Composites. *Constr. Build. Mater.* **2022**, *318*, 126027. [CrossRef]
65. Marsh, A.; Heath, A.; Patureau, P.; Evernden, M.; Walker, P. Phase Formation Behaviour in Alkali Activation of Clay Mixtures. *Appl. Clay Sci.* **2019**, *175*, 10–21. [CrossRef]
66. Wang, Y.; Xu, M.; Zhao, J.; Xin, A. Nano-ZnO Modified Geopolymer Composite Coatings for Flame-Retarding Plywood. *Constr. Build. Mater.* **2022**, *338*, 127649. [CrossRef]
67. Perumal, P.; Hasnain, A.; Luukkonen, T.; Kinnunen, P.; Illikainen, M. Role of Surfactants on the Synthesis of Impure Kaolin-Based Alkali-Activated, Low-Temperature Porous Ceramics. *Open Ceram.* **2021**, *6*, 100097. [CrossRef]
68. Gao, H.; Liu, H.; Liao, L.; Mei, L.; Zhang, F.; Zhang, L.; Li, S.; Lv, G. A Bifunctional Hierarchical Porous Kaolinite Geopolymer with Good Performance in Thermal and Sound Insulation. *Constr. Build. Mater.* **2020**, *251*, 118888. [CrossRef]
69. Dietel, J.; Warr, L.N.; Bertmer, M.; Steudel, A.; Grathoff, G.H.; Emmerich, K. The Importance of Specific Surface Area in the Geopolymerization of Heated Illitic Clay. *Appl. Clay Sci.* **2017**, *139*, 99–107. [CrossRef]

70. Okoye, F.N.; Durgaprasad, J.; Singh, N.B. Mechanical Properties of Alkali Activated Flyash/Kaolin Based Geopolymer Concrete. *Constr. Build. Mater.* **2015**, *98*, 685–691. [CrossRef]
71. Ceylantekin, R.; Başar, R. Solid Solution Limit of Fe_2O_3 in Mullite Crystals, Produced from Kaolin by Solid State Reactions. *Ceram. Int.* **2018**, *44*, 7599–7604. [CrossRef]
72. Dedzo, G.K.; Detellier, C. Functional Nanohybrid Materials Derived from Kaolinite. *Appl. Clay Sci.* **2016**, *130*, 33–39. [CrossRef]
73. Prasad, M.S.; Reid, K.J.; Murray, H.H. Kaolin: Processing, Properties and Applications. *Appl. Clay Sci.* **1991**, *6*, 87–119. [CrossRef]
74. Monsif, M.; Rossignol, S.; Allali, F.; Zeroual, A.; Idrissi Kandri, N.; Joussein, E.; Tamburini, S.; Bertani, R. The Implementation of Geopolymers Materials from Moroccan Clay, within the Framework of the Valorization of the Local Natural Resources. *J. Mater. Environ. Sci.* **2017**, *8*, 2704–2721.
75. Borges, P.H.R.; Bhutta, A.; Bavuzo, L.T.; Banthia, N. Effect of SiO_2/Al_2O_3 Molar Ratio on Mechanical Behavior and Capillary Sorption of MK-Based Alkali-Activated Composites Reinforced with PVA Fibers. *Mater. Struct.* **2017**, *50*, 148. [CrossRef]
76. Belmokhtar, N.; Ammari, M.; Brigui, J. Comparison of the Microstructure and the Compressive Strength of Two Geopolymers Derived from Metakaolin and an Industrial Sludge. *Constr. Build. Mater.* **2017**, *146*, 621–629. [CrossRef]
77. Lahoti, M.; Narang, P.; Tan, K.H.; Yang, E.-H. Mix Design Factors and Strength Prediction of Metakaolin-Based Geopolymer. *Ceram. Int.* **2017**, *43*, 11433–11441. [CrossRef]
78. Kwasny, J.; Soutsos, M.N.; Mcintosh, J.A.; Cleland, D.J. Comparison of the Effect of Mix Proportion Parameters on Behaviour of Geopolymer and Portland Cement Mortars. *Constr. Build. Mater.* **2018**, *187*, 635–651. [CrossRef]
79. Matalkah, F.; Aqel, R.; Ababneh, A. Enhancement of the Mechanical Properties of Kaolin Geopolymer Using Sodium Hydroxide and Calcium Oxide. *Procedia Manuf.* **2020**, *44*, 164–171. [CrossRef]
80. Nnaemeka, O.F.; Singh, N.B. Durability Properties of Geopolymer Concrete Made from Fly Ash in Presence of Kaolin. *Mater. Today Proc.* **2019**, *29*, 781–784. [CrossRef]
81. Derouiche, R.; Baklouti, S. Phosphoric Acid Based Geopolymerization: Effect of the Mechanochemical and the Thermal Activation of the Kaolin. *Ceram. Int.* **2021**, *47*, 13446–13456. [CrossRef]
82. Cheng-Yong, H.; Yun-Ming, L.; Abdullah, M.M.A.B.; Hussin, K. Thermal Resistance Variations of Fly Ash Geopolymers: Foaming Responses. *Sci. Rep.* **2017**, *7*, 45355. [CrossRef]
83. Diffo, B.B.K.; Elimbi, A.; Cyr, M.; Manga, J.D.; Kouamo, H.T. Effect of the Rate of Calcination of Kaolin on the Properties of Metakaolin-Based Geopolymers. *Integr. Med. Res.* **2015**, *3*, 130–138. [CrossRef]
84. MacKenzie, K.J.D.; Brew, D.R.M.; Fletcher, R.A.; Vagana, R. Formation of Aluminosilicate Geopolymers from 1:1 Layer-Lattice Minerals Pre-Treated by Various Methods: A Comparative Study. *J. Mater. Sci.* **2007**, *42*, 4667–4674. [CrossRef]
85. Liew, Y.M.; Kamarudin, H.; Al Bakri, A.M.M.; Bnhussain, M.; Luqman, M.; Nizar, I.K.; Ruzaidi, C.M.; Heah, C.Y. Optimization of Solids-to-Liquid and Alkali Activator Ratios of Calcined Kaolin Geopolymeric Powder. *Constr. Build. Mater.* **2012**, *37*, 440–451. [CrossRef]
86. Prasanphan, S.; Wannagon, A.; Kobayashi, T.; Jiemsirilers, S. Reaction Mechanisms of Calcined Kaolin Processing Waste-Based Geopolymers in the Presence of Low Alkali Activator Solution. *Constr. Build. Mater.* **2019**, *221*, 409–420. [CrossRef]
87. Gao, B.; Jang, S.; Son, H.; Lee, H.J.; Lee, H.J.; Yang, J.J.; Bae, C.J. Study on Mechanical Properties of Kaolin-Based Geopolymer with Various Si/Al Ratio and Aging Time. *J. Korean Ceram. Soc.* **2020**, *57*, 709–715. [CrossRef]
88. Alonso, M.M.; Gascó, C.; Morales, M.M.; Suárez-Navarro, J.A.; Zamorano, M.; Puertas, F. Olive Biomass Ash as an Alternative Activator in Geopolymer Formation: A Study of Strength, Durability, Radiology and Leaching Behaviour. *Cem. Concr. Compos.* **2019**, *104*, 103384. [CrossRef]
89. Naghsh, M.; Shams, K. Synthesis of a Kaolin-Based Geopolymer Using a Novel Fusion Method and Its Application in Eff Ective Water Softening. *Appl. Clay Sci.* **2017**, *146*, 238–245. [CrossRef]
90. Sornlar, W.; Wannagon, A.; Supothina, S. Stabilized Homogeneous Porous Structure and Pore Type Effects on the Properties of Lightweight Kaolinite-Based Geopolymers. *J. Build. Eng.* **2021**, *44*, 103273. [CrossRef]
91. Majdoubi, H.; Haddaji, Y.; Mansouri, S.; Alaoui, D.; Tamraoui, Y.; Semlal, N.; Oumam, M.; Manoun, B.; Hannache, H. Thermal, Mechanical and Microstructural Properties of Acidic Geopolymer Based on Moroccan Kaolinitic Clay. *J. Build. Eng.* **2021**, *35*, 102078. [CrossRef]
92. Merabtene, M.; Kacimi, L.; Clastres, P.; Eco-mat, L. Elaboration of Geopolymer Binders from Poor Kaolin and Dam Sludge Waste. *Heliyon* **2019**, *5*, e01938. [CrossRef]
93. Villaquirán-Caicedo, M.A.; De Gutiérrez, R.M. Synthesis of Ceramic Materials from Ecofriendly Geopolymer Precursors. *Mater. Lett.* **2018**, *230*, 300–304. [CrossRef]
94. Khoury, H.; Salhah, Y.A.; Al Dabsheh, I.; Slaty, F.; Alshaaer, M.; Rahier, H.; Esaifan, M.; Wastiels, J. Geopolymer Products from Jordan for Sustainability of the Environment. *Adv. Mater. Sci. Environ. Nucl. Technol. II* **2011**, *227*, 289–300.
95. Hubadillah, S.K.; Othman, M.H.D.; Matsuura, T.; Ismail, A.F.; Rahman, M.A.; Harun, Z.; Jaafar, J.; Nomura, M. Fabrications and Applications of Low Cost Ceramic Membrane from Kaolin: A Comprehensive Review. *Ceram. Int.* **2018**, *44*, 4538–4560. [CrossRef]
96. Tchakouté, H.K.; Melele, S.J.K.; Djamen, A.T.; Kaze, C.R.; Kamseu, E.; Nanseu, C.N.P.; Leonelli, C.; Rüscher, C.H. Microstructural and Mechanical Properties of Poly(Sialate-Siloxo) Networks Obtained Using Metakaolins from Kaolin and Halloysite as Aluminosilicate Sources: A Comparative Study. *Appl. Clay Sci.* **2020**, *186*, 105448. [CrossRef]

97. Bouguermouh, K.; Bouzidi, N.; Mahtout, L.; Pérez-Villarejo, L.; Martínez-Cartas, M.L. Effect of Acid Attack on Microstructure and Composition of Metakaolin-Based Geopolymers: The Role of Alkaline Activator. *J. Non-Cryst. Solids* **2017**, *463*, 128–137. [CrossRef]
98. Wu, W.; Tian, L. Formulation and Morphology of Kaolin-Filled Rubber Composites. *Appl. Clay Sci.* **2013**, *80–81*, 93–97. [CrossRef]
99. Shilar, F.A.; Ganachari, S.V.; Patil, V.B.; Khan, M.Y.; Dawood, S.; Khadar, A.; Ash, F. Molarity Activity Effect on Mechanical and Microstructure Properties of Geopolymer Concrete: A Review. *Case Stud. Constr. Mater.* **2022**, *16*, e01014. [CrossRef]
100. Ding, Y.; Shi, C.; Li, N. Fracture Properties of Slag/Fly Ash-Based Geopolymer Concrete Cured in Ambient Temperature. *Constr. Build. Mater.* **2018**, *190*, 787–795. [CrossRef]
101. Nenadović, S.S.; Ferone, C.; Nenadović, M.T.; Cioffi, R.; Mirković, M.M.; Vukanac, I.; Kljajević, L.M. Chemical, Physical and Radiological Evaluation of Raw Materials and Geopolymers for Building Applications. *J. Radioanal. Nucl. Chem.* **2020**, *325*, 435–445. [CrossRef]
102. Mohammed, B.S.; Haruna, S.; Mubarak bn Abdul Wahab, M.; Liew, M.S. Optimization and Characterization of Cast In-Situ Alkali-Activated Pastes by Response Surface Methodology. *Constr. Build. Mater.* **2019**, *225*, 776–787. [CrossRef]
103. Reddy, M.S.; Dinakar, P.; Rao, B.H. A Review of the Influence of Source Material's Oxide Composition on the Compressive Strength of Geopolymer Concrete. *Microporous Mesoporous Mater.* **2016**, *234*, 12–23. [CrossRef]
104. Hosan, A.; Haque, S.; Shaikh, F. Compressive Behaviour of Sodium and Potassium Activators Synthetized Fly Ash Geopolymer at Elevated Temperatures: A Comparative Study. *J. Build. Eng.* **2016**, *8*, 123–130. [CrossRef]
105. Shahmansouri, A.A.; Yazdani, M.; Ghanbari, S.; Akbarzadeh Bengar, H.; Jafari, A.; Farrokh Ghatte, H. Artificial Neural Network Model to Predict the Compressive Strength of Eco-Friendly Geopolymer Concrete Incorporating Silica Fume and Natural Zeolite. *J. Clean. Prod.* **2021**, *279*, 123697. [CrossRef]
106. Ababneh, A.; Matalkah, F.; Aqel, R. Synthesis of Kaolin-Based Alkali-Activated Cement: Carbon Footprint, Cost and Energy Assessment. *J. Mater. Res. Technol.* **2020**, *9*, 8367–8378. [CrossRef]
107. Martins, L.; Geraldo, N.; Almeida, S.; Houmard, M.; Roberto, P.; Jorge, G.; Silva, B.; Teresa, M.; Aguilar, P. Influence of the Addition of Amorphous and Crystalline Silica on the Structural Properties of Metakaolin-Based Geopolymers. *Appl. Clay Sci.* **2021**, *215*, 106312. [CrossRef]
108. Chouia, F.; Belhouchet, H.; Sahnoune, F.; Bouzrara, F. Reaction Sintering of Kaolin-Natural Phosphate Mixtures. *Ceram. Int.* **2015**, *41*, 8064–8069. [CrossRef]
109. Heah, C.Y.; Kamarudin, H.; Mustafa Al Bakri, A.M.; Bnhussain, M.; Luqman, M.; Khairul Nizar, I.; Ruzaidi, C.M.; Liew, Y.M. Study on Solids-to-Liquid and Alkaline Activator Ratios on Kaolin-Based Geopolymers. *Constr. Build. Mater.* **2012**, *35*, 912–922. [CrossRef]
110. Kamseu, E.; Kaze, C.R.; Fekoua, J.N.N.; Melo, U.C.; Rossignol, S.; Leonelli, C. Ferrisilicates Formation during the Geopolymerization of Natural Fe-Rich Aluminosilicate Precursors. *Mater. Chem. Phys.* **2020**, *240*, 122062. [CrossRef]
111. Marfo, K.K.; Dodoo-Arhin, D.; Agyei-Tuffou, B.; Nyankson, E.; Obada, D.O.; Damoah, L.N.W.; Annan, E.; Yaya, A.; Onwona-Agyeman, B.; Bediako, M. The Physico-Mechanical Influence of Dehydroxylized Activated Local Kaolin: A Supplementary Cementitious Material for Construction Applications. *Case Stud. Constr. Mater.* **2020**, *12*, e00306. [CrossRef]
112. Guo, X.; Shi, H.; Wei, X. Pore Properties, Inner Chemical Environment, and Microstructure of Nano-Modified CFA-WBP (Class C Fly Ash-Waste Brick Powder) Based Geopolymers. *Cem. Concr. Compos.* **2017**, *79*, 53–61. [CrossRef]
113. Singh, B.; Ishwarya, G.; Gupta, M.; Bhattacharyya, S.K. Geopolymer Concrete: A Review of Some Recent Developments. *Constr. Build. Mater.* **2015**, *85*, 78–90. [CrossRef]
114. Galvão Souza Azevedo, A.; Strecker, K. Kaolin, Fly-Ash and Ceramic Waste Based Alkali-Activated Materials Production by the "One-Part" Method. *Constr. Build. Mater.* **2021**, *269*, 121306. [CrossRef]
115. Jaya, N.A.; Mustafa, M.; Bakri, A. Correlation between Na2SiO3/NaOH and NaOH Molarity to Flexural Strength of Geopolymer Ceramic Correlation between Na2SiO3/NaOH and NaOH Molarity to Flexural Strength of Geopolymer Ceramic. *Appl. Mech. Mater.* **2015**, *754*, 152–156. [CrossRef]
116. Duxson, P.; Mallicoat, S.W.; Lukey, G.C.; Kriven, W.M.; van Deventer, J.S.J. The Effect of Alkali and Si/Al Ratio on the Development of Mechanical Properties of Metakaolin-Based Geopolymers. *Colloids Surf. A Physicochem. Eng. Asp.* **2007**, *292*, 8–20. [CrossRef]
117. Pelisser, F.; Guerrino, E.L.; Menger, M.; Michel, M.D.; Labrincha, J.A. Micromechanical Characterization of Metakaolin-Based Geopolymers. *Constr. Build. Mater.* **2013**, *49*, 547–553. [CrossRef]
118. Mathivet, V.; Jouin, J.; Parlier, M.; Rossignol, S. Control of the Alumino-Silico-Phosphate Geopolymers Properties and Structures by the Phosphorus Concentration. *Mater. Chem. Phys.* **2021**, *258*, 123867. [CrossRef]
119. Jia, D.; Li, Y.; He, P.; Fu, S.; Duan, X.; Sun, Z.; Cai, D.; Li, D.; Yang, Z.; Zhou, Y. In-Situ Formation of Bulk and Porous h-AlN/SiC-Based Ceramics from Geopolymer Technique. *Ceram. Int.* **2019**, *45*, 24727–24733. [CrossRef]
120. Pacheco-torgal, F.; Moura, D.; Ding, Y.; Jalali, S. Composition, Strength and Workability of Alkali-Activated Metakaolin Based Mortars. *Constr. Build. Mater.* **2011**, *25*, 3732–3745. [CrossRef]
121. Selmani, S.; Sdiri, A.; Bouaziz, S.; Joussein, E.; Rossignol, S. Effects of Metakaolin Addition on Geopolymer Prepared from Natural Kaolinitic Clay. *Appl. Clay Sci.* **2017**, *146*, 457–467. [CrossRef]
122. Coudert, E.; Deneele, D.; Russo, G.; Vitale, E.; Tarantino, A. Microstructural Evolution and Mechanical Behaviour of Alkali Activated Fly Ash Binder Treated Clay. *Constr. Build. Mater.* **2021**, *285*, 122917. [CrossRef]

123. Singaravel, A.; Nathan, C.; Tah, R.; Balasubramanium, M.K. Evaluation of Fracture Toughness of Zirconia Silica Nano-Fi Bres Reinforced Feldespathic Ceramic. *J. Oral Biol. Craniofacial Res.* **2018**, *8*, 221–224. [CrossRef] [PubMed]
124. Zawrah, M.F.; Farag, R.S.; Kohail, M.H. Improvement of Physical and Mechanical Properties of Geopolymer through Addition of Zircon. *Mater. Chem. Phys.* **2018**, *217*, 90–97. [CrossRef]
125. Assaedi, H.; Shaikh, F.U.A.; Low, I.M. Influence of Mixing Methods of Nano Silica on the Microstructural and Mechanical Properties of Flax Fabric Reinforced Geopolymer Composites. *Constr. Build. Mater.* **2016**, *123*, 541–552. [CrossRef]
126. Gilmore, C. *Materials Science and Engineering Properties*; Cengage Learning: Boston, MA, USA, 2014; ISBN 1305178173.
127. Okada, M.; Taketa, H.; Hara, E.S.; Torii, Y.; Irie, M.; Matsumoto, T. Improvement of Mechanical Properties of Y-TZP by Thermal Annealing with Monoclinic Zirconia Nanoparticle Coating. *Dent. Mater.* **2019**, *35*, 970–978. [CrossRef]
128. Egilmez, F.; Ergun, G.; Cekic-Nagas, I.; Vallittu, P.K.; Lassila, L.V.J. Factors Affecting the Mechanical Behavior of Y-TZP. *J. Mech. Behav. Biomed. Mater.* **2014**, *37*, 78–87. [CrossRef]
129. Phair, J.W.; Van Deventer, J.S.J.; Smith, J.D. Mechanism of Polysialation in the Incorporation of Zirconia into Fly Ash-Based Geopolymers. *Ind. Eng. Chem. Res.* **2000**, *39*, 2925–2934. [CrossRef]
130. Kenawy, S.H.; Awaad, M.; Awad, H. In-Situ Mullite–Zirconia Composites from Kaolin. *Am. Ceram. Soc. Bull.* **2016**, *85*, 9401-U16.

Article

Influence of Replacing Cement with Waste Glass on Mechanical Properties of Concrete

Özer Zeybek [1], Yasin Onuralp Özkılıç [2,*], Memduh Karalar [3], Ali İhsan Çelik [4], Shaker Qaidi [5,6], Jawad Ahmad [7], Dumitru Doru Burduhos-Nergis [8,*] and Diana Petronela Burduhos-Nergis [8]

[1] Department of Civil Engineering, Faculty of Engineering, Mugla Sitki Kocman University, Mugla 48000, Turkey
[2] Department of Civil Engineering, Faculty of Engineering, Necmettin Erbakan University, Konya 42000, Turkey
[3] Department of Civil Engineering, Faculty of Engineering, Zonguldak Bulent Ecevit University, Zonguldak 67100, Turkey
[4] Department of Construction, Tomarza Mustafa Akincioglu Vocational School, Kayseri University, Kayseri 38940, Turkey
[5] Department of Civil Engineering, College of Engineering, University of Duhok, Duhok 42001, Iraq
[6] Department of Civil Engineering, College of Engineering, Nawroz University, Duhok 42001, Iraq
[7] Department of Civil Engineering, Military College of Engineering (NUST), Risalpur 24080, Pakistan
[8] Faculty of Materials Science and Engineering, Gheorghe Asachi Technical University of Iasi, 700050 Iasi, Romania
* Correspondence: yozkilic@erbakan.edu.tr (Y.O.Ö.); doru.burduhos@tuiasi.ro (D.D.B.-N.)

Citation: Zeybek, Ö.; Özkılıç, Y.O.; Karalar, M.; Çelik, A.İ.; Qaidi, S.; Ahmad, J.; Burduhos-Nergis, D.D.; Burduhos-Nergis, D.P. Influence of Replacing Cement with Waste Glass on Mechanical Properties of Concrete. *Materials* 2022, *15*, 7513. https://doi.org/10.3390/ma15217513

Academic Editor: F. Pacheco Torgal

Received: 28 September 2022
Accepted: 24 October 2022
Published: 26 October 2022

Publisher's Note: MDPI stays neutral with regard to jurisdictional claims in published maps and institutional affiliations.

Copyright: © 2022 by the authors. Licensee MDPI, Basel, Switzerland. This article is an open access article distributed under the terms and conditions of the Creative Commons Attribution (CC BY) license (https://creativecommons.org/licenses/by/4.0/).

Abstract: In this study, the effect of waste glass on the mechanical properties of concrete was examined by conducting a series of compressive strength, splitting tensile strength and flexural strength tests. According to this aim, waste glass powder (WGP) was first used as a partial replacement for cement and six different ratios of WGP were utilized in concrete production: 0%, 10%, 20%, 30%, 40%, and 50%. To examine the combined effect of different ratios of WGP on concrete performance, mixed samples (10%, 20%, 30%) were then prepared by replacing cement, and fine and coarse aggregates with both WGP and crashed glass particles. Workability and slump values of concrete produced with different amounts of waste glass were determined on the fresh state of concrete, and these properties were compared with those of plain concrete. For the hardened concrete, 150 mm × 150 mm × 150 mm cubic specimens and cylindrical specimens with a diameter of 100 mm and a height of 200 mm were tested to identify the compressive strength and splitting tensile strength of the concrete produced with waste glass. Next, a three-point bending test was carried out on samples with dimensions of 100 × 100 × 400 mm, and a span length of 300 mm to obtain the flexure behavior of different mixtures. According to the results obtained, a 20% substitution of WGP as cement can be considered the optimum dose. On the other hand, for concrete produced with combined WGP and crashed glass particles, mechanical properties increased up to a certain limit and then decreased owing to poor workability. Thus, 10% can be considered the optimum replacement level, as combined waste glass shows considerably higher strength and better workability properties. Furthermore, scanning electron microscope (SEM) analysis was performed to investigate the microstructure of the composition. Good adhesion was observed between the waste glass and cementitious concrete. Lastly, practical empirical equations have been developed to determine the compressive strength, splitting tensile strength, and flexure strength of concrete with different amounts of waste glass. Instead of conducting an experiment, these strength values of the concrete produced with glass powder can be easily estimated at the design stage with the help of proposed expressions.

Keywords: eco-friendly concrete; waste glass; workability; compressive strength; splitting tensile strength; flexural strength

1. Introduction

The use of ordinary Portland cement (OPC) by replacing it with recycled cement in certain proportions is an interesting issue for sustainable environmental awareness [1,2]. During cement production, a high amount of energy is consumed, and a high amount of carbon dioxide (CaO_2) is released into the atmosphere [3]. Therefore, cement additives are of interest to reduce cement production [4,5]. Waste is generally defined as residues from industrial production processes, residues arising from the transfer, or product residues that have completed their economic life [6–14]. With the development of modern cities, the decrease in natural resources, climate change, and increasing awareness of environmental protection, there is an urgent need to develop building materials that will reduce greenhouse gas emissions [2,15]. Waste glass powder (WGP), which is increasing with the effect of industrialization and the increase in urban transformation, attracts the attention of researchers as a concrete additive material due to its economic and mechanical performance effects [4,5,16–21]. Since there is no regular storage area for waste glass, it increases the risk of soil and water pollution due to its oxidation effect. Therefore, the use of recycled glass in concrete production will provide significant contributions to reducing environmental problems [2,3,22,23].

According to ASTM C618-19, 2019, since waste glass consists of a large amount of calcium and an amorphous structure, it can be ground into powder to obtain pozzolanic material or cement additive [24–26]. Therefore, WGP can be used in concrete production by replacing it with cement in certain proportions [2]. Recent studies show that replacing 15–25% glass powder with cement increases the mechanical properties of concrete [6,15,21]. Aliabdo et al. used 25% substitution with cement to investigate the mechanical effect of WGP in concrete as a cement substitute. According to the results obtained, it was observed that the void ratio and density of the samples decreased, while the tensile and compressive strengths increased [27]. Al Saffar et al. in their study to test the interaction of WGP, added cement mortar with other materials and found that the compressive strength of the samples increased as the glass powder addition rate increased. They obtained the highest strength with the addition of 25% glass powder [15]. Elaqra et al. (2019) added 4% by weight to the mix to investigate the effect of WGP on fresh and hardened concrete. They found that as the amount of WGP increased, the machinability increased, and the maximum compressive strength was reached with the addition of 20% WGP at 28 days of cure [28]. It is well known that in the case of fine grinding of soda–lime glass, the reactivity of pozzolanic increases as the particle size decreases [29]. Zhang et al. in their study to examine the effect of WGP particle size on the mechanical and microstructures of concrete, found that the particle size distribution of WGP has a significant effect on the properties of WGP-based concrete [30]. Shao et al. in their experimental study to observe the effect of finely ground WGP on the compressive strength of concrete found that the pozzolanic reactivity increases as the WGP particle size decreases. They found that WGP, with a particle size of 38 μm, was replaced by 30% with cement, resulting in 4.1 MPa greater compressive strength [31]. It was found that WGP pozzolanic reactivity was replaced by 0, 15, 30, 45, and 60% of weight cement, while below 30% the concrete compressive strength did not decrease due to the pozzolanic reaction between WGP cement hydration products. In fact, with the addition of 60% WGP, the resistance to chloride ion and water penetration increased continuously, while the concrete compressive strength increased by around 85% [32]. In the study by Peril and Sangle, it was observed that when WGP was mixed with 30% cement, an increase in compressive strength between <38 μm and <75 μm was between 20% and 10% [33]. Khatib et al. found in their study that there was a 1.2% increase in concrete compressive strength with the addition of 10% WGP [34]. In a similar study, Madandoust and Ghavidel found that the compressive strength of the control sample was higher than the concrete with glass powder added at every stage, the compressive strength of both samples increased with aging [35]. In their study, Tejaswi et al. stated that the compressive strength increased by 1.5% with the addition of 10% WGP. However, they found that the compressive strengths were at equal levels when 20% was substituted [36]. Vasudevan et al.

observed a 1.05% increase in concrete compressive strength when 20% WFP addition was <90 μm [37]. Schwarz et al. found that the addition of 20% WGP increased the compressive strength by 19% when <75 μm [38]. They state that while the strength decreases as a result of the 7-day compressive strength test, the compressive strength does not change as a result of the 28- and 91-day tests <75 μm. In addition to strength performance, the failure processes of brittleness materials are also important. Some researchers have studied the failure processes of brittleness materials [39,40]. Since the risk of fracture will increase under sudden loads, it is beneficial to take measures to increase elastic behavior.

Dust and particles from soda glass were mostly used in previous studies. However, its usability in concrete was investigated by replacing fine aggregate in self-compacting concrete with waste particles obtained from cathode ray tubes. The results had a positive effect on the durability properties of concrete [41]. Past studies have shown that ground glass powder can increase the pozzolanic reactivity of secondary cementitious materials. Therefore, the granule size of the ground glass powder has important effects [42,43]. Ahmet et al. examined different methods of using waste glass in concrete and stated that particle size, substitution ratio, and chemical composition have important effects on the mechanical durability of concrete. For example, as the grain size decreases, workability becomes more difficult but pozzolanic and strength increase [44]. The particle size of the waste glass, which must be taken into account during the mix design, may affect the active silica reaction depending on the rate of substitution.

Solid waste management is an important issue for most developing countries [45]. Instead of storing or disposing of waste materials such as glass, plastic, and metal, reusing or recycling has become a more attractive option. Waste glass has started to be widely preferred for concrete production in civil engineering applications in recent years. Since the employing of waste glass in concrete can assist to reduce environmental pollution, protect natural resources and produce low-cost concrete, WGP can be preferred instead of either natural aggregates or cement due to its pozzolanic effect. Although many studies have been conducted on this subject, there are differences in the results obtained from the literature. Thus, there still remains a need to investigate the mechanical behavior of concrete with partial substitution of waste glass and the ideal dosage of it. Based on this motivation, an experimental study was carried out on some test specimens. Analytical solution proposals have been developed according to the data obtained from the experimental study. The proposed formulas will serve as a guide for researchers and manufacturers and will accelerate future studies to be more effective.

2. Experimental Program

In this study, the main aim is to investigate the effect of glass powder when it is replaced with cement. Furthermore, the effects of all replacements with glass are investigated. Nine mixes including the reference were designed. C represents cement replacement, MIX represents cement, fine aggregate, and coarse aggregate replacement. Table 1 summarizes the sample properties. For the MIX design, each type of material was replaced with certain amounts of recycled glass. The size of fine aggregates is 1 mm to 4 mm and the corresponding waste glass was 1.7 mm to 4 mm. Size of coarse aggregate size was selected as 5–12 mm and the corresponding waste glass was also 5–12 mm. The particle size of cement particles was between 0.02 mm to 0.1 mm and the size of glass waste powder was 0.1–0.2 mm. Figure 1 demonstrates the used aggregates and cement and also their replacements which are glass powder and glass grains. Figure 2 demonstrates the recycled glass in concrete.

Table 1. Sample Properties.

REF	No Glass Was Used
C10%	10% cement was replaced with waste glass powder
C20%	20% cement was replaced with waste glass powder
C30%	30% cement was replaced with waste glass powder
C40%	40% cement was replaced with waste glass powder
C50%	50% cement was replaced with waste glass powder
MIX10%	10% cement, 10% fine and 10% coarse aggregates were replaced with waste glass
MIX20%	20% cement, 20% fine and 20% coarse aggregates were replaced with waste glass
MIX30%	30% cement, 30% fine and 30% coarse aggregates were replaced with waste glass

Figure 1. Recycled glass, cement, and aggregates.

Figure 2. Recycled glass in concrete.

Cement was selected as CEM I 32.5 Portland cement. The water-to-cement ratio was chosen as 0.5. Figure 3 demonstrates the slump test results. The workability of the concrete produced by adding WGP increased as the waste rate increased. Accordingly, there was an increase in slump values. It is seen that the slump value increases as the recycled glass ratio increases. On the other hand, when all ingredients were replaced with glass, the slump value significantly decreases. Madandous and Ghavidel obtained an 80 mm slump in their study with the addition of 5–20% WGP. In this study, it is seen that the slump value is 200 mm, since the additional WCP is 50%. It can be said that the results are similar to the literature [35].

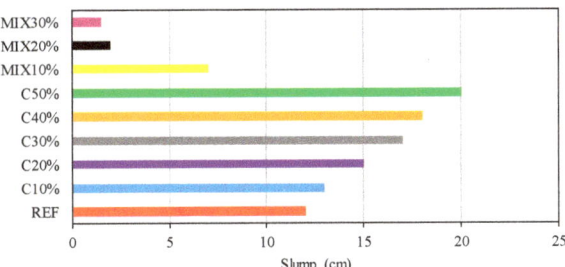

Figure 3. Slump test.

Three types of tests were performed in order to evaluate the performance of the concrete with recycled glass. A compression test was performed on $15 \times 15 \times 15$ cm cubic samples while splitting tests were conducted on 10×20 cm cylinder samples. Furthermore, bending tests were performed using $10 \times 10 \times 40$ cm samples. Three repetitions were completed for each mix and each test. Compressive ability is the capability of a specimen to decline load under pressure. The CST was performed according to ASTM C39/C39M (C39&C39M A, 2003). In this test, specimens have dimensions of 150 mm $\times$ 300 mm. In this experiment, concrete specimens were subjected to constraint lengthwise load at a level in the offered as concerns the specimen fractures. At that point, the compressive capability was estimated from the critical failure strength separated by the part of the specimen. To provide the tensile strength of concrete where compactor load is performed in anticipation of examples unravel caused by expansion of tensile force in concrete as ASTM (Designation, 1976). In this way, cylindrical examples are divided through the vertical dimension.

3. Experimental Results and Discussions

3.1. Compressive Strength (CS)

In this part, to carry through the compressive strength test (CST), concrete examples were arranged without any adulteration. Figure 4 expresses the CST consequences at 28 days period for low w/c ratio combinations formed with 100% reference and numerous ratios of recycled aggregate substituting reference aggregates, with and lacking waste glass. The CS values of the reference concrete without waste glass series at 28 days were found as 18.9 MPa, respectively. The lowest CS values were also found to be 7.10 MPa in the concrete produced with cement including 50% waste glass for 28 days periods, respectively. As observed in Figure 4, it has been detected that the CS of concrete combinations including waste glass utilization as a fractional replacement for cement were lesser than those of the corresponding concrete mixes lacking waste glass. As observed in Figure 4, statistical investigation of test values shows the noteworthy impact of waste glass (as a fractional replacement for cement) on the CS of concrete.

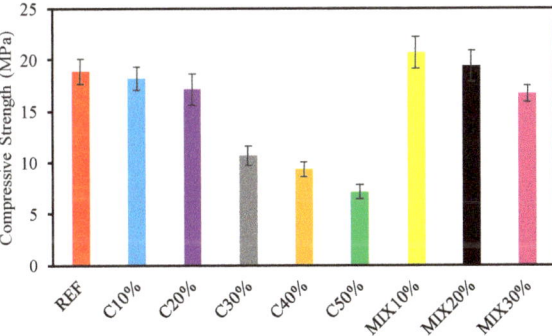

Figure 4. Results of CS.

In the comparison of the cement that was changed with waste glass powder, the CS of C10% (10% cement replaced with waste glass) is 6% greater than that of C20%. Correspondingly, the CS of C20% (20% cement replaced with waste glass) is 60% greater than that of C30%. Comparing the CS of C30% (30% of the cement was replaced with waste glass powder) it can be observed that the value is 15% greater than that of C40%. Finally, the comparison of CS of C40% (40% of the cement was substituted with waste glass powder) is 31% greater than that of C50%. Consequently, it is observed that the CS of C10% (10% cement was replaced with waste glass powder) is similar to that of reference concrete. As shown in Figure 4, while fine and coarse aggregate for 10%, 20%, and 30% (MIX10%, MIX20%, and MIX30%) were exchanged with waste glass, this trend was reversed as fractional replacement of cement with waste glass promoted the CS of concrete. On the other hand, at the comparison of CS of MIX10%, MIX20%, and MIX30%, the CST consequences for the great waste glass ratio continue trends comparable to those for the small waste glass ratio. In this situation too, the CS of MIX10% is 24% greater than the corresponding mix (MIX30%). While waste glass is used instead of cement, fine and coarse aggregate, remarkable improvements in strength are observed when compared with the reference concrete. Therefore, it was observed that by replacing cement with glass powder, a significant reduction in CS will be obtained.

3.2. Splitting Tensile Strength (STS)

A comparison of relative STS of the various concrete percentage combinations formed with 100% reference and numerous ratios of recycled aggregate replacing reference aggregates, with and without waste glass are shown in Figure 5. As presented in Figure 5, the results of STS commonly pursue a similar tendency as the compressive strength. As observed from Figure 5, the remarkable improvement in concrete tensile strength is up to 10% with waste glass used as a fractional replacement for cement. Upon altering the amount of waste glass, consequences of tensile strength are influenced similarly to the CS. As shown in Figure 5, tensile strength optimal norms are found at 10% waste glass obligating maximum divided tensile. A comparative evaluation was performed where controller concrete tensile strength was deliberate as reference strength, from which concrete of the changing ratio of waste glass is collated. At 10% substitution of waste glass, STS is only 3% lower than the reference mix. The correlation of CS in competition with STS is provided in Figure 6. As shown in Figure 6, the regression model among the compressive and STS appeared to be flat. Furthermore, as observed in Figure 6, the regression stroke represents a strong relationship between CS contrasted with STS having an R2 worth of more than 94 percent.

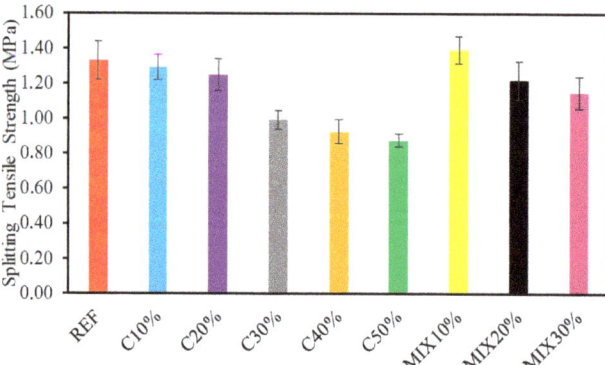

Figure 5. Results of STS.

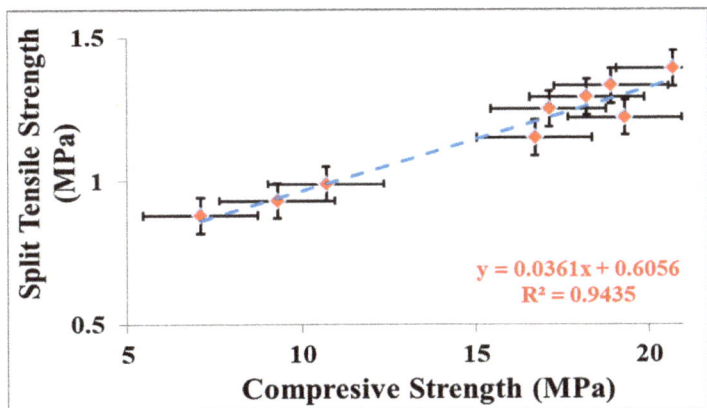

Figure 6. STS versus CS.

3.3. Flexural Performance

Flexure strength (FS) of investigational examples was established on examples after CST. The acquired norms of strength ranged between 6.6 MPa and 3.9 MPa. The consequence of the FS of the examples is presented in Figure 7. In Figure 7, it is detected FS with variable ratios of waste glass. Related to CS, FS at the initial phase failures with the incorporation of waste glass. It was detected that with the addition of waste glass at 10%, 20%, 30%, 40%, and 50% of cement weight, the decrease in FS was 6.7%, 12.5%, and 21.1%, 46.5%, and 61.5% correspondingly in proportion to the reference sample (6.3 MPa). Figure 8 demonstrates a rectilinear relationship between the tensile FS of the example and the substance of waste glass addition. On the other hand, if fine and coarse aggregates in cement are replaced with waste glass, it was noticed that the waste glass was replaced with the cement, fine and coarse aggregate at 10%, the increase in FS was 4.7% correspondingly in proportion to the reference sample (6.3 MPa). Nevertheless, if the quantity of waste glass is occupied relative to the cement content, it may be detected that the association between FS and waste glass content is related (the slope of the curves is approximate (Figure 8)).

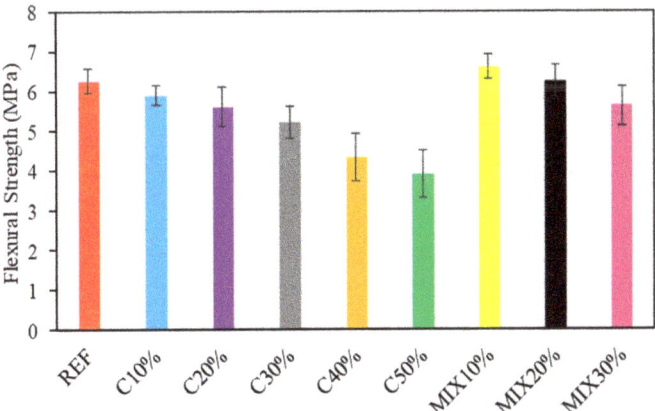

Figure 7. Results of FS.

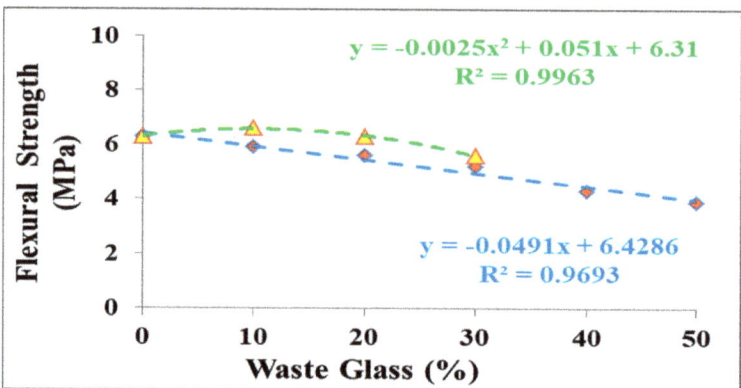

Figure 8. Results of tensile flexural strength tests of concrete examples with altered waste glass substances.

3.4. Comparison of Findings of this Study with Other Existing Studies

The effect of waste glass on the structural performance of concrete was investigated by many research teams. Effects of the use of waste glass as cement replacement on some mechanical properties such as CS, STS, and FS have been investigated. In this part of the study, values of CS, STS, and FS for plain concrete and concrete produced from different glass powder amounts have been collected from the research publications [46–63]. Then, measured strength values of concrete produced with waste glass were first normalized by plain concrete strengths. These normalized strength values were plotted in Figures 9–11, respectively, as a function of the waste glass content.

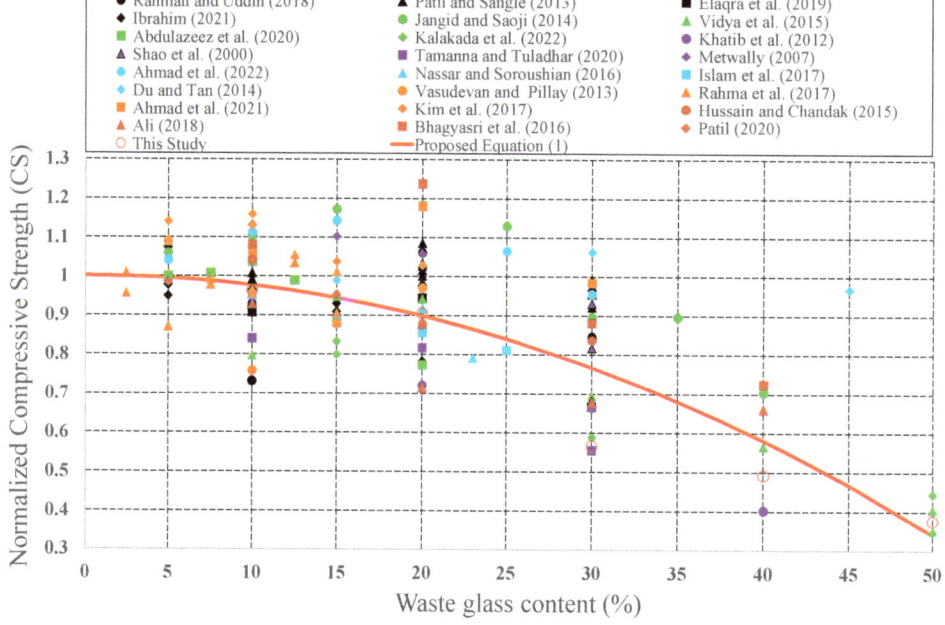

Figure 9. Variation of the normalized CS of the concrete produced with waste glass.

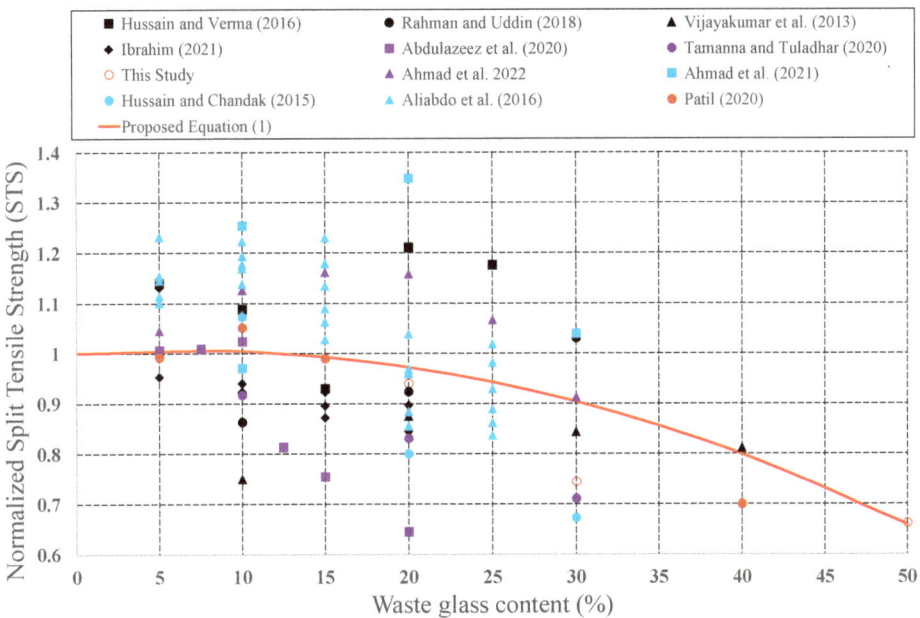

Figure 10. Variation of the STS of the concrete produced with waste glass.

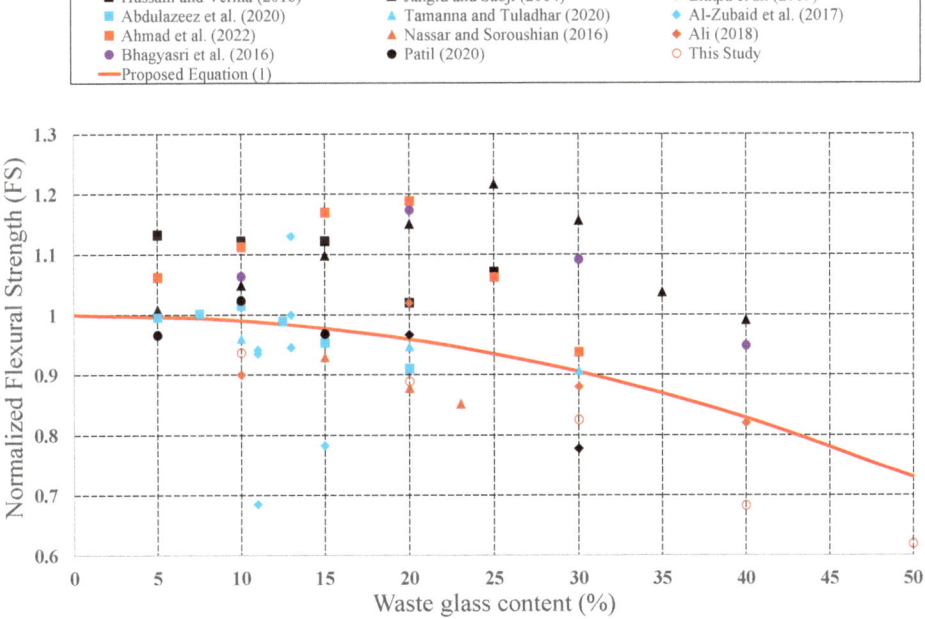

Figure 11. Variation of the FS of the concrete produced with waste glass.

As shown in these figures, with the addition of waste glass, the strength values of concrete produced with glass powder generally decrease after a certain value. The maximum decrease in compressive strength was observed in the study by Kalakada et. al., 2022. When the cement was partially replaced with glass powder at 50%, there is a 65%

reduction in compression strength of the plain concrete specimen. In the same manner, a 20% addition of the waste glass leads to a 36% reduction in the splitting tensile strength of the plain concrete specimen in the study of Abdulazeez et al., (2020). With the addition of 50% glass powder, a 38% reduction in flexural strength was observed in our experimental study. Thus, a rational design expression was developed to consider these reductions in the strength values as follows:

$$f = [1 + c_1 \times (WGPR) + c_2 \times (WGPR)^2] \times f' \qquad (1)$$

where f is strength values as follows: f_c: compressive strength f_t: splitting tensile strength; f_f: flexure strength; $WGPR$: waste glass power ratio (0 < WGPR < 50). c_1 and c_2 are coefficients given in Table 2; f' is strength values of the plain concrete.

Table 2. Constants for Equation (1).

Strength Values (f)	c_1	c_2
f_c	0.00033	−0.00027
f_t	0.00217	−0.00018
f_f	0.00017	−0.00011

As shown in Equation (1), CS, STS, and FS values of concrete produced with waste glass were expressed as a function of the amount of the waste glass ($WGPR$). These expressions can be easily used in the design stages.

3.5. Scanning Electron Microscope (SEM) Analysis

Scanning electron microscope (SEM) analysis was performed from the sample pieces taken after the compressive strength test from the concrete samples produced with recycled glass waste powder. SEM analysis is carried out to show the typical morphology of the surface tissue of OPC and WGP. The particles of both OPC and WGP are composed of glassy structures and irregular shapes with sharp edges. OPC particles consist of sharper edges and shapes, while WGP particle appears on smoother surfaces, sharper prismatic edges, and denser content [2]. The best images were selected to observe the effect of glass dust and glass particles in OPC concrete production. Figure 12a–f contains images where WGP is replaced by cement. Images of the mix design samples from Figure 12g–j contain the details of the change in glass particles with cement and aggregates. In order for WGP to be processed as a filler by grinding in electric mills without any treatment, 74% of the particles must be passed through a 36 μm sieve [64]. The key findings of the inner image details at 500 times magnification are shown in detail in Figure 12. Judging from the general view in Figure 12a, it can be said that glass powder provides good bonding with cement and aggregates. A good interlocking is observed in terms of surface quality, but some flat zones have cracks due to fracture. In Figure 12b, the appearance of glass powder particles like sharp fibers is remarkable. Although there is a homogeneous distribution, gaps are seen in some regions. Voids can be removed by better mixing, shaking, and skewering [64–66]. The image in Figure 12c shows material connections more closely, such as a finely woven bee comb. Figure 12d shows the concrete surface bonding and spacing. Figure 12e,f shows the gelled state of the concrete surface. In terms of surface quality, it can be said that the glass powder is in a good correlation, but it has some holes and gaps. In the mixed model, glass powder was used as a binder with cement, and broken glass particles were used by replacing fine and coarse aggregates in certain proportions. Calcium silicate hydrate (C-S-H), an amorphous phase, is defined as weak crystals [67]. The fine glass particles can be seen in Figure 12g. They are seen as a honeycomb in the C-S-H phase. Since this honeycomb structure provides a good pozzolanic effect, it increases compressive strength [68,69]. Figure 12h shows how large and small glass particles are bonded with binders in concrete. It is seen in Figure 12i with good fillings of binders between large pieces of glass. Figure 12j shows the homogeneous distribution and connection of the glass particles in the concrete.

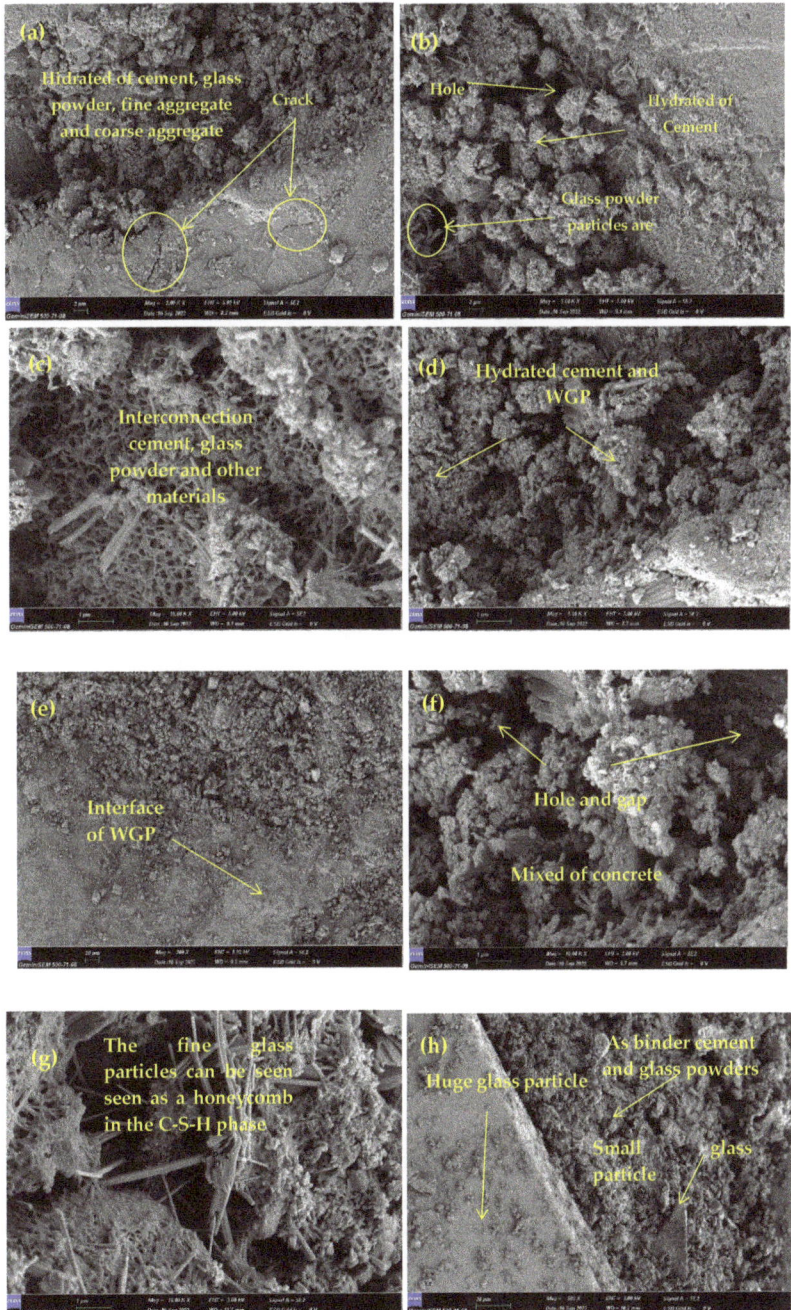

Figure 12. *Cont.*

Figure 12. SEM micrographs of the obtained samples.

4. Conclusions and Summary

Waste management has gained great importance due to the increase in the amount of waste. The reuse or recycling approach of these waste products, especially in the construction sector, has become a more attractive option. In this study, the applicability of waste glass on concrete production was investigated by considering two different conditions, i.e., replacing cement with waste glass powder and replacing cement, fine and coarse aggregate with waste glasses. A series of tests were conducted on concrete samples produced with different amounts of waste glass to determine engineering properties both in the fresh and hardened cases. In the fresh case of concrete, workability and slump properties were explored. In the case of hardened concrete, compression strength, splitting tensile strength, and flexural strength of the produced test specimens were investigated. These properties were then compared with those of plain concrete. Then, SEM analysis was carried out to explore the interaction between glass powder, cement, and aggregates. Furthermore, practical equations were developed to identify the compressive strength, splitting tensile strength, and flexure strength of concrete produced with WGP. Based on our study, the following conclusions can be drawn:

- Based on the slump test, the slump value decreases as the amount of waste glass increases. In the same manner, the workability decreases when the rate of glass powder increases.
- The compressive test results indicated that the usage of WGP as a replacement of cement adversely affects the compressive strength of concrete. Compared to the reference sample, 10%, 20%, 30%, 40% and 50% replacement of WPG reduced the compressive strength by 3%, 6%, 37%, 13% and 23%, respectively.
- On the other hand, when cement, fine and coarse aggregates were replaced with waste glasses, there was an increase in the compression strength, the flexural and splitting tensile strength values up to a certain value of the amount of the waste. However, further increasing the waste glass addition led to a decrease in the compression value.
- According to the results of the split tensile and bending strength tests, the strength values decrease as the value of the substitution of cement with waste glass increases. Compared to plain concrete, 50% replacement of WPG reduced the splitting tensile strength by 34%, respectively. On the other hand, this replacement reduced flexural strength by 38%.
- When cement, fine and coarse aggregates were replaced with waste glasses there was an increase in the flexural and splitting tensile strength values up to a certain value of the amount of the waste glass. However, further increasing the waste glass addition resulted in a decrease in the strength values. Replacing the waste glass by 10% resulted in a 5% increase in splitting tensile strength while replacing the waste glass by 30% resulted in a 14% decrease in splitting tensile strength. On the other hand, a 10% substitute of waste glass resulted in a 5% increase in flexural strength, while a 30% substitute of waste glass resulted in an 11% decrease in flexural strength.

- It can be concluded that when waste glass is used instead of cement, fine and coarse aggregate, significant increases in strength are observed.
- SEM analysis results showed that there was a good bond between the glass powder and cementitious concrete.
- The proposed equations for the compressive strength, flexural, and splitting tensile strength are quite general and they can be suitable for direct adoption into design specifications of the concretes produced with waste glass.

In this study, it was observed that waste glass can be used as a partial cement substitute, or as cement, fine and coarse aggregate in concrete. Optimum waste glass dosage has also been identified for design purposes. However, lower strength values were observed when waste glass was used as a partial replacement for cement. The current study defines waste glass as suitable, reachable in large amounts, a local eco-material, cheap, that can be selected for concrete construction, in a point of view between economically and environmentally responsive. To mitigate this shortcoming, future studies will concentrate on the use of waste glass with other recycling fibers.

Author Contributions: Conceptualization, Y.O.Ö. and D.D.B.-N.; methodology, Y.O.Ö. and D.D.B.-N.; data curation, Ö.Z., M.K., D.P.B.-N. and Y.O.Ö.; investigation, A.İ.Ç., Y.O.Ö., Ö.Z., M.K.; writing—original draft preparation, A.İ.Ç., Y.O.Ö., Ö.Z., S.Q., J.A., D.D.B.-N. and D.P.B.-N.; writing—review and editing, Y.O.Ö., Ö.Z., M.K., S.Q., J.A., D.D.B.-N. and D.P.B.-N.; funding acquisition, D.D.B.-N. and D.P.B.-N. All authors have read and agreed to the published version of the manuscript.

Funding: This work was supported by Gheorghe Asachi Technical University of Iași—TUIASI-Romania, Scientific Research Funds, FCSU-2022.

Institutional Review Board Statement: Not applicable.

Informed Consent Statement: Not applicable.

Data Availability Statement: Not applicable.

Acknowledgments: This paper was financially supported by the Project "Network of excellence in applied research and innovation for doctoral and postdoctoral programs"/InoHubDoc, project co-funded by the European Social Fund financing agreement no. POCU/993/6/13/153437. This paper was also supported by "Gheorghe Asachi" Technical University from Iași (TUIASI), through the Project "Performance and excellence in postdoctoral research 2022".

Conflicts of Interest: The authors declare no conflict of interest.

References

1. Batayneh, M.; Marie, I.; Asi, I. Use of selected waste materials in concrete mixes. *Waste Manag.* **2007**, *27*, 1870–1876. [CrossRef] [PubMed]
2. Jiang, X.; Xiao, R.; Bai, Y.; Huang, B.; Ma, Y. Influence of waste glass powder as a supplementary cementitious material (SCM) on physical and mechanical properties of cement paste under high temperatures. *J. Clean. Prod.* **2022**, *340*, 130778. [CrossRef]
3. Bilondi, M.P.; Toufigh, M.M.; Toufigh, V. Experimental investigation of using a recycled glass powder-based geopolymer to improve the mechanical behavior of clay soils. *Constr. Build. Mater.* **2018**, *170*, 302–313. [CrossRef]
4. Tayeh, B.A.; Al Saffar, D.M.; Aadi, A.S.; Almeshal, I. Sulphate resistance of cement mortar contains glass powder. *J. King Saud. Univ. Eng. Sci.* **2020**, *32*, 495–500. [CrossRef]
5. Tayeh, B.A.; Almeshal, I.; Magbool, H.M.; Alabduljabbar, H.; Alyousef, R. Performance of sustainable concrete containing different types of recycled plastic. *J. Clean. Prod.* **2021**, *328*, 129517. [CrossRef]
6. Wang, H.-Y.; Huang, W.-L. Durability of self-consolidating concrete using waste LCD glass. *Constr. Build. Mater.* **2010**, *24*, 1008–1013. [CrossRef]
7. Wang, H.-Y.; Zeng, H.-H.; Wu, J.-Y. A study on the macro and micro properties of concrete with LCD glass. *Constr. Build. Mater.* **2014**, *50*, 664–670. [CrossRef]
8. Aksoylu, C.; Özkılıç, Y.O.; Hadzima-Nyarko, M.; Işık, E.; Arslan, M.H. Investigation on Improvement in Shear Performance of Reinforced-Concrete Beams Produced with Recycled Steel Wires from Waste Tires. *Sustainability* **2022**, *14*, 13360. [CrossRef]
9. Ali, E.E.; Al-Tersawy, S.H. Recycled glass as a partial replacement for fine aggregate in self compacting concrete. *Constr. Build. Mater.* **2012**, *35*, 785–791. [CrossRef]
10. Qaidi, S.; Tayeh, B.A.; Zeyad, A.M.; de Azevedo, A.R.; Ahmed, H.U.; Emad, W. Recycling of mine tailings for the geopolymers production: A systematic review. *Case Stud. Constr. Mater.* **2022**, e00933. [CrossRef]

11. Ahmed, S.N.; Sor, N.H.; Ahmed, M.A.; Qaidi, S.M. Thermal conductivity and hardened behavior of eco-friendly concrete incorporating waste polypropylene as fine aggregate. *Mater. Today Proc.* **2022**, *57*, 818–823. [CrossRef]
12. Ahmad, J.; Martínez-García, R.; De-Prado-Gil, J.; Irshad, K.; El-Shorbagy, M.A.; Fediuk, R.; Vatin, N.I. Concrete with Partial Substitution of Waste Glass and Recycled Concrete Aggregate. *Materials* **2022**, *15*, 430. [CrossRef] [PubMed]
13. Qaidi, S.; Najm, H.M.; Abed, S.M.; Özkılıç, Y.O.; Al Dughaishi, H.; Alosta, M.; Sabri, M.M.S.; Alkhatib, F.; Milad, A. Concrete Containing Waste Glass as an Environmentally Friendly Aggregate: A Review on Fresh and Mechanical Characteristics. *Materials* **2022**, *15*, 6222. [CrossRef] [PubMed]
14. Karalar, M.; Özkılıç, Y.O.; Deifalla, A.F.; Aksoylu, C.; Arslan, M.H.; Ahmad, M.; Sabri, M.M.S. Improvement in Bending Performance of Reinforced Concrete Beams Produced with Waste Lathe Scraps. *Sustainability* **2022**, *14*, 12660. [CrossRef]
15. Giannopoulou, I.; Dimas, D.; Maragkos, I.; Panias, D. Utilization of metallurgical solid by-products for the development of inorganic polymeric construction materials. *Glob. NEST J.* **2009**, *11*, 127–136.
16. Al Saffar, D.M.A.R. Experimental investigation of using ultra-fine glass powder in concrete. *Int. J. Eng. Res. Appl.* **2017**, *7*, 33–39.
17. Almeshal, I.; Al-Tayeb, M.M.; Qaidi, S.M.; Abu Bakar, B.; Tayeh, B.A. Mechanical properties of eco-friendly cements-based glass powder in aggressive medium. *Mater. Today: Proc.* **2022**, *58*, 1582–1587. [CrossRef]
18. Ez-Zaki, H.; El Gharbi, B.; Diouri, A. Development of eco-friendly mortars incorporating glass and shell powders. *Constr. Build. Mater.* **2018**, *159*, 198–204. [CrossRef]
19. Jubeh, A.I.; Al Saffar, D.M.; Tayeh, B.A. Effect of recycled glass powder on properties of cementitious materials contains styrene butadiene rubber. *Arab. J. Geosci.* **2019**, *12*, 39. [CrossRef]
20. Mousa, M.; Cuenca, E.; Ferrara, L.; Roy, N.; Tagnit-Hamou, A. Tensile Characterization of an "Eco-Friendly" UHPFRC with Waste Glass Powder and Glass Sand. In Proceedings of the International Conference on Strain-Hardening Cement-Based Composites, Dresden, Germany, 18–20 September 2017; pp. 238–248. [CrossRef]
21. Tayeh, B.A. Effects of marble, timber, and glass powder as partial replacements for cement. *J. Civ. Eng. Constr.* **2018**, *7*, 63–71. [CrossRef]
22. Tan, K.H.; Du, H. Use of waste glass as sand in mortar: Part I—Fresh, mechanical and durability properties. *Cem. Concr. Compos.* **2013**, *35*, 109–117. [CrossRef]
23. Xiao, R.; Polaczyk, P.; Zhang, M.; Jiang, X.; Zhang, Y.; Huang, B.; Hu, W. Evaluation of Glass Powder-Based Geopolymer Stabilized Road Bases Containing Recycled Waste Glass Aggregate. *Transp. Res. Rec. J. Transp. Res. Board* **2020**, *2674*, 22–32. [CrossRef]
24. C09; Committee Specification for Coal Fly Ash and Raw or Calcined Natural Pozzolan for Use in Concrete. ASTM International: West Conshohocken, PA, USA, 2013. [CrossRef]
25. Omran, A.; Tagnit-Hamou, A. Performance of glass-powder concrete in field applications. *Constr. Build. Mater.* **2016**, *109*, 84–95. [CrossRef]
26. Vijayakumar, G.; Vishaliny, H.; Govindarajulu, D. Studies on glass powder as partial replacement of cement in concrete production. *Int. J. Emerg. Technol. Adv. Eng.* **2013**, *3*, 153–157.
27. Aliabdo, A.A.; Abd Elmoaty, A.E.M.; Aboshama, A.Y. Utilization of waste glass powder in the production of cement and concrete. *Constr. Build. Mater.* **2016**, *124*, 866–877. [CrossRef]
28. Elaqra, H.A.; Haloub, M.A.A.; Rustom, R. Effect of new mixing method of glass powder as cement replacement on mechanical behavior of concrete. *Constr. Build. Mater.* **2019**, *203*, 75–82. [CrossRef]
29. Zheng, K. Pozzolanic reaction of glass powder and its role in controlling alkali–silica reaction. *Cem. Concr. Compos.* **2016**, *67*, 30–38. [CrossRef]
30. Zhang, Y.; Xiao, R.; Jiang, X.; Li, W.; Zhu, X.; Huang, B. Effect of particle size and curing temperature on mechanical and microstructural properties of waste glass-slag-based and waste glass-fly ash-based geopolymers. *J. Clean. Prod.* **2020**, *273*, 122970. [CrossRef]
31. Shao, Y.; Lefort, T.; Moras, S.; Rodriguez, D. Studies on concrete containing ground waste glass. *Cem. Concr. Res.* **2000**, *30*, 91–100. [CrossRef]
32. Du, H.; Tan, K.H. Waste Glass Powder as Cement Replacement in Concrete. *J. Adv. Concr. Technol.* **2014**, *12*, 468–477. [CrossRef]
33. Patil, D.M.; Sangle, K.K. Experimental investigation of waste glass powder as partial replacement of cement in concrete. *Int. J. Adv. Technol. Civ. Eng.* **2013**, *2*, 2231–5721. [CrossRef]
34. Khatib, J.M.; Negim, E.M.; Sohl, H.S.; Chileshe, N. Glass powder utilisation in concrete production. *Eur. J. Appl. Sci.* **2012**, *4*, 173–176. [CrossRef]
35. Madandoust, R.; Ghavidel, R. Mechanical properties of concrete containing waste glass powder and rice husk ash. *Biosyst. Eng.* **2013**, *116*, 113–119. [CrossRef]
36. Tejaswi, S.S.; Rao, R.C.; Vidya, B.; Renuka, J. Experimental investigation of waste glass powder as partial replacement of cement and sand in concrete. *IUP J. Struct. Eng.* **2015**, *8*, 14.
37. Vasudevan, G.; Pillay, S.G.K. Performance of using waste glass powder in concrete as replacement of cement. *Am. J. Eng. Res.* **2013**, *2*, 175–181.
38. Schwarz, N.; Cam, H.; Neithalath, N. Influence of a fine glass powder on the durability characteristics of concrete and its comparison to fly ash. *Cem. Concr. Compos.* **2008**, *30*, 486–496. [CrossRef]
39. Qiu, H.; Zhu, Z.; Wang, F.; Wang, M.; Zhou, C.; Luo, C.; Wang, X.; Mao, H. Dynamic behavior of a running crack crossing mortar-rock interface under impacting load. *Eng. Fract. Mech.* **2020**, *240*, 107202. [CrossRef]

40. Wang, F.; Wang, M.; Zhu, Z.; Deng, J.; Nezhad, M.M.; Qiu, H.; Ying, P. Rock Dynamic Crack Propagation Behaviour and Determination Method with Improved Single Cleavage Semi-circle Specimen Under Impact Loads. *Acta Mech. Solida Sin.* **2020**, *33*, 793–811. [CrossRef]
41. Sua-Iam, G.; Makul, N. Use of Limestone Powder to Improve the Properties of Self-Compacting Concrete Produced Using Cathode Ray Tube Waste as Fine Aggregate. *Appl. Mech. Mater.* **2012**, *193–194*, 472–476. [CrossRef]
42. Alexander, K.M. Reactivity of ultrafine powders produced from siliceous rocks. *J. Proc.* **1960**, *57*, 557–570.
43. Vizcayno, C.; DE Gutierrez, R.M.; Castello, R.; Rodriguez, E.; Guerrero, C. Pozzolan obtained by mechanochemical and thermal treatments of kaolin. *Appl. Clay Sci.* **2010**, *49*, 405–413. [CrossRef]
44. Ahmad, J.; Zhou, Z.; Usanova, K.I.; Vatin, N.I.; El-Shorbagy, M.A. A Step towards Concrete with Partial Substitution of Waste Glass (WG) in Concrete: A Review. *Materials* **2022**, *15*, 2525. [CrossRef] [PubMed]
45. Çelik, A.I.; Özkılıç, Y.O.; Zeybek, Ö.; Özdöner, N.; Tayeh, B.A. Performance Assessment of Fiber-Reinforced Concrete Produced with Waste Lathe Fibers. *Sustainability* **2022**, *14*, 11817. [CrossRef]
46. Ahmad, J.; Aslam, F.; Martinez-Garcia, R.; De-Prado-Gil, J.; Qaidi, S.M.A.; Brahmia, A. Effects of waste glass and waste marble on mechanical and durability performance of concrete. *Sci. Rep.* **2021**, *11*, 1–17. [CrossRef]
47. Rahma, A.; El Naber, N.; Ismail, I.S. Effect of glass powder on the compression strength and the workability of concrete. *Cogent Eng.* **2017**, *4*, 1373415. [CrossRef]
48. AbdulAzeez, A.S.; Idi, M.A.; Kolawole, M.A.; Hamza, B. Effect of Waste Glass Powder as A Pozzolanic Material in Concrete Production. *Int. J. Eng. Res.* **2020**, *9*, 589–594. [CrossRef]
49. Kim, S.K.; Kang, S.T.; Kim, J.K.; Jang, I.Y. Effects of Particle Size and Cement Replacement of LCD Glass Powder in Concrete. *Adv. Mater. Sci. Eng.* **2017**, *2017*, 1–12. [CrossRef]
50. Rahman, S.; Uddin, M. Experimental Investigation of Concrete with Glass Powder as Partial Replacement of Cement. *Civ. Eng. Arch.* **2018**, *6*, 149–154. [CrossRef]
51. Jangid, B.J.; Saoji, A. Experimental investigation of waste glass powder as the partial replacement of cement in concrete production. In Proceedings of the International Conference on Advances in Engineering & Technology, Singapore, 29–30 March 2014; pp. 55–60.
52. Hussain, G.; Verma, G. Experimental investigation on glass powder as partial replacement of cement for M-30 concrete. *Int. J. Sci. Res. Sci. Eng. Technol.* **2016**, *2*, 846–853.
53. Nassar, R.-U.; Soroushian, P. Field investigation of concrete incorporating milled waste glass. *J. Solid Waste Technol. Manag.* **2011**, *37*, 307–319. [CrossRef]
54. Metwally, I.M. Investigations on the Performance of Concrete Made with Blended Finely Milled Waste Glass. *Adv. Struct. Eng.* **2007**, *10*, 47–53. [CrossRef]
55. Ibrahim, K.I.M. Recycled waste glass powder as a partial replacement of cement in concrete containing silica fume and fly ash. *Constr. Build. Mater.* **2021**, *15*, e00630. [CrossRef]
56. Islam, G.S.; Rahman, M.; Kazi, N. Waste glass powder as partial replacement of cement for sustainable concrete practice. *Int. J. Sustain. Built Environ.* **2017**, *6*, 37–44. [CrossRef]
57. Tamanna, N.; Tuladhar, R. Sustainable Use of Recycled Glass Powder as Cement Replacement in Concrete. *Open Waste Manag. J.* **2020**, *13*, 1–13. [CrossRef]
58. Ali, I.; IFTM University. Behavior of Concrete by using Waste Glass Powder and Fly Ash as a Partial Replacement of Cement. *Int. J. Eng. Res.* **2015**, *4*, 1238–1243. [CrossRef]
59. Kumar, V.; Sood, H. Effect of Waste Glass Powder in Concrete by Partial Replacement of Cement. *Int. J. Civ. Eng.* **2017**, *4*, 13–22. [CrossRef]
60. Bhagyasri, T.; Prabhavathi, U.; Vidya, N. Role of Glass Powder in Mechanical strength of concrete. *Int. J. Adv. Mech. Civ. Eng.* **2016**, *3*, 74–78.
61. Patil, B.H. Utilization of Waste Glass Powder as a Replacement of Cement in Concrete. *Int. J. Res. Eng. Sci. Manag.* **2020**, *3*, 466–468.
62. Hussain, M.V.; Chandak, R. Strength properties of concrete containing waste glass powder. *Int. J. Eng. Res. Appl.* **2015**, *5*, 1–4.
63. Al-Zubaid, A.B.; Shabeeb, K.M.; Ali, A.I. Study The Effect of Recycled Glass on The Mechanical Properties of Green Concrete. *Energy Procedia* **2017**, *119*, 680–692. [CrossRef]
64. Ramdani, S.; Guettala, A.; Benmalek, M.; Aguiar, J.B. Physical and mechanical performance of concrete made with waste rubber aggregate, glass powder and silica sand powder. *J. Build. Eng.* **2019**, *21*, 302–311. [CrossRef]
65. Nergis, D.D.B.; Vizureanu, P.; Ardelean, I.; Sandu, A.V.; Corbu, O.C.; Matei, E. Revealing the Influence of Microparticles on Geopolymers' Synthesis and Porosity. *Materials* **2020**, *13*, 3211. [CrossRef] [PubMed]
66. Hao, D.L.C.; Razak, R.A.; Kheimi, M.; Yahya, Z.; Abdullah, M.M.A.B.; Nergis, D.D.B.; Fansuri, H.; Ediati, R.; Mohamed, R.; Abdullah, A. Artificial Lightweight Aggregates Made from Pozzolanic Material: A Review on the Method, Physical and Mechanical Properties, Thermal and Microstructure. *Materials* **2022**, *15*, 3929. [CrossRef] [PubMed]
67. Kim, J.; Yi, C.; Zi, G. Waste glass sludge as a partial cement replacement in mortar. *Constr. Build. Mater.* **2015**, *75*, 242–246. [CrossRef]

68. Najad, A.A.A.-J.; Kareem, J.H.; Azline, N.; Ostovar, N. Waste glass as partial replacement in cement—A review. *IOP Conf. Series Earth Environ. Sci.* **2019**, *357*, 012023. [CrossRef]
69. Ansari, W.S. Porosity analysis using Image J. In *Proceedings of the 8th Graduate-Student Forum on Building Materials*; Southeast University: Nanjing, China, 2019.

Article

Biochar Produced from Saudi Agriculture Waste as a Cement Additive for Improved Mechanical and Durability Properties—SWOT Analysis and Techno-Economic Assessment

Kaffayatullah Khan [1,2,*], Muhammad Arif Aziz [3], Mukarram Zubair [4,*] and Muhammad Nasir Amin [1,2]

1. Department of Civil and Environmental Engineering, College of Engineering, King Faisal University, Al-Ahsa 31982, Saudi Arabia; mgadir@kfu.edu.sa
2. Al Bilad Bank Scholarly Chair for Food Security in Saudi Arabia, The Deanship of Scientific Research, The Vice Presidency for Graduate Studies and Scientific Research, King Faisal University, Al-Ahsa 31982, Saudi Arabia
3. Department of Civil and Construction Engineering, College of Engineering, Imam Abdulrahman Bin Faisal University, Dammam 31451, Saudi Arabia; maaahmed@iau.edu.sa
4. Department of Environmental Engineering, College of Engineering, Imam Abdulrahman Bin Faisal University, Dammam 31451, Saudi Arabia
* Correspondence: kkhan@kfu.edu.sa (K.K.); mzzubair@iau.edu.sa (M.Z.)

Citation: Khan, K.; Aziz, M.A.; Zubair, M.; Amin, M.N. Biochar Produced from Saudi Agriculture Waste as a Cement Additive for Improved Mechanical and Durability Properties—SWOT Analysis and Techno-Economic Assessment. *Materials* **2022**, *15*, 5345. https://doi.org/10.3390/ma15155345

Academic Editor: Dumitru Doru Burduhos Nergis

Received: 14 May 2022
Accepted: 19 July 2022
Published: 3 August 2022

Publisher's Note: MDPI stays neutral with regard to jurisdictional claims in published maps and institutional affiliations.

Copyright: © 2022 by the authors. Licensee MDPI, Basel, Switzerland. This article is an open access article distributed under the terms and conditions of the Creative Commons Attribution (CC BY) license (https:// creativecommons.org/licenses/by/ 4.0/).

Abstract: The Kingdom of Saudi Arabia generates an enormous amount of date palm waste, causing severe environmental concerns. Green and strong concrete is increasingly demanded due to low carbon footprints and better performance. In this research work, biochar derived from locally available agriculture waste (date palm fronds) was used as an additive to produce high-strength and durable concrete. Mechanical properties such as compressive and flexural strength were evaluated at 7, 14, and 28 days for control and all other mixes containing biochar. In addition, the durability properties of the concrete samples for the mixes were investigated by performing electric resistivity and ultra-sonic pulse velocity testing. Finally, a SWOT (strengths, weaknesses, opportunities, and threats) analysis was carried out to make strategic decisions about biochar's use in concrete. The results demonstrated that the compressive strength of concrete increased to 28–29% with the addition of 0.75–1.5 wt% of biochar. Biochar-concrete containing 0.75 wt% of biochar showed 16% higher flexural strength than the control specimen. The high ultrasonic pulse velocity (UPV) values (>7.79 km/s) and low electrical resistivity (<22.4 kΩ-cm) of biochar-based concrete confirm that the addition of biochar resulted in high-quality concrete free from internal flaws, cracks, and better structural integrity. SWOT analysis indicated that biochar-based concrete possessed improved performance than ordinary concrete, is suitable for extreme environments, and has opportunities for circular economy and applications in various construction designs. However, cost and technical shortcomings in biochar production and biochar-concrete mix design are still challenging.

Keywords: green concrete; biochar; compressive strength; durability properties; SWOT; techno-economic analysis

1. Introduction

The construction industry is one of the most rapidly growing industries globally. The substantial growth of the construction industry led to the increasing demand for concrete. The population in urban regions is likely to grow from around 3.4 billion in 2009 to 6.5 billion by 2050. It is estimated that the yearly concrete production withstood at 10,000 Mt and is expected to rise twice in the next 40 years [1]. As a result, with time, there is an increasing demand for green building materials and eco-friendly construction practices to reduce the environmental impact of the concrete industry [2,3]. Carbon dioxide (CO_2) emission has always been a severe worldwide concern in cement manufacturing. Moreover, activities related to the processing and transportation of cement are also responsible for greenhouse gas emissions, which are considered a severe environmental threat. Concrete contributes to approximately 8% of the entire world's production of CO_2 during the production, processing, and preparation phase [4,5]. As a result of the high CO_2 emissions and environmental issues, it has become necessary to implement sustainable CO_2 reduction methods relevant to cement-based materials (CBM) in the environment [2]. Environmentally friendly cementitious materials (CBM) can entirely or partially replace cement to reduce the negative environmental impacts of concrete production [6]. In recent years, various waste materials such as fly ash, silica fume, glass, rubber and tires, steel slag, etc., have been utilised in concrete to improve performance and lower carbon footprint [7–9]. The use of these materials has a two-fold benefit. It reduces carbon emissions by reducing the use of cement in concrete, but it also diverts waste materials away from landfills, which helps increase sustainability.

Biowaste, including municipal, industrial, agricultural, and other forms, is widely generated worldwide. This waste can be dumped into landfills and cause severe environmental consequences [10]. Transforming waste into value-added products for various applications via various approaches is a step toward a circular economy and an effective waste management strategy [11]. The pyrolysis technique is most commonly adopted to convert various biomass sources such as organic industrial and household waste, wood, and agricultural waste into biochar, biogas, and bio-oil [12,13]. Biochar is a carbon-rich material with high porosity obtained via thermochemical conversion of biomass in the absence of oxygen [14]. Recent studies showed that a pyrolysis temperature of > 500 °C releases all the organic components from biochar leading to high surface area biochars [2,15]. The biochar's excellent mechanical and thermal stability, high surface area, and porosity proved it to be favourable to use as a cement replacement as an admixture in concrete [16,17]. For instance, Choi et al. [18] partially replaced the cement by adding biochar in a mortar and reported replacing 5% biochar showed a 10% increase in the compressive strength. Correspondingly, substantial improvement in mechanical strength (16–20%), water penetration (40%), and water absorption (35–60%) by the addition of biochar produced from various feedstock were reported [3,19]. Furthermore, Restuccia et al. [16] stated that adding biochar to cement paste could enhance its fracture energy and modulus of rupture. Wang et al. [20] reported that biochar in concrete enhances the mechanical properties by reducing the microcracks in the concrete and improving cement's hydration. Nevertheless, there is still a considerable gap in investigating the effect of biochar from various feedstock on concrete's mechanical and durability properties. The Kingdom of Saudi Arabia (KSA) is one of the largest producers of date palm trees, with about 35 million trees. These trees generate large agricultural waste, either carried to landfills or burned in the open areas. This has a significant detrimental impact on humans and the ecosystem [14]. Walid et al. [21] stated an efficient and valuable use of ash obtained from date palm waste as a partial replacement of Portland cement in concrete structures. Though, the addition of biochar derived from date palm wastes for construction application has not been studied yet.

The primary aim of the presented research is to study the impact of biochar derived from date palm fronds as an additive in concrete to produce high-strength and durable concrete. The control and all other biochar-containing mixes assessed compressive and flexural strengths after 7, 14, and 28 days. Additionally, the durability properties of the concrete samples for the mixes were evaluated by measuring their electric resistivity and ultrasonic pulse velocity. Finally, the SWOT analysis and techno-economic assessment of the biochar-concrete system were performed to provide detailed insight into date palm derived biochar's potential application on a commercial level.

2. Materials and Methods

2.1. Materials

Type-1 Ordinary Portland cement (OPC) from a Saudi cement factory was used as a primary binder in all concrete specimens [22]. The cement's specific gravity and maximum particle size were recorded as 3.14 and 0.072 mm, respectively, provided by the local manufacturer. Biochar was produced using date palm fronds at 500 °C with a heating rate of 10 °C/min for 2 h. Industrially available sand was utilised as a finer aggregate (FA) in all concrete mixtures. A 0.49% absorption capacity and 2.6% specific gravity were recorded in FA [23]. The particle size distribution investigation showed that particles passed 100% from the ASTM sieve size #4 (2 mm), 100% from sieve size #8 (4 mm), 100% from sieve size #16 (600 μm), 74% from sieve size #30 (425 μm), 12% from sieve size #50 (212 μm), 6% from sieve size #100 (150 μm) and 5% from sieve size #200 (75 μm). Pulverised limestone was added as a coarser aggregate (CA) in all mixtures. Absorption, bulk specific gravity, and maximum particle size was recorded as 1.19%, 2.57, and 18.5 mm. Tap water is used as a mix of water in fabricating all concrete specimens.

2.2. Preparation of Biochar-Concrete Specimens

A total of 6 mix ratios were established to calculate the impact of biochar in standard strength concrete; 16 cylinders and two beams were cast for each mix. Five mixes contain various dosages of biochar, and one mix consists of ordinary Portland Cement (OPC) only. Biochar was added to the mix at a rate of 0.25%, 0.50%, 0.75%, 1.00%, and 1.50 wt% of the overall volume of concrete (Table 1), representing one mix with OPC only and five mixes with different percentages of biochar addition, as described by Akhtar et al. [2]. The water to cement ratio was kept constant at 0.45. MasterGlenium 110M was utilised as a superplasticiser. Initially, materials were dry mixed for about 2 min in a rotary mixer containing special blades. Then, water was added and mixed for another 3–4 min to achieve a homogenous mix. All specimens were cast as per ASTM C192 [24]. Beam specimens (100 mm × 100 mm × 500 mm) and cylindrical specimens (100 mm × 200 mm) were fabricated after putting the concrete mixture in the individual moulds in two consecutive layers and compacted each layer for 10 sec. Afterwards, it was levelled and covered the surface for 24 ± 2 h for setting and then submerged in a curing tank at room temperature 26 ± 2 °C for 28 days. Water curing helps to minimise the loss of mixing water from the concrete's surface, and the additional water accelerates the strength gain. After 28 days, the samples are taken out from the curing tank to determine the mechanical properties of the concrete (Figure 1).

Table 1. Mix design composition of Biochar-concrete specimen.

Specimen	Cement (kg/m^3)	Sand (kg/m^3)	Gravel (kg/m^3)	Biochar (kg/m^3)	w/c Ratio
Control	466.56	685	934	-	0.45
0.25 wt% BC	465.39	685	934	1.166	0.45
0.50 wt% BC	466.22	685	934	2.33	0.45
0.75 wt% BC	463.06	685	934	3.49	0.45
1.00 wt% BC	461.89	685	934	4.66	0.45
1.50 wt% BC	459.56	685	934	6.99	0.45

Figure 1. Concrete cylinder and beam specimens.

2.3. Experimental Tests on Concrete Specimens

2.3.1. Compressive Strength

After 7, 14, and 28 days of water curing, the compressive strength of cylindrical specimens was determined in three-time intervals. From each mix, three specimens were tested on a compression testing machine (CTM) with a loading rate of 2.4 kN/s, and the average value was noted according to [25] (Figure 2).

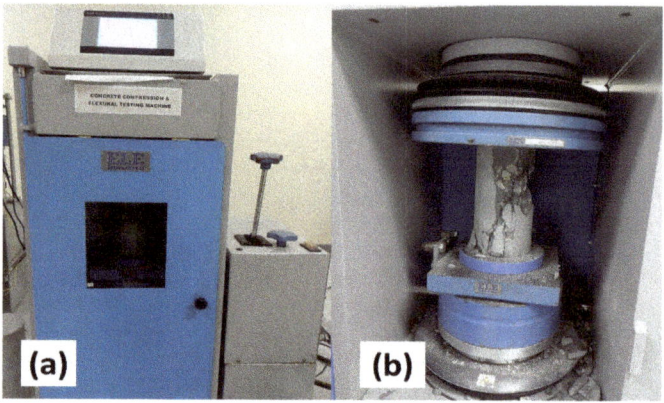

Figure 2. (**a**) Compressive Testing machine (**b**) Crushed cylindrical specimen.

2.3.2. Flexural Strength

Beams with (100 mm × 100 mm × 500 mm) were utilised to calculate the flexural strength of concrete beams after 7, 14, and 28 days of water curing. Two beam specimens from every mix were assessed on a flexural testing machine with a consistent loading rate of 0.05 kN/s, and the mean value was noted according to (ASTM C 78) [26] (Figure 3).

Figure 3. Setup for flexural testing of beam specimen.

2.3.3. Electric Resistivity

The electrical resistivity of concrete was determined using nondestructive test equipment RESIPOD as per (ASTM C1876) standard [27] (Figure 4). This test measures the bulk electrical resistivity of moulded specimens or cored segments of hardened concrete after 28 days of curing. The samples were tested after 28 days of water curing based on the practice of Su et al., 2002 [28].

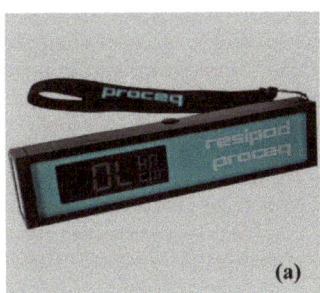

Figure 4. (a) RESIPOD electrical resistivity apparatus (b) Test setup to calculate electrical resistivity.

2.3.4. Ultrasonic Pulse Velocity

The ultrasound pulse velocity of concrete specimens from each mix was determined using a portable ultrasonic nondestructive digital indicating tester (PUNDIT), according to (ASTM C597) standard [29], after 28 days of curing, the same practice was adopted by Aziz et al. [30] previously. The setup of equipment for PUNDIT is shown in Figure 5.

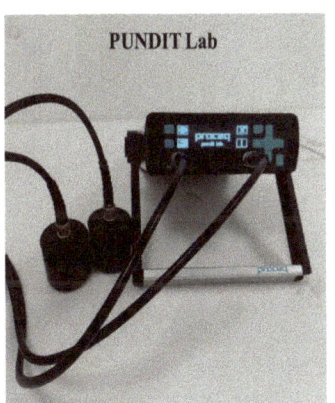
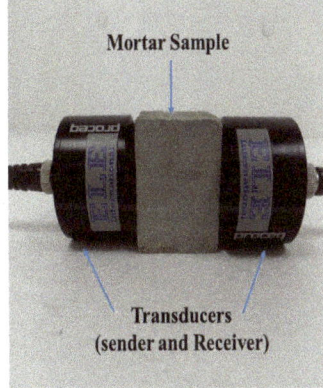

Figure 5. Ultrasonic pulse velocity test setup.

3. Results and Discussion

3.1. Compressive Strength

Figure 6 represents the compressive strength improvement at 7, 14 and 28 days of curing cylindrical specimens made with date palm fronds biochar and compared with a biochar-free control mix. The compressive strength of the control mix was noted to be 28.2, 36.6, and 43.5 MPa at an interval of 7, 14, and 28 days, respectively. The incorporation of biochar showed a linear rise in compressive strength. For instance, adding biochar with a dosage of 0.75%, 1.00%, and 1.50% increased compressive strength by 11%, 12%, and 14%, respectively, while at 0.25% and 0.50%, biochar addition represents similar strength to control mix. A similar result on compressive strength due to biochar addition was also indicated at the 14-day and 28-day ages of concrete. It is observed that the addition of 0.50%, 0.75%, 1.00% and 1.50% of biochar indicated the strength improvement of 17%, 23%, 24% and 28%, respectively, at 14-day age and 16%, 28%, 26% and 29%, respectively, at 28-day age. However, a 0.25% biochar addition represents no significant change in strength compared to the control mix. The increase in compressive strength due to biochar's addition was mainly associated with the biochar's high surface area, porosity, and water retention capability [3,18]. Dry biochar particles absorbed some of the mixing water during concrete mixing, resulting in a reduced free water–cement ratio in the concrete matrix. The presence of capillary water causes the cementitious matrix to have a high capillary porosity, which has a negative impact on strength development [31]. During the initial hardening of concrete, the water absorption through porous biochar resulted in the increasing density of the cement matrix by lowering the available water in the pores. The water absorbed in the pores of the biochar was eventually provided internally to assist cement hydration via internal curing, which contributed to the cementitious matrix's strength development [4]. Additionally, the smaller particle size of biochar exhibited a filler effect, which helps to minimise voids and gaps between cement particles and aggregates [3]. The findings discovered that the optimum biochar dosage was 0.75% and 1.50%, representing compressive strength improvement at 7, 14, and 28-day age compared to control specimens.

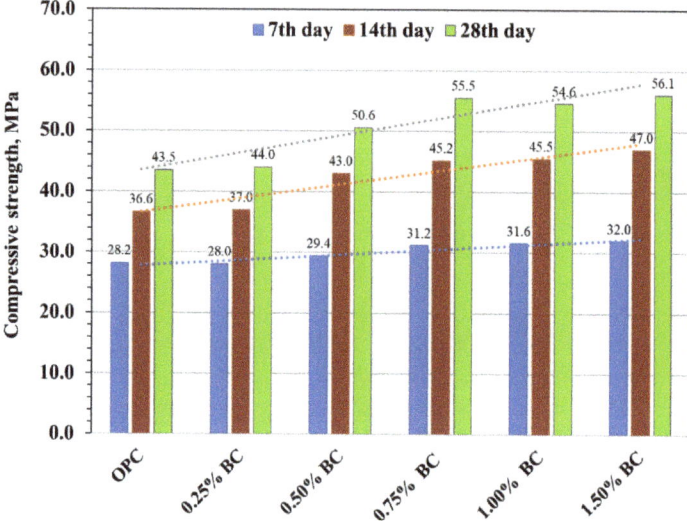

Figure 6. Compressive strength of concrete cylinder specimens.

3.2. Flexural Strength

The flexural strength biochar concrete specimens are displayed in Figure 7 at 7, 14, and 28 days of testing. The results indicate that, unlike compressive strength, adding biochar, even at minimal dosages (0.25–0.5 wt%), was favourable to enhance the flexural strength of the concrete beam. The flexural strength of the control mix was recorded as 4.75, 4.95, and 5.06 MPa at 7, 14, and 28 days, respectively. The flexural strength of all biochar-concrete mixes showed higher flexural strength than the control specimen. Adding biochar from 0.25 to 0.75 wt% in concrete showed a linear increase in flexural strength. However, with high biochar loading above 0.75 wt%, the flexural strength was not significantly enhanced. For instance, at a 7-day age, 0.25%, 0.50% and 0.75% biochar indicated a 9%, 13% and 16% improvement in flexural strength as compared to the control mix. Consequently, an increase in biochar loading, i.e., 1.00% and 1.5%, caused only 14% and 13% improvement in flexural strength, which was almost similar to 0.75% biochar-concrete but still better than the control mix. Likewise, at a 14-day and 28-day age, similar trends were noticed, 0.25%, 0.50% and 0.75% biochar resulted in around 7%, 10%, and 12% increase, respectively, at 14 days and 8%, 10%, and 11% increase, respectively, at 28 days while 1.00% and 1.50% biochar showed 10% and 9% increment at 14 days and 9% and 8% increment at 28 days in flexural strength than that of control mix. The substantial improvement in flexural strength due to the addition of biochar could be due to the flexibility provided by biochar in concrete, which functions as a link between biochar particles and hydrated cement, preventing premature fracture. Maljaee et al. [32] concluded that biochar-based mortar's flexural strength improved due to the addition of biochar. The concrete becomes dense and tough due to the addition of porous biochar, contributing micro-reinforcement effect. This ultimately resists the crack propagation, deflects the crack path, and increases flexural strength [2]. However, a negative impact was noticed after adding biochar by greater than 0.75%. A large amount of biochar in the cement matrix may cause the aggregation of biochar particles leading to an increase in inhomogeneity in the tensile plane of the cement-biochar matrix. Similar behaviour was also reported by Ahmed et al. when using biochar in a cement composite [33]. The results indicate that the optimum biochar dosage was 0.75%, and the flexural strength produced by 0.75% biochar was improved up to 16%, 12%, and 11% at 7, 14, and 28 days compared to the control specimen.

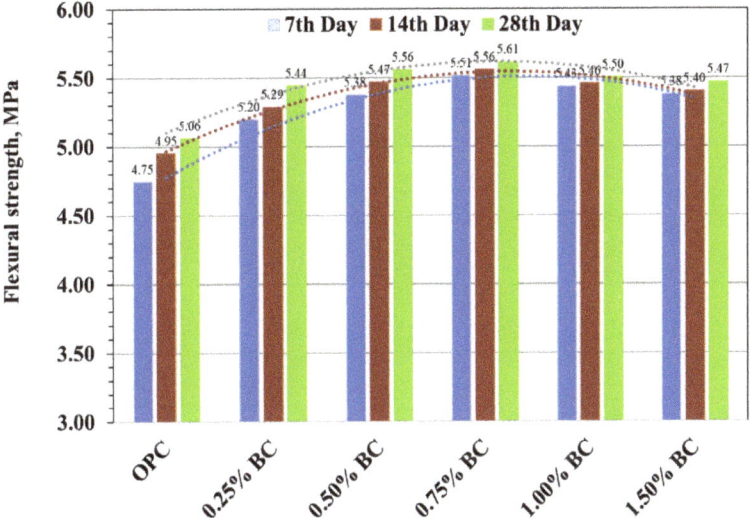

Figure 7. Flexural strength of concrete beam specimens.

3.3. Ultrasonic Pulse Velocity (UPV)

In contrast to destructive testing, the research group focused on determining the mechanical properties of biochar-based concrete using non-destructive testing. Figure 8 shows the ultrasonic pulse velocity (UPV) testing results for control and biochar concrete. According to ASTM C 597 [29], the test technique has been validated and standardised. It indicates that the values of UPV were around 7.54 and 8.04 km/s, with a mean of 7.79 km/s, according to the given experimental data. However, compared to control concrete, the biochar-concrete samples had higher UPVs. The development in UPV results within biochar-concrete can be attributed to the pozzolanic activity due to the incorporation of biochar. Similar conclusions have been observed by other researchers [34,35]. It is worth mentioning that the UPV values for all the biochar-concrete samples were over 7.79 km/s, indicating that the quality of concrete is exceptional [36]. According to the literature and standard, it can be concluded that the biochar-concrete with UPV value > 7.79 km/s indicates that it is free from internal flaws, large voids, cracks, and segregation leading to reduced structural integrity and good concrete quality in terms of density, uniformity, homogeneity [37,38].

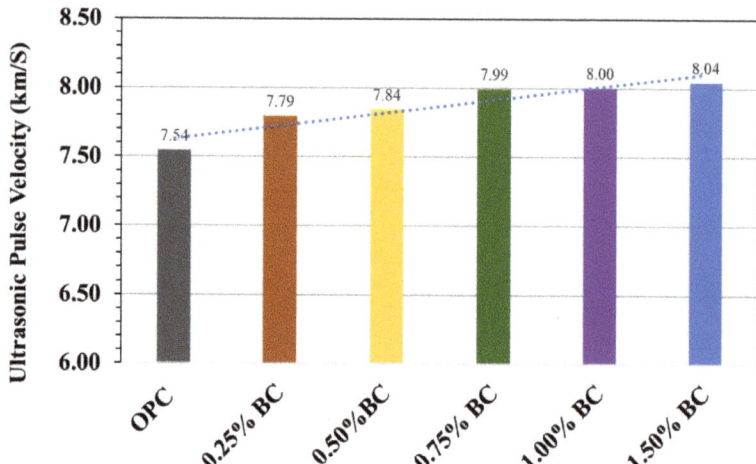

Figure 8. Ultrasonic pulse velocity for various concrete specimens.

3.4. Electrical Resistivity ρ (kΩ-cm)

The electrical resistivity of biochar-concrete and the control specimen is displayed in Figure 9. It is indicated that the electric resistivity ρ decreases gradually with increased biochar content. The value of electric resistivity was recorded as 26.1 kΩ-cm for the control mix. The values of electric resistivity noticed for 0.25%0.50%,0.75%,1.00%,1.50% biochar-concrete specimen was 22.4,22.2, 20.9,19.6,16.5kΩ-cm, respectively, which indicate linear decrease in the values of electric resistivity. Any material's electrical resistivity (ρ) is described as its ability to resist the ions transfer exposed to an electrical field. It mainly relies on the microstructure elements associated with the shape of interconnection and pore size [39]. A finer pore network with fewer connections results in lower permeability, leading to increased electrical resistivity [40–42].

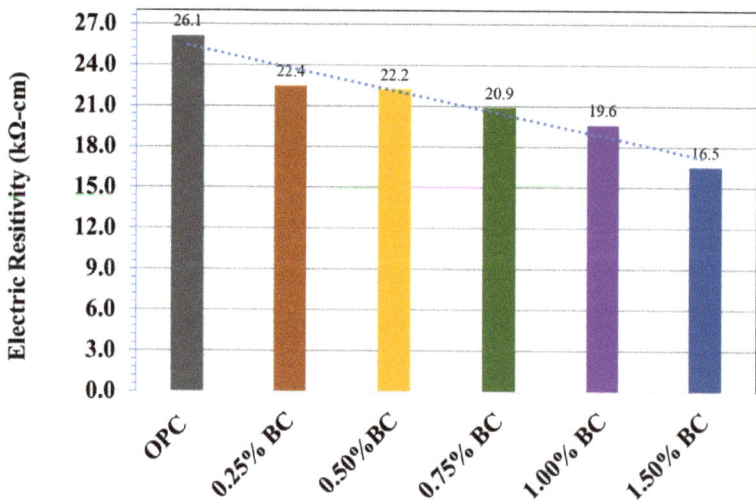

Figure 9. Electric resistivity of different concrete samples.

3.5. SWOT Analysis

Concrete is considered an essential building material globally and widely used for various construction applications. However, concrete manufacturing accounts for substantial greenhouse gas emissions associated with cement production. Therefore, innovative approaches toward green concrete building materials reduce environmental and climate impact and promote sustainable societal development. Recently, biochar-based concrete gained increasing attention due to its sustainability and improved mechanical and durable properties compared to ordinary Portland concrete [2,43–45] However, to critically evaluate its potential for real-time application, it is necessary to summarise its merits, demerits, and limitations. Therefore, in this section, the SWOT analysis is carried out as a sustainable approach focusing on business strengths, weaknesses, profiting from opportunities, and potential identified threats of date palm derived biochar-based concrete to gain insight into and guide the relevance of the adoption of biochar in the construction industry. Table 2 summarises the main components of the SWOT analysis of date palm derived biochar-concrete construction, which is discussed below.

Table 2. SWOT analysis of biochar-based concrete.

Strengths	Weaknesses
<ul><li>Stronger and rigid concrete</li><li>Dense concrete matrix</li><li>Internal curing agent</li><li>Reduce carbon footprint</li><li>Reduce hydration rate</li><li>High-quality concrete</li><li>Low electrical resistivity</li></ul>	Biochar production is an energy-intensive process Biochar dispersion in concrete is not homogeneous Biochar possessed varied surface morphology Limited research High biochar production cost Lower acceptability
Opportunities	**Threats**
<ul><li>Effective use of biowaste</li><li>Other products formation</li><li>Biochar composite industries</li><li>Biochar insulation materials</li><li>Carbon sequestering material</li><li>Global climate change</li></ul>	High energy consumption in biochar production Limited technology advancements

3.5.1. Strengths

The major strengths of using date palm fronds derived biochar-based concrete as building materials are listed in Table 2. Accordingly, date palm-derived biochar-concrete possessed desirable characteristics to develop a sustainable and green concrete material without compromising its mechanical and durable properties. The date palm-derived biochar exhibits a porous graphite carbon structure and high surface area, which facilitates the formation of denser concrete, ultimately improving the compressive and flexural strength to 29% and 16%, respectively. Previous studies revealed that biochar-based concrete demonstrated comparatively better compressive strength than ordinary concrete. It was reported that the compressive strength increased to around 31% using paper sludge-derived biochar concrete after curing for 28 days [2]. In another study, adding biochar (0.08 wt% of cement) improved the compressive strength to 85 MPa and 100 MPa. Similarly, adding coarse-sized biochar (140 μm) particles to concrete may enhance the flexural strength of concrete. It was observed that 0.5 wt% of coarse biochar-based concrete, after curing for seven days, indicated a 51% higher flexural strength (3.34 MPa) than the reference concrete [46]. The date palm-derived biochar concrete demonstrated higher UPVs (7.79 km/s), indicating improved concrete durability, which is attributed to the reduction of large voids, and internal cracks in the concrete matrix. The biochar concrete exhibited a low electrical resistivity value (ρ), which is 47% lesser than the control concrete, suggesting that the biochar-concrete matrix consists of a heterogeneous structure with a strongly connected pore network.

Additionally, studies also confirmed that the high thermal stability of biochar-based cement composites is another essential factor demonstrating its applicability compared to ordinary concrete. It was reported that biochar-based mortar specimens consisting of different proportions (5%, 10%, and 20% of cement weight) when subjected to different heating environments (200 °C, 450 °C, and 700 °C), showed minimal % loss in strength compared to ordinary mortar [47]. The study reported that adding 5 wt% biochar retained nearly 88%, 76%, and 38% of compressive strength when exposed to high temperatures (200 °C, 450 °C, and 700 °C). Biochar produced at high pyrolysis temperature exhibits high thermal stability, significantly improving concrete fire resistance [40]. The fire stability characteristic attracts applications in concrete structures used in mines and tunnels, reducing human risks and substantial damage. Moreover, the highly porous structure of biochar serves as a thermal insulator in a concrete matrix. Generally, the low interfacial adhesion of biochar with cement matrix leads to poor heat transfer, leading to decreased thermal conductivity. It was reported that biochar derived from the peach shell and apricot, when added to concrete, showed low thermal conductivity of 0.40 and 0.34 $\times$ 10^{-6} m^2/s, respectively [48,49]. Therefore, using biochar in concrete as cement replacement improves the mechanical, durable, and thermal properties of concrete and reduces the CO_2 emissions of the concrete industry. The study evaluated the impact of various governing factors, including raw materials, methods, synthesis, and transportation of biochar-concrete systems on the environment. It was estimated that approximately 0 to 20 wt% of biochar additions might expect to reduce 0.15–0.20 kg of cement in concrete. Therefore, using low cement amounts in biochar-concrete may be expected to reduce greenhouse gas emissions, ozone depletion, climate change, and hazardous biowaste management [50].

3.5.2. Weakness

While other agricultural wastes such as rice husk ash, bagasse ash, palm oil fuel ash, etc., have been widely used to replace OPC in cementitious materials, there is little knowledge and availability on the properties of biochar made from date palm fronds and cementitious materials from it in most regions of the world. As a result of this lack of understanding, date palm fronds biochar application is very limited in the construction industry due to the lack of high confidence in the material. Furthermore, the effect of biochar on composite cement performance still requires various experiments to be completed to draw a more accurate conclusion. It is imperative to conduct an imminent study that examines the properties of date palm fronds biochar and its impact on cementitious materials' long-term performance to promote the practical application of date palm fronds.

The use of date palm fronds biochar in cementitious materials also has a weakness: its lower strength at high volume percentages. However, higher date palm fronds biochar content (1.5%) in the cementitious mix reduces its engineering performance, limiting its application as a binder component in cementitious materials [51,52]. Conversely, other agricultural wastes, such as rice husk ash, wheat straw ash, palm oil fuel ash, etc., are effective even at high dosages (up to 20%) in cementitious materials [53,54]. Combining date palm fronds biochar with high-reactive materials such as nano- and micro-silica makes it possible to use high dosages of date palm fronds biochar without affecting its engineering properties [55,56]. However, the date palm fronds biochar amount in the cementitious matrix must be carefully controlled since it can reduce free water and, consequently, the fluidity of concrete, increasing the demand for superplasticisers.

Similarly, cementitious materials that contain date palm fronds biochar have a longer setting time, making them less suitable for applications requiring shorter setting times. A chemical additive such as an accelerator can be added to cementitious materials incorporating DPFA as a replacement for OPC to shorten their initial and final set times [57].

3.5.3. Opportunities

Growing sustainability awareness in the construction industry has led to the search for sustainable materials that can replace OPC in cementitious materials. Due to its chemical properties and the fact that it is derived from agricultural waste, date palm frond biochar is an excellent alternative source of revenue for developing sustainable concrete. Saudi Arabia ranks among the world's leading date-producing countries. As a result of the high production of dates, the date palm industry produces a tremendous amount of agricultural waste. If these wastes are improperly disposed of in the environment, they could pose a fire and safety risk. In addition, valuable land spaces could be depleted, and the aesthetics of the environment might be impacted. The processing of these date palm fronds into biochar would provide an opportunity to efficiently manage these waste materials and use them as a raw material for making green concrete. Further, converting these wastes into valuable resources would entail a monetary value for date palm factories, opening up another source of revenue.

The application of date palm fronds biochar in building materials has been shown to improve the mechanical properties of building materials and enhance the durability of composites under extreme environmental conditions. The improved properties suggest that biochar-containing building materials perform equally or even better than those without [4,58–61]. Therefore, there is a massive opportunity for date palm fronds derived from biochar-based concrete to be used for various building applications for a better design life. Due to its wide range of applications and vast production, concrete has a substantial carbon footprint, contributing to 8% of the global carbon dioxide emissions [5]. Therefore, it is imperative to look for pathways for reducing emissions within the cement and concrete industry to reduce its environmental impact [4]. The use of date palm fronds-derived biochar in building materials has the potential to reduce carbon footprints and mitigate climate change [62]. The ability of biochar to sequester carbon in stable forms and capture CO_2 directly from the atmosphere in building materials and the addition of CO_2-saturated biochar have a vital role. Even with a lack of studies on this topic, building materials that contain date palm fronds derived from biochar have superb possibilities for reducing carbon footprints and mitigating climate change [63].

3.5.4. Threats

Various potential barriers that may restrict date palm derived biochar-concrete commercialisation of biochar-concrete are listed in Table 2. Although biochar-concrete knowledge is progressively expanding, biochar production's cost and engineering shortcoming is still challenging. Biochar production is an energy-intensive process that may increase biochar cost compared to cement. Therefore, an efficient design for biochar production is needed, which is economically feasible, improves biochar quality, and reduces net greenhouse gas emissions. Additionally, extensive research and development are required to identify sustainable and cost-efficient alternative production approaches to mitigate energy use. The new techniques would ensure biochar strength and market potential as sustainable future materials in the concrete industry.

3.6. Technical and Economic Feasibility of the Biochar-Concrete System

The possible emissions of GHGs to the environment because of the decay/decomposition of the biomass are avoided by the process of valorising the biomass to produce biochar. Reducing GHG emissions of CO_2-eq./kg in the biochar life cycle using different biomass was estimated. In addition to reducing the net emission of GHG, the use of biochar in concrete has played a vital role in improving its chemical and mechanical properties. Most studies in the past showed significant improvements in concrete's compressive and flexural strength [3,64–68]. Other mechanical properties of concrete, such as toughness, flexibility, elongation, permeability, thermal stability, and thermal conductivity, were also observed [19,64,69,70]. Despite a relatively high cost of production of biochar [71], compared to natural filler materials such as sand, it is still considered a better construction material due to its associated environmental benefits of reduced CO_2-eq./kg as well as the generation of other value-added products such as syngas, bio-oil production in pyrolysis.

Apart from the environmental and other technical benefits of using biochar in concrete, its economic viability is crucial in deploying it in the construction industry. Kung et al. [72] reported a higher and more feasible feedstock value of 10.98 $/t for biochar production by slow pyrolysis compared to a correspondingly lower value of 2.85 $/t pyrolysis. A lesser value of biochar production in fast pyrolysis is due to higher net losses of feedstock and is considered unviable both in economic and environmental profits. A more excellent feedstock value in slow pyrolysis is a viable solution for biochar production. As reported by several researchers, [73] biochar production from forest residue using a portable system, the cost of biochar production can be further reduced, equaling 470 $/t of oven-dried by technologically improving the portable system.

4. Conclusions

In this research, biochar derived from date palm waste was used as an additive to concrete at the different mass compositions of 0.25 wt% to 1.5 wt%. The performance of biochar-concrete specimens derived from date palm waste was examined by the fresh concrete specimen's representative mechanical and durability characteristics. The following conclusions can be drawn based on the outcomes:

1. The compressive strength of biochar-concrete increased with increasing biochar content and showed a maximum 28%, 26%and 29% improvement in power at 28-day age with the incorporation of 0.75%, 1.00%, and 1.50% of biochar. The biochar-concrete containing 0.75 wt% biochar loading indicated 16% higher flexural strength than the control mix. The increased surface area, small particle size, and water retention capability of porous biochar lead to a denser concrete matrix, formation of cement hydrates, and filler effect resulting in stronger concrete.
2. Biochar-concrete showed high values (>7.79 km/s) of UPV demonstrating high-performance concrete. The electrical resistivity reduced linearly with the incorporation of biochar. This confirmed the formation homogeneous and denser biochar-concrete network resulting in lower permeable concrete.
3. The SWOT and techno-economic assessment analysis further corroborates that the biochar-concrete system possessed the high potential to be commercially adopted as green and sustainable material despite the economic and engineering challenges.
4. In general, it is suggested that biochar derived from Saudi agriculture waste can be used as a beneficial product for infrastructure designs requiring high-performance and durable building materials to attain technical and environmental benefits.

Author Contributions: Conceptualization, K.K. and M.Z.; Data curation, M.A.A. and M.N.A.; Formal analysis, M.A.A., M.Z. and M.N.A.; Funding acquisition, K.K.; Investigation, M.A.A. and M.Z.; Methodology, M.A.A. and M.Z.; Project administration, K.K.; Resources, K.K.; Supervision, K.K.; Validation, M.A.A., M.Z. and M.N.A.; Visualization, M.A.A. and M.N.A.; Writing—original draft, K.K., M.A.A., M.Z. and M.N.A.; Writing—review & editing, K.K. and M.Z. All authors have read and agreed to the published version of the manuscript.

Funding: This work was supported by the Al Bilad Bank Scholarly Chair for Food Security in Saudi Arabia, the Deanship of Scientific Research, Vice Presidency for Graduate Studies and Scientific Research, King Faisal University, Saudi Arabia [Grant No. CHAIR73]. The APC was funded by the same [GRANT No. CHAIR73].

Institutional Review Board Statement: Not applicable.

Informed Consent Statement: Not applicable.

Data Availability Statement: The data used in this research has been appropriately cited and reported in the main text.

Acknowledgments: The authors acknowledge the Al Bilad Bank Scholarly Chair for Food Security in Saudi Arabia, the Deanship of Scientific Research, Vice Presidency for Graduate Studies and Scientific Research, King Faisal University, Saudi Arabia, for the financial support (Grant No. CHAIR73).

Conflicts of Interest: The authors declare no conflict of interest.

References

1. Torgal, F.P.; Mistretta, M.; Kaklauskas, A.; Granqvist, C.G.; Cabeza, L.F. *Nearly Zero Energy Building Refurbishment: A Multidisciplinary Approach*; Springer Science & Business Media: Berlin, Germany, 2014; ISBN 9781447155232.
2. Akhtar, A.; Sarmah, A.K. Novel biochar-concrete composites: Manufacturing, characterization and evaluation of the mechanical properties. *Sci. Total Environ.* **2018**, *616–617*, 408–416. [CrossRef] [PubMed]
3. Gupta, S.; Kua, H.W.; Koh, H.J. Application of biochar from food and wood waste as green admixture for cement mortar. *Sci. Total Environ.* **2018**, *619–620*, 419–435. [CrossRef] [PubMed]
4. Winters, D.; Boakye, K.; Simske, S. Toward carbon-neutral concrete through biochar–cement–calcium carbonate composites: A critical review. *Sustainability* **2022**, *14*, 4633. [CrossRef]
5. Liu, B.; Qin, J.; Shi, J.; Jiang, J.; Wu, X.; He, Z. New perspectives on utilization of CO_2 sequestration technologies in cement-based materials. *Constr. Build. Mater.* **2021**, *272*, 121660. [CrossRef]
6. Nasir, M.; Aziz, M.A.; Zubair, M.; Ashraf, N.; Hussein, T.N.; Allubli, M.K.; Manzar, M.S.; Al-Kutti, W.; Al-Harthi, M.A. Engineered cellulose nanocrystals-based cement mortar from office paper waste: Flow, strength, microstructure, and thermal properties. *J. Build. Eng.* **2022**, *51*, 104345. [CrossRef]
7. Abdalla, L.B.; Ghafor, K.; Mohammed, A. Testing and modeling the young age compressive strength for high workability concrete modified with PCE polymers. *Results Mater.* **2019**, *1*, 100004. [CrossRef]
8. Yang, K.H.; Jung, Y.B.; Cho, M.S.; Tae, S.H. Effect of supplementary cementitious materials on reduction of CO2 emissions from concrete. *J. Clean. Prod.* **2015**, *103*, 774–783. [CrossRef]
9. Batalin, B.S.; Saraikina, K.A. Interaction of glass fiber and hardened cement paste. *Glass Ceram.* **2014**, *71*, 294–297. [CrossRef]
10. Amrul, N.F.; Kabir Ahmad, I.; Ahmad Basri, N.E.; Suja, F.; Abdul Jalil, N.A.; Azman, N.A. A review of organic waste treatment using black soldier fly (hermetia illucens). *Sustainability* **2022**, *14*, 4565. [CrossRef]
11. Elbeshbishy, E.; Dhar, B.R. Processes for bioenergy and resources recovery from biowaste. *Processes* **2020**, *8*, 1005. [CrossRef]
12. Ihsanullah, I.; Khan, M.T.; Zubair, M.; Bilal, M.; Sajid, M. Removal of pharmaceuticals from water using sewage sludge-derived biochar: A review. *Chemosphere* **2022**, *289*, 133196. [CrossRef]
13. Vieira, F.R.; Romero Luna, C.M.; Arce, G.L.A.F.; Ávila, I. Optimization of slow pyrolysis process parameters using a fixed bed reactor for biochar yield from rice husk. *Biomass Bioenergy* **2020**, *132*, 105412. [CrossRef]
14. Zubair, M.; Mu'azu, N.D.; Jarrah, N.; Blaisi, N.; Aziz, H.A.; Al-Harthi, M.A. Adsorption behavior and mechanism of methylene blue, crystal violet, eriochrome black T, and methyl orange dyes onto biochar-derived date palm fronds waste produced at different pyrolysis conditions. *Water Air Soil Pollut.* **2020**, *231*, 240. [CrossRef]
15. Mu'azu, N.D.; Zubair, M.; Ihsanullah, I. Process optimization and modeling of phenol adsorption onto sludge-based activated carbon intercalated MgAlFe ternary layered double hydroxide composite. *Molecules* **2021**, *26*, 4266. [CrossRef]
16. Restuccia, L.; Ferro, G.A. Promising low cost carbon-based materials to improve strength and toughness in cement composites. *Constr. Build. Mater.* **2016**, *126*, 1034–1043. [CrossRef]
17. Tan, K.; Pang, X.; Qin, Y.; Wang, J. Properties of cement mortar containing pulverized biochar pyrolyzed at different temperatures. *Constr. Build. Mater.* **2020**, *263*, 120616. [CrossRef]
18. Choi, W.C.; Yun, H.D.; Lee, J.Y. Mechanical properties of mortar containing bio-char from pyrolysis. *J. Korea Inst. Struct. Maint. Insp.* **2012**, *16*, 67–74. [CrossRef]
19. Restuccia, L.; Ferro, G.A.; Suarez-Riera, D.; Sirico, A.; Bernardi, P.; Belletti, B.; Malcevschi, A. Mechanical characterization of different biochar-based cement composites. *Procedia Struct. Integr.* **2020**, *25*, 226–233. [CrossRef]
20. Wang, L.; Chen, L.; Tsang, D.C.W.; Kua, H.W.; Yang, J.; Ok, Y.S.; Ding, S.; Hou, D.; Poon, C.S. The roles of biochar as green admixture for sediment-based construction products. *Cem. Concr. Compos.* **2019**, *104*, 103348. [CrossRef]
21. Al-Kutti, W.; Saiful Islam, A.B.M.; Nasir, M. Potential use of date palm ash in cement-based materials. *J. King Saud Univ. Eng. Sci.* **2019**, *31*, 26–31. [CrossRef]
22. *ASTM C150*; Standard Specification for Portland Cement. ASTM International: West Conshohocken, PA, USA, 2019.
23. *ASTM C 128*; Standard Test Method for Density, Relative Density (Specific Gravity), and Absorption of Fine Ag-gregate. Annual Book of ASTM Standards. ASTM International: West Conshohocken, PA, USA, 2003.
24. *ASTM C192*; Standard Practice for Making and Curing Concrete Test Specimens in the Laboratory. ASTM International: West Conshohocken, PA, USA, 2016.
25. *ASTM C39 ASTM Standard C39/C39M-16*; Standard Test Method for Compressive Strength of Cylindrical Concrete Specimens. ASTM International: West Conshohocken, PA, USA, 2016.
26. *American Society for Testing and Materials (ASTM) Astm C78/C78M-02*; Standard Test Method for Flexural Strength of Concrete (Using Simple Beam with Third-Point Loading). ASTM International: West Conshohocken, PA, USA, 2002.

27. *ASTM C1876*; Standard Test Method for Bulk Electrical Resistivity or Bulk Conductivity of Concrete. ASTM International: West Conshohocken, PA, USA, 2012.
28. Su, J.; Yang, C.; Wu, W.; Huang, R. Effect of moisture content on concrete resistivity measurement. *J. Chin. Inst. Eng.* **2002**, *25*, 117–122. [CrossRef]
29. *ASTM C 597-09*; Standard Test Method for Pulse Velocity Through Concrete. ASTM International: West Conshohocken, PA, USA, 2010.
30. Aziz, M.A.; Zubair, M.; Saleem, M. Development and testing of cellulose nanocrystal-based concrete. *Case Stud. Constr. Mater.* **2021**, *15*, e00761. [CrossRef]
31. Chen, X.; Wu, S.; Zhou, J. Influence of porosity on compressive and tensile strength of cement mortar. *Constr. Build. Mater.* **2013**, *40*, 869–874. [CrossRef]
32. Maljaee, H.; Paiva, H.; Madadi, R.; Tarelho, L.A.C.; Morais, M.; Ferreira, V.M. Effect of cement partial substitution by waste-based biochar in mortars properties. *Constr. Build. Mater.* **2021**, *301*, 124074. [CrossRef]
33. Ahmad, S.; Tulliani, J.M.; Ferro, G.A.; Khushnood, R.A.; Restuccia, L.; Jagdale, P. Crack path and fracture surface modifications in cement composites. *Frat. Ed Integrità Strutt.* **2015**, *9*, 34. [CrossRef]
34. Chen, T.T.; Wang, W.C.; Wang, H.Y. Mechanical properties and ultrasonic velocity of lightweight aggregate concrete containing mineral powder materials. *Constr. Build. Mater.* **2020**, *258*, 119550. [CrossRef]
35. Zhang, Y.; Aslani, F. Compressive strength prediction models of lightweight aggregate concretes using ultrasonic pulse velocity. *Constr. Build. Mater.* **2021**, *292*, 123419. [CrossRef]
36. *IS 13311 (Part 1)*; Method of Non-Destructive Testing of Concret, Part 1: Ultrasonic Pulse Velocity. Bureau of Indian Standards: Manak Bhawan, Old Delhi, 1992.
37. Zaheer, M.M.; Hasan, S.D. Mechanical and durability performance of carbon nanotubes (CNTs) and nanosilica (NS) admixed cement mortar. *Mater. Today Proc.* **2021**, *42*, 1422–1431. [CrossRef]
38. Sabbağ, N.; Uyanık, O. Prediction of reinforced concrete strength by ultrasonic velocities. *J. Appl. Geophys.* **2017**, *141*, 13–23. [CrossRef]
39. Layssi, H.; Ghods, P.; Alizadeh, A.R.; Salehi, M. Electrical Resistivity of Concrete. *Concr. Int.* **2016**, *37*, 41–46.
40. Yang, H.M.; Zhang, S.M.; Wang, L.; Chen, P.; Shao, D.K.; Tang, S.W.; Li, J.Z. High-ferrite Portland cement with slag: Hydration, microstructure, and resistance to sulfate attack at elevated temperature. *Cem. Concr. Compos.* **2022**, *130*, 104560. [CrossRef]
41. Wang, J.; He, T.; Zhou, Y.; Tang, S.; Tan, J.; Liu, Z.; Su, J. The influence of fiber type and length on the cracking resistance, durability and pore structure of face slab concrete. *Constr. Build. Mater.* **2021**, *282*, 122706. [CrossRef]
42. Huang, J.; Li, W.; Huang, D.; Wang, L.; Chen, E.; Wu, C.; Wang, B.; Deng, H.; Tang, S.; Shi, Y.; et al. Fractal analysis on pore structure and hydration of magnesium oxysulfate cements by first principle, thermodynamic and microstructure-based methods. *Fractal Fract.* **2021**, *5*, 164. [CrossRef]
43. Sirico, A.; Bernardi, P.; Belletti, B.; Malcevschi, A.; Dalcanale, E.; Domenichelli, I.; Fornoni, P.; Moretti, E. Mechanical characterization of cement-based materials containing biochar from gasification. *Constr. Build. Mater.* **2020**, *246*, 118490. [CrossRef]
44. Aamar Danish, M.; Usama Salim, T.A. Trends and developments in green cement "a sustainable approach". *Sustain. Struct. Mater.* **2019**, *2*, 45–60. [CrossRef]
45. Mrad, R.; Chehab, G. Mechanical and microstructure properties of biochar-based mortar: An internal curing agent for PCC. *Sustainability* **2019**, *11*, 2491. [CrossRef]
46. Restuccia, L.; Ferro, G.A. Influence of filler size on the mechanical properties of cement-based composites. *Fatigue Fract. Eng. Mater. Struct.* **2018**, *41*, 797–805. [CrossRef]
47. Navaratnam, S.; Wijaya, H.; Rajeev, P.; Mendis, P.; Nguyen, K. Residual stress-strain relationship for the biochar-based mortar after exposure to elevated temperature. *Case Stud. Constr. Mater.* **2021**, *14*, e00540. [CrossRef]
48. Drzymała, T.; Jackiewicz-Rek, W.; Gałaj, J.; Šukys, R. Assessment of mechanical properties of high strength concrete (HSC) after exposure to high temperature. *J. Civ. Eng. Manag.* **2018**, *24*, 138–144. [CrossRef]
49. Carević, I.; Baričević, A.; Štirmer, N.; Šantek Bajto, J. Correlation between physical and chemical properties of wood biomass ash and cement composites performances. *Constr. Build. Mater.* **2020**, *256*, 119450. [CrossRef]
50. Campos, J.; Fajilan, S.; Lualhati, J.; Mandap, N.; Clemente, S. Life cycle assessment of biochar as a partial replacement to Portland cement. *IOP Conf. Ser. Earth Environ. Sci.* **2020**, *479*, 12025. [CrossRef]
51. Cuthbertson, D.; Berardi, U.; Briens, C.; Berruti, F. Biochar from residual biomass as a concrete filler for improved thermal and acoustic properties. *Biomass Bioenergy* **2019**, *120*, 77–83. [CrossRef]
52. Gupta, S.; Kua, H.W.; Pang, S.D. Biochar-mortar composite: Manufacturing, evaluation of physical properties and economic viability. *Constr. Build. Mater.* **2018**, *167*, 874–889. [CrossRef]
53. Khan, K.; Ullah, M.F.; Shahzada, K.; Amin, M.N.; Bibi, T.; Wahab, N.; Aljaafari, A. Effective use of micro-silica extracted from rice husk ash for the production of high-performance and sustainable cement mortar. *Constr. Build. Mater.* **2020**, *258*, 119589. [CrossRef]
54. Amin, M.N.; Murtaza, T.; Shahzada, K.; Khan, K.; Adil, M. Pozzolanic potential and mechanical performance of wheat straw ash incorporated sustainable concrete. *Sustainability* **2019**, *11*, 519. [CrossRef]
55. Dixit, A.; Gupta, S.; Pang, S.D.; Kua, H.W. Waste Valorisation using biochar for cement replacement and internal curing in ultra-high performance concrete. *J. Clean. Prod.* **2019**, *238*, 117876. [CrossRef]

56. Gupta, S.; Kua, H.W. Combination of biochar and silica fume as partial cement replacement in mortar: Performance evaluation under normal and elevated temperature. *Waste Biomass Valoriz.* **2020**, *11*, 2807–2824. [CrossRef]
57. Gupta, S.; Kua, H.W. Carbonaceous micro-filler for cement: Effect of particle size and dosage of biochar on fresh and hardened properties of cement mortar. *Sci. Total Environ.* **2019**, *662*, 952–962. [CrossRef]
58. Tan, K.; Qin, Y.; Wang, J. Evaluation of the properties and carbon sequestration potential of biochar-modified pervious concrete. *Constr. Build. Mater.* **2022**, *314*, 125648. [CrossRef]
59. Praneeth, S.; Saavedra, L.; Zeng, M.; Dubey, B.K.; Sarmah, A.K. Biochar admixtured lightweight, porous and tougher cement mortars: Mechanical, durability and micro computed tomography analysis. *Sci. Total Environ.* **2021**, *750*, 142327. [CrossRef]
60. Gupta, S.; Muthukrishnan, S.; Kua, H.W. Comparing influence of inert biochar and silica rich biochar on cement mortar–Hydration kinetics and durability under chloride and sulfate environment. *Constr. Build. Mater.* **2021**, *268*, 121142. [CrossRef]
61. Dixit, A.; Verma, A.; Pang, S.D. Dual waste utilization in ultra-high performance concrete using biochar and marine clay. *Cem. Concr. Compos.* **2021**, *120*, 104049. [CrossRef]
62. Suarez-Riera, D.; Restuccia, L.; Ferro, G.A. The use of biochar to reduce the carbon footprint of cement-based. *Procedia Struct. Integr.* **2020**, *26*, 199–210. [CrossRef]
63. Roberts, K.G.; Gloy, B.A.; Joseph, S.; Scott, N.R.; Lehmann, J. Life cycle assessment of biochar systems: Estimating the energetic, economic, and climate change potential. *Environ. Sci. Technol.* **2010**, *44*, 827–833. [CrossRef]
64. Gupta, S.; Kua, H.W.; Tan Cynthia, S.Y. Use of biochar-coated polypropylene fibers for carbon sequestration and physical improvement of mortar. *Cem. Concr. Compos.* **2017**, *83*, 171–187. [CrossRef]
65. Das, O.; Sarmah, A.K.; Bhattacharyya, D. A novel approach in organic waste utilization through biochar addition in wood/polypropylene composites. *Waste Manag.* **2015**, *38*, 132–140. [CrossRef]
66. Asadi Zeidabadi, Z.; Bakhtiari, S.; Abbaslou, H.; Ghanizadeh, A.R. Synthesis, characterization and evaluation of biochar from agricultural waste biomass for use in building materials. *Constr. Build. Mater.* **2018**, *181*, 301–308. [CrossRef]
67. Gupta, S.; Kua, H.W.; Low, C.Y. Use of biochar as carbon sequestering additive in cement mortar. *Cem. Concr. Compos.* **2018**, *87*, 110–129. [CrossRef]
68. Gupta, S.; Kashani, A. Utilization of biochar from unwashed peanut shell in cementitious building materials—Effect on early age properties and environmental benefits. *Fuel Process. Technol.* **2021**, *218*, 106841. [CrossRef]
69. Gupta, S.; Kua, H.W. Effect of water entrainment by pre-soaked biochar particles on strength and permeability of cement mortar. *Constr. Build. Mater.* **2018**, *159*, 107–125. [CrossRef]
70. Jeon, J.; Park, J.H.; Yuk, H.; Kim, Y.U.; Yun, B.Y.; Wi, S.; Kim, S. Evaluation of hygrothermal performance of wood-derived biocomposite with biochar in response to climate change. *Environ. Res.* **2021**, *193*, 110359. [CrossRef]
71. Alhashimi, H.A.; Aktas, C.B. Life cycle environmental and economic performance of biochar compared with activated carbon: A meta-analysis. *Resour. Conserv. Recycl.* **2017**, *118*, 13–26. [CrossRef]
72. Kang, K.; Nanda, S.; Sun, G.; Qiu, L.; Gu, Y.; Zhang, T.; Zhu, M.; Sun, R. Microwave-assisted hydrothermal carbonization of corn stalk for solid biofuel production: Optimization of process parameters and characterization of hydrochar. *Energy* **2019**, *186*, 115795. [CrossRef]
73. Shabangu, S.; Woolf, D.; Fisher, E.M.; Angenent, L.T.; Lehmann, J. Techno-economic assessment of biomass slow pyrolysis into different biochar and methanol concepts. *Fuel* **2014**, *117*, 742–748. [CrossRef]

Review

The Present State of the Use of Waste Wood Ash as an Eco-Efficient Construction Material: A Review

Rebeca Martínez-García [1], P. Jagadesh [2], Osama Zaid [3,*], Adrian A. Șerbănoiu [4], Fernando J. Fraile-Fernández [1], Jesús de Prado-Gil [1], Shaker M. A. Qaidi [5] and Cătălina M. Grădinaru [4]

1. Department of Mining Technology, Topography and Structures, Campus of Vegazana s/n, University of León, 24071 León, Spain; rmartg@unileon.es (R.M.-G.); fjfraf@unileon.es (F.J.F.-F.); jesusdepradogil@gmail.com (J.d.P.-G.)
2. Department of Civil Engineering, Coimbatore Institute of Technology, Coimbatore 641014, Tamil Nadu, India; jaga.86@gmail.com
3. Department of Civil Engineering, Swedish College of Engineering and Technology, Wah Cantt 47080, Pakistan
4. Faculty of Civil Engineering and Building Services, Gheorghe Asachi Technical University of Iași, 700050 Iași, Romania; serbanoiu.adrian@tuiasi.ro (A.A.Ș.); catalina.gradinaru@tuiasi.ro (C.M.G.)
5. Department of Civil Engineering, College of Engineering, University of Duhok, Duhok 42001, Iraq; shaker.abdal@uod.ac
* Correspondence: osama.zaid@scetwah.edu.pk

Citation: Martínez-García, R.; Jagadesh, P.; Zaid, O.; Şerbănoiu, A.A.; Fraile-Fernández, F.J.; de Prado-Gil, J.; Qaidi, S.M.A.; Grădinaru, C.M. The Present State of the Use of Waste Wood Ash as an Eco-Efficient Construction Material: A Review. *Materials* 2022, 15, 5349. https://doi.org/10.3390/ma15155349

Academic Editor: Dumitru Doru Burduhos Nergis

Received: 7 July 2022
Accepted: 2 August 2022
Published: 3 August 2022

Publisher's Note: MDPI stays neutral with regard to jurisdictional claims in published maps and institutional affiliations.

Copyright: © 2022 by the authors. Licensee MDPI, Basel, Switzerland. This article is an open access article distributed under the terms and conditions of the Creative Commons Attribution (CC BY) license (https://creativecommons.org/licenses/by/4.0/).

Abstract: A main global challenge is finding an alternative material for cement, which is a major source of pollution to the environment because it emits greenhouse gases. Investigators play a significant role in global waste disposal by developing appropriate methods for its effective utilization. Geopolymers are one of the best options for reusing all industrial wastes containing aluminosilicate and the best alternative materials for concrete applications. Waste wood ash (WWA) is used with other waste materials in geopolymer production and is found in pulp and paper, wood-burning industrial facilities, and wood-fired plants. On the other hand, the WWA manufacturing industry necessitates the acquisition of large tracts of land in rural areas, while some industries use incinerators to burn wood waste, which contributes to air pollution, a significant environmental problem. This review paper offers a comprehensive review of the current utilization of WWA with the partial replacement with other mineral materials, such as fly ash, as a base for geopolymer concrete and mortar production. A review of the usage of waste wood ash in the construction sector is offered, and development tendencies are assessed about mechanical, durability, and microstructural characteristics. The impacts of waste wood ash as a pozzolanic base for eco-concreting usages are summarized. According to the findings, incorporating WWA into concrete is useful to sustainable progress and waste reduction as the WWA mostly behaves as a filler in filling action and moderate amounts of WWA offer a fairly higher compressive strength to concrete. A detail study on the source of WWA on concrete mineralogy and properties must be performed to fill the potential research gap.

Keywords: geopolymer concrete; waste wood ash; environmental impact mechanical properties; durability

1. Introduction

Globally, 0.74 kg of solid waste is generated per capita per day, with national rates varying between 0.11 and 4.54 kg per capita per day depending on urbanization rates and income levels [1–3]. The Europe and Central Asia regions, with 20% (392 million tons per year), rank second in solid waste generation [4–6]. The overall composition of waste mainly corresponds to organic and green waste (44%); paper and cardboard (19%); other materials (14%); plastics (12%); glass (5%); metal (4%); wood (2%); and rubber and leather (2%). As for waste treatment, it mainly focuses on recycling (20%) and incineration (17.8%), providing the possibility of giving a new useful life to the materials after their use and ensuring adequate final disposal [7]. This is in line with the adaptation of a circular economy as a novelty and eco-friendly production model. In the specific case of the construction industry, part of the environmental impact is due to the demolition of structures, which generates different types of solid waste. On the other hand, the use of cement in the production of bricks/block and concrete, which is used in the latter to make it more resistant [8], implies a significant anthropic emission of carbon dioxide (CO_2) of 5–8% worldwide, which could increase, according to projections, to 27% by the year 2050, especially taking into account that one cubic meter of concrete is produced annually per person [8–11]. Based on this reality and the projected scenario, the cement and concrete industry has been developing a series of strategies and innovations to reduce CO_2 emissions. One of these innovations is the production of geopolymers to be used as alternative materials to replace all or part of the ordinary Portland cement used in construction, which is obtained either from metakaolin or from industrial, forestry, and agricultural waste with a high aluminosilicates content [12–15]. A geopolymer is a binder of mineral origin (inorganic) obtained from the dissolution [16–19] and subsequent polycondensation of ashes rich in aluminosilicates in the presence of an alkaline solution (hydroxides and silicates of alkali metals, Na and K) [13,20,21]. Additionally, the use of mixed geopolymers, which are generated by the combination of two or more types of chemically stabilized industrial wastes or ashes, has been considered [22]. The use of this type of materials can reduce CO_2 production by up to 90%, while preserving or even improving their mechanical properties (e.g., porosity, structure, compressive strength, water absorption, and durability) [12,23].

Several researchers have devoted themselves to using different raw materials for the production of concrete, for example, agricultural residues such as rice husk ash and palm oil ash [24], sugar cane bagasse [25,26], and corn cob ash [27], finding good results in the properties of concrete [28]. On the other hand, wood waste ashes [29,30] have emerged as a good option for the fractional replacement of binder and kaolin used in the formation of geopolymers, since in addition to increasing workability, porosity, and drying shrinkage, these wastes are given an alternative use, and potential environmental pollution [29–33] is reduced by their entry into the environment, contributing directly to sustainable development [34,35]. Ekaputri reported [36] obtaining a concrete (geopolymer) with high compressive strength (48.5 MPa to 48.5 MPa) from class F ash with 10 mol/L NaOH due to the generation of hydroxide ions that significantly influence the dissolution of the Si and Al atoms of the source material. Despite the advantages of using high concentrations of alkali (NaOH, between 8–10 M) to obtain a high compression strength product of 104.5 MPa and 71 MPa for the paste and mortar, as well as a lower change in length due to temperature and water evaporation that have the lowest shrinkage percentage [37], it has been proven that the use of ashes from forest biomass (wood) can decrease the requirements of alkaline activators by up to 20% without the loss of properties [38–41]. However, when the substitution level of these ashes is higher than 10% by mass, the mechanical properties of the geopolymer are affected [42–45], proportionally reducing the compressive and flexural strength of the mortars, for all curing times [35]. Likewise, it is highlighted that different conditions can be used during the process of obtaining geopolymers, such as the type of curing, humidity control, temperature, concentration and proportions of alkaline activators, type and quantity of raw material or proportions of starting materials (in case of mixtures), which will influence the properties of the final product. Among the findings, it can be mentioned that the increase in SiO_2/Al_2O_3 ratios positively influences the mechanical compressive strength of geopolymers [15], and it was found that the inclusion of 5–15% wood ash in the process can generate greater strength and durability depending on the age (aging time) of 3–7 days as a consequence of the formation of gels and minerals that increase alkalinity [12]. Research has also been conducted on the effects of the solid–liquid ratio and the alkaline activator in the synthesis of pure geopolymers. Alves et al. [46] used as precursor material ground blast furnace slag with a solid–liquid ratio between 1.5 and 2.2, and as activator solutions (a) a sodium hydroxide/sodium silicate/water mixture and (b) a potassium hydroxide/potassium silicate/water mixture, finding that the resulting geopolymer possessed high compressive strength depending on the solid–liquid ratio and the percentage of water added to the mixture, which is further impacted by the composition of the activating solution. They also noticed that the strength increases with aging [46]. Currently, the addition of plastics to the optimized wood ash-based geopolymer is being tested; for example, in the case of polypropylene (PP), it has been reported that the addition of 1% PP fiber generates an increase in compressive, tensile, and flexural strength by 3.7%, 15.6%, and 10%, respectively [47]. Other types of materials are also being developed. Kristály et al. [48] produced a composite of geopolymer foam and glass to obtain a lightweight and environmentally friendly concrete from waste materials (secondary raw material), which is a valuable building material useful for thermal and acoustic insulation of walls that is also heat-, fire-, and acid-resistant [48].

Cement consumption in the world currently amounts to approximately 3 billion tons, which translates into 1.5 billion tons of carbon dioxide emitted into the environment [34]. According to the United Nations, the world population has increased in recent years, from 5300 million inhabitants in 1990 to 7300 million inhabitants in 2015; with a projected increase by the year 2050 of 24.74%, the requirements for cement, concrete, and other types of construction materials will increase significantly [49]. In this sense, the development of new and better alternative materials for the efficient substitution of cement for other materials at a global level will reduce production costs while reducing emissions, contributing to goals 11 and 13 of the 2030 Agenda for Sustainable Development, "Make cities and human settlements inclusive, safe, resilient and sustainable" and "Climate action", respectively. This literature review focuses on the approach to the processes for obtaining geopolymers from the use of wood ash, as well as the physical and chemical effects that take place under different production conditions. As per the authors' best knowledge, no significant review study exists on the physical, chemical, strength, durability, and microstructural analysis of concrete, which points to the originality of present work.

2. Environmental Impact of WWA

2.1. Air Pollution

Energy extracted from burning the wood results in the formation of WWA. WWA is very fine, which results in the ease of pollution causing respiratory problems for human beings and animals around the site of WWA production [50]. Loose ash has a high possibility for harmful influence on ground vegetation [51], predominantly to the cover and certain kinds of moss groups [52].

2.2. Land Pollution

WWA is problematic if spread regularly and requires slow delivery rates from spreaders [51]. Because of the huge variety of WWA quality, reliance on the sort of chemical structure of WWA is needed before demonstrating the direction of management as agricultural or forest-related systems [53]. WWA recycling to agricultural or forests appears a decent environmental solution, but there are a lot of possible difficulties related to its use in systems, which are more multifaceted [51].

2.3. pH Increase

The topsoil of the system is affected by pH differences and its blocks the crop or tree to obtain enough amount of nutrition from the soil. The delivery rate of calcium to soils is reliant on the primary shape of the ash, with loose ash such as WWA possibly instigating a temporary quick increase in pH in the soil [54]. For the first 7 years, the soil under 100 mm depth had a very minor change in pH value after WWA application, but after 16 years, an increase in pH value was observed [51]. The land dumping of WWA results in the slow transfer of pH from the topsoil to bottom soil, which can be observed over time. There is an increase in the pH of runoff water over the same period where WWA is applied, as observed by Fransman et al. [55].

2.4. Higher Production Rate

Approximately 2.5 kilo-tons of WWA are annually discarded in lands, as of 2006 [51], but it may increase at a high rate and a decrease in forest land has been observed. In several countries across the globe, 90% of WWA is sent to landfill and the balance part goes as land applied purpose, co-composed with sewage sludge [51]. Apart from the several environmental effects discussed above, Pitman et al. (2006) [51] studied specifically soil properties and soil vegetation.

2.5. PH Affects the Nutrition (Phosphorus, Nitrogen, and Potassium) Addition of Soil

When WWA is in contact with water, the pH solution becomes higher as the hydroxides and oxides in the WWA are dissolved and hydroxide ions are developed. WWA has a liming impact when introduced into soils and could be utilized to neutralize acidity. Three tons of WWA have a liming effect equal to one ton of quicklime. The solubility of different nutrition elements in the WWA varies considerably. Generally, the solubility of the nutrients elements are in the order of potassium > magnesium > calcium oxide > phosphorus [56].

2.6. Heavy Metal Contamination of Soil

pH, organic material content, and hydrous oxide play the main roles in the adsorption of heavy metals from the soil [57]. When WWA dissolves in an acid environment such as soils in forests, the alkalinity of WWA is consumed and the metals are exposed to a pH far lower than that of the ash, causing higher solubility [56]. WWA could also have high concentrations of heavy metals due to the fuel, which is contaminated. Wood from and wood preservers and demolition in waste wood generally comprise higher proportions of heavy metals. As a result of the relatively low volatilization temperatures for many of the heavy metals, they become enhanced by WWA. In the combustion of untreated wood [58], the concentrations of lead and antimony are one order of magnitude higher, while the concentrations of arsenic, cadmium, chromium, copper, nickel, and zinc are approximately twice as high.

2.7. Soil Water Leachate

Williams et al. (1996) [59] observed that there are amplified concentrations of both calcium and potassium in groundwaters and soils, with some movement of aluminum and magnesium in it. In a long-term experiment, soil leachate at a 20 cm depth taken from mineral soils displayed elevated levels of calcium, magnesium, and potassium, but no significant impact on nitrate concentration, pH or Cd, Cu, Cr, and Pb levels [51]. The storage of moistened WWA in the air led to an adverse effect: it increased potassium leaching. The leaching of phosphorous, magnesium, and metal species from the ash matrix is generally low with a high pH prevailing in the water phase during short-term leaching [60].

3. Source and Production of WWA

As per Grau et al. [61], the WWA obtained from the total amount of available quantity is from 0.4 to 2.1%.

4. Physical and Chemical Compositions

4.1. Physical Properties

As per the report from Etiegni and Campbell et al. [62], 80% of WWA consists of particles size less than 1.0 mm and the balance is unburned wood particles with different sizes. Specifically, WWA consists of 25.4% of fine particles, which are less than 75 μm. Wood ash shows a relatively high specific surface area that provides good absorption. The specific gravity of WWA is 2.41, pH 12.57, average particle diameter d_{50} (mm) is 0.223, bulk density (kg/m^3) is 663–997, and specific surface area (m^2/kg) is 4200–100,600 (Grau et al., 2015) [61]. An increase in the particle size increases the concentrations of aluminum, arsenic, barium, and copper, and decreases the concentrations of boron, cadmium, manganese, lead, zinc, potassium, magnesium, and calcium [63].

4.2. Chemical Properties

Wood combustion produces highly alkaline ash (pH varies from 9 to 12) [62]. Ash yield is decreased approximately to 45% with an increase in burning temperature from 538 °C to 1093 °C. The metal contents in ash increase with the increase in the burning temperature. With the increase in the burning temperature, elements such as calcium, iron, magnesium, manganese, and phosphorus increase, and elements such as zinc, potassium, and sodium decrease [62].

The alkalinity of WWA depends on the carbonate, bicarbonate, and hydroxide content in it. WWA composition also varies during storage and under different environmental conditions as carbon dioxide and moisture react with WWA to form carbonates, bicarbonates, and hydroxides [62]. An increase in potassium, sodium, and manganese concentration increases linearly with ash concentration. The leaching of these elements increased to result in a decrease in pH. Table 1 shows the chemical elements of different types of ashes.

Table 1. Chemical elements of the ashes (ppm) [53].

Type of Ash	P	K	Ca	S	Cu	Fe	Mn	Zn	Ni	Cr	Pb	As
Birch wood	20,853	71,290	132,583	5631	97.10	6518	17,585	212.67	34.91	39.07	40.48	1.01
Pine wood	18,618	116,436	201,109	7142	196	3665	10,693	193.13	45.84	62.04	28.89	1.59
Oak wood	15,071	57,331	156,738	5107	190.67	9256	10,114	169.33	125.67	89.87	54.49	1.91
Horen beam wood	16,548	69,905	249,050	3956	140.67	8598	18,587	155.0	158.67	10.65	40.20	1.13
Ash wood	17,967	70,442	279,785	3077	121.00	5758	10,545	183.0	24.84	30.66	15.31	0.78
Wood residue chips—forest	17,680	69,104	203,935	1546	188.0	3403	6920	171.0	110.33	95.64	50.67	1.44
Wood residue chips—municipal	32,039	108,081	245,075	8464	181.0	4678	2815	320.33	176.33	25	12.69	0.13
Poplar wood	6419	64,985	173,872	5015	96.92	4612	549.67	81.41	26.19	20.57	9.65	0.18
Willow	3342	37,339	135,981	4732	123.5	2662	910	394.0	32.0	45.97	8.93	0.34
Acacia wood	2679	38,799	227,225	1826	158.0	6156	794.3	244.0	59.72	36.31	15.83	0.49
Average (%)	15,121.6	70,371.2	200,535.3	4649.6	149.286	5530.6	7951.297	212.387	79.45	45.578	27.714	0.9

Zajac et al. [53] divided ash components into three categories based on their concentration: 1. macro-elements: phosphorus, potassium, calcium, and sulfur; 2. micro-elements: manganese, iron, copper, and zinc; 3. toxic elements: chromium, nickel, arsenic, and lead. The quantity and quality of WWA content depend upon the organic, inorganic, and impurity elements present in it. The chemical and physical characteristics of WWA depend upon on the sampling point, the sort of biomass, plant kind, growth process, growth circumstances, plant age, fertilization, the applied dosage of plant protection products, harvesting conditions, and process of burning (preparation of fuel, burning method used, and circumstances) [51,53].

Szakova et al. [64] determined the chemical composition of WWA (wood chips and wood waste) using the XRF technique. Different elements analyzed (in ppm) were P: 5300–10,800; S: 1200–11,100; K: 38,000–58,000; Ca: 78,000–159,000; Cr: 118; Mn: 6200–10,700; Fe: 29,300–34,800; N: 28.9; Cu: 153; Zn: 300–1100; As: 9.8; and Pb: 313. Tarun et al. (2003) revealed the subsequent elements in wood ash: C (5% to 30%), Ca (5% to 30%), carbon (7% to 33%), K (3% to 4%), Mg (1% to 2%), P (0.3% to 1.4%), and Na (0.2% to 0.5%). Elemental arrangement varies for WWA because ashes derived from branches and roots are rich in many elements than those derived from stem wood [51].

The following compound composition limits were also reported: titanium dioxide (0% to 1.5%), sulfur trioxide (0.1% to 15%), silica (4% to 60%), aluminum oxide (5% to 20%), ferric oxide (10% to 90%), magnesium oxide (0.7% to 5%), potassium oxide (0.4% to 14%), calcium oxide (2% to 37%), loss of ignition (0.1% to 33%), moisture content (0.1% to 22%), and available alkalis (0.4% to 20%). Table 2 shows the chemical compounds in waste wood ash that were obtained in past research.

WWA is usually very low in nitrogen because it evaporates during incineration. Trace elements such as boron (B), molybdenum (Mo), copper (Cu), and zinc (Zn) have been observed in WWA, which are called micronutrients [65].

Table 2. Chemical compounds of WWA from past studies.

SiO_2	Al_2O_3	Fe_2O_3	CaO	MgO	K_2O	NaO	L.O.I	Ref.
31.8	28	2.34	10.53	9.32	10.38	6.5	1.13	[66]
32.8	27.0	2.2	11.7	9.1	10.5	6.7	0.7	[61]
30.4	26.5	1.9	12.8	9.4	11.4	5.9	1.7	[67]
32.4	27.4	2.1	10.3	9.4	10.85	6.4	1.15	[68]
31.8	28.2	2.4	10.6	9.2	10.80	6.0	1.00	[69]
33.6	27.8	2.6	11.2	8.3	10.2	5.7	0.60	[70]
32.7	26.4	2.2	10.25	9.45	10.45	7.21	1.34	[71]
31.5	26.3	2.6	10.4	9.3	10.72	8.2	0.98	[72]

5. Influence of Waste Wood Ash on Hardened Concrete

5.1. Strength Properties

Waste wood ash is an easily available agricultural discarded material that enhances the workability, quality of microstructure, and improves the strength characteristics of concrete samples. The key characteristics of concrete can be enhanced, and most importantly, the time for hydration is reduced due to the pozzolanic effect [70]. The distinct proportioning of the mix can be achieved by substituting cement with waste wood ash and slag and other supplementary cementitious materials (SCMs). The different characteristics of concrete from the previous literature that were assessed through different forms of strength and durability testing are presented in Table 3.

The properties of various mixes have been studied with different water to binder ratios. The optimal dose was 20% WWA with rice husk ash, which improved the compression strength significantly and also showed enhanced durability [73]. Because of the low amount of silica in waste wood ash, a low water to cement ratio of 0.4 was selected. Ramos et al. noted that waste wood ash seemed a potential fractional substitute pozzolanic material for cement because it improves strength and durability characteristics and also assists in making concrete sustainable [68].

Table 3. Influence of waste wood ash on the characteristics of concrete.

Level of Substitution	Observed Properties	Results	Discussion	Ref.
10–35% (25% optimum dosage)	Specific gravity Bulk density Initial setting time Final setting time Compression strength Slump value Water demand	2.21 755 kg/m^3 221 min 547 min 7.5–23.2 MPa at 56 days 40–55 mm 134–140 mL	When the water to cement ratio was kept at 0.55, the maximum slump was 55 mm for 25% and 30% WWA, and maximum strength was 23.2 MPa for 25% WWA at 56 days and then reduced	[66]
0–25% (20% optimum dosage)	Compression strength Water absorption Weight loss	40–48 MPa at 90 days (before acid test), 29–41 MPa (after acid test) 2.2%–2.64% 6–10.5%	When the water to cement ratio was kept at 0.45 and utilizing 10% sulfuric acid, the highest loss in strength was 29 MPa 90 days for; 20% WWA loss in weight was minimum with only 6%	[74]
5–25% (15–20% optimum dosage)	Compression strength Flexural strength Slump value Water absorption	17 MPa to 29 MPa at 28 days 4 MPa to 6.25 MPa at 7 and 28 days 0–15 mm 0.20% to 1.70%	At a water to binder ratio of 0.50, samples with 15 and 20% WWA had maximum compressive strength with 17 and 29 MPa at 28 days, and then strength began to decrease	[67]
10–30% (20% optimum dosage)	Split tensile strength Compression strength	4.25 MPa to 6.70 MPa at 28 and 90 days 42 to 49 at 28 days and 47 to 55 MPa at 90 days	At a water to binder ratio of 0.48, at 20% WWA, the strength was slightly less than the reference sample due to WWA acting as a filler, not a binder, but microstructure was enhanced	[72]
0–20% (20% optimum dosage)	Compression strength Flexural strength Alkali silica reaction Carbonation	39 to 54 MPa at 28 to 90 days 7 to 9 MPa at 56 days Expansion of ASR at 28 days was 0.17% at 20% WWA The average depth was 3.75 mm at 20% WWA	The highest compression strength was obtained at 54 MPa at 90 days with water to cement ratio of 0.52 with 20% WWA; the same w/c led to a sample with low ASR levels and carbonation depth	[68]
0–40% (25% optimum dosage)	Compression strength Slump value	12–15 MPa at 21 days without admixture, with admixture the strength 28 MPa, With w/c 0.55, the slump was 40 mm	Sample with w/c of 0.55 had the highest slump value 40 mm at 15% WWA, utilizing admixture enhanced compression strength significantly with 45% more strength at 25% WWA	[75]

Table 3. Cont.

Level of Substitution	Observed Properties	Results	Discussion	Ref.
10–25% with 5% silica fume (20% + 5% SF optimum dosage)	SEM analysis Compression strength	The creation of pores in mortars was considerably impacted because of the substitution of the binder with WWA and SF 20–42 MPa at 28 days	With constant w/c of 0.44, mix with 20% WWA and 5% SF had the highest mechanical strength, and further adding of WWA led to the development of pores in the matrix	[76]
10–35% (20% optimum dosage)	Compression strength Compaction factor	29.5–54 MPa at 90 days 0.741	At later ages, the concrete strength improved considerably, because the water absorption from the blend by WWA reduced the workability slowly	[77]
0–30% (25% optimum dosage)	Pressure resistance Water absorption	2.9–3.8 9–11.5	With a 25% dose of WWA, the samples had the least absorption and maximum pressure resistance in comparison to the reference sample.	[78]
0–25% (20% optimum dosage)	Slump value Compression strength Sieve analysis	45 mm with 20% WWA 10.57 to 35.47 MPa at 20% WWA for 28 days Size ranged from 0.059 to 32.5 mm	With a w/c of 0.55, the optimal mechanical strength was 35.4 MPa at 90 days with 20% WWA as a partial substitute for cement, and workability was in an acceptable range	[79]
5–20% (15% optimum dosage)	Chemical and physical analysis Compression strength Flexural strength Split tensile strength X-ray diffraction spectra	Comprised 70.5% silica, alumina, and ferric that was similar to class F type pozzolanic material and mean size, bulk density, and specific gravity of WWA were 170 microns, 720 kg/m^3, and 2.21, respectively For w/b of 0.40, the strength was 36.3 MPa at 28 days For w/b of 0.40, the strength was 6.52 MPa at 28 days For w/b of 0.40, the strength was 2.37 MPa at 28 days WWA comprised silica both in crystal and formless shapes with the highest peak at 29 degrees against 2-theta	At a w/b ratio of 0.40, with 15% of waste wood ash as a partial substitute of cement, the highest compression at 28 days was 36.3 MPa, which was more than the control sample	[80]

5.2. Effect of Waste Wood Ash (WWA) on the Durability Characteristics of Concrete

5.2.1. Acid Resistance Test

Dashibil and Udoeyo (2002) [81] examined the capability of waste wood ash concrete to withstand acid tests. Two groups of samples had a similar content of aggregates, water, and binder. The only difference was that the first group had only cement as the primary binder and the other group had 15% waste wood ash as a fractional binder substitute. The concrete samples were dipped in strong acid (sulfuric acid) for 54 days. It was revealed that concrete samples that had waste wood ash had a less decrease in their mass loss in comparison to concrete with no waste wood ash at all.

Ejeh and Elinwa (2004) [82] investigated the impact of adding WWA in samples for acid tests against the possibility of corrosion. Two sorts of acids were tried; one was sulfuric acid and the other was nitric acid at 20% concentration. One group of samples had 10% WWA utilized as a fractional binder substitute and the other group of samples was the same as the previous mixes but without WWA. Both groups of samples were dipped in both sorts of acids for 35 days. It was noticed that, in samples with 10% WWA, their resistance to nitric acid was much more enhanced because the loss in mass was lower in comparison to the samples with no WWA, as shown in Figure 1. However, samples with 10% WWA had less resistance to sulfuric acid in comparison to samples with no WWA. This is because of a higher loss in the weight of 10% WWA concrete in comparison to the control sample when dipped in 20% H_2SO_4, as shown in Figure 2.

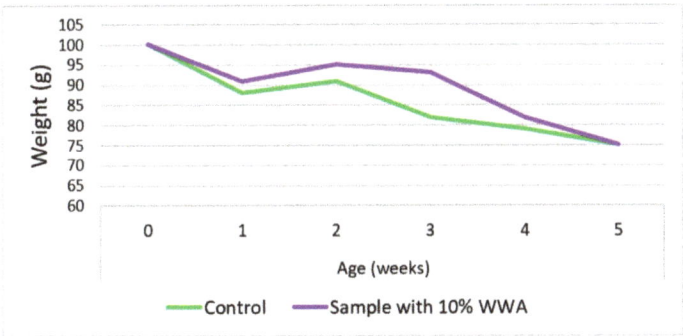

Figure 1. Change in concrete mass with a period of dipping samples in nitric acid (data from reference [82]).

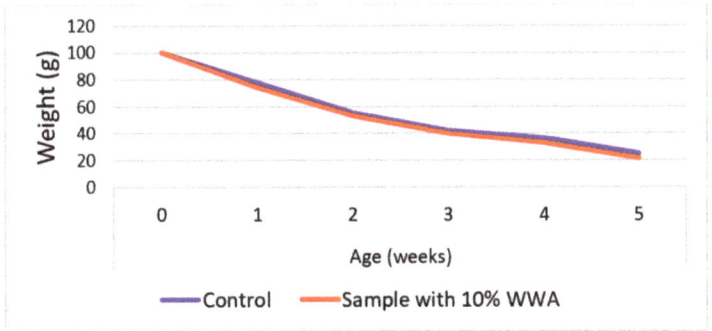

Figure 2. Change in concrete mass with a period of dipping samples in sulfuric acid (data from reference [82]).

5.2.2. Water Absorption

Ejeh and Elinwa (2004) [82] inspected the influence of the inclusion of WWA as a fractional binder substitute in mortar blends on the properties of water absorption. Two groups of mortars were developed with similar mixing content except for cement; one blend had only cement as a binder and the other one had 15% WWA as a fractional substitute of cement in the blend. It was revealed that the addition of WWA as a binder substitute at 15% of cement weight assisted in decreasing the water absorption of the developed blends. The mean water absorption of the blends with 15% WWA and no WWA proportions were noted to be 0.75% and 1.30%, correspondingly, but both of the blends were still lower than 10% of the highest water absorption criteria.

Udoeyo et al. (2006) [67] studied the characteristics of water absorption with WWA as a fractional binder substitute material. Sample blends with proportions of WWA ranging from 5 to 30% at an interval of 5% were developed to evaluate the water absorption properties. The water absorption of the sample with WWA as a fractional binder substitute was noted to rise steadily from 0.15 to 1.10% with a rise in the proportion of binder substitution from 5% to 30%, as displayed in Figure 3. At proportions of binder substitution by WWA up to 30%, the developed sample had still reasonable values of water absorption under 10%, which is a tolerable criterion for all of the materials that are used for construction.

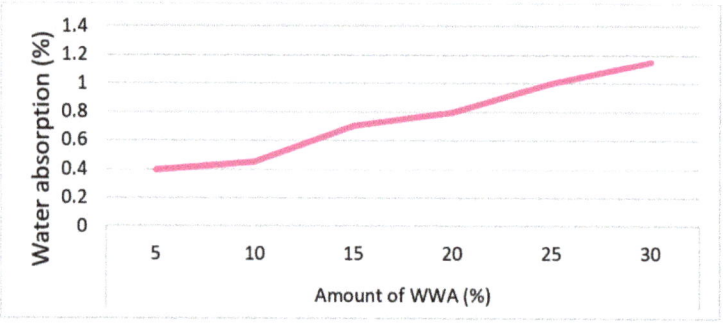

Figure 3. Co-relation of water with WWA in concrete (data from reference [82]).

5.2.3. Permeability of Chloride Test

Wang et al. (2008) [83] examined the resistance against the permeability of chloride of air entrained in the sample with a fractional substitution of the binder with wood/coal fly ash (WCFA) and wood fly ash (WFA). Proportions of binder substitution by different sorts of FA were utilized as a fractional substitution of binder, such as Class F fly ash, class C fly as, ash from the combusted wood, and coal fly ash. All the concrete specimens were placed in water for 56 days before placing in the chloride permeability test, and the chloride permeability test was conducted per ASTM C 1202 [84]. From the test outcome, it was revealed that the inclusion of WWA at a 25% substitution of cement in the sample had no adverse effect from the chloride on the concrete. The usage of class F/coal mixed and wood ash in fractional replacement of cement had considerable help in dropping the permeability of chloride property of the sample. A minor rise in the permeability of chloride in the sample mix with 25% WWA as cement substitute was noted in comparison to the control sample, perhaps ascribed to the coarse size of WWA particles (30 to 130 microns).

Horsakulthai et al. (2010) [73] investigated the impact of adding very fine ash from the combustion of rice husk, wood, and sugarcane waste from bagasse as a fractional binder substitute on the permeability of chloride of a developed blend of concrete. To assess the concrete permeability of chloride, an accelerated salt ponding technique was utilized for two distinct grades of concrete (grades 20 and 35) developed by the inclusion of ash from rice husk, wood, and bagasse at cement substitution proportions of 0, 10, 20, and 40% of cement weight. The test outcome revealed that the inclusion of fine size ash from rice husk, wood, and bagasse as a fractional replacement of binder in the sample led to the improvement in permeability against chloride and also reduced the coefficient of chloride diffusion. The existence of fine size ash from rice husk, wood, and bagasse in a blend at binder substitutions of 10, 20, and 40% caused a decrease in the coefficient of chloride diffusion by 35–45%, 65–75%, and 80% correspondingly, as compared to the reference mix with only Portland cement as a binder. The inclination of a steady decrease in the coefficient of chloride diffusion for the two distinct grades of concrete was evaluated. The raising dose of binder substitution by ash from rice husk, wood, and bagasse is displayed in Figure 4. The term "PC" in Figure 4 denotes plain concrete; BRWA denotes co-combination of bagasse, rice and waste wood ash; and the numbers after them denotes their percentage added in the mix.

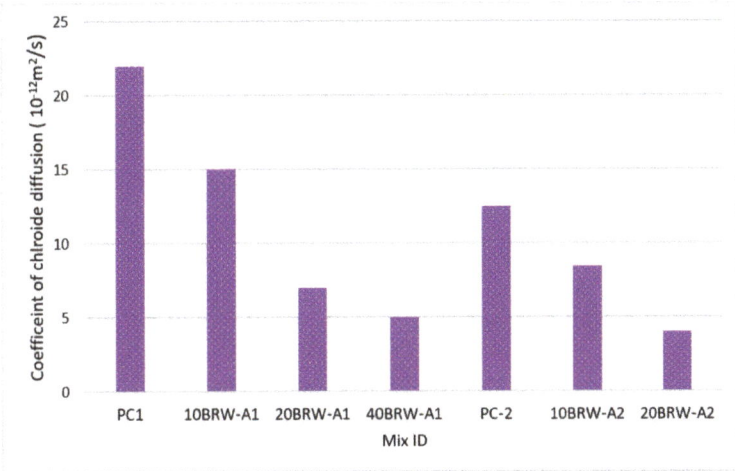

Figure 4. Coefficient of the chloride diffusion of samples at 28 days (data from reference [82]).

5.2.4. Alkali Silica Reaction (ASR)

Baxter and Wang (2007) [85] studied the conduct of expansion in mortar blends due to ASR comprising an opal aggregate, which is a very reactive, highly alkaline cement, and three distinct sorts of fly ash (FA). The different sorts of FA were acquired from heating the class C coal. Four groups of mortar blends with the same proportions of ingredients were arranged. The first group of mortar had only Portland cement as a binder and the remaining three groups of mortar blend had three distinct sorts of FA utilized at a uniform dose of binder substitution with 35% of cement by weight. The test outcome revealed that coal fly ash had a higher quantity of alkaline matter as compared to class C FA. The utilization of coal FA in the blend of mortar was revealed to be capable of decreasing the expansion of the alkali-silica reaction at 180 days under 0.1% (highest expansion stated by ASTM C33) from 0.27%. This happened with the reference blend of mortar having Portland cement as the only binder. Between the different sorts of fly ash that were tested, coal fly ash was revealed to have optimal behavior in the mitigating expansion of the alkali-silica reaction.

5.2.5. Shrinkage Test of WWA Concrete

Naik et al. (2003) [86] examined the dry shrinkage characteristics of sample mixes developed by the inclusion of WWA as a fractional binder substitution material. For blends developed in the research, WWA was utilized at cement replacement levels of 0, 5, 8, and 12%. A variation in length in the formed concrete samples was observed up to 240 days. It was revealed that the value of shrinkage for the reference concrete samples was 0.009% for 7 days and 0.05% for 240 days. The values of shrinkage for sample mixes with 5, 8, 12%

were observed to be 0.01–0.02%, 0.0139–0.014%, and 0.049–0.044%. From the test outcomes of dry shrinkage, it was noted that the incorporation of WWA considerably helped in the decrease in the extent of concrete upon drying. This is an essential property that could reduce the development of micro-cracks within the sample upon drying.

6. Microstructural Study

Garcia et al. [71] studied the microstructure properties with the help of a scan electron microscope (SEM) of ground waste wooden ash, obtained from forest regions surrounding a power plant in Portugal. From the SEM images, it was observed that ground waste wooden ash has two governing properties of particles and fibers in layers. The SEM micrographs of the ground waste wooden ash at higher magnification and electron dispersive X-ray (EDX) spectra are shown in Figure 5.

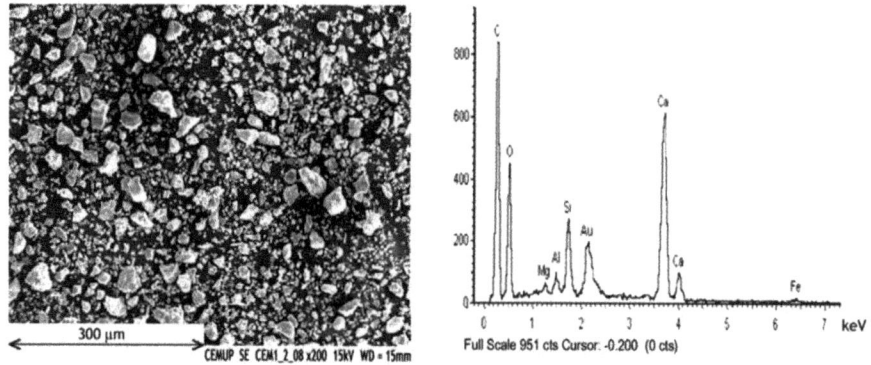

Figure 5. SEM micrograph and EDX analysis of waste wood ash (used with permission from Elsevier [71]).

Awolusi et al. [87] investigated the microstructure characteristics of OPC mortar with waste wood ash from sawdust. The mortars were developed with WWA ranging from 0 to 10% with a water to binder ratio of 0.6. SEM micrograph displayed maximum inter-spatial distance between the particles of WWA in comparison to the binder, which was dense with each other [87]. The SEM outcomes of a composite made from corn cob–polypropylene–WWA displayed that the pores present in matrix become smaller as the proportion of WWA was raised. This can be attributed to the pores between the corn cob and polypropylene being filled by the waste wooden ash. Due to the increased dose of waste wooden ash in the mix, the stress concentration stretched, and the shape of WWA was revealed to be considerably small; thus, the distances became very less. The bridging behavior of waste wooden ash could lead to a maximum wrap among corn cobs [88].

It can be observed from the SEM image shown in Figure 6 that waste wood ash with silica fume enhanced the matrix when the ash was dispersed uniformly. Figure 6 shows the study of pores enclosed in the reference sample and sample with different proportions of waste wood ash and silica fume. The doses of waste wood ash and silica fume added to the mix considerably impacted the pores shaped in concrete mortars. The proof of this effect is displayed in the formation of larger pores in the reference specimen. Concrete mixes utilized different proportions of silica fume and waste wood ash by substituting the binder with 15% reduced estimate of pores. This positive test outcome was due to the rich silica in the silica fumes [76].

Chowdhury et al. [80] studied the X-ray diffraction of waste wood ash concrete. Figure 7 displays the X-ray diffraction spectra of waste wood ash concrete. The hump specifies that the specimen was formless, with the peaks of silica demonstrating a crystalline behavior. Thus, waste wood ash comprises silica in crystalline and formless shapes.

Crystals of silica present high peaks at 29 degrees at 2-theta. Formless silica concentrates in the mixture as a suitable binder substitution material due to its pozzolanic behavior [80]. Similar observations were also noted by Elahi et al. [89] from the X-ray diffraction spectra of waste wood ash, which displayed the existence of formless silica, although only in low proportions. A study of the waste wood ash concrete microstructure showed that the inclusion of waste wood ash impacted the pore estimations, and a significant reduction in porosity occurred [90]. The succeeding samples showed a microstructure that was very dense with a lower permeability [90].

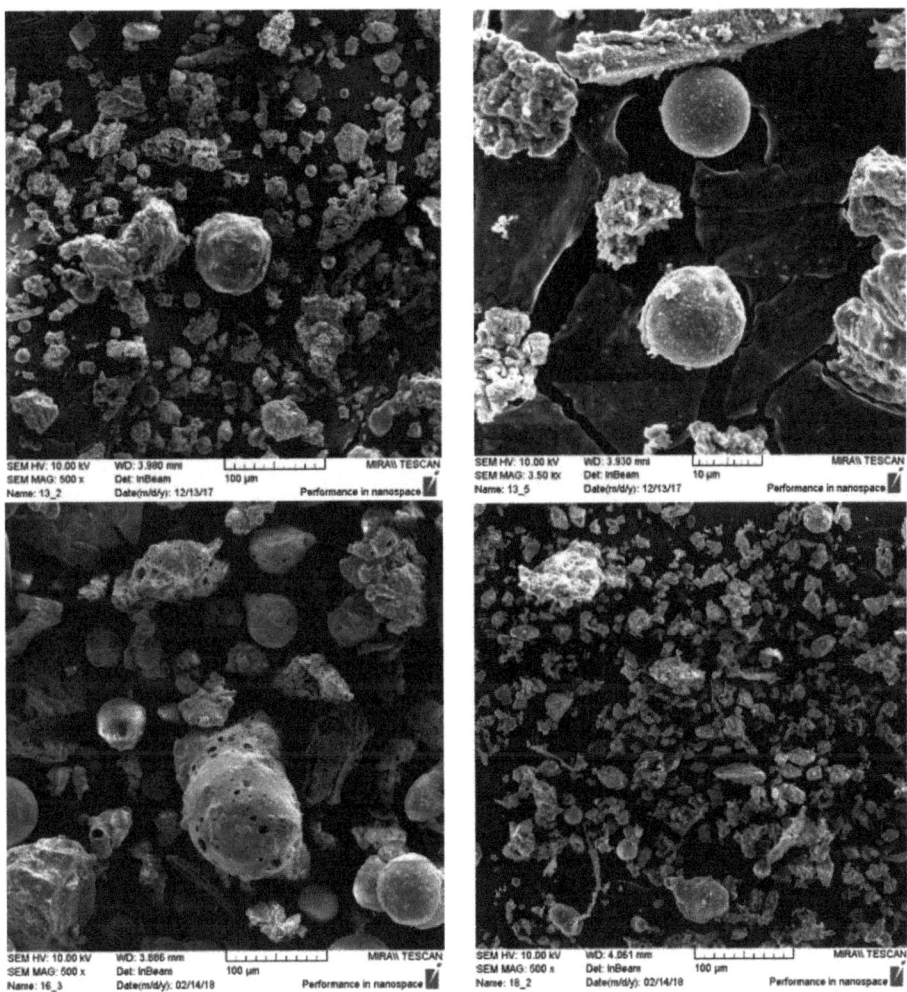

Figure 6. Pores in the reference concrete and reference sample with different proportions of WWA and SF (used from an open source journal of MDPI [91]).

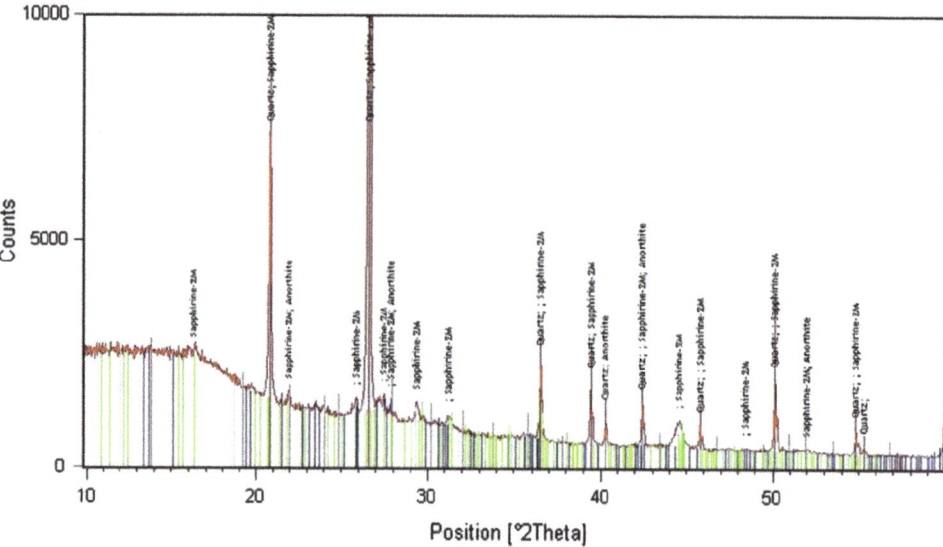

Figure 7. XRD analysis of waste wood ash (used with the permission from Elsevier [71]).

WWA consists of particles sizes that vary between 10 μm and 200 μm, as is observed in Figure 8. The shape and size of particles vary for WWA with small particles adhering to the surface, as is observed. WWA consists of particles with an irregular shape due to the inorganic particles present in it (Etiegni and Campbell, 1991) [62].

 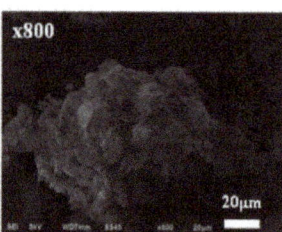 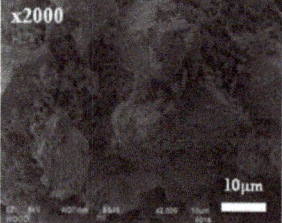

Figure 8. SEM image of WWA (Grau et al., 2015) (Used from an open source journal of MDPI [61]).

7. Effect of WWA Concrete on the Environment

WWA can have a positive influence in developing cement from an environmental standpoint [92]. The mineralogical and physical assembly of burning ash, such as WWA ash, and the availability of metals depend on the treatment of temperature and feed material [93]. The last method for using waste wood ash must be appropriately sustained due to the fineness of particles and lenience of air pollution, which can lead to breathing issues for the public near the manufacturing places [50]. In addition to this impact, the impact of waste wood ash on acidic material can result in releasing heavy metals into the environment [67]. Thus, issues have to be raised and more studies must be performed on the environmental impacts of waste wood ash concrete. Udoeyo et al. [67] utilized waste wood ash as a mineral material to research its effect on the environment and revealed that, if the waste wood ash is discarded in lands, then acidic rain releases heavy metals to the surrounding area. Thus, the usage of waste wood ash decreases pollution by reducing the necessity of discarding [73].

Gorpade et al. [72] assessed the impact of the inclusion of waste wooden ash from 0% to 30% in concrete. It was revealed that up to 10% of cement can be efficiently substituted with waste wood ash. If waste wood ash is utilized as a mineral material in concrete, the amount of discarded WWA and its adverse effects on the environment can be diminished significantly. Adding waste wood ash as a substitute for cement decreases the usage of cement in concrete, which reduces the manufacturing cost of binders and its related outflow of harmful gases. Waste wood ash is a very fine material compared to cement and it can fill all the voids within the microstructure of concrete, which makes it hard for the outside chlorides or salts to enter the concrete. Making concrete buildings near the sea areas with waste wood ash concrete is recommended to avert structure catastrophes from heavy salt-oriented climates [94].

Waste wood ash seems to be an auspicious pozzolanic material for the partial replacement of binder, with no reduction in concrete strength, with enhanced durability of sample, and contributing significantly to the sustainability of the construction industry [68]. Waste wood ash in samples help to makes an eco-efficient substitute cementitious material, which is efficient and cost-friendly [80]. Studies [70,95] have been performed to evaluate the creation of sustainable construction material by providing waste wood ash as a binder replacement material. The natural influence of the usage of waste wood ash in cement mortar, carbon impression, and the degree of consumed energy was taken as a major parameters that can be used as a quantitative limitation to signify the likely recompences of waste wood ash applications in cementitious materials. The study was carried out according to the exclusive method shown by Pavlikova et al. [69]. Waste wood ash can be an actual pozzolanic additive used as a partial substitute of cement to assist in the environmentally friendly concrete construction of buildings.

8. Conclusions

The quality and quantity of waste wood ash are dependent on different features, specifically, the temperature of the burning of waste wood and the type of burning technique utilized for waste wood. Thus, the appropriate classification of WWA is obligatory before its usage as an ingredient material in the development of geopolymer concrete mixes.

- The distribution of wood ash particles is usually grainier as compared to cement. However, the specific surface of WWA is moderately smoother than that of Portland cement because of the higher irregularity in wood particles and their permeable behavior.
- The chemical arrangement of WWA differs considerably within types of trees from which the biomass of wood is obtained, but it is usually rich in CaO and SiO_2 elements.
- Binders blended with WWA as a fractional substitute have higher initial and final times and high standard consistency. Geopolymer mixes having WWA are inclined to have a low heat of hydration.
- A considerable amount of ettringite crystals is shaped within a paste of binder upon the hydration of OPC–WWA geopolymer samples, specifically at high doses of binder replacement with WWA.
- Geopolymer mixes of mortar and concrete comprising WWA as a fractional substitution of the binder have more water requirements to obtain a desirable level of slump value in comparison to similar geopolymer mixtures with no WWA.
- The addition of WWA as fractional binder substitution in mixes of mortar and concrete at a high dose of binder substitution could lead to a steady decrease in the bulk density of hard mixes of geopolymer mortar and concrete.
- Usually, the inclusion of WWA as fractional binder substitution in the preparation of geopolymer concrete blend decreases the compression, flexural, and split tensile strength of geopolymer concrete. However, there are hopeful outcomes as the addition of WWA at a low dosage level of binder substitution truly assisted in the improvement of the compression strength of the developed mixes of geopolymer concrete. WWA as a fractional substitute for binder at a substitution level of 10% by binder weight can make geopolymer mortar or concrete, which can be produced and utilized in building applications with suitable strength and durability characteristics.

- Metakaolin can be utilized as an activator for making geopolymer concrete or mortar with WWA as a fractional substitute of binder to improve the mechanical strength of geopolymer concrete or mortar.
- Geopolymer concrete blends comprising WWA as fractional binder substitute display more resistance against rusting when exposed to strong acids in comparison to mixes with no WWA.
- Geopolymer concrete blends having more quantity of WWA as a partial substitute of binder can have a high degree of water absorption.
- The utilization of WWA as fractional replacement of binder in geopolymer concrete blends at substitution levels of up to 25% by binder weight does not have detrimental impacts on the resistance of the geopolymer concrete against chloride ion diffusion. Furthermore, the utilization of 80% fly ash and 20% WWA in geopolymer concrete considerably improves the sample's capability to resist chloride ions diffusion.
- The advantage of present study is that the addition of very fine size WWA, made from burning of rice husk, wood, bagasse, assists considerably in enhancing the durability characteristics of the geopolymer sample in terms of ASR and resistance against chloride ions. The existence of WWA in geopolymer concrete had a considerable role, as it reduced the extent of the geopolymer sample's drying shrinkage considerably.
- The disadvantage of present study is that, to make WWA, it needs a considerable higher degree of fire to burn the waste wood, which will need a lot of energy and resources, and finding naturally burnt WWA is highly difficult.

9. Recommendations

Waste wood ash has the possibility of being a substitute construction material for sustainability purposes, as a fractional substitute of binder and aggregates. The usage of waste wood ash in large volumes is conceivable. Some research has been conducted on this and some hopeful outcomes have been observed, as WWA can be utilized as an eco-efficient material with little to no compromise on the properties of geopolymer concrete samples. However, for now, WWA has been utilized in a limited amount in the development of samples. This extensive review study of WWA on geopolymer concrete as a fractional substitute of binder showed that the shape, size, source, method of making WWA, and chemical and physical composition of WWA have a significant impact on the strength and durability properties of the sample in which WWA is utilized. Thus, waste wood ash is suitable as a binder replacement, and if it is used in construction as a building material, it will reduce the demand for cement, which will reduce the outflow of greenhouse gases from the production of cement and also preserve the natural reserves of limestone used in the making of cement, thus helping the environment and assisting the construction industry by increasing its sustainability.

Author Contributions: Conceptualization, R.M.-G. and P.J.; methodology, R.M.-G. and O.Z.; software, P.J. and S.M.A.Q.; validation, C.M.G., J.d.P.-G. and A.A.Ș.; formal analysis, F.J.F.-F.; investigation, A.A.Ș.; resources, F.J.F.-F.; data curation, C.M.G.; writing—original draft preparation, R.M.-G. and O.Z.; writing—review and editing, O.Z.; visualization, S.M.A.Q.; supervision, R.M.-G.; project administration, A.A.Ș.; funding acquisition, J.d.P.-G. All authors have read and agreed to the published version of the manuscript.

Funding: This research received no external funding.

Institutional Review Board Statement: Not applicable.

Informed Consent Statement: Not applicable.

Data Availability Statement: Data can be provided upon request from the corresponding author.

Conflicts of Interest: The authors declare no conflict of interest.

References

1. Zaid, O.; Mukhtar, F.M.; García, R.M.; El Sherbiny, M.G.; Mohamed, A.M. Characteristics of high-performance steel fiber reinforced recycled aggregate concrete utilizing mineral filler. *Case Stud. Constr. Mater.* **2022**, *16*, e00939. [CrossRef]
2. Althoey, F.; Zaid, O.; de-Prado-Gil, J.; Palencia, C.; Ali, E.; Hakeem, I.; Martínez-García, R. Impact of sulfate activation of rice husk ash on the performance of high strength steel fiber reinforced recycled aggregate concrete. *J. Build. Eng.* **2022**, *54*, 104610. [CrossRef]
3. Zaid, O.; Zamir Hashmi, S.R.; Aslam, F.; Alabduljabbar, H. Experimental Study on Mechanical Performance of Recycled Fine Aggregate Concrete Reinforced With Discarded Carbon Fibers. *Front. Mater.* **2021**, *8*, 481. [CrossRef]
4. Zaid, O.; Ahmad, J.; Siddique, M.S.; Aslam, F. Effect of Incorporation of Rice Husk Ash Instead of Cement on the Performance of Steel Fibers Reinforced Concrete. *Front. Mater.* **2021**, *8*, 14–28. [CrossRef]
5. Aslam, F.; Zaid, O.; Althoey, F.; Alyami, S.H.; Qaidi, S.M.A.; de Prado Gil, J.; Martínez-García, R. Evaluating the influence of fly ash and waste glass on the characteristics of coconut fibers reinforced concrete. *Struct. Concr.* **2022**. [CrossRef]
6. Zaid, O.; Ahmad, J.; Siddique, M.S.; Aslam, F.; Alabduljabbar, H.; Khedher, K.M. A step towards sustainable glass fiber reinforced concrete utilizing silica fume and waste coconut shell aggregate. *Sci. Rep.* **2021**, *11*, 12822. [CrossRef] [PubMed]
7. Kaza, S.; Lisa, Y.; Perinaz, B.-T.; Van Woerden, F. *What a Waste 2.0: A Global Snapshot of Solid Waste Management to 2050*; Urban Development Series; World Bank: Washington, DC, USA, 2018.
8. Lim, J.; Raman, S.N.; Lai, F.-C.; Mohd Zain, M.F.; Hamid, R. Synthesis of Nano Cementitious Additives from Agricultural Wastes for the Production of Sustainable Concrete. *J. Clean. Prod.* **2017**, *171*, 1150–1160. [CrossRef]
9. Malhotra, V.M. Introduction: Sustainable Development and Concrete Technology. *Concr. Int.* **2002**, *24*, 22.
10. Hh, M.; Al-Sulttani, A.; Abbood, I.; Hanoon, A. Emissions Investigating of Carbon Dioxide Generated by the Iraqi Cement Industry. *IOP Conf. Ser. Mater. Sci. Eng.* **2020**, *928*, 22041. [CrossRef]
11. Tripathi, N.; Hills, C.; Singh, R.; Singh, J.S. Offsetting anthropogenic carbon emissions from biomass waste and mineralised carbon dioxide. *Sci. Rep.* **2020**, *10*, 958. [CrossRef] [PubMed]
12. Abdulkareem, O.A.; Matthews, J.; Abdullah, M.M.A.B. Strength and Porosity Characterizations of Blended Biomass Wood Ash-fly Ash-Based Geopolymer Mortar. *AIP Conf. Proc.* **2018**, *2045*, 20096.
13. Davidovits, J. *Geopolymer Chemistry and Applications*, 5th ed.; Institut Géopolymère, Geopolymer Institute: Saint-Quentin, France, 2008; Volume 171, ISBN 9782954453118.
14. Ekinci, E.; Kazancoglu, Y.; Mangla, S.K. Using system dynamics to assess the environmental management of cement industry in streaming data context. *Sci. Total Environ.* **2020**, *715*, 136948. [CrossRef] [PubMed]
15. De Rossi, A.; Simão, L.; Ribeiro, M.; Hotza, D.; Moreira, R. Study of cure conditions effect on the properties of wood biomass fly ash geopolymers. *J. Mater. Res. Technol.* **2020**, *9*, 7518–7528. [CrossRef]
16. Maglad, A.M.; Zaid, O.; Arbili, M.M.; Ascensão, G.; Şerbănoiu, A.A.; Grădinaru, C.M.; García, R.M.; Qaidi, S.M.A.; Althoey, F.; de Prado-Gil, J. A Study on the Properties of Geopolymer Concrete Modified with Nano Graphene Oxide. *Buildings* **2022**, *12*, 1066. [CrossRef]
17. Zaid, O.; Martínez-García, R.; Abadel, A.A.; Fraile-Fernández, F.J.; Alshaikh, I.M.H.; Palencia-Coto, C. To determine the performance of metakaolin-based fiber-reinforced geopolymer concrete with recycled aggregates. *Arch. Civ. Mech. Eng.* **2022**, *22*, 114. [CrossRef]
18. He, X.; Yuhua, Z.; Qaidi, S.; Isleem, H.F.; Zaid, O.; Althoey, F.; Ahmad, J. Mine tailings-based geopolymers: A comprehensive review. *Ceram. Int.* **2022**, *48*, 24192–24212. [CrossRef]
19. Qaidi, S.M.A.; Mohammed, A.S.; Ahmed, H.U.; Faraj, R.H.; Emad, W.; Tayeh, B.A.; Althoey, F.; Zaid, O.; Sor, N.H. Rubberized geopolymer composites: A comprehensive review. *Ceram. Int.* **2022**, *48*, 24234–24259. [CrossRef]
20. Ismail, I.; Bernal, S.A.; Provis, J.L.; San Nicolas, R.; Hamdan, S.; van Deventer, J.S.J. Modification of phase evolution in alkali-activated blast furnace slag by the incorporation of fly ash. *Cem. Concr. Compos.* **2014**, *45*, 125–135. [CrossRef]
21. Salih, A.P.D.M.; Ali, A.; Farzadnia, N. Characterization of mechanical and microstructural properties of palm oil fuel ash geopolymer cement paste. *Constr. Build. Mater.* **2014**, *65*, 592–603. [CrossRef]
22. Cheah, C.; Ken, P.; Ramli, M. The hybridizations of coal fly ash and wood ash for the fabrication of low alkalinity geopolymer load bearing block cured at ambient temperature. *Constr. Build. Mater.* **2015**, *88*, 41–55. [CrossRef]
23. Li, Z.; Ding, Z.; Zhang, Y. Development of sustainable cementitious materials. In Proceedings of the International Workshop on Sustainable development and Concrete Technology, Beijing, China, 20–21 May 2004.
24. Jamil, M.; Khan, M.N.N.; Karim, M.; Kaish, A.B.M.; Zain, M.F.M. Physical and chemical contributions of Rice Husk Ash on the properties of mortar. *Constr. Build. Mater.* **2016**, *128*, 185–198. [CrossRef]
25. Sales, A.; Bessa, S. Use of Brazilian sugarcane bagasse ash in concrete as sand replacement. *Waste Manag.* **2010**, *30*, 1114–1122. [CrossRef] [PubMed]
26. Payá, J.; Monzo, J.; Borrachero, M.; Díaz-Pinzón, L.; Ordonez, L.M. Sugar-cane bagasse ash (SCBA): Studies on its properties for reusing in concrete production. *J. Chem. Technol. Biotechnol.* **2002**, *77*, 321–325. [CrossRef]
27. Adesanya, D.A. Evaluation of blended cement mortar, concrete and stabilized earth made from ordinary Portland cement and corn cob ash. *Constr. Build. Mater.* **1996**, *10*, 451–456. [CrossRef]

28. Rangasamy, G.; Mani, S.; Senathipathygoundar Kolandavelu, S.K.; Alsoufi, M.S.; Mahmoud Ibrahim, A.M.; Muthusamy, S.; Panchal, H.; Sadasivuni, K.K.; Elsheikh, A.H. An extensive analysis of mechanical, thermal and physical properties of jute fiber composites with different fiber orientations. *Case Stud. Therm. Eng.* **2021**, *28*, 101612. [CrossRef]
29. El-Kassas, A.; Elsheikh, A.H. A new eco-friendly mechanical technique for production of rice straw fibers for medium density fiberboards manufacturing. *Int. J. Environ. Sci. Technol.* **2020**, *18*, 979–988. [CrossRef]
30. Elsheikh, A.H.; Panchal, H.; Shanmugan, S.; Muthuramalingam, T.; El-Kassas, A.M.; Ramesh, B. Recent progresses in wood-plastic composites: Pre-processing treatments, manufacturing techniques, recyclability and eco-friendly assessment. *Clean. Eng. Technol.* **2022**, *8*, 100450. [CrossRef]
31. Elsheikh, A.H.; Abd Elaziz, M.; Ramesh, B.; Egiza, M.; Al-Qaness, M.A.A. Modeling of drilling process of GFRP composite using a hybrid random vector functional link network/parasitism-predation algorithm. *J. Mater. Res. Technol.* **2021**, *14*, 298–311. [CrossRef]
32. Showaib, E.A.; Elsheikh, A.H. Effect of surface preparation on the strength of vibration welded butt joint made from PBT composite. *Polym. Test.* **2020**, *83*, 106319. [CrossRef]
33. Anand Raj, M.K.; Muthusamy, S.; Panchal, H.; Mahmoud Ibrahim, A.M.; Alsoufi, M.S.; Elsheikh, A.H. Investigation of mechanical properties of dual-fiber reinforcement in polymer composite. *J. Mater. Res. Technol.* **2022**, *18*, 3908–3915. [CrossRef]
34. Danraka, M.; Aziz, F.; Jaafar, M.; Mohd Nasir, N.; Abdulrashid, S. Application of Wood Waste Ash in Concrete Making: Revisited. In Proceedings of the Global Civil Engineering Conference (GCEC 2017), Kuala Lumpur, Malaysia, 25–28 July 2017; pp. 69–78.
35. Candamano, S.; De Luca, P.; Frontera, P.; Crea, F. Production of Geopolymeric Mortars Containing Forest Biomass Ash as Partial Replacement of Metakaolin. *Environments* **2017**, *4*, 74. [CrossRef]
36. Ekaputri, J.J.; Triwulan Damayanti, O. The Influence of Alkali Activator Concentration to Mechanical Properties of Geopolymer Concrete with Trass as a Filler. *Mater. Sci. Forum* **2015**, *803*, 125–134.
37. Ekaputri, J. Geopolymer Grout Material. *Mater. Sci. Forum* **2015**, *841*, 40–47. [CrossRef]
38. Smirnova, O.; Menéndez-Pidal, I.; Alekseev, A.; Petrov, D.; Popov, M. Strain Hardening of Polypropylene Microfiber Reinforced Composite Based on Alkali-Activated Slag Matrix. *Materials* **2022**, *15*, 1607. [CrossRef] [PubMed]
39. Smirnova, O. Development of classification of rheologically active microfillers for disperse systems with Portland cement and superplasticizer. *Int. J. Civ. Eng. Technol.* **2018**, *9*, 1966–1973.
40. Smirnova, O.M. Low-Clinker Cements with Low Water Demand. *J. Mater. Civ. Eng.* **2020**, *32*, 6020008. [CrossRef]
41. Smirnova, O.M.; de Navascués, I.; Mikhailevskii, V.R.; Kolosov, O.I.; Skolota, N.S. Sound-Absorbing Composites with Rubber Crumb from Used Tires. *Appl. Sci.* **2021**, *11*, 7347. [CrossRef]
42. Yakovlev, G.; Polyanskikh, I.; Gordina, A.; Pudov, I.; Černý, V.; Gumenyuk, A.; Smirnova, O. Influence of Sulphate Attack on Properties of Modified Cement Composites. *Appl. Sci.* **2021**, *11*, 8509. [CrossRef]
43. Saidova, Z.; Yakovlev, G.; Smirnova, O.; Gordina, A.; Kuzmina, N. Modification of Cement Matrix with Complex Additive Based on Chrysotyl Nanofibers and Carbon Black. *Appl. Sci.* **2021**, *11*, 6943. [CrossRef]
44. Smirnova, O.; Kazanskaya, L.; Koplík, J.; Tan, H.; Gu, X. Concrete Based on Clinker-Free Cement: Selecting the Functional Unit for Environmental Assessment. *Sustainability* **2021**, *13*, 135. [CrossRef]
45. Smirnova, O. Compatibility of shungisite microfillers with polycarboxylate admixtures in cement compositions. *ARPN J. Eng. Appl. Sci.* **2019**, *14*, 600–610.
46. Alves, L.; Leklou, N.; de Barros, S. A comparative study on the effect of different activating solutions and formulations on the early stage geopolymerization process. *MATEC Web Conf.* **2020**, *322*, 1039. [CrossRef]
47. Kumar, A.; Muthukannan, M.; Babu, A.; Hariharan, A.; Muthuramalingam, T. Effect on addition of Polypropylene fibers in wood ash-fly ash based geopolymer concrete. *IOP Conf. Ser. Mater. Sci. Eng.* **2020**, *872*, 12162. [CrossRef]
48. Kristály, F.; Szabo, R.; Madai, F.; Ákos, D.; Mucsi, G. Lightweight composite from fly ash geopolymer and glass foam. *J. Sustain. Cem. Mater.* **2020**, *10*, 1–22. [CrossRef]
49. Ali, B.; Raza, S.; Kurda, R.; Alyousef, R. Synergistic effects of fly ash and hooked steel fibers on strength and durability properties of high strength recycled aggregate concrete. *Resour. Conserv. Recycl.* **2021**, *168*, 105444. [CrossRef]
50. Aprianti, E.; Shafigh, P.; Bahri, S.; Farahani, J.N. Supplementary cementitious materials origin from agricultural wastes—A review. *Constr. Build. Mater.* **2015**, *74*, 176–187. [CrossRef]
51. Pitman, R. Wood ash use in forestry—A review of the environmental impacts. *Forestry* **2006**, *79*, 563–588. [CrossRef]
52. Kurda, R.; de Brito, J.; Silvestre, J.D. Influence of recycled aggregates and high contents of fly ash on concrete fresh properties. *Cem. Concr. Compos.* **2017**, *84*, 198–213. [CrossRef]
53. Zając, G.; Szyszlak-Bargłowicz, J.; Gołębiowski, W.; Szczepanik, M. Chemical Characteristics of Biomass Ashes. *Energies* **2018**, *11*, 2885. [CrossRef]
54. Kahl, J.; Fernandez, I.; Rustad, L.; Peckenham, J. Threshold Application Rates of Wood Ash to an Acidic Forest Soil. *J. Environ. Qual.* **1996**, *25*, 220–227. [CrossRef]
55. Fransman, B.; Nihlgård, B.J. Water chemistry in forested catchments after topsoil treatment with liming agents in South Sweden. *Water. Air. Soil Pollut.* **1995**, *85*, 895–900. [CrossRef]
56. Eriksson, J. *Dissolution of Hardened Wood Ash in Forest Soils Studies in a Column Experiment*; Swedish National Board for Industrial and Technical Development: Stockholm, Sweden, 1996.

57. Baker, A. *Heavy Metals in Soils*, 2nd ed.; Alloway, B.J., Ed.; Blackie Academic & Professional: London, UK, 1995; Volume 90, p. 269, ISBN 0-7514-0198-6.
58. Lanzerstorfer, C. Chemical composition and physical properties of filter fly ashes from eight grate-fired biomass combustion plants. *J. Environ. Sci.* **2015**, *30*, 191–197. [CrossRef]
59. Williams, T.M.; Hollis, C.A.; Smith, B.R. Forest Soil and Water Chemistry following Bark Boiler Bottom Ash Application. *J. Environ. Qual.* **1996**, *25*, 955–961. [CrossRef]
60. Steenari, B.-M.; Karlsson, L.G.; Lindqvist, O. Evaluation of the leaching characteristics of wood ash and the influence of ash agglomeration. *Biomass Bioenergy* **1999**, *16*, 119–136. [CrossRef]
61. Grau, F.; Choo, H.; Hu, J.W.; Jung, J. Engineering Behavior and Characteristics of Wood Ash and Sugarcane Bagasse Ash. *Materials* **2015**, *8*, 6962–6977. [CrossRef] [PubMed]
62. Etiégni, L.; Campbell, A.G. Physical and chemical characteristics of wood ash. *Bioresour. Technol.* **1991**, *37*, 173–178. [CrossRef]
63. Lanzerstorfer, C. Fly Ash from the Combustion of Post-Consumer Waste Wood: Distribution of Heavy Metals by Particle Size. *Int. J. Environ. Sci.* **2017**, *2*, 438–442.
64. Szakova, J.; Ochecova, P.; Hanzlicek, T.; Perna, I.; Tlustos, P. Viability of total and mobile element contents in ash derived from biomass combustion. *Chem. Pap.* **2013**, *67*, 1376–1385. [CrossRef]
65. Karltun, E.; Saarsalmi, A.; Ingerslev, M.; Mandre, M.; Andersson, S.; Gaitnieks, T.; Ozolinčius, R.; Varnagiryte-Kabasinskiene, I. Wood Ash Recycling—Possibilities And Risks. In *Sustainable Use of Forest Biomass for Energy*; Springer: Dordrecht, The Netherlands, 2008; pp. 79–108, ISBN 978-1-4020-5053-4.
66. Abdullahi, M. Characteristics of wood ash/OPC concrete. *Leonardo Electron. J. Pract. Technol.* **2006**, *8*, 9–16.
67. Udoeyo, F.; Inyang, H.; Young, D.; Oparadu, E. Potential of Wood Waste Ash as an Additive in Concrete. *J. Mater. Civ. Eng.* **2006**, *18*, 605–611. [CrossRef]
68. Ramos, T.; Matos, A.M.; Sousa-Coutinho, J. Mortar with wood waste ash: Mechanical strength carbonation resistance and ASR expansion. *Constr. Build. Mater.* **2013**, *49*, 343–351. [CrossRef]
69. Pavlíková, M.; Zemanová, L.; Pokorny, J.; Záleská, M.; Jankovský, O.; Lojka, M.; Sedmidubský, D.; Pavlik, Z. Valorization of wood chips ash as an eco-friendly mineral admixture in mortar mix design. *Waste Manag.* **2018**, *80*, 89–100. [CrossRef] [PubMed]
70. Chowdhury, S.; Mishra, M.; Suganya, O. The incorporation of wood waste ash as a partial cement replacement material for making structural grade concrete: An overview. *Ain Shams Eng. J.* **2015**, *6*, 429–437. [CrossRef]
71. Da Luz Garcia, M.; Sousa-Coutinho, J. Strength and durability of cement with forest waste bottom ash. *Constr. Build. Mater.* **2013**, *41*, 897–910. [CrossRef]
72. Ghorpade, V.G. Effect of wood waste ash on the strength characteristics of concrete. *Nat. Environ. Pollut. Technol.* **2012**, *11*, 121–124.
73. Horsakulthai, V.; Phiuvanna, S.; Kaenbud, W. Investigation on the corrosion resistance of bagasse-rice husk-wood ash blended cement concrete by impressed voltage. *Constr. Build. Mater.* **2011**, *25*, 54–60. [CrossRef]
74. Sashidhar, C.; Rao, S. Durability Studies On Concrete with Wood Ash Additive. In Proceedings of the 35th Conference on Our World in Concrete & Structures, Singapore, 25–27 August 2010.
75. Okeyinka, O.M.; Oladejo, O.A. The Influence of Calcium Carbonate as an Admixture on the Properties of Wood Ash Cement Concrete. *Int. J. Emerg. Technol. Adv. Eng.* **2014**, *4*, 432–437.
76. Mydin, M.A.; Shajahan, M.F.; Ganesan, S.; Md Sani, N. Laboratory Investigation on Compressive Strength and Micro-structural Features of Foamed Concrete with Addition of Wood Ash and Silica Fume as a Cement Replacement. *MATEC Web Conf.* **2014**, *17*, 01004. [CrossRef]
77. Kusuma, S. Studies on strength characteristics of fibre reinforced concrete with wood waste ash. *Int. Res. J. Eng. Technol.* **2015**, *2*, 181–187.
78. Prabagar, S.; Subasinghe, K.; Fonseka, W. Wood ash as an effective raw material for concrete blocks. *Int. J. Res. Eng. Technol.* **2015**, *4*, 228–233. [CrossRef]
79. Fapohunda, C.; Bolatito, A.; Akintoye, O. A Review of the Properties, Structural Characteristics and Application Potentials of Concrete Containing Wood Waste as Partial Replacement of one of its Constituent Material. *YBL J. Built Environ.* **2018**, *6*, 63–85. [CrossRef]
80. Chowdhury, S.; Maniar, A.; Suganya, O.M. Strength development in concrete with wood ash blended cement and use of soft computing models to predict strength parameters. *J. Adv. Res.* **2015**, *6*, 907–913. [CrossRef] [PubMed]
81. Udoeyo, F.; Dashibil, P. Sawdust Ash as Concrete Material. *J. Mater. Civ. Eng.* **2002**, *14*, 173–176. [CrossRef]
82. Elinwa, A.U.; Ejeh, S.P. Effects of the Incorporation of Sawdust Waste Incineration Fly Ash in Cement Pastes and Mortars. *J. Asian Archit. Build. Eng.* **2004**, *3*, 1–7. [CrossRef]
83. Wang, S.; Llamazos, E.; Baxter, L.; Fonseca, F. Durability of biomass fly ash concrete: Freezing and thawing and rapid chloride permeability tests. *Fuel* **2008**, *87*, 359–364. [CrossRef]
84. ASTM C 1202; Standard Test Method for Electrical Induction of Concrete, Stability to Resist Chloride Ion Penetration. American Society for Testing and Materials International: West Conshohocken, PA, USA, 2009.
85. Wang, S.; Baxter, L. Comprehensive study of biomass fly ash in concrete: Strength, microscopy, kinetics and durability. *Fuel Process. Technol.* **2007**, *88*, 1165–1170. [CrossRef]

86. Naik, T.; Kraus, R.N.; Siddique, R. Controlled low-strength materials containing mixtures of coal ash and new pozzolanic material. *ACI Mater. J.* **2003**, *100*, 208–215.
87. Awolusi, T.F.; Sojobi, A.O.; Afolayan, J.O. SDA and laterite applications in concrete: Prospects and effects of elevated temperature. *Cogent Eng.* **2017**, *4*, 1387954. [CrossRef]
88. Wan, Y.; Wu, H.; Huang, L.; Zhang, J.; Tan, S.; Cai, X. Preparation and characterization of corn cob/polypropylene composite reinforced by wood ash. *Polym. Bull.* **2018**, *75*, 2125–2138. [CrossRef]
89. Elahi, M.; Qazi, A.; Yousaf, M.; Akmal, U. Application of wood ash in the production of concrete. *Sci. Int.* **2015**, *27*, 1277–1280. [CrossRef]
90. Cheah, C.B.; Ramli, M. The implementation of wood waste ash as a partial cement replacement material in the production of structural grade concrete and mortar: An overview. *Resour. Conserv. Recycl.* **2011**, *55*, 669–685. [CrossRef]
91. Gabrijel, I.; Jelčić Rukavina, M.; Štirmer, N. Influence of Wood Fly Ash on Concrete Properties through Filling Effect Mechanism. *Materials* **2021**, *14*, 7164. [CrossRef] [PubMed]
92. Yin, K.; Ahamed, A.; Lisak, G. Environmental perspectives of recycling various combustion ashes in cement production—A review. *Waste Manag.* **2018**, *78*, 401–416. [CrossRef] [PubMed]
93. Vollprecht, D.; Berneder, I.; Capo Tous, F.; Stöllner, M.; Sedlazeck, P.; Schwarz, T.; Aldrian, A.; Lehner, M. Stepwise treatment of ashes and slags by dissolution, precipitation of iron phases and carbonate precipitation for production of raw materials for industrial applications. *Waste Manag.* **2018**, *78*, 750–762. [CrossRef]
94. Manikanta, B.; Vummaneni, R.R.; Achyutha Kumar Reddy, M. Performance of wood ash blended reinforced concrete beams under acid (HCl), base (NaOH) and salt (NaCl) curing conditions. *Int. J. Eng. Technol.* **2018**, *7*, 1045–1048. [CrossRef]
95. Siddique, R. Utilization of wood ash in concrete manufacturing. *Resour. Conserv. Recycl.* **2012**, *67*, 27–33. [CrossRef]

Review

Artificial Lightweight Aggregates Made from Pozzolanic Material: A Review on the Method, Physical and Mechanical Properties, Thermal and Microstructure

Dickson Ling Chuan Hao [1], Rafiza Abd Razak [1,2,*], Marwan Kheimi [3], Zarina Yahya [1,2], Mohd Mustafa Al Bakri Abdullah [2,4], Dumitru Doru Burduhos Nergis [5,*], Hamzah Fansuri [6], Ratna Ediati [6], Rosnita Mohamed [2] and Alida Abdullah [2]

1. Faculty of Civil Engineering Technology, Universiti Malaysia Perlis (UniMAP), Arau 02600, Perlis, Malaysia; dlch6179@gmail.com (D.L.C.H.); zarinayahya@unimap.edu.my (Z.Y.)
2. Centre of Excellence Geopolymer and Green Technology (CEGeoGTech), Universiti Malaysia Perlis (UniMAP), Arau 02600, Perlis, Malaysia; mustafa_albakri@unimap.edu.my (M.M.A.B.A.); rosnitamohamed21@gmail.com (R.M.); alida@unimap.edu.my (A.A.)
3. Department of Civil Engineering, Faculty of Engineering—Rabigh Branch, King Abdulaziz University, Jeddah 21589, Saudi Arabia; mmkheimi@kau.edu.sa
4. Faculty of Chemical Engineering Technology, Universiti Malaysia Perlis (UniMAP), Arau 02600, Perlis, Malaysia
5. Faculty of Materials Science and Engineering, Gheorghe Asachi Technical University of Iasi, 700050 Iasi, Romania
6. Department of Chemistry, Institut Teknologi Sepuluh Nopember, Surabaya 60115, Indonesia; h.fansuri@chem.its.ac.id (H.F.); rediati@chem.its.ac.id (R.E.)
* Correspondence: rafizarazak@unimap.edu.my (R.A.R.); doru.burduhos@tuiasi.ro (D.D.B.N.)

Abstract: As the demand for nonrenewable natural resources, such as aggregate, is increasing worldwide, new production of artificial aggregate should be developed. Artificial lightweight aggregate can bring advantages to the construction field due to its lower density, thus reducing the dead load applied to the structural elements. In addition, application of artificial lightweight aggregate in lightweight concrete will produce lower thermal conductivity. However, the production of artificial lightweight aggregate is still limited. Production of artificial lightweight aggregate incorporating waste materials or pozzolanic materials is advantageous and beneficial in terms of being environmentally friendly, as well as lowering carbon dioxide emissions. Moreover, additives, such as geopolymer, have been introduced as one of the alternative construction materials that have been proven to have excellent properties. Thus, this paper will review the production of artificial lightweight aggregate through various methods, including sintering, cold bonding, and autoclaving. The significant properties of artificial lightweight aggregate, including physical and mechanical properties, such as water absorption, crushing strength, and impact value, are reviewed. The properties of concrete, including thermal properties, that utilized artificial lightweight aggregate were also briefly reviewed to highlight the advantages of artificial lightweight aggregate.

Keywords: artificial lightweight aggregate; geopolymer; aggregate crushing value; aggregate impact value; thermal properties; pozzolanic materials

1. Introduction

Aggregates are widely used in the field of construction, specifically in the manufacture of concrete. The increasing demand for concrete will result in an increase in the requirements of all concrete ingredients, including aggregates, which represent 60–70% of the total volume of the concrete [1]. Most of the aggregates used are obtained from natural resources, including rocks. The concrete industry has an impact on the global environmental problem due to the utilization of large amounts of natural resources. Natural resources are decreasing at a quicker rate because of the high demand for usage in the manufacturing of concrete. The development of lightweight materials, such as lightweight aggregate, will help to minimize the use of natural resources.

In the construction field, lightweight aggregate that is used in concrete production is a type of material that is very environmentally friendly. Lightweight aggregate is dramatically different from conventional aggregate. The modifications bring advantages for the designers for a number of reasons other than weight reduction, such as decreased early cracking, decreased permeability, and longer lifespan [2]. Lightweight aggregates can be grouped into the following categories [3]:

1. Materials that naturally occur and require further processing, such as expanded clay, shale and slate, and vermiculite;
2. Industrial waste by-products, such as sintered pulverized fuel ash (fly ash), foamed or expanded-blast-furnace slag, and hemalite;
3. Materials that naturally occur, such as pumice, foamed lava, volcanic tuff, and porous limestone.

Sintered fly ash aggregate, also known as Lytag, is the most widely used artificial aggregate in the construction field [4]. According to Nadesan and Dinakar [5], sintered fly ash aggregate is capable of producing concrete with high strength performance. In addition, the application of lightweight aggregate in concrete has increased in popularity due to its low density, good thermal conductivity, being environmentally friendly, and many economic advantages [6]. Concrete with lightweight aggregate also has low thermal conductivity [6]. The concrete with lightweight aggregate had lower thermal conductivity [2]. The objective of this paper is to review the manufacturing method of lightweight aggregate for the production of lightweight concrete, as well as the properties of lightweight aggregate. The properties include physical properties, such as specific gravity and water absorption, and mechanical properties, such as crushing strength and aggregate impact value. The mechanical and thermal properties of concrete consisting of lightweight aggregate that have been reported previously are also reported in this review.

2. Aggregate

The main component of concrete is aggregates, which occupy around 70% to 80% of the total volume, with fine aggregate accounting for 25% to 30% and coarse aggregate accounting for 40% to 50% [7]. The coarse aggregate that is usually utilized in construction work comes from various types of resources, such as rock, crushed stone, and gravel [8]. Crushed rocks are commonly used as coarse aggregate and river sand as fine aggregate, both of which can be found naturally. A large amount of natural aggregate, such as sand, gravel, or crushed rock, is mined for concrete manufacturing, and the world's aggregate usage is estimated to be in excess of 40 billion tonnes per year, with concrete accounting for 64% to 75% of all mined aggregate [9]. The massive use of raw materials for aggregates is expected to reach 62.9 billion metric tonnes per year by 2024 with global construction aggregates consumption, and this can deplete natural resources, as it creates an immediate risk to the environment [10]. The construction projects need a considerable amount of natural aggregate for the production of concrete, which increases the depletion of natural aggregate resources and makes the sustainability of the construction projects more challenging [11]. Due to the increasing expansion of building construction, natural aggregate supplies are rapidly decreasing, resulting in a shortage of resources. This shortage of resources requires proper utilization for sustainable growth [7]. The natural sand can be replaced by using by-products of coarse aggregate, such as copper mine waste rocks, as it reduces the manufacturing cost, reduces CO_2 emissions from the industrial process of producing natural sand, and water consumption for sand washing would be minimized [12]. Manufactured sands are essentially a waste product from the production of coarse aggregate, which are generally available, have a cheaper cost, and would reduce natural sand mining [13]. In addition, manufactured sand had become a popular choice to be used to replace fine aggregate, as it is generally mined from stream beds, and harvesting sand is thought to be environmentally detrimental [14]. In addition, manufactured sand which is made from hard granite rocks can be produced locally by lowering the expense of shipping from a distant river sand bank, and it is dust-free, which is readily regulated to suit the needed grading for the construction [15]. Furthermore, the application of lightweight aggregate in concrete has been developed in order to resolve the depletion of natural aggregate. Due to its advantages in decreasing load bearing and also improving performance in thermal insulation, the development of lightweight concrete by including lightweight aggregates has caught researchers' attention [16].

3. Lightweight Aggregate

A lightweight aggregate (LWA) is a solid substance having a particle density of less than 2.0 g/cm^3 and a loose bulk density of less than 1.2 g/cm^3 (BS EN 13055-1, 2002) [17]. LWAs are porous and granular materials that have been widely used in architecture, landscaping, and geotechnics [18]. In addition, it can provide better sound absorption and thermal insulation [9]. Lightweight aggregates are ecologically friendly construction materials made from a variety of wastes and frequently produced through high-temperature burning [19]. There are two types of lightweight aggregate, which are natural lightweight aggregate and artificial lightweight aggregate. The following are the two types of lightweight aggregate:

1. Aggregates that occur naturally and can only be used after mechanical treatment, such as pumice and scoria aggregates [20].
2. Artificial aggregates are made up of waste materials, such as fly ash, husks, or volcanic form and ground granulated blast-furnace slag (GGBS) [21]

Various types of lightweight aggregate have been widely utilized as construction materials in the construction field. Considering aggregate makes up approximately 70% of the concrete mixture, substituting natural aggregate with lightweight aggregate manufactured from waste materials will be an efficient method to minimize nonrenewable resource usage [22]. Lightweight aggregate has been discovered as significant in the formation of lightweight concrete by lowering greenhouse emissions in buildings and decreasing the self-weight of the structure [23]. Application of LWA in concrete will enhance thermal insulation characteristics, decrease structural dead load, allowing larger structures to be built with the same foundation size, and lead to lower CO_2 emissions [24]. Furthermore, LWA is a critical component in the construction of earthquake-resistant buildings [25]. Due to the larger amount of internal pores in lightweight aggregate, absorption of moisture from cement paste is more rapid than in normal-weight aggregate, which makes the concrete less workable and lower in strength performance than the concrete prepared with normal aggregate [26]. Regardless of the performance when compared to natural aggregate concrete, LWA is worthwhile to be explored specifically for enhancing the performance towards minimizing environmental problems, alongside maintaining long-term sustainability by improving water quality or as a growth medium for green roofs to mitigate the urban heat island effect [27].

Lightweight Aggregate with Inclusion of Geopolymer

The inclusion of geopolymer in lightweight aggregate has become more concerning due to its advantages in improving the properties of lightweight aggregate. The strength of activated fly ash-based artificial lightweight aggregate by inclusion of geopolymer is comparable to that of commercialized expanded clay lightweight aggregate [22]. In addition, the inclusion of geopolymer in lightweight aggregate produced from fluidized bed combustion (FBC) fly ashes and mine tailings showed excellent mechanical properties in mortar and concrete as compared to the application of commercialized aggregate (LECA) [28]. The inclusion of geopolymer in the lightweight aggregate manufactured from recycled silt and palm oil fuel ash meets the demand for high-strength lightweight concrete and can be utilized for lightweight construction or insulating concrete [23]. In addition, the inclusion of geopolymer in lightweight aggregate that is produced from the combination of fly ash and silica fume can be used for heavy-duty floors due to its high strength [29]. Therefore, the inclusion of geopolymer in lightweight aggregate brings advantages in the construction field, especially in the structural components. However, the use of geopolymer in lightweight aggregate is still limited, and more research is needed to identify lightweight aggregate properties.

4. Manufacturing of Lightweight Aggregate

The manufacturing process of artificial aggregate consists of three stages, which are the mixing of raw materials, pelletization, and hardening. In the first stage, which is mixing, the well-proportioned ingredients are mixed until the mixture achieves consistency. In a disc-based pelletizer machine, the mixture of the raw materials undergoes the pelletization process by the agglomeration of the fine particles using a suitable binder. Some previous studies used pozzolanic materials as binders, such as metakaolin and bentonite [5,26,27]. Meanwhile, alkaline activators are commonly used as binders for the production of geopolymer aggregate [18,27–29]. Depending on the angle of the disc, the speed of the pelletizer, and the moisture content, the appropriate size of pellets will be collected in the disc. The hardening of the fresh pellets can be accomplished by using sintering, cold bonding, or autoclaving in order to gain the strength of the aggregate. The flow chart of the production of lightweight aggregates can be illustrated as in Figure 1. Meanwhile, the current research on lightweight aggregate is summarized as in Table 1.

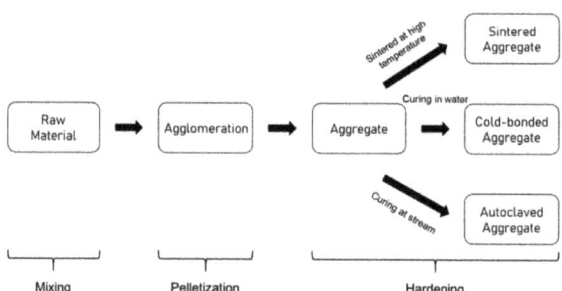

Figure 1. Flow chart of producing lightweight aggregate.

Table 1. Previous studies on the type of method to manufacture lightweight aggregate.

Researcher	Method	Raw Materials	Additives	Significant Finding
Kwek et al. (2022) [16]	Sintering	Palm Oil Fuel Ash and Silt	Alkaline activator (NaOH and Na_2SiO_3) and lime	- The bulk density can obtain as low as 1.18 kg/m^3 - The individual crushing strength is almost the same as commercialized aggregate used in lightweight concrete
Kwek and Awang (2021) [23]	Sintering	Palm Oil Fuel Ash	Alkaline activator (NaOH and Na_2SiO_3) and lime	- The strength of the aggregate achieved to be utilized for lightweight constructions or insulating concrete
Li et al. (2020) [30]	Sintering	Sewage sludge	Waste glass powder	- The addition of waste glass powder helps in reduction of water absorption
Ren et al. (2020) [31]	Sintering	Fly ash and clay	Coke particles	- Coke particles reduce the apparent density

Table 1. Cont.

Researcher	Method	Raw Materials	Additives	Significant Finding
Chien et al. (2020) [19]	Sintering	Industrial sludge and marine clay	Na_2CO_3	- The Na_2CO_3 can reduce the specific gravity and the firing temperature required for production of lightweight aggregate
Abdullah et al. (2021) [32]	Cold bonding	Fly ash	Alkaline activator (NaOH and Na_2SiO_3)	- NaOH molarity will affect the strength of the aggregate - 12 M of NaOH provide the optimum mix design of geopolymer aggregate
Risdanareni et al. (2020) [22]	Cold bonding	Fly ash	Sodium hydroxide (NaOH) solution	- 6 M of NaOH brought a positive impact to aggregate strength - The highest compressive strength at 8 M of NaOH
Ul Rehman et al. (2020) [33]	Cold bonding	Fly ash, Slag	Cement (Cement based) and Alkaline activator (Geopolymer based)	- Lightweight aggregate produced from cement based is strongest - Aggregate with hot water curing shows good properties as it can withstand half of the compressive load as compared to normal aggregate
Vali and Murugan (2020) [34]	Cold bonding	Fly ash, GGBS, hydrated lime	glass fibers	- Lightweight aggregate produced achieved the requirement for structural components
Patel et al. (2019) [35]	Cold bonding	Fly ash	Styrene-butadiene rubber	- Compressive strength increase compared to normal lightweight aggregate concrete
Tang et al. (2019) [36]	Cold bonding	Concrete slurry waste (CSW) and fine incineration bottom ash (IBA)	Cement and ground granulated blast furnace slag (GGBS)	- Addition of cement or GGBS as the additives in the manufacturing process will increase the strength of the aggregate
Wang et al. (2022) [37]	Autoclaved	Quartz tailings, fly ash, cement	Alkaline activator (NaOH and Na_2SiO_3)	- The strength of aggregate increase from 7.61 MPa to 10.20 MPa when increasing the autoclaved pressure - The water absorption decreases from 2.22% to 1.83% when increasing the autoclaved pressure
Wang et al. (2020) [38]	Autoclaved	Quartz tailings, fly ash	Quicklime	- The compressive strength is high - Higher structure efficiency for quartz tailing aggregate concrete

In the study of Punlert et al. (2017) [39], lightweight concrete was manufactured using fly ash lightweight aggregates instead of coarse aggregates, resulting in a much lower density and good strength compared to conventional concrete. Furthermore, when sintered at around 1100 °C, lightweight aggregate made from sewage sludge and river sediment achieved high density, low water absorption, and high strength. However, the existence of air voids in fly ash lightweight aggregates, which are crucial for absorbency, leads to difficulty in producing lightweight aggregate concrete, especially in the mix design, which requires further work to enhance the properties [40]. For this purpose, additional binders or additives are introduced as one of the alternatives towards improving the properties of lightweight aggregate.

From the previous results, the salt additives (NaCl) resulted in less viscosity and produced wider internal pores, which allowed the production of ultralight aggregates. However, the use of Na_2CO_3 as an additive, which is low cost and low corrosion hazard, allows the creation of ultra-lightweight aggregates [19]. In the study reported by Ren et al. [31], addition of coke particles in the manufacturing of lightweight aggregate will help to reduce the apparent density of the aggregate produced. The use of styrene butadiene rubber (SBR) improves the microstructure of lightweight aggregate, thus improving the aggregate's mechanical properties [35]. In addition, the inclusion of waste glass powder causes the pozzolanic material to inflate, resulting in a more efficient lightweight aggregate by enhancing the porosity of lightweight aggregate and decreasing water absorption [30].

In addition, pozzolanic materials with high SiO_2, Al_2O_3, and CaO content have a high potential to be utilized in producing artificial aggregates with the addition of an alkaline activator [23]. The usage of alkaline activator as an additive for pozzolanic materials will help to influence the formation of C-S-H binding gel and sodium aluminosilicate hydrates during the geopolymerization process [41]. The formation of geopolymer aggregate by mixing pozzolanic material with alkaline activator will decrease the porosity of aggregate and improve the strength due to the extra C-S-H and calcium reaction during the reaction process alongside the denser microstructure produced [33]. Furthermore, the NaOH molarity will affect the strength of the geopolymer aggregate. For instance, low sodium hydroxide content leads to improper dissolution of fly ash, thus causing the inter-particle spaces of the participating gels to not be entirely filled [35]. It is critical to investigate the optimization of mix designation for each kind of material utilized in the manufacturing of lightweight aggregate-based geopolymer.

To summarize, recent research has shown that fly ash is the common material that had been chosen to produce the lightweight aggregate due to their excellent properties. In the future, alternative pozzolanic materials should be studied to establish their suitability in the manufacturing of lightweight aggregate. In addition, it showed that additional additives will improve the properties of lightweight aggregate in terms of specific gravity, strength, and water absorption. Furthermore, the use of geopolymer in lightweight aggregate has been shown to increase the porosity and strength of the aggregate, making it a viable option for lightweight aggregate manufacturing. According to Table 1, various methods can be used to manufacture lightweight aggregate, which are sintering, cold bonding, and autoclaving, which have been reported previously. The sintering and cold bonding methods are the methods that have been used wisely due to the excellent properties of aggregate. Despite the fact that the sintering technique consumes a lot of energy, the quality of the lightweight aggregate generated is excellent, with good strength at a low density. Because of the considerable energy required in the sintering procedure, cold bonding has grown popular because it does not require any additional sintering or heating. This process can create various grade aggregates, depending on the density of the aggregate produced. Generally, aggregate with high density will have a high strength. As a result, the use of a foaming agent may be necessary to make the aggregate lighter. There is relatively limited research on autoclaving methods in the current literature, necessitating more inquiry to assess the possible application of autoclaving methods in the manufacturing of lightweight aggregate. The strength and water absorption capabilities of lightweight aggregate generated by the autoclaving technique were good, and the incorporation of geopolymer improved the properties. As a result, greater research into autoclaving procedures is needed, particularly in terms of simplifying the process so that it can be commercialized.

4.1. Sintering

Sintering is a process that consumes high energy to produce artificial aggregate with enhanced properties. As reported by Sun et al. (2021) [34], raw materials with a high amount of SiO_2 and Al_2O_3 commonly use sintering. When the pellets in the disc pelletizer are shaped, the pellets will dry for a day before undergoing the sintering process at a temperature of between 1180 °C and 1200 °C [42,43]. In some cases, some of the pellets are fused at a temperature above 1200 °C for optimum properties [21,44]. Most of the previous research used a similar drying method prior to sintering [45,46]. Chen et al. [47] also reported a similar method where the pellets undergo a drying stage, followed by preheating at 500 °C expanding temperature of a temperature between 1100 °C and 1150 °C for the sintering process.

Meanwhile, according to Grygo and Pranevich [48], the aggregate produced through the sintering process is lighter and has high strength performance. The sintering process is a popular application for mass manufacture of lightweight aggregates that does not require a long-term curing process [49]. Lytag, Pollytag, LECA, and liapour are some of the commercially sintered lightweight aggregates around the world. The factory that manufactured LECA has three lightweight aggregate production lines with a total capacity of 750,000 cubic meters per year [50]. Sintered artificial lightweight aggregate is one of the possible materials to make concrete lighter than the standard aggregate concrete [5]. In addition, Tian et al. (2021) [51] state that sintering aggregates with the help of geopolymerization reactions can have higher aggregate strength and low density. However, the sintering process involves a high level of energy during pelletization, which results in higher manufacturing costs [44]. Aside from that, the sintering process generates a large amount of pollutants, which will cause environmental problems. The sintering method needed a lot of energy regardless of its potential engineering properties with respective mix design applied [52].

In short, the lightweight aggregate produced at the temperature of 1200 °C provides the best properties of the aggregate. To acquire the best features of lightweight aggregate, the suggested sintering temperature for metakaolin is 900 °C, 1100 °C for sewage sludge and river sediment, and 900 °C for fly ash. As a result, materials such as metakaolin and fly ash are more advantageous due to the energy savings at the lowest sintering temperature required to manufacture lightweight aggregate. The sintering method will also be able to produce lightweight aggregate in a shorter time, at which it is suitable to be used to replace natural aggregate. Nevertheless, the sintering method will require high energy during the production, and this will increase the price of the production. The usage of sintered lightweight aggregate in the construction field will increase the overall cost of construction. As a result, new approaches, such as cold bonding, are being studied to address the problems that the sintering method encountered.

4.2. Cold Bonding

Cold bonding is a process of enhancing fine particles, either by pressure or non-pressure agglomeration methods, forming larger particles. In the cold bonding process, cement or alkaline activator will be chosen as the binder. The cold bonding method has been noted as a cost-effective method as it agglomerates at room temperature [47]. Furthermore, the cold bonding method tends to minimize energy usage when compared to other production processes [21]. For cold bonding, the pellets will be dried at room temperature for 24 h once the shape of the pellets is formed. The pellets are then sealed in the bag until the testing day [22,35,43,47,53,54]. According to Jiang et al. [55], normal curing at room temperature was required for cold-bonded artificial aggregate to achieve the strength. However, aggregates produced using the cold bonding method required curing at room temperature or in an enclosed chamber with steam until the required strength was attained [56]. The major challenge for cold-bonded aggregate is the requirement for a longer hardening period, as it is normally required to cure for 28 days before being discharged and used as construction materials [34].

From both economic and environmental viewpoints, the cold bonding process is fulfilling, as it involves low energy consumption. Wastewater treatment sludge, ground granulated blast furnace slag, rice husk ash, and fly ash are some of the common materials used to produce cold bonding lightweight aggregate. In addition, lightweight aggregate produced by using cold bonding instead of sintering is considered to have a strong effect on customer acceptance, as it reduces the environmental impact [22].

Meanwhile, the addition of nano SiO_2 from 0.5% to 1.5% during the production of artificial lightweight aggregate leads to increasing water absorption from 12.5% to 30.1% [57]. In addition, the utilization of foaming agents in lightweight aggregate causes more pores and makes the aggregate lighter than cold-bonded artificial lightweight aggregate. This was supported by the high water absorption ranging from 28.7% to 33.5% when compared to cold-bonded artificial lightweight aggregate, which had a water absorption ranging from 15.1% to 18.9% [58].

In a nutshell, the cold bonding method is considered a low-cost method, as it can be hardened at room temperature. Cold bonding has been recognized as a major step forward in the production of lightweight aggregate. Moreover, the cold bonding method is more likely to be adopted by society, as it does not require additional energy during the process. However, in order to improve the properties of the cold-bonded lightweight aggregate, it needs considerable treatment, particularly the use of a foaming agent during the manufacturing process. Another challenge is the curing day for cold bonding lightweight aggregate, which should be reduced to achieve acceptance in the construction industry.

4.3. Autoclaving

Autoclaving is a process that involves the addition of chemicals, such as lime or gypsum, in the agglomeration stage. In addition, autoclaving produced aggregates with little binding material and low curing time [21]. For autoclaving, the pellets will be hardened by the autoclave pressure and temperature to gain strength. A previous study by Wan et al. [38] reported autoclaving for the production of aggregates. The quartz tailing aggregate was cured at room temperature for 24 h, followed by curing at a temperature of up to 195 °C for 3 h with an autoclaved pressure of 1.38 MPa. The aggregates were further cured at 195 °C for another 10 h without autoclaved pressure before cooling at room temperature. The cured aggregate was then dried in order to achieve the desired weight of less than 1100 kg/m^3. The autoclaving method produced aggregates very quickly and it required little binding material and curing time [21]. Moreover, lightweight aggregate can be made with a considerable number of industrial solid wastes utilizing autoclave technology, which not only reduces the curing time (to only 4 h) to maximize space utilization, but also meets commercial environmental and economic requirements [37].

However, there are still limited studies available on autoclaving. This is because the autoclaving method requires an autoclaved machine with the required temperature and pressure to harden the aggregate. In addition, the autoclaved machine is very expensive and requires high power consumption and large production facilities to complete the process.

Generally, the sintering method has been widely used around the world with some popular commercial products, such as LECA and Lytag. LECA is one of the most popular artificial lightweight aggregates that has been commercialized in the market used to replace natural aggregate. The production rate can be up to 200,000 m^3 per year, depending on local LECA requirements and capital available. LECA is a new revolutionary material, and its manufacturing rate compared to standard aggregate is still dependent on customer demand. However, due to the requirement of high sintering temperature to produce sintered aggregate, the cold bonding method has been introduced as an initiative towards saving energy. The lightweight aggregate produced through the cold bonding method has the potential to be applied in concrete production due to its comparable properties to other methods. Moreover, cold bonding also showed promising properties, such as high compressive strength when applied to the concrete. Cold bonding also contributes to low pollutant production and low operating expenses. In addition, the autoclaving method is also another method that can be used to produce artificial lightweight aggregate. However, an autoclaved pressure machine is required for this method, which is very costly. The autoclaving method also required longer curing time to achieve the strength of the aggregate. Therefore, among all of the commonly reported methods, the cold bonding method with proper optimization of mix design is noted to be advantageous to the construction field. Furthermore, variations in manufacturing methods, as well as mix designation, were known to have a significant impact on the properties of lightweight aggregate, particularly on physical and mechanical properties.

5. Physical and Mechanical Properties of Lightweight Aggregate

5.1. Specific Gravity

The specific gravity of the aggregate varies depending on the type of raw material used. The cold-bonded fly ash aggregate that used different molarities of alkaline activator had a specific gravity of between 1.8 and 1.85 [22]. In addition, mixing 90% of fly ash with 10% of cement by using cold bonding gives a specific gravity of 1.76 [56]. The artificial aggregate that was made up of fly ash by using the cold bonding method had a specific gravity of 1.63 as compared to normal coarse aggregate at 2.71 [54]. The specific gravity was found to be increased from 1.84 to 1.91 when the styrene–butadiene rubber (SBR) was added from 1% to 3% to the lightweight aggregate [35]. In addition, the aggregate produced by mixing bentonite and metakaolin together with fly ash has a specific gravity of 1.8 to 1.93 and 1.95 to 1.99 [59].

The sintered fly ash aggregate had a specific gravity between 1.41 and 1.44, with a size that varied from 2 mm to 12 mm [5]. The specific gravity of aggregate that was made from water treatment residual increased from 1.21 to 1.78 when the sintering temperature was increased from 1000 °C to 1100 °C [42]. The sintered dredged sludge lightweight aggregate had a specific gravity of 1.00 to 1.38 for the particle size of 4.75 mm to 12.5 mm [47]. The sintered fly ash aggregate with bentonite added had a specific gravity of 1.57, while the sintered fly ash aggregate with glass powder added had a specific gravity of 1.60 at the temperature at 1200 °C [54]. Meanwhile, coarse aggregate manufactured from bentonite and water glass has a specific gravity of 1.63 at a temperature of 800 °C [46].

In comparison to natural aggregates, the specific gravity of artificial geopolymer aggregates formed by sintering is quite low [21]. For instance, Kamal and Mishra (2020) [60] reported on the specific gravity of the fly ash aggregates as well as raw materials, including fly ash and binder, and the amount of void space in the aggregate. In addition, whenever cold-bonded aggregate was combined with other pozzolanic binding materials, such as GGBS, the specific gravity was found to be as high as 2.42, in which the hydrated lime acts as a primary binder [61]. Previous research on the determination of specific gravity of aggregates can be summarized as in Table 2.

Table 2. Previous studies on specific gravity of lightweight aggregate.

Researcher	Aggregate	Specific Gravity
Shahane and Patel (2021) [62]	Cold-bonded Fly ash aggregate at 75 °C	2.1
	Cold-bonded Fly ash aggregate at 65 °C	2.2
Risdanareni et al. (2020) [22]	Cold-bonded Fly ash-based artificial lightweight aggregate	1.8–1.85
Kamal and Mishra (2020) [60]	Sintered Fly ash aggregate	1.52–1.9
	Sintered Fly ash aggregate (bentonite as binder)	1.61–1.65
	Sintered Fly ash aggregate (glass powder as binder)	1.64–1.67
Satpathy et al. (2019) [63]	Sintered fly ash lightweight aggregate	1.89
Nadesan and Dinakar (2018) [5]	Sintered fly ash lightweight aggregate	1.41–1.44
Abbas et al. (2018) [46]	Sintered lightweight aggregate produced from bentonite clay and water glass (sodium silicate)	1.63
Shivaprasad and Das (2018) [64]	Cold-bonded fly ash aggregate (heat cured)	1.94–2.03

From Table 2, the specific gravity of lightweight aggregate was found to be in the range from 1.41 to 2.2. Based on BS EN 13055-1 [17], the specific gravity of lightweight aggregate should be less than 2.0. The wide range of the specific gravity of lightweight aggregate values can be explained by the influence factors, including type of material used, type of method used, and type of binder used during the manufacturing process. The specific gravity of lightweight aggregate is often increased by adding additives. Furthermore, the specific gravity of lightweight aggregate is affected by the curing temperature, with a greater curing temperature resulting in a lower specific gravity. The addition of a foaming agent will aid in lowering the specific gravity. In addition, the sintering method always was proven to have lower specific gravity compared to other methods due to the formation of voids at higher temperatures. However, the low specific gravity can be achieved by the cold bonding and autoclaving methods through addition of additive, such as protein-based foaming agent.

5.2. Water Absorption

Water absorption provides an indication of the internal aggregate structure. Higher water absorption of aggregates indicates the large number of pores in nature and usually gives drawbacks to the aggregates. For instance, expanded perlite powder (EPP) was noted as being high in porosity, thus causing the water absorption to increase from 33% to 52% when the replacement of fly ash increased to 30% [53]. Meanwhile, when sintering is applied, increasing the sintering temperature was found to decrease water absorption of all aggregates due to the increment in the fusion of material, which led to less water surface permeability [48]. In the study conducted by Sun et al. (2021) [65], it was found that the sintered aggregate made up of red mud and municipal solid waste incineration bottom ash with the temperature increased from 1010 °C to 1090 °C caused the porosity of the aggregate to increase and the water absorption to decrease significantly until 1070 °C. Furthermore, Liu et al. (2018) [66] also found that lower water absorption can be obtained when the lightweight aggregate was sintered at a temperature of around 1100 °C. Lightweight aggregate made up of drill cuttings containing synthetic-based mud, when sintered at 1180 °C, had water absorption of 3.6% [24]. In addition, the water absorption of metakaolin artificial lightweight aggregate increases at the sintering temperature over 900 °C. The pores formed during the sintering process were found to be closed pores, thus causing a reduction in permeability to water [67].

In addition, the addition of the waste glass powder to lightweight aggregates was found to significantly reduce water absorption from 7.73% to 0.5% [30]. Meanwhile, due to the porosity of the hydrated calcium silicate, the quartz tailing aggregate (QTA) also possessed high water absorption, which varies from 13.77% to 21.93% [38]. In another study, the water absorption was reduced from 12.1% to 8.58% when styrene–butadiene rubber (SBR) was added from 1% to 3% to the lightweight aggregate, which proved the minimizing of voids when the SBR increased in the pellets [35]. Furthermore, it was found that the lightweight aggregate produced from different ratio palm oil fuel ash and silt causes high water absorption of 32.2% when 90% of clay is used [8]. This is due to the high pozzolanic reaction rate within the mixture, thus causing higher water absorption through the capillary of silt. Moreover, water absorption for aggregates incorporated with 10% cement was lowered by 13.97% due to a stronger hydration reaction and a denser microstructure with more C-S-H and CH products, thus leading to an increase in the strength of the aggregate [68].

In addition, the utilization of alkaline activator as a binder for the production of lightweight aggregate was found to increase the water absorption from 22% to 23% [22]. Meanwhile, the geopolymer lightweight aggregate sintered using microwave radiation had water absorption of 18.98%, as it was affected by the high density of the aggregate [29]. In another study, it was found that the pores in the fly ash geopolymer aggregate were reduced after the geopolymerization process, thus giving a denser microstructure, which resulted in lower water absorption at 10.05% [32]. Meanwhile, increasing the $Na_2SiO_3/NaOH$ ratio from 1.5 to 4.0 caused the water absorption values for geopolymer lightweight aggregate to steadily increase from 15.2% to 19% due to the foaming activity of Na_2SiO_3 [58]. Furthermore, the metakaolin geopolymer aggregate sintered at 600 °C will have lower water absorption as the greater the sintering temperature, the more voids created, and, hence, the higher the water absorption of lightweight aggregate [69]. Moreover, the fly ash geopolymer aggregates had higher water absorption when curing at 80 °C due to the water in the aggregates participating in the geopolymerization process, which improves the strength of the pellets [64].

From Table 3, the water absorption for artificial lightweight aggregate was proven to be higher than that of the natural aggregate, which can be explained by the effect of porosity due to pore formation. In addition, the water absorption of the lightweight aggregate can be affected by influence factors, including the type of materials used, the type of curing, and the type of binder used. The sintering method demonstrated that, when the temperature rises, the water absorption of lightweight aggregate decreases because it contains closed pores. The majority of research has indicated that the cold bonding procedure will have higher water absorption and will require additional treatment to eliminate this problem. Furthermore, adding additives to the lightweight aggregate will aid in water absorption reduction. The addition of geopolymer to lightweight aggregate will increase water absorption significantly. Water absorption will be reduced by the presence of a vitrified shell around the artificial lightweight aggregate. As a result, the water absorption of lightweight aggregate has an impact on mechanical qualities and should be assessed before using it in concrete.

Table 3. Previous studies on water absorption of lightweight aggregate.

Researcher	Aggregate	Water Absorption (%)
Ren et al. (2020) [31]	Artificial aggregate (fly ash + clay)	8.7
	Artificial aggregate (fly ash + clay + coke particles)	22.97
	Artificial aggregate (fly ash + clay + sodium carbonate solution)	8.84
Risdanareni et al. (2020) [22]	Alkali activated Fly ash-based artificial lightweight aggregate (4 M NaOH)	23.92
	Alkali activated Fly ash-based artificial lightweight aggregate (6 M NaOH)	23.23
	Alkali activated Fly ash-based artificial lightweight aggregate (8 M NaOH)	22.08
Rehman et al. (2020) [68]	Lightweight aggregate (fly ash + GGBS + 10% cement)	14.53
	Lightweight aggregate (fly ash + GGBS + 20% cement)	12.50
Vali and Murugan (2019) [57]	Cold-bonded artificial aggregate (fly ash + hydrated lime + cement + nano SiO_2)	22.9–30.1
	Cold-bonded artificial aggregate (fly ash + hydrated lime + metakaolin + nano SiO_2)	20.7–28.2
	Cold-bonded artificial aggregate (fly ash + hydrated lime + slag + nano SiO_2)	12.5–23.8
Narattha and Chaipanich (2018) [70]	Cold-bonded fly ash lightweight aggregates	14.08
	Cold-bonded fly ash lightweight aggregates (Additional of Portland cement)	13.34–16.90
	Cold-bonded fly ash lightweight aggregates (Calcium hydroxide)	14.10–18.46
Mohamad Ibrahim et al. (2018) [44]	Cold-bonded lightweight aggregate (cured at room temperature)	22.1–39.8
	Cold-bonded lightweight aggregate (cured under water at room temperature)	21.1–39.0
	Cold-bonded lightweight aggregate (cured at oven)	26.5–41.3
	Cold-bonded lightweight aggregate (cured under water at oven)	24.5–39.5

5.3. Mechanical Properties

In terms of mechanical properties, fly ash with expanded perlite powder (EPP) has a higher crushing strength due to increasing pozzolanic activity [53]. Sintered sediment lightweight aggregate with a bulk density of 859 kg/m^3 had the highest crushing strength of 13.4 MPa. This occurrence proves that the aggregate strength increases with increasing bulk density [47]. The crushing strength decreased as the silt content increased due to the binding failure of palm oil fuel ash (POFA) with the silt content [8]. Meanwhile, the fly-ash metakaolin binder aggregate showed high crushing strength when curing under high temperatures [59]. For instance, when the sintering temperature exceeds 900 °C, the sintered aggregate made up of metakaolin and alkaline activator has high aggregate impact value, which cause decreasing aggregate strength due to the increasing amounts of pore space in the aggregate [67]. Moreover, combination of fly ash and clay with 10% of sodium carbonate and sintering temperature of 1220 °C leads to a higher pellet strength of 4.25 MPa [31].

In addition, regardless of methods applied, the cold-bonded fly ash aggregate, sintered fly ash aggregate, as well as autoclaved aggregates were found to have a high impact value of 9.56%, 10.2%, and 11.46%, respectively [71]. According to research of Kamal and Mishra (2020) [60], the addition of binder is noted as effective due to the binder's role, which is to wrap the pellets, therefore causing the voids to have better resistance to compression. Meanwhile, the addition of styrene–butadiene rubber (SBR) to lightweight aggregate leads to a lower impact value, which makes the aggregate stronger [35]. The impact resistance of cement-based fly ash aggregate was enhanced by the cement content because of the increased hydration reaction. In addition, the curing temperature will also increase the impact resistance of artificial aggregate [33]. The high porosity of phosphogypsum-based cold-bonded aggregates with 90% of phosphogypsum accounts for high water absorption with 13.6%, as it holds fewer binders and allows it to absorb more water [72]. The inclusion of cement enhanced the pellet strength from 1 MPa to 2.3 MPa when compared to the pellet strength of the cold-bonded lightweight aggregate made with only concrete slurry waste and fine incinerator bottom ash [36].

In addition, the lightweight aggregate with the lowest water absorption has better sustainability towards the impact of the load. In a previous study, it was discovered that increasing the maximum amount of fly ash replacement with 10% cement or 5% calcium hydroxide increased cold-bonded aggregate strength with decreasing water absorption [70]. On the other hand, curing at higher temperatures causes the impact value of artificial lightweight aggregate to improve by 12.5% to 14.75% and the crushing strength by 28.2% to 39.7% [64]. According to a study by Rehman et al. (2020) [68], the aggregate with the lowest water absorption of 12.5% had the lowest aggregate impact value of 22.12%, thus proving the stronger microstructure and lower porosity had led to high resistance to crack penetration and increasing strength. Meanwhile, according to Ghosh (2018) [73], the autoclaved aggregate made by using fly ash and cement can be used to replace the gravel as the crushing value and impact value due to the similar value.

From the previous studies in Table 4, it is shown that the mechanical properties of lightweight aggregate mainly depend on the type of material used. Furthermore, adding additives to the lightweight aggregate can aid to improve its strength. The majority of the researchers concluded that adding geopolymer to the aggregate leads to an improvement in strength performance. The method of curing, on the other hand, will have an impact on the strength of the lightweight aggregate, as a higher curing temperature will result in greater strength. The mechanical properties of the lightweight aggregate can be affected by the microstructure of the lightweight aggregate.

Table 4. Previous studies on mechanical properties of lightweight aggregate.

Researcher	Aggregate	Crushing Strength (MPa)	Aggregate Impact Value, AIV (%)
Ding et al. (2022) [72]	Phosphogypsum-based cold-bonded aggregates	8.11–11.04	-
Zafar et al. (2021) [58]	Foam Lightweight Aggregate Geopolymer Lightweight Aggregate	0.35–0.83 3.69–4.14	-
Aslam et al. (2020) [74]	Geopolymer Lightweight Aggregate (fly ash, silica fume, baking soda)	3.34–4.54	10.03
Saleem et al. 2020 [29]	Geopolymer lightweight aggregates sintered by microwave radiations	3.08–3.96	22.1–35.7
Saad et al. (2019) [75]	Artificial granular lightweight aggregates (bottom ash + cement)	4.0–7.13	25.5–42.5
Taijra et al. (2018) [53]	Core-shell structured lightweight aggregate (expanded perlite powder + fly ash)	2.04–2.66	-
Shivaprasad and Das (2018) [64]	Fly ash aggregate (Ambient Cured) Fly ash aggregate (60 °C Cured) Fly ash aggregate (80 °C Cured)	2.87 3.68 4.01	27.57 24.10 23.50
Mohamad Ibrahim et al. (2018) [44]	Cold-bonded lightweight aggregate (cured at room temperature) Cold-bonded lightweight aggregate (cured under water at room temperature) Cold-bonded lightweight aggregate (cured at oven) Cold-bonded lightweight aggregate (cured under water at oven)	-	17.2–57.9 15.4–55.7 25.4–61.0 22.1–58.5
Abdullah et al. (2018) [76]	Fly ash geopolymer artificial aggregate (fly ash/alkaline activator = 2.0) Fly ash geopolymer artificial aggregate (fly ash/alkaline activator = 2.5) Fly ash geopolymer artificial aggregate (fly ash/alkaline activator = 3.0) Fly ash geopolymer artificial aggregate (fly ash/alkaline activator = 3.5)	-	23.19 23.14 19.6 25.56

5.4. Morphology

The morphology of lightweight aggregate can be observed through scanning electron microscopy (SEM). The artificial lightweight aggregate that was made from calcining coal ash and dredged soil was observed through SEM, and the morphology can be shown as in Figure 2. The observation of voids from the morphology proved that the formation of voids contributes to lower specific density and loose bulk density.

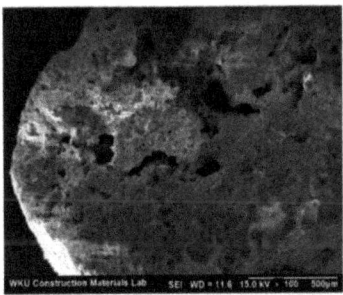

Figure 2. SEM of artificial lightweight aggregate made up of calcining coal ash and dredged soil [77].

Figure 3 shows SEM images of sintered lightweight aggregate at 1180 °C. The increasing amount of waste glass powder in the sample allows the tiny voids to be filled up, thus causing less porosity of the sintered lightweight aggregate with increasing particle density [30]. Besides that, excess gas may be created when using the sintering method. The formation of pores will occur continuously when the temperature applied is too high [78]. In addition, for the sintering method, the lightweight aggregate sintered at 1100 °C had a smooth and thick surface with solitary and round pores with widths ranging from 10 to 20 μm which minimizes porosity. This will lead to enhanced densification by creating samples with minimal water absorption and high compressive strength [66]. As shown in Figure 4, the pores in the aggregates LWA1 and LWA2 are significantly larger because organic substance components derived from sewage sludge release gases that aid in the development of pores and thus form a porous aggregate structure [18].

In addition, alkaline activators such as sodium hydroxide and sodium silicate have been used previously as liquid precursors and mixed with aluminosilicate materials such as fly ash and rice husk ash to create a cold bonded lightweight aggregate known as geopolymer aggregate, and the result can be depicted as in Figure 5. The denser matrix that is shown in Figure 5 for geopolymer aggregate with the solid-to-liquid (fly ash/alkaline activator) ratio of 3.0 results in a lower AIV value, where the strength of the geopolymer aggregate is higher. The geopolymer aggregates indicated that excellent solidification of the fly ash with alkaline activators occurred by geopolymerization reaction as the alkaline activator dissolved most of the fly ash particles [79]. Furthermore, the SEM of quarry tailings autoclaved aggregate in Figure 6 showed that high stream curing had a denser structure, which increased the strength alongside with reduced water absorption of the aggregate [37].

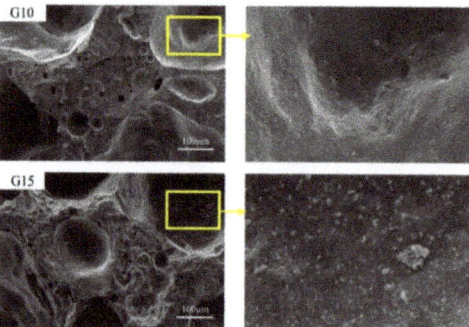

Figure 3. SEM of sintered lightweight aggregate samples with 10% and 15% of water glass content at 1180 °C [30].

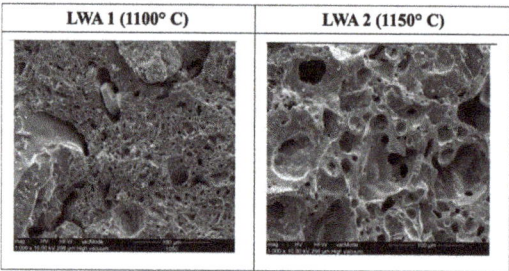

Figure 4. SEM of sintered lightweight aggregate made up of clay and sewage sludge at 1100 °C and 1150 °C [18].

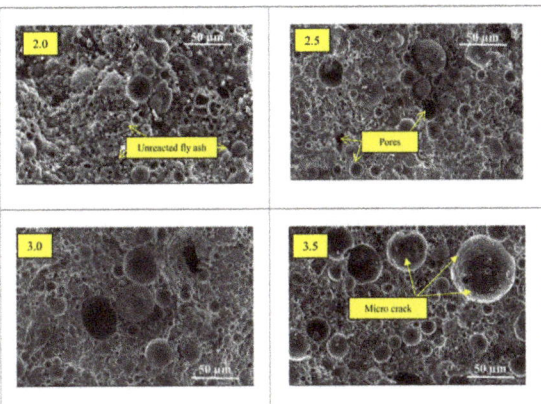

Figure 5. SEM of geopolymer aggregate at solid/liquid ratio of 2.0, 2.5, 3.0, and 3.5 [76].

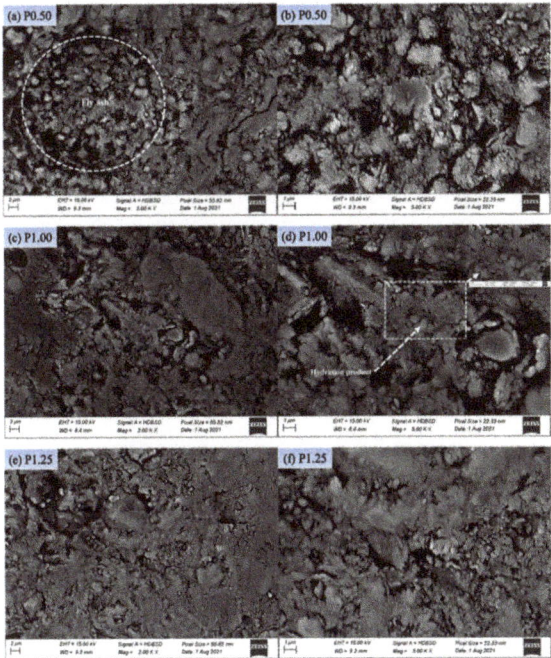

Figure 6. SEM of autoclaved lightweight aggregate with different curing pressure: (**a**,**b**) P0.50, (**c**,**d**) P1.00, (**e**,**f**) P1.25 [37].

From the morphology of lightweight aggregate, the pore of the lightweight aggregate can be observed through SEM. The formation of pores observed from the morphology is significant towards proving the increment and decrement in some properties of aggregates, such as specific gravity and density. Based on the previous study, aggregate that was made up of fly ash through geopolymerization showed the highest distribution of pores. The connected pores lead to higher water absorption than disconnected pores due to the ability to absorb more water into the aggregate. Meanwhile, the denser structure that was observed through SEM can provide good properties of lightweight aggregate, which can bring benefits in the application of concrete.

6. Lightweight Aggregate Concrete

6.1. Mechanical Properties

The utilization of lightweight aggregate also affects the significant properties of the concrete produced, including mechanical properties, thermal conductivity, and ultrasonic pulse velocity. The mechanical properties of lightweight aggregate concrete determine the suitability of artificial lightweight aggregate used in the concrete. The interfacial zone (ITZ) between the coarse aggregate and paste is a critical factor that will affect the mechanical properties of the concrete [80].

Compressive strength is the ability of the structure to resist compression, and this is the main design variable for engineers. The compressive strength of mortar containing fly ash aggregate will increase slowly at the beginning of the hardening stage, but it will increase rapidly after 14 days. In addition, the heavier the mortar, the higher the compressive strength. For instance, mortar of fly ash with an 8 M concentration of NaOH shows the heaviest bulk density with the highest compressive strength [22]. The compressive strength for quartz tailings aggregate (QTA) concrete reaches 74 MPa, which is considered high strength concrete. This is because the cement binders are readily penetrated to form a later mechanical interlocking structure around the aggregates to strengthen the bonding between aggregate and cement paste since the shell of QTA is a porous, fibrous, and needle-flake tobermorite structure [38]. The increasing amount of styrene-butadiene rubber (SBR) in pellets will cause an increase in the compressive strength of SBR modified lightweight aggregate (SLWA) concrete. This is due to the microstructure of the SLWA, which generates a strong bond between the aggregate and the cement paste [35]. Due to the impact on the restriction of the spread of cracking occurring, the beneficial effect of the addition of fibres to the lightweight aggregate cement mix gives a higher compressive strength (over 40 MPa) and can be used as a structural application in the construction field [81]. Table 5 shows the compressive strength of lightweight aggregate concrete reported by past researches. According to Table 5, it can be concluded that the artificial lightweight aggregate concrete had achieved the compressive strength for structure concrete which is 17 MPa based on ACI 318M-14.

In general, the concrete that used the autoclaved quartz tailing lightweight aggregate had a compressive strength of up to 74 MPa and alkali-activated fly-ash-based artificial lightweight aggregate achieved 64 MPa of compressive strength at 28 days. In addition, when lightweight aggregate has been used in concrete, it achieved early strength compared to conventional concrete. The additives added to the lightweight aggregate enhanced the aggregate's strength, which will also increase the lightweight concrete's strength. The use of lightweight aggregate in concrete met the minimal compressive strength requirements for the use of structural components. As a result, artificial lightweight aggregate has been an innovative material that can be used to manufacture lightweight structural components that have been highlighted in the recent construction field to minimize the structure's dead load and protect it from earthquakes.

Table 5. Previous studies on compressive strength of lightweight aggregate concrete.

Researcher	Aggregate	28 Days of Compressive Strength (MPa)
Risdanareni et al. (2020) [22]	Alkali-activated fly-ash-based artificial lightweight aggregate	64.0
Sahoo et al. (2020) [81]	Sintered fly ash aggregate with synthetic fiber	46.0
Wang et al. (2020) [38]	Autoclaved quartz tailing lightweight aggregate	74.0
Patel et al. (2019) [35]	SBR-modified lightweight aggregate	42.0
Abbas et al. (2018) [46]	Sintered fly ash aggregate	35.8
Lau et al. (2018) [82]	Sintered lime-treated sewage sludge and palm oil fuel ash	50.4

6.2. Thermal Conductivity

The capacity of a substance to transport heat is measured by thermal conductivity, which is significant in determining the insulation of materials. Lightweight concrete normally has low thermal conductivity compared to conventional concrete [3]. The thermal conductivity of lightweight concrete is 0.9567 W/m·K, whereas the thermal conductivity of normal weight aggregate is 1.98–2.94 W/m·K [46]. The low thermal conductivity of lightweight concrete can be due to the type of material used to produce lightweight aggregate. According to Tajra et al. [53], the low conductivity of lightweight concrete is due to the use of expanded perlite as the core structure, which has a low thermal conductivity of about 0.05 W/m·K, as well as the use of expanded perlite powder in the shell structure, which improved its thermal properties. Concrete with lightweight coarse and fine aggregate has the lowest thermal properties, 0.0703 W/m·K, when compared to conventional concrete, which has thermal properties of 1.736 W/m·K [2]. Figure 7 shows the comparison between lightweight aggregate concrete and conventional concrete. Based on Figure 7, the lightweight aggregate concrete had decreased by 95.95% and 67.46% as compared to normal aggregate concrete. The application of sintered expanded slate aggregate in coarse and fine aggregate showed that it had the lowest thermal conductivity. Therefore, it can be concluded that the application of sintered lightweight aggregate in concrete is more suitable to be used as thermal insulation material. It would be fascinating to look into finding the thermal conductivity of concrete using cold-bonded lightweight aggregate and autoclaved lightweight aggregate.

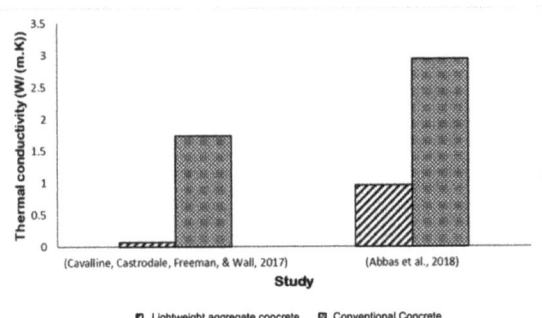

Figure 7. Previous studies' comparison on thermal conductivity between lightweight aggregate concrete and conventional concrete.

6.3. Ultrasonic Pulse Velocity

Ultrasonic pulse velocity (UPV) testing is used to verify the integrity and quality of structural concrete alongside voiding, honeycombing, cracking, and other defects. According to BS 1881-203 [83], the ultrasonic pulse velocity, which is greater or equal to 4.5 km/s, is considered as excellent concrete quality, while less than 2.0 km/s is classified as very weak concrete quality. Based on the study conducted by Tanaka et al. (2020) [4], the pulse velocity is 3.5 km/s to 4.4 km/s, which is slower as compared to the control sample due to the large amount of air voids in the artificial aggregate in the concrete, which affects the quality of the concrete. In study of Rehman et al. (2020) [68], it was found that geopolymer-based lightweight concrete has a higher ultrasonic pulse velocity, which is 2936 m/s to 3016 m/s as compared to cement-based lightweight concrete, which is 2601 m/s to 2835 m/s. This can be explained by the higher strength and more complex microstructure of the geopolymer-based concrete. Further, the occurrence of large amounts of micro-cracks in the concrete can cause the ultrasonic pulse velocity to be lowered, which is 3.42–4.51 km/s when the replacement of coarse and fine aggregates is increased with artificial lightweight aggregate (Satpathy et al., 2019) [63]. For geopolymer concrete specimens containing artificial lightweight aggregate, the ultrasonic pulse velocity value vary from 4.15 km/s to 4.35 km/s, which can be considered good quality concrete (Abbas et al., 2018) [46]. Furthermore, the UPV of the concrete mixes decreased from 4.29 km/s to 3.58 km/s as the lightweight expanded clay aggregate (LECA) and expanded perlite aggregate (EPA) replacement percentage increased, and this could be due to the existence of voids in LECA and EPA lengthening the travel path of the ultrasonic pulse and resulting in a lower UPV value [84].

From Table 6, it can be concluded that the lightweight aggregate concrete will provide good ultrasonic pulse velocity to prevent the defects of the concrete. It is also possible to deduce that the majority of the ultrasonic pulse velocity of concrete is determined by the use of commercialized lightweight aggregate, such as LECA, which is manufactured using the sintering method. As a result, the ultrasonic pulse velocity for concrete with cold-bonded lightweight aggregate and autoclaved lightweight aggregate is limited, and more research is needed.

Table 6. Previous studies on ultrasonic pulse velocity of lightweight aggregate concrete.

Researcher	Ultrasonic Pulse Velocity	Concrete Quality Based on BS:118-203	Material
Othman et al. (2020) [84]	3.58 km/s to 4.21 km/s	Good	Lightweight Expanded Clay Aggregate (LECA) and Expanded Perlite Aggregate (EPA)
Tanaka et al. (2020) [4]	3.5 km/s to 4.4 km/s	Good	Lightweight artificial aggregate from industrial by-product
Satpathy et al. (2019) [63]	3.42 km/s to 4.51 km/s	Good, 4.51 (Excellent)	Fly ash cenosphere and sintered fly ash aggregate
Abbas et al. (2018) [46]	4.15 km/s to 4.35 km/s	Good	Sintered fly ash aggregate

7. Conclusions

Based on the review, the cold bonding method and the autoclaving method can be considered as alternative ways to produce the lightweight aggregate due to the comparable properties to the sintering method. However, the aggregate produced from the cold bonding method and the autoclaving method are still limited and require further exploration in order to be commercialized as a sintering method. Other than features such as water absorption and specific gravity, the curing days for aggregate generated should be researched further for the cold bonding process, as this approach requires the same curing days as OPC. The construction sector will accept shorter curing days since they are more practical to commercialize. The performance of the aggregate produced by the autoclave process is good; however, the low temperature during autoclave pressure may need to be researched more in the future to lower production costs and, thus, make it a viable alternative to commercially available good-grade aggregate.

Further, lightweight aggregate having a density of less than 2.0 can be utilized extensively in lightweight concrete to lower the structure's dead load. The specific gravity of lightweight aggregate will be mostly affected by the type of material and binder used. In addition, the artificial lightweight aggregate showed lower water absorption than normal aggregate. The water absorption of lightweight aggregate is increased when geopolymer is used in the manufacturing process, and it can be further improved by utilizing additional treatments, such as vacuum impregnation or coating. Furthermore, due to the fact that the pozzolanic activity with the inclusion of geopolymer in the lightweight aggregate is higher, mechanical properties of lightweight aggregate are improvable. Moreover, the morphology of lightweight aggregate can help to determine the microstructure of lightweight aggregate. Lightweight aggregate with a denser microstructure and lower porosity was found to be more resistant to crack penetration and have higher strength. The properties of lightweight aggregate can be improved by a greater distribution of internal aggregate holes and fewer continuous pores observed on the microstructure.

On the other hand, the interfacial zone (ITZ) between coarse aggregate and paste is a significant factor that will affect the compressive strength of the concrete. Thus, more research into the bonding mechanism between aggregate and cement matrix is critical. The compressive strength of lightweight aggregate concrete can be increased by adding additives, such as synthetic fibers. Higher compressive strength can be obtained by using lightweight aggregate in order to construct lightweight structural components to protect it from earthquake resistance. Additionally, lightweight aggregate with geopolymer demonstrates early strength development in lightweight concrete. The low thermal conductivity of lightweight aggregate concrete can be applied in the construction field as a thermal insulation material. This can assist the building to be more comfortable while using less energy. Furthermore, the application of lightweight aggregate in the concrete will produce excellent quality structural concrete, which can reduce defects when applied in the construction field. Therefore, the creation of lightweight aggregate can be a unique material that can be used in a variety of applications to minimize the usage of natural aggregate and bring benefits to the environment, such as reducing the dead load of the structure, enhancing thermal insulation properties, and reducing CO_2 emissions.

Funding: This research was funded by Ministry of Higher Education Malaysia, Fundamental Research Grant Scheme (FRGS), and grant number of 9003-00747 FRGS/1/2019/TK06/UNIMAP/02/1. Authors gratefully acknowledge technical and financial support from King Abdulaziz University, DSR, Jeddah, Saudi Arabia.

Institutional Review Board Statement: Not applicable.

Informed Consent Statement: Not applicable.

Data Availability Statement: Not applicable.

Conflicts of Interest: The authors declare no conflict of interest.

References

1. Alqahtani, F.K.; Zafar, I. Characterization of processed lightweight aggregate and its effect on physical properties of concrete. *Constr. Build. Mater.* **2019**, *230*, 116992. [CrossRef]
2. Cavalline, T.L.; Castrodale, R.W.; Freeman, C.; Wall, J. Impact of Lightweight Aggregate on Concrete Thermal Properties. *ACI Mater. J.* **2017**, *114*, 945–956. [CrossRef]
3. Kumar, P.S.; Babu, M.J.R.K.; Kumar, K.S. Experimental Study on Lightweight Aggregate. *Int. J. Civ. Eng. Res.* **2010**, *1*, 65–74.
4. Tanaka, C.J.; Mohd Sam, A.R.; Abdul Shukor Lim, N.; Awang, A.Z.; Hamzah, N.; Loo, P. Properties of Concrete Containing Blended Cement and Lightweight Artificial Aggregate. *Malays. J. Civ. Eng.* **2020**, *32*, 59–68.
5. Nadesan, M.S.; Dinakar, P. Influence of type of binder on high-performance sintered fly ash lightweight aggregate concrete. *Constr. Build. Mater.* **2018**, *176*, 665–675. [CrossRef]
6. Pramusanto, P.; Nurrochman, A.; E Mamby, H.; Nugraha, P. High strength lightweight concrete with expandable perlite as the aggregate. *IOP Conf. Ser. Mater. Sci. Eng.* **2020**, *830*, 42040. [CrossRef]
7. Kanojia, A.; Jain, S.K. Performance of coconut shell as coarse aggregate in concrete. *Constr. Build. Mater.* **2017**, *140*, 150–156. [CrossRef]
8. Ying, K.S.; Awang, H. Performance of Aggregate Incorporating Palm Oil Fuel Ash (Pofa) and Silt. *Int. J. Eng. Adv. Technol.* **2019**, *9*, 1218–1223. [CrossRef]
9. Eziefula, U.; Ezeh, J.C.; Eziefula, B.I. Properties of seashell aggregate concrete: A review. *Constr. Build. Mater.* **2018**, *192*, 287–300. [CrossRef]
10. Salahuddin, H.; Nawaz, A.; Maqsoom, A.; Mehmood, T.; Zeeshan, B.U.A. Effects of elevated temperature on performance of recycled coarse aggregate concrete. *Constr. Build. Mater.* **2019**, *202*, 415–425. [CrossRef]
11. Zhou, C.; Chen, Z. Mechanical properties of recycled concrete made with different types of coarse aggregate. *Constr. Build. Mater.* **2017**, *134*, 497–506. [CrossRef]
12. Benahsina, A.; El Haloui, Y.; Taha, Y.; Elomari, M.; Bennouna, M.A. Natural sand substitution by copper mine waste rocks for concrete manufacturing. *J. Build. Eng.* **2021**, *47*, 103817. [CrossRef]
13. Altuki, R.; Ley, M.T.; Cook, D.; Gudimettla, M.J.; Praul, M. Increasing sustainable aggregate usage in concrete by quantifying the shape and gradation of manufactured sand. *Constr. Build. Mater.* **2022**, *321*, 125593. [CrossRef]

14. Suleman, S.; Needhidasan, S. Utilization of manufactured sand as fine aggregates in electronic plastic waste concrete of M30 mix. *Mater. Today Proc.* **2020**, *33*, 1192–1197. [CrossRef]
15. Needhidasan, S.; Ramesh, B.; Prabu, S.J.R. Experimental study on use of E-waste plastics as coarse aggregate in concrete with manufactured sand. *Mater. Today Proc.* **2019**, *22*, 715–721. [CrossRef]
16. Kwek, S.Y.; Awang, H.; Cheah, C.B.; Mohamad, H. Development of sintered aggregate derived from POFA and silt for lightweight concrete. *J. Build. Eng.* **2022**, *49*, 104039. [CrossRef]
17. BS EN 13055-1:2002; Lightweight Aggregates—Part 1: Lightweight Aggregates for Concrete, Mortar and Grout. European Committee for Standardization: Brussels, Belgium, 2002.
18. Franus, M.; Barnat-Hunek, D.; Wdowin, M. Utilization of sewage sludge in the manufacture of lightweight aggregate. *Environ. Monit. Assess.* **2015**, *188*, 10. [CrossRef]
19. Chien, C.-Y.; Show, K.-Y.; Huang, C.; Chang, Y.-J.; Lee, D.-J. Effects of sodium salt additive to produce ultra lightweight aggregates from industrial sludge-marine clay mix: Laboratory trials. *J. Taiwan Inst. Chem. Eng.* **2020**, *111*, 105–109. [CrossRef]
20. Durga, J.; Kumar, C.; Arunakanthi, E. The Use of Light Weight Aggregates for Precast Concrete Structural Members. *Int. J. Appl. Eng. Res.* **2018**, *13*, 7779–7787.
21. George, G.K.; Revathi, P. Production and Utilisation of Artificial Coarse Aggregate in Concrete—A Review. *IOP Conf. Ser. Mater. Sci. Eng.* **2020**, *936*, 12035. [CrossRef]
22. Risdanareni, P.; Schollbach, K.; Wang, J.; De Belie, N. The effect of NaOH concentration on the mechanical and physical properties of alkali activated fly ash-based artificial lightweight aggregate. *Constr. Build. Mater.* **2020**, *259*, 119832. [CrossRef]
23. Kwek, S.; Awang, H. Utilisation of Recycled Silt from Water Treatment and Palm Oil Fuel Ash as Geopolymer Artificial Lightweight Aggregate. *Sustainability* **2021**, *13*, 6091. [CrossRef]
24. Ayati, B.; Molineux, C.; Newport, D.; Cheeseman, C. Manufacture and performance of lightweight aggregate from waste drill cuttings. *J. Clean. Prod.* **2018**, *208*, 252–260. [CrossRef]
25. Agrawal, Y.; Gupta, T.; Sharma, R.; Panwar, N.; Siddique, S. A Comprehensive Review on the Performance of Structural Lightweight Aggregate Concrete for Sustainable Construction. *Constr. Mater.* **2021**, *1*, 3. [CrossRef]
26. Hunag, L.-J.; Wang, H.-Y.; Wu, Y.-W. Properties of the mechanical in controlled low-strength rubber lightweight aggregate concrete (CLSRLC). *Constr. Build. Mater.* **2016**, *112*, 1054–1058. [CrossRef]
27. Liu, R.; Coffman, R. Lightweight Aggregate Made from Dredged Material in Green Roof Construction for Stormwater Management. *Materials* **2016**, *9*, 611. [CrossRef]
28. Yliniemi, P.; Ferreira, T.; Illikainen. Development and incorporation of lightweight waste-based geopolymer aggregates in mortar and concrete. *Constr. Build. Mater.* **2017**, *131*, 784–792. [CrossRef]
29. Saleem, N.; Rashid, K.; Fatima, N.; Hanif, S.; Naeem, G.; Aslam, A.; Fatima, M.; Aslam, K. Appraisal of geopolymer lightweight aggregates sintered by microwave radiations. *J. Asian Concr. Fed.* **2020**, *6*, 37–49. [CrossRef]
30. Li, X.; He, C.; Lv, Y.; Jian, S.; Liu, G.; Jiang, W.; Jiang, D. Utilization of municipal sewage sludge and waste glass powder in production of lightweight aggregates. *Constr. Build. Mater.* **2020**, *256*, 119413. [CrossRef]
31. Ren, Y.; Ren, Q.; Huo, Z.; Wu, X.; Zheng, J.; Hai, O. Preparation of glass shell fly ash-clay based lightweight aggregate with low water absorption by using sodium carbonate solution as binder. *Mater. Chem. Phys.* **2020**, *256*, 123606. [CrossRef]
32. Abdullah, A.; Hussin, K.; Abdullah, M.; Yahya, Z.; Sochacki, W.; Razak, R.; Błoch, K.; Fansuri, H. The Effects of Various Concentrations of NaOH on the Inter-Particle Gelation of a Fly Ash Geopolymer Aggregate. *Materials* **2021**, *14*, 1111. [CrossRef]
33. Rehman, M.-U.; Rashid, K.; Haq, E.U.; Hussain, M.; Shehzad, N. Physico-mechanical performance and durability of artificial lightweight aggregates synthesized by cementing and geopolymerization. *Constr. Build. Mater.* **2019**, *232*, 117290. [CrossRef]
34. Vali, K.; Murugan, S. Influence of industrial by-products in artificial lightweight aggregate concrete: An Environmental Benefit Approach. *Ecol. Environ. Conserv.* **2020**, *26*, S233–S241.
35. Patel, J.; Patil, H.; Patil, Y.; Vesmawala, G. Strength and transport properties of concrete with styrene butadiene rubber latex modified lightweight aggregate. *Constr. Build. Mater.* **2018**, *195*, 459–467. [CrossRef]
36. Tang, P.; Xuan, D.; Poon, C.S.; Tsang, D.C. Valorization of concrete slurry waste (CSW) and fine incineration bottom ash (IBA) into cold bonded lightweight aggregates (CBLAs): Feasibility and influence of binder types. *J. Hazard. Mater.* **2019**, *368*, 689–697. [CrossRef]
37. Wang, S.; Yu, L.; Yang, F.; Zhang, W.; Xu, L.; Wu, K.; Tang, L.; Yang, Z. Resourceful utilization of quarry tailings in the preparation of non-sintered high-strength lightweight aggregates. *Constr. Build. Mater.* **2022**, *334*. [CrossRef]
38. Wang, D.; Cui, C.; Chen, X.-F.; Zhang, S.; Ma, H. Characteristics of autoclaved lightweight aggregates with quartz tailings and its effect on the mechanical properties of concrete. *Constr. Build. Mater.* **2020**, *262*, 120110. [CrossRef]
39. Punlert, S.; Laoratanakul, P.; Kongdee, R.; Suntako, R. Effect of lightweight aggregates prepared from fly ash on lightweight concrete performances. *J. Phys. Conf. Ser.* **2017**, *901*, 12086. [CrossRef]
40. Nadesan, M.S.; Dinakar, P. Mix design and properties of fly ash waste lightweight aggregates in structural lightweight concrete. *Case Stud. Constr. Mater.* **2017**, *7*, 336–347. [CrossRef]
41. Kwek, S.; Awang, H.; Cheah, C. Influence of Liquid-to-Solid and Alkaline Activator (Sodium Silicate to Sodium Hydroxide) Ratios on Fresh and Hardened Properties of Alkali-Activated Palm Oil Fuel Ash Geopolymer. *Materials* **2021**, *14*, 4253. [CrossRef]
42. Huang, C.; Pan, J.R.; Liu, Y. Mixing Water Treatment Residual with Excavation Waste Soil in Brick and Artificial Aggregate Making. *J. Environ. Eng.* **2005**, *131*, 272–277. [CrossRef]

43. Güneyisi, E.; Gesoğlu, M.; Pürsünlü, Ö.; Mermerdaş, K. Durability aspect of concretes composed of cold bonded and sintered fly ash lightweight aggregates. *Compos. Part B Eng.* **2013**, *53*, 258–266. [CrossRef]
44. Ibrahim, N.M.; Ismail, K.N.; Amat, R.C.; Ghazali, M.I.M. Properties of cold-bonded lightweight artificial aggregate containing bottom ash with different curing regime. *E3S Web Conf.* **2018**, *34*, 1038. [CrossRef]
45. Lau, P.; Teo, D.; Mannan, M. Characteristics of lightweight aggregate produced from lime-treated sewage sludge and palm oil fuel ash. *Constr. Build. Mater.* **2017**, *152*, 558–567. [CrossRef]
46. Abbas, W.; Khalil, W.; Nasser, I. Production of lightweight Geopolymer concrete using artificial local lightweight aggregate. *MATEC Web Conf.* **2018**, *162*, 2024. [CrossRef]
47. Chen, H.-J.; Yang, M.-D.; Tang, C.-W.; Wang, S.-Y. Producing synthetic lightweight aggregates from reservoir sediments. *Constr. Build. Mater.* **2012**, *28*, 387–394. [CrossRef]
48. Grygo, R.; Pranevich, V. Lightweight sintered aggregate as construction material in concrete structures. *MATEC Web Conf.* **2018**, *174*, 2008. [CrossRef]
49. Özkan, H.; Kabay, N.; Miyan, N. Properties of Cold-Bonded and Sintered Aggregate Using Washing Aggregate Sludge and Their Incorporation in Concrete: A Promising Material. *Sustainability* **2022**, *14*, 4205. [CrossRef]
50. The Third Line of the Biggest LECA Lightweight Aggregate Production in the Region Was Lunched—Leca Asia. Available online: https://leca.asia/coming-soon-the-biggest-leca-lightweight-aggregate-production-in-the-region/ (accessed on 16 May 2022).
51. Tian, K.; Wang, Y.; Hong, S.; Zhang, J.; Hou, D.; Dong, B.; Xing, F. Alkali-activated artificial aggregates fabricated by red mud and fly ash: Performance and microstructure. *Constr. Build. Mater.* **2021**, *281*, 122552. [CrossRef]
52. Mohan, A.B.; Vasudev, R. Artificial Lightweight Aggregate Through Cold Bonding Pelletization of Fly Ash: A Review. *Int. Res. J. Eng. Technol.* **2018**, *5*, 778–783.
53. Tajra, F.; Elrahman, M.A.; Chung, S.-Y.; Stephan, D. Performance assessment of core-shell structured lightweight aggregate produced by cold bonding pelletization process. *Constr. Build. Mater.* **2018**, *179*, 220–231. [CrossRef]
54. Kockal, N.U.; Ozturan, T. Durability of lightweight concretes with lightweight fly ash aggregates. *Constr. Build. Mater.* **2010**, *25*, 1430–1438. [CrossRef]
55. Jiang, Y.; Ling, T.-C.; Shi, M. Strength enhancement of artificial aggregate prepared with waste concrete powder and its impact on concrete properties. *J. Clean. Prod.* **2020**, *257*, 120515. [CrossRef]
56. Güneyisi, E.; Gesoğlu, M.; Altan, İ.; Öz, H.Ö. Utilization of cold bonded fly ash lightweight fine aggregates as a partial substitution of natural fine aggregate in self-compacting mortars. *Constr. Build. Mater.* **2015**, *74*, 9–16. [CrossRef]
57. Vali, K.S.; Murugan, B. Impact of Nano SiO_2 on the Properties of Cold-bonded Artificial Aggregates with Various Binders. *Int. J. Technol.* **2019**, *10*, 897. [CrossRef]
58. Zafar, I.; Rashid, K.; Ju, M. Synthesis and characterization of lightweight aggregates through geopolymerization and microwave irradiation curing. *J. Build. Eng.* **2021**, *42*, 102454. [CrossRef]
59. Gomathi, P.; Sivakumar, A. Fly ash based lightweight aggregates incorporating clay binders. *Indian J. Eng. Mater. Sci.* **2014**, *21*, 227–232.
60. Kamal, J.; Mishra, U.K. Influence of Fly Ash Properties on Characteristics of Manufactured Angular Fly Ash Aggregates. *J. Inst. Eng. Ser. A* **2020**, *101*, 735–742. [CrossRef]
61. Vali, K.S.; Murugan, B. Effect of different binders on cold-bonded artificial lightweight aggregate properties. *Adv. Concr. Constr.* **2020**, *9*, 183–193.
62. Shahane, H.A.; Patel, S. Influence of curing method on characteristics of environment-friendly angular shaped cold bonded fly ash aggregates. *J. Build. Eng.* **2020**, *35*, 101997. [CrossRef]
63. Satpathy, H.; Patel, S.; Nayak, A. Development of sustainable lightweight concrete using fly ash cenosphere and sintered fly ash aggregate. *Constr. Build. Mater.* **2019**, *202*, 636–655. [CrossRef]
64. Shivaprasad, K.N.; Das, B.B. Effect of Duration of Heat Curing on the Artificially Produced Fly Ash Aggregates. *IOP Conf. Ser. Mater. Sci. Eng.* **2018**, *431*, 92013. [CrossRef]
65. Sun, Y.; Li, J.-S.; Chen, Z.; Xue, Q.; Sun, Q.; Zhou, Y.; Chen, X.; Liu, L.; Poon, C.S. Production of lightweight aggregate ceramsite from red mud and municipal solid waste incineration bottom ash: Mechanism and optimization. *Constr. Build. Mater.* **2021**, *287*, 122993. [CrossRef]
66. Liu, M.; Wang, C.; Bai, Y.; Xu, G. Effects of sintering temperature on the characteristics of lightweight aggregate made from sewage sludge and river sediment. *J. Alloy. Compd.* **2018**, *748*, 522–527. [CrossRef]
67. Risdanareni, P.; Ekaputri, J.J.; Triwulan. The effect of sintering temperature on the properties of metakaolin artificial lightweight aggregate. *AIP Conf. Proc.* **2017**, *1887*, 20045. [CrossRef]
68. Rehman, M.-U.; Rashid, K.; Zafar, I.; Alqahtani, F.K.; Khan, M.I. Formulation and characterization of geopolymer and conventional lightweight green concrete by incorporating synthetic lightweight aggregate. *J. Build. Eng.* **2020**, *31*, 101363. [CrossRef]
69. Risdanareni, P.; Choiri, A.A.; Djatmika, B.; Puspitasari, P. Effect of the Use of Metakaolin Artificial Lightweight Aggregate on the Properties of Structural Lightweight Concrete. *Civ. Eng. Dimens.* **2017**, *19*, 86–92. [CrossRef]
70. Narattha, C.; Chaipanich, A. Phase characterizations, physical properties and strength of environment-friendly cold-bonded fly ash lightweight aggregates. *J. Clean. Prod.* **2018**, *171*, 1094–1100. [CrossRef]
71. Dash, S.K.; Kar, B.B.; Mukherjee, P.S.; Mustakim, S.M. A Comparison Among the Physico-Chemical-Mechanical of Three Potential Aggregates Fabricated from Fly Ash. *J. Civ. Environ. Eng.* **2016**, *6*, 243.

72. Ding, C.; Sun, T.; Shui, Z.; Xie, Y.; Ye, Z. Physical properties, strength, and impurities stability of phosphogypsum-based cold-bonded aggregates. *Constr. Build. Mater.* **2022**, *331*, 127307. [CrossRef]
73. Ghosh, B. Autoclaved Fly-Ash Pellets as Replacement of Coarse Aggregate in Concrete Mixture. *AMC Indian J. Civ. Eng.* **2018**, *1*, 36. [CrossRef]
74. Aslam, A.; Rashid, K.; Naeem, G.; Saleem, N.; Hanif, S. Microwave Assisted Synthesis and Experimental Exploration of Geopolymer Lightweight Aggregate. *Tech. J. Univ. Eng. Technol.* **2020**, *25*, 8–16.
75. Saad, M.; Baalbaki, O.; Khatib, J.; El Kordi, A.; Masri, A. Manufacturing of Lightweight Aggregates from Municipal Solid Waste Incineration Bottom Ash and their Impacts on Concrete Properties. In Proceedings of the 2nd International Congress on Engineering and Architecture, Marmaris, Turkey, 22–24 April 2019; pp. 1638–1649.
76. Abdullah, A.; Abdullah, M.M.A.B.; Hussin, K.; Junaidi, S.; Tahir, M.F.M. Effect of fly ash/alkaline activator ratio on fly ash geopolymer artificial aggregate. *AIP Conf. Proc.* **2018**, *2045*, 20102. [CrossRef]
77. Choi, S.-J.; Kim, J.-H.; Bae, S.-H.; Oh, T.-G. Strength, Drying Shrinkage, and Carbonation Characteristic of Amorphous Metallic Fiber-Reinforced Mortar with Artificial Lightweight Aggregate. *Materials* **2020**, *13*, 4451. [CrossRef]
78. Ayati, B.; Ferrándiz-Mas, V.; Newport, D.; Cheeseman, C. Use of Clay in the Manufacture of Lightweight Aggregate. *Constr. Build. Mater.* **2018**, *162*, 124–131. [CrossRef]
79. Karyawan, I.D.; Ekaputri, J.J.; Widyatmoko, I.; Ariatedja, E. The effects of Na2SiO3/NaOH ratios on the volumetric properties of fly ash geopolymer artificial aggregates. *Mater. Sci. Forum* **2019**, *967*, 228–235. [CrossRef]
80. Razak, R.; Abdullah, M.M.A.B.; Hussin, K.; Ismail, K.N.; Hardjito, D.; Yahya, Z. Performances of Artificial Lightweight Geopolymer Aggregate (ALGA) in OPC Concrete. *Key Eng. Mater.* **2016**, *673*, 29–35. [CrossRef]
81. Sahoo, S.; Selvaraju, A.K.; Prakash, S.S. Mechanical characterization of structural lightweight aggregate concrete made with sintered fly ash aggregates and synthetic fibres. *Cem. Concr. Compos.* **2020**, *113*, 103712. [CrossRef]
82. Lau, P.; Teo, D.; Mannan, M. Mechanical, durability and microstructure properties of lightweight concrete using aggregate made from lime-treated sewage sludge and palm oil fuel ash. *Constr. Build. Mater.* **2018**, *176*, 24–34. [CrossRef]
83. *BS 1881-203*; Testing Concrete—Part 203: Recommendations for Measurement of Velocity of Ultrasonic Pulses in Concrete. European Committee for Standardization: Brussels, Belgium, 1986.
84. Othman, M.; Sarayreh, A.; Abdullah, R.; Sarbini, N.; Yassin, M.; Ahmad, H. Experimental Study on Lightweight Concrete using Lightweight Expanded Clay Aggregate (LECA) and Expanded Perlite Aggregate (EPA). *J. Eng. Sci. Technol.* **2020**, *15*, 1186–1201.

Review

Mapping Research Knowledge on Rice Husk Ash Application in Concrete: A Scientometric Review

Muhammad Nasir Amin [1,*], Waqas Ahmad [2], Kaffayatullah Khan [1] and Mohamed Mahmoud Sayed [3]

1. Department of Civil and Environmental Engineering, College of Engineering, King Faisal University, Al-Ahsa 31982, Saudi Arabia; kkhan@kfu.edu.sa
2. Department of Civil Engineering, COMSATS University Islamabad, Abbottabad 22060, Pakistan; waqasahmad@cuiatd.edu.pk
3. Architectural Department, Faculty of Engineering and Technology, Future University in Egypt, New Cairo 11845, Egypt; mohamed.mahmoud@fue.edu.eg
* Correspondence: mgadir@kfu.edu.sa; Tel.: +966-13-589-5431; Fax: +966-13-581-7068

Abstract: This study aimed to carry out a scientometric review of rice husk ash (RHA) concrete to assess the various aspects of the literature. Conventional review studies have limitations in terms of their capacity to connect disparate portions of the literature in a comprehensive and accurate manner. Science mapping, co-occurrence, and co-citation are a few of the most difficult phases of advanced research. The sources with the most articles, co-occurrences of keywords, the most prolific authors in terms of publications and citations, and areas actively involved in RHA concrete research are identified during the analysis. The Scopus database was used to extract bibliometric data for 917 publications that were then analyzed using the VOSviewer (version: 1.6.17) application. This study will benefit academics in establishing joint ventures and sharing innovative ideas and strategies because of the statistical and graphical representation of contributing authors and countries.

Keywords: rice husk ash; concrete; supplementary cementitious material; waste management; scientometric analysis; eco-friendly construction material

1. Introduction

Increased greenhouse gas (GHG) discharges have caused the melting of the Antarctic and Arctic polar ice caps. This has resulted in significant environmental problems on Earth [1]. The manufacture and transportation of building materials, as well as the installation and construction of structures, require considerable energy and produce significant volumes of GHG. In the European Union's member states, buildings use around 50% of the total energy consumption and contribute to almost 50% of the CO_2 emissions in the environment over their life cycle, which includes construction, operation, and destruction [2,3]. The building sector is still experiencing an increase in demand for concrete [4–9]. Ordinary Portland cement (OPC) is a critical component of concrete that contributes considerably to GHG emissions [10–13]. OPC production causes around 5–8% of worldwide CO_2 emissions [14–17]. Annual cement usage is over 4000 million tons and is predicted to reach approximately 6000 million tons by 2060 [18]. These GHG emissions have been a significant contributor to climate change [19–21]. In recent years, there has been a rise in the figure of thorough studies on the many triggers of climate change (natural and man-made), their effects on living conditions, and possible adaptation and mitigation techniques [22–27]. Blended cement manufacturing demands the use of a number of different cementitious components because of the higher energy and emission issues associated with OPC production [28]. Industrial waste utilization as supplementary cementitious materials (SCMs) is one of the methods that might cause a significant reduction in the usage of OPC, while also eliminating the risks connected with the disposal of waste materials from varied sectors [29–33]. Therefore, the most efficient technique for reducing the carbon footprint of the construction industry is to replace OPC with suitable alternative SCMs [34–37]. There are several binders that might be utilized in concrete to decrease GHG emissions from the concrete industry [38,39]. Utilizing recycled/waste materials in concrete is a viable method of mitigating the impact of environmental challenges [40]. This not only meets the increasing need for concrete, but also significantly reduces the direct danger to society [41]. Numerous researchers in the building sector have focused on the utilization of waste resources, particularly SCMs [42,43]. The production of environmentally friendly concrete has been critical to decreasing GHG emissions [44]. Agriculture wastes such as rick husk ash (RHA), sugarcane bagasse ash, olive oil ash, etc., as well as industrial wastes, are being utilized to partially replace OPC in the manufacture of sustainable concrete [45–49]. By polluting air and water systems, dumping these waste materials in the open ground creates a major environmental threat [50]. Globally, rice husk is produced by nearly 110 million tons and RHA by 22 million tons [51]. Rice husk is effectively and extensively used as a fuel in numerous nations for rice paddy milling operations and electricity production facilities [52]. This procedure results in the formation of a pozzolanic substance known as RHA, which contains more than 75% silica by weight (after incineration, 20% of the rice husk remains in the form of RHA) [23]. The ash formed by this operation is often dumped into water flows, contaminating the water and causing ecological damage [53]. Utilizing waste materials in concrete might enhance the durability and strength of the material owing to the pozzolanic effect [54]. This decreases industrial demand for OPC, lowering the expense of producing concrete and mitigating the negative impacts of CO_2 discharges during the OPC production process [28]. Given RHA's advantageous characteristics as an SCM, its use is not limited to cementitious concrete, but may also include geopolymer concrete, self-compacting concrete, fiber-reinforced concrete, pavement blocks, bricks, and high-performance nanocomposites [55–61].

The key properties of SCMs are their compatibility with aggregates (similar to OPC) and their better pozzolanic nature [62,63]. The application of RHA in concrete has sparked tremendous interest in the usage of sustainable and environmentally friendly SCM [64–67]. RHA has amorphous nature, high surface area, and compatibility with OPC-concrete, which results in outstanding pozzolanic capabilities [55,68–70]. Each kilogram of rice milled yields 0.28 kg of rice husk [71]. As a result, an enormous quantity of waste is generated annually. These rice husks are utilized as a fuel source in a variety of sectors to generate heat energy, including incineration and combustion units [72–74]. After the complete burning of rice husk, around 20–25% RHA by weight is formed [56]. A very small quantity of the RHA is subsequently employed as a field fertilizer, and sadly, most of it is thrown in open landfills [73,75]. RHA includes amorphous silica and calcium oxide and so may be utilized efficiently as an SCM in concrete [76–78]. Utilizing RHA in concrete results in better durability and strength, reduced material expenditures owing to OPC savings, and ecological advantages associated with waste material disposal [48]. RHA has been employed in recent studies as a partial replacement for OPC as well as fine aggregate in concrete mixes [79–81]. The properties of RHA concrete vary by the amount of OPC or fine aggregate replaced, the RHA grain size, the chemical characteristics of RHA independent of the water-cement ratio, and aggregate size/shape in the matrix [82]. However, for optimum strength growth, it is advised that around 10–25% of OPC be replaced [55,56]. The use of RHA in concrete has a number of benefits, as depicted in Figure 1. RHA has been researched for its possible use in cement-based composites as SCM or fine aggregate replacement. Also, natural aggregate extraction uses substantial energy and leads to increased CO_2 discharges [83]. As a result, issues about the manufacturing and use of OPC may be reduced, while natural resources can be conserved. Thus, including RHA into cementitious materials reduces the demand for OPC and fine aggregate and results in an ecologically beneficial building material. Furthermore, waste management issues can be alleviated by the use of RHA in construction materials.

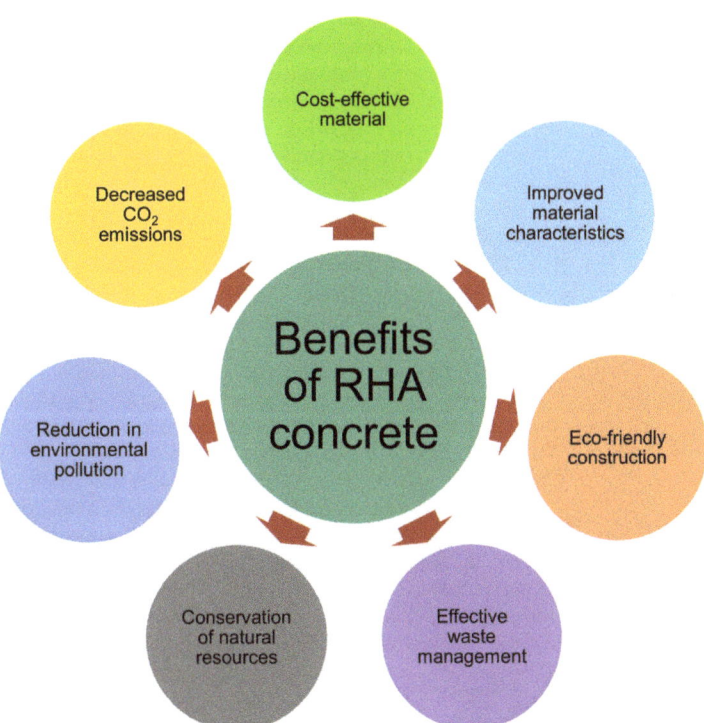

Figure 1. Benefits of RHA concrete.

As research on RHA concrete develops in response to the expanding environmental concerns, scientists face information constraints that may stymie creative investigation and scholarly collaboration. As a result, it is vital to create and apply a method that enables researchers to obtain critical information from the most reliable sources feasible. A scientometric method may assist in overcoming this shortcoming via the software application. The intention of this work is to conduct a scientometric analysis of bibliographic records published on RHA concrete up to 2021. Using a proper software tool, a scientometric analysis may undertake a quantitative examination of massive bibliometric data. Conventional review studies are weak in their ability to connect diverse sections of the literature in a complete and accurate manner. Science mapping, co-occurrence, and co-citation are a few of the most demanding parts of modern-day exploration [84–86]. The scientometric analysis identifies sources with the most articles, keyword co-occurrence, the most prolific authors in terms of papers and citations, and areas actively engaged in RHA concrete research. The Scopus database was utilized to extract bibliometric data for 917 relevant articles, which were then evaluated using the VOSviewer program. As a consequence of the statistical and graphical depiction of authors and countries, this study will aid academics in forming joint ventures and exchanging novel ideas and methods.

2. Methods

For the quantitative evaluation of the various features of the bibliographic data, this study carried out the scientometric analysis of the bibliographic data [87–89]. Numerous papers have been written on the issue, and it is critical to use a search engine that is reputable. Scopus and Web of Science are two very accurate search engines that are particularly well-suited for this purpose [90,91]. The bibliographic data for this study on RHA concrete were gathered using Scopus, which comes highly recommended by academics [92,93]. As of March 2022, a Scopus search for "rice husk ash concrete" found 1234 articles. Numerous filter preferences were employed to eradicate superfluous documents. The document types "journal article", "conference paper", "journal review", and "conference review" were selected. "Journal" and "conference proceeding" were chosen as the "source type". The "publication year" restriction was set to "2021", and the "language" constraint was set to "English". For further examination, the "subject areas" of "engineering", "material science", and "environmental science" were selected. A total of 917 records were kept following the application of these requirements. Numerous researchers have likewise reported on the same technique [94–96].

Scientometric investigations employ scientific mapping, a technique developed by academics for the purpose of analyzing bibliometric information [97]. Scopus records were saved in the Comma Separated Values (CSV) (see Supplementary Materials) files for further evaluation using appropriate computer software. VOSviewer (version: 1.6.17) was employed to generate the scientific visualization and quantitative assessment of the literature from the retrieved records. VOSviewer is an easily available and open-source mapping tool that is broadly employed across a range of areas and is well-suggested by academics [98–101]. As a result, the current study's goals were satisfied through the use of the VOSviewer. The obtained CSV files were loaded into the VOSviewer, and additional assessment was performed while retaining data integrity and consistency. During the bibliographic assessment, the sources of publications, the highly regularly appearing keywords, the scholars with the most publications and citations, and the country's participation were all assessed. The many facets, their relationships, and co-occurrence were shown graphically, while their statistical figures were reported in tables. The flowchart of the scientometric strategy is depicted in Figure 2.

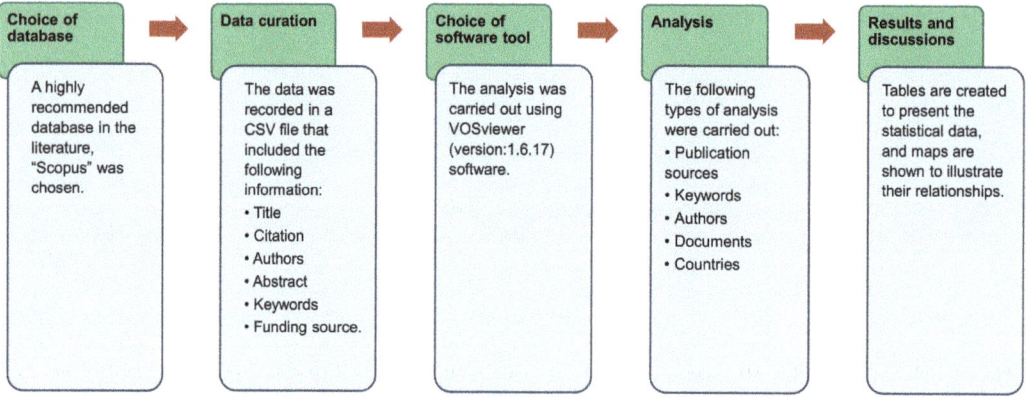

Figure 2. Sequence of the research methods.

3. Analysis of Results

3.1. Relevant Subject Areas and Yearly Publications

The Scopus analyzer was employed to carry out this analysis to discover the most pertinent study fields. Engineering, materials science, and environmental science were found to be the leading three document-producing areas, with around 39, 27, and 10% of documents, respectively, accounting for a total 76% of contributions based on document count, as seen in Figure 3. Additionally, as seen in Figure 4, the kind of paper was evaluated in the searched term in the Scopus database. According to this research, journal articles, conference papers, journal reviews, and conference reviews accounted for almost 66, 25, 7, and 2% of total documents, respectively. The yearly trend in publications in the present research area from 1977 to 2021 is depicted in Figure 5, since the first document on the subject research field was discovered in 1977. In the research of RHA concrete, a slow increase in the amount of publications was seen, with an average of roughly three papers per year up to 2000. Following this, there was a continuous increase in publications, with an average of roughly 20 papers each year from 2001 to 2016. The quantity of publications increased significantly during the previous five years (2017–2021), averaging approximately 110 each year.

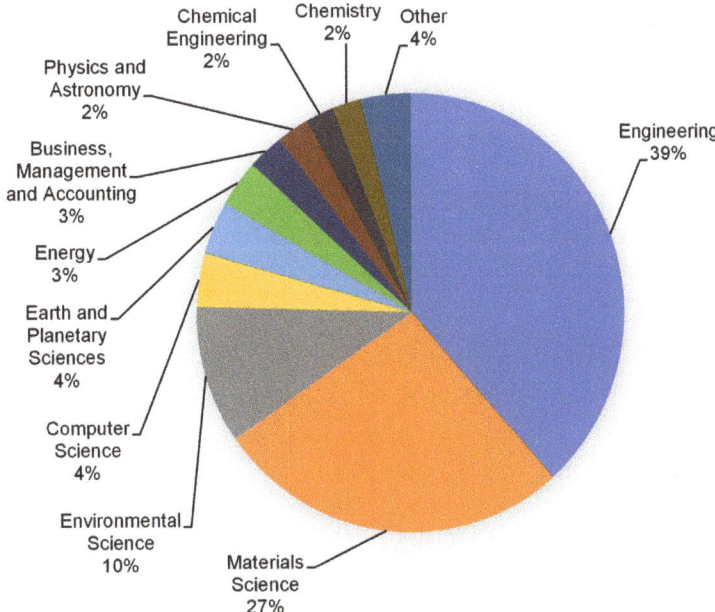

Figure 3. The subject area of articles.

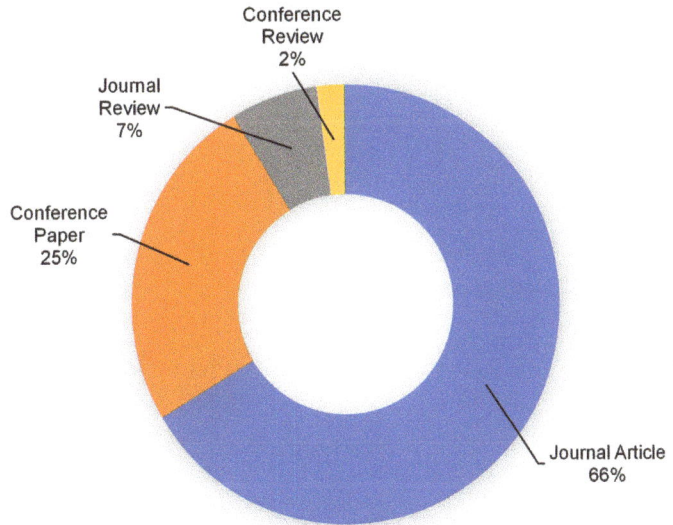

Figure 4. Various types of documents published in the related study field.

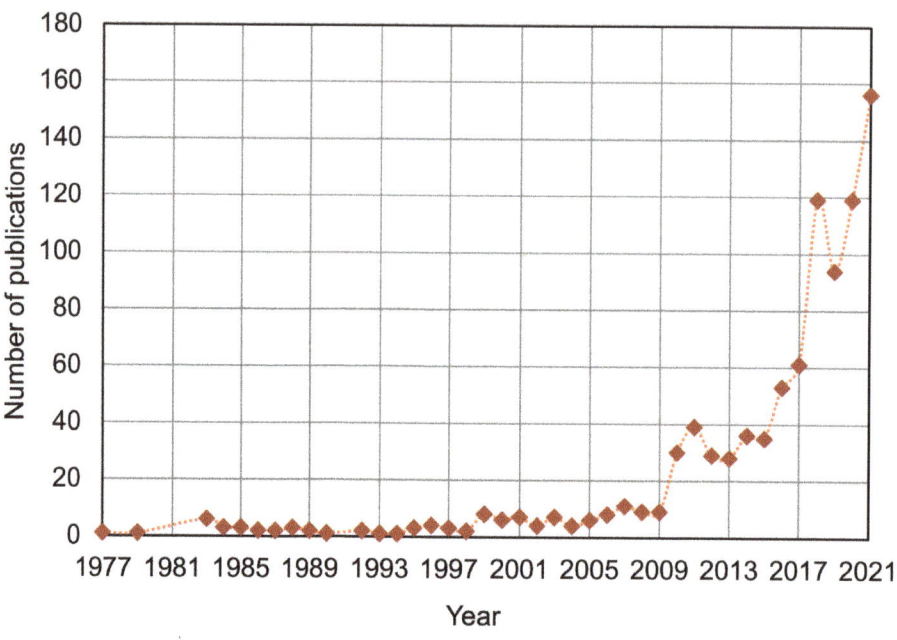

Figure 5. Annual publication trend of articles.

3.2. Sources of Publications

The assessment of publication sources was carried out using the VOSviewer on the collected bibliographic data. During the analysis, "bibliographic coupling" was selected as the "kind of analysis", while "sources" were retained as the "unit of analysis". At least ten papers per source restraint were set, and 14 of the 265 publication sources met these criteria. Table 1 shows the publishing sources that published a minimum of ten documents, providing data on RHA concrete, up to 2021, together with the amount of citations obtained during that time period. The main three sources/journals based on paper count are "Construction and building materials", "IOP conference series: materials science and engineering", and "Materials today: proceedings", with 110, 48, and 45 papers, respectively. Moreover, the top three sources based on the overall citations are "Construction and building materials" with 6797, "Cement and concrete composites" with 2268, and "Journal of cleaner production" with 1579. Remarkably, this exploration would provide a basis for upcoming scientometric investigations in the research of RHA concrete. In addition, prior traditional reviews were unable to generate scientific visualization maps.

Figure 6 illustrates a map of journals that have published a minimum of ten documents. The box size is proportional to the journal's impact on the current research area's document quantity; a bigger box dimension implies a superior impact. As an example, "Construction and building materials" has a bigger box than the others, implying that it is a source of considerable importance in that field. Five clusters were created, each of which is represented in the artwork by a different hue (red, blue, green, yellow, and purple). Clusters are formed on the basis of the research source's extent or the frequency with which they are co-cited in a similar article [102]. The VOSviewer created clusters of journals based on their co-citation patterns in published papers. For instance, the red cluster consists of six sources that have been co-cited several times in identical works. Additionally, nearly spaced frames (journals) in a cluster have stronger relationships than widely distributed frames. For instance, "Construction and building materials" is more strongly correlated with "Materials today: proceedings" than with "Journal of cleaner production".

Table 1. Publication sources with at least ten publications in the related research field up to 2021.

S/N	Publication Source	Number of Publications	Total Number of Citations
1	Construction and building materials	110	6797
2	IOP conference series: materials science and engineering	48	110
3	Materials today: proceedings	45	227
4	American concrete institute, ACI special publication	35	261
5	International journal of civil engineering and technology	32	54
6	IOP conference series: earth and environmental science	28	22
7	Journal of cleaner production	27	1579
8	Cement and concrete composites	21	2268
9	Journal of materials in civil engineering	17	424
10	Cement and concrete research	14	1539
11	materials	13	215
12	Journal of building engineering	10	144
13	International journal of applied engineering research	10	96
14	International journal of innovative technology and exploring engineering	10	10

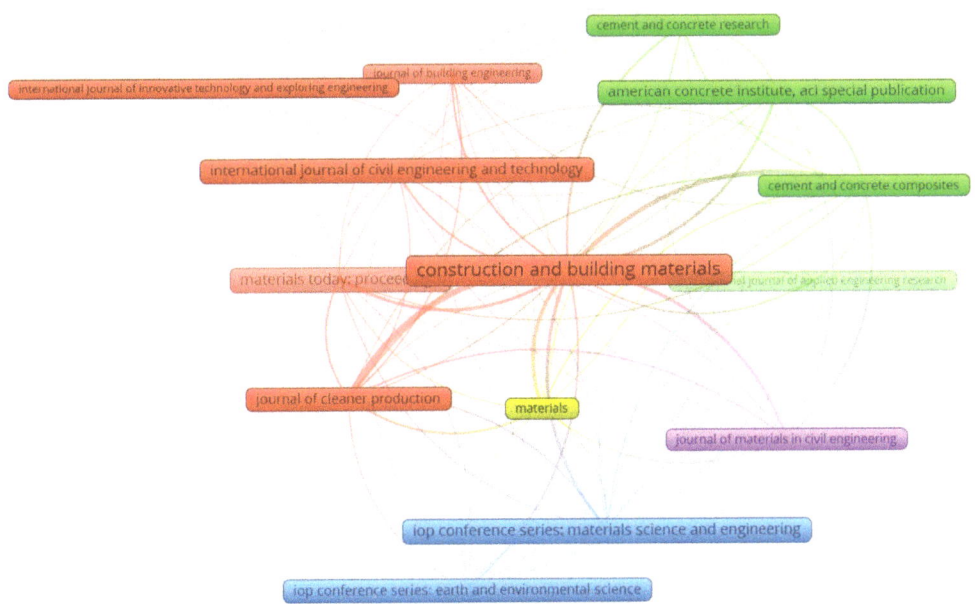

Figure 6. Scientific visualization of publication sources with at least ten publications in the related research area.

3.3. Keywords

Keywords are important in research because they define and highlight the study domain's fundamental subject [103]. The "analysis type" was set to "co-occurrence" and the "analysis unit" to "all keywords" for the evaluation. The least repetition constraint for a keyword was maintained at 20, and 96 of the 4185 keywords were retained. The leading 20 keywords most commonly used in published articles in the topic area are listed in Table 2. Rice husk ash, compressive strength, concretes, fly ash, and cements are the five most-often appearing keywords in the subject research area. According to the keyword analysis, RHA has been studied primarily as an SCM in normal concrete, self-compacting concrete, and high-performance concrete, as well as a precursor material in geopolymers. Figure 7 depicts a visualization map of keywords in terms of co-occurrences, linkages, and the density related to their frequency of occurrence. In Figure 7a, the size of a keyword circle implies its frequency, whereas its position implies its co-occurrence in articles. Also, the graph illustrates that the leading keywords have wider circles than the others, implying that they are critical terms for RHA concrete research. Clusters of keywords have been highlighted in the graph in a way that reflects their co-occurrence across a range of publications. The color-coded clustering is based on the co-occurrence of numerous keywords in published publications. The existence of four clusters is indicated by distinct colors (blue, red, green, and yellow) (Figure 7a). As seen in Figure 7b, different colors indicate varying concentrations of keyword density. The colors red, yellow, green, and blue are organized, corresponding to their density concentrations, with red indicating the highest and blue indicating the lowest density concentration. Compressive strength, rice husk ash, and concretes all exhibit red signs implying a higher concentration of density. This discovery will assist aspiring authors in choosing keywords that will facilitate the identification of published data in a certain field.

Table 2. The leading 20 frequently employed keywords in the research of RHA concrete.

S/N	Keyword	Occurrences
1	Rice husk ash	460
2	Compressive strength	402
3	Concretes	284
4	Fly ash	214
5	Cements	185
6	Concrete	173
7	Portland cement	154
8	Durability	146
9	Silica fume	126
10	Mechanical properties	106
11	Silica	103
12	Tensile strength	83
13	Water absorption	83
14	Slags	81
15	Chlorine compounds	78
16	Concrete mixtures	75
17	Agricultural wastes	72
18	Supplementary cementitious material	70
19	Aggregates	66
20	High performance concrete	64

3.4. Authors

Citations indicate a researcher's influence within a certain study domain [104]. For the evaluation of authors, the "kind of analysis" was chosen "co-authorship", and the "unit of analysis" was chosen "authors". The minimal paper restrictions for a writer were kept at 5, and 50 of the 2226 authors met this condition. Table 3 summarizes the most prolific authors in terms of publications and citations in the research of RHA concrete, as determined by data obtained from the Scopus search engine. The average citations for each author were calculated by dividing the total citations by the total publications. It will be difficult to quantify a scientist's efficacy when all factors such as the number of publications, total citations, and average citations are included. In contrast, the writer's assessment will be determined independently of each factor, i.e., total publications, total citations, and average citations. Nuruddin M.F. is the leading author with 16, followed by Zain M.F.M. and Mahmud H.B. with 14 each, and Shafiq N. with 13 publications. Jaturapitakkul C. leads the field in terms of total citations with 973, Zain M.F.M. is second with 738, and Chindaprasirt P. is third with 668 total citations in the current study area. Furthermore, when comparing average citations, the following writers stand out: Jaturapitakkul C. has around 97, Chindaprasirt P. has approximately 84, and Bui D.D. has approximately 82 average citations. Figure 8 illustrates the relationship between authors who have published at least ten publications and the most eminent authors. It was noticed that the largest set of connected authors based on citations are 6 of the 60 authors. This study revealed that a small number of writers are connected by citations in the research of RHA concrete.

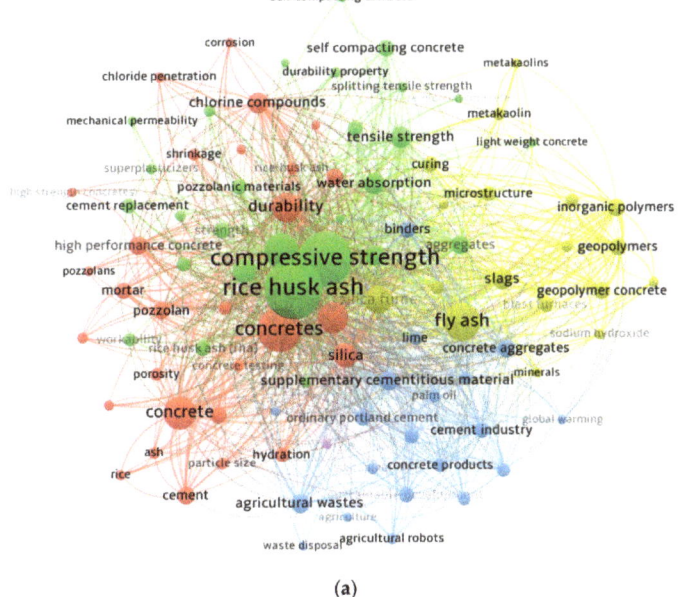

(a)

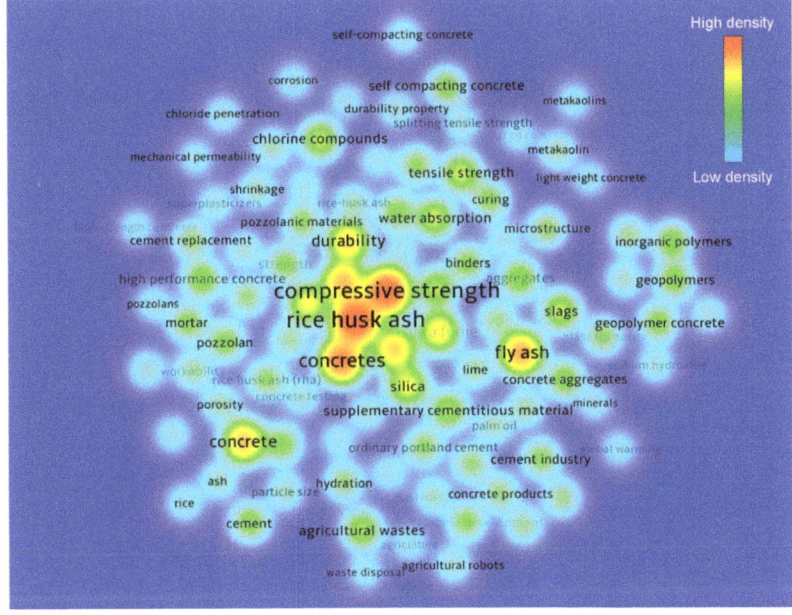

(b)

Figure 7. Keywords analysis: (**a**) scientific visualization; (**b**) density visualization.

Table 3. Authors with at least five publications in the research of RHA concrete up to 2021.

S/N	Author	Number of Publications	Total Number of Citations	Average Citations
1	Nuruddin M.F.	16	309	19
2	Zain M.F.M.	14	738	53
3	Mahmud H.B.	14	397	28
4	Shafiq N.	13	260	20
5	Jaturapitakkul C.	10	973	97
6	Makul N.	10	335	34
7	Isaia G.C.	9	635	71
8	Gastaldini A.L.G.	9	581	65
9	Stroeven P.	9	357	40
10	Rüscher C.H.	9	196	22
11	Ramadhansyah P.J.	9	90	10
12	Bahurudeen A.	9	56	6
13	Chindaprasirt P.	8	668	84
14	Siddique R.	8	487	61
15	Karim M.R.	8	380	48
16	Sua-Iam G.	8	334	42
17	Tchakouté H.K.	8	189	24
18	Nimityongskul P.	8	59	7
19	Jamil M.	7	441	63
20	Alengaram U.J.	7	222	32
21	Hainin M.R.	7	44	6
22	Mohamad N.	7	31	4
23	Bui D.D.	6	492	82
24	Ludwig H.-M.	6	437	73
25	Cordeiro G.C.	6	290	48
26	Soudki K.A.	6	248	41
27	Kamseu E.	6	214	36
28	Leonelli C.	6	214	36
29	Raman S.N.	6	213	36
30	Alyousef R.	6	73	12
31	Jaya R.P.	6	68	11
32	Murthi P.	6	24	4
33	Islam M.N.	5	327	65
34	Le H.T.	5	308	62
35	Safiuddin M.	5	247	49
36	West J.S.	5	247	49
37	Giaccio G.	5	229	46
38	Zerbino R.	5	229	46
39	Sugita S.	5	209	42
40	Jumaat M.Z.	5	167	33
41	Gobinath R.	5	64	13
42	Wan Ibrahim M.H.	5	60	12
43	Alabduljabbar H.	5	40	8
44	Fediuk R.	5	39	8
45	Hossain Z.	5	39	8
46	Samad A.A.A.	5	28	6
47	Jaini Z.M.	5	22	4
48	Hadipramana J.	5	10	2
49	Riza F.V.	5	10	2
50	Fang G.	5	0	0

3.5. Documents

The amount of citations a document obtains reflects its influence on a certain area of research. Papers with a high citation count are recognized as pioneers in their respective fields of research. For the assessment of documents, the "kind of analysis" was set to "bibliographic coupling" and "unit of analysis" to "documents". The least citations requirement for a document was 50, and 121 of 917 documents satisfied this requirement. The top ten papers in the area of RHA concrete by citations are included in Table 4, along with their writers and citation information. Ganesan K. [105] received 346 citations for their article "Rice husk ash blended cement: Assessment of optimal level of replacement for strength and permeability properties of concrete". G.C. Isaia [106] and D.-Y. Yoo [107] received 329 and 228 citations, respectively, for their publications and were positioned in the leading three. However, up until 2021, only 18 publications received more than 200 citations. In addition, Figure 9 illustrates the map of linked papers based on citations, as well as the density of those documents in the current study subject. The analysis revealed that 112 of 121 papers were linked by citations. Figure 9a illustrates the citation-based mapping of connected articles. Also, the density mapping (Figure 9b) reveals the top articles' enhanced density concentration.

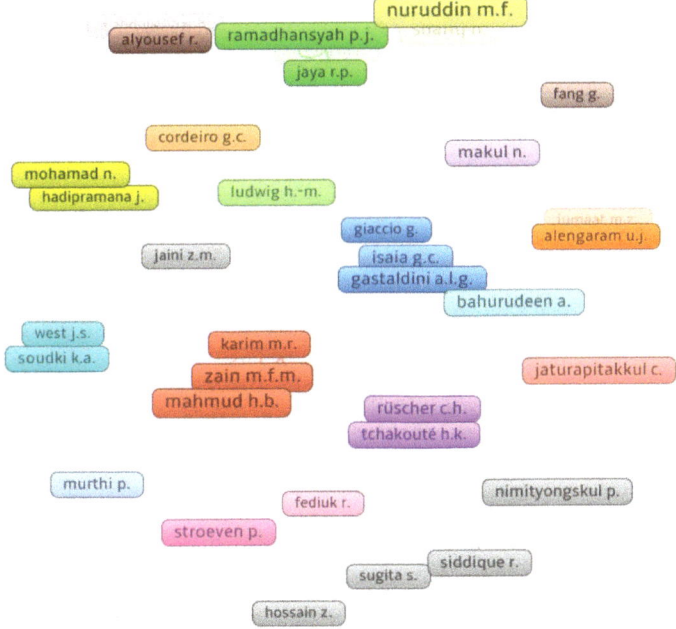

Figure 8. Scientific visualization of authors that published articles in the related research area.

Table 4. The top ten highly cited published articles up to 2021 in the research of RHA concrete.

S/N	Article	Title	Total Number of Citations Received
1	Ganesan K. [105]	Rice husk ash blended cement: Assessment of optimal level of replacement for strength and permeability properties of concrete	346
2	Isaia G.C. [106]	Physical and pozzolanic action of mineral additions on the mechanical strength of high-performance concrete	329
3	Yoo D.-Y. [107]	Mechanical properties of ultra-high-performance fiber-reinforced concrete: A review	288
4	Bui D.D. [108]	Particle size effect on the strength of rice husk ash blended gap-graded Portland cement concrete	283
5	Sata V. [109]	Influence of pozzolan from various by-product materials on mechanical properties of high-strength concrete	278
6	Zhang M.-H. [110]	High-performance concrete incorporating rice husk ash as a supplementary cementing material	271
7	Rodríguez De Sensale G. [53]	Strength development of concrete with rice-husk ash	261
8	Nehdi M. [111]	Performance of rice husk ash produced using a new technology as a mineral admixture in concrete	257
9	Paris J.M. [112]	A review of waste products utilized as supplements to Portland cement in concrete	232
10	Wongpa J. [113]	Compressive strength, modulus of elasticity, and water permeability of inorganic polymer concrete	223

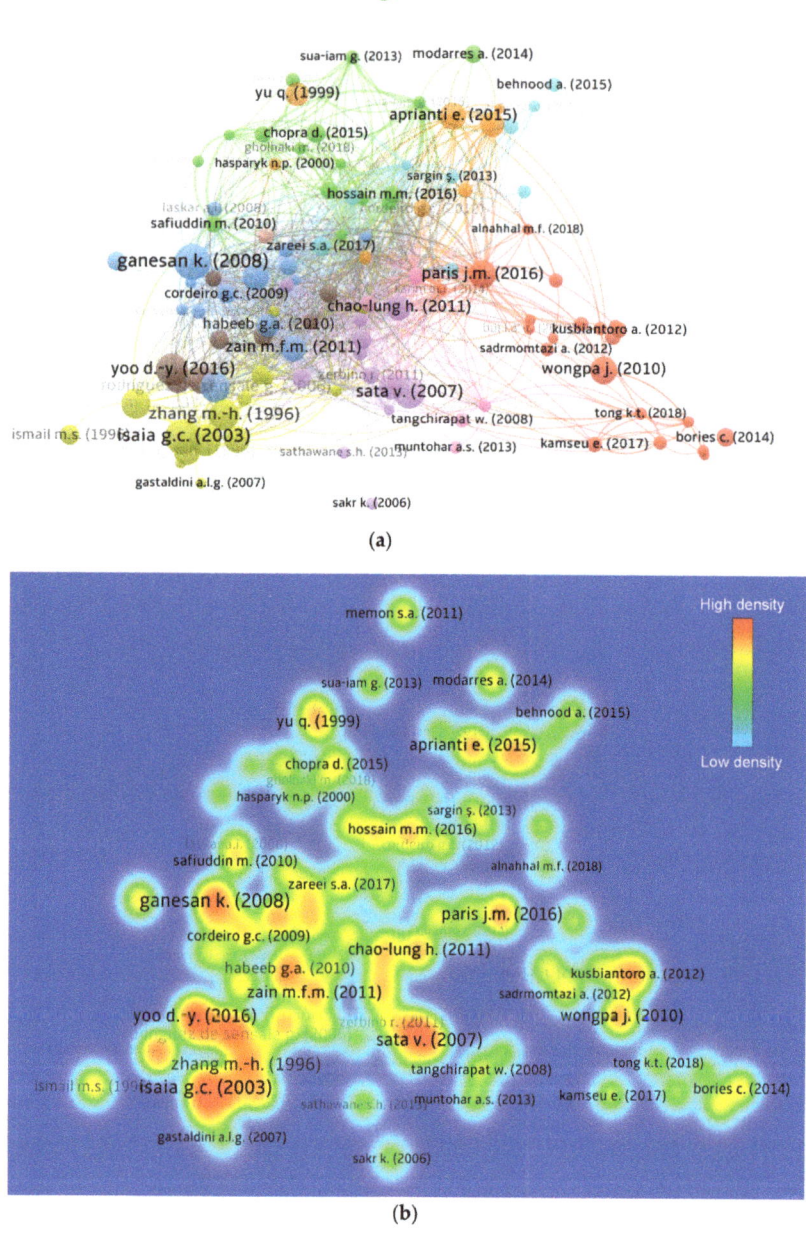

Figure 9. Scientific mapping of published articles in the related research area up to 2021; (**a**) connected articles in terms of citations, (**b**) density of connected articles.

3.6. Countries

Several countries have contributed more to current research than others have and are expected to contribute further. The network map was created to allow readers to view areas committed to the research of RHA concrete. "Bibliographic coupling" was selected as the "kind of analysis", and "countries" as the "unit of analysis". The minimum document limit for a nation was set at 10, and 27 countries met this requirement. The nations listed in Table 5 have published at least ten documents in the present study field. India, Malaysia, and Thailand presented the most papers with 293, 133, and 48 documents. Moreover, these nations received the most citations, with Malaysia receiving 3104, India receiving 3098, and Thailand receiving 2049 citations. Figure 10 illustrates the visualization of the science mapping as well as the density of nations connected via citations. The size of a box is proportional to a nation's effect on the subject research (Figure 10a). The nations with the most engagement had a higher density, as indicated by the density visualization (Figure 10b). The statistical and graphical analysis of the contributing states will aid emerging researchers in establishing scientific alliances, forming joint ventures, and exchanging innovative techniques and ideas. Researchers from nations interested in promoting research on RHA concrete can work with experts in the field and profit from their experience.

Table 5. Leading countries based on published documents in the present research area until 2021.

S/N	Country	Number of Publications	Total Number of Citations
1	India	293	3098
2	Malaysia	133	3104
3	Thailand	48	2049
4	Indonesia	46	271
5	Iran	44	1528
6	Brazil	39	1294
7	United States	39	1180
8	China	39	818
9	Pakistan	37	728
10	Nigeria	28	360
11	Germany	25	729
12	Canada	22	1270
13	Vietnam	21	1334
14	Australia	21	459
15	Saudi Arabia	21	225
16	United Kingdom	19	394
17	Bangladesh	18	481
18	Netherlands	14	834
19	Japan	14	450
20	Russian Federation	13	186
21	Iraq	12	225
22	Spain	12	153
23	Turkey	11	358
24	Cameroon	11	295
25	South Korea	10	736
26	Egypt	10	240
27	Colombia	10	223

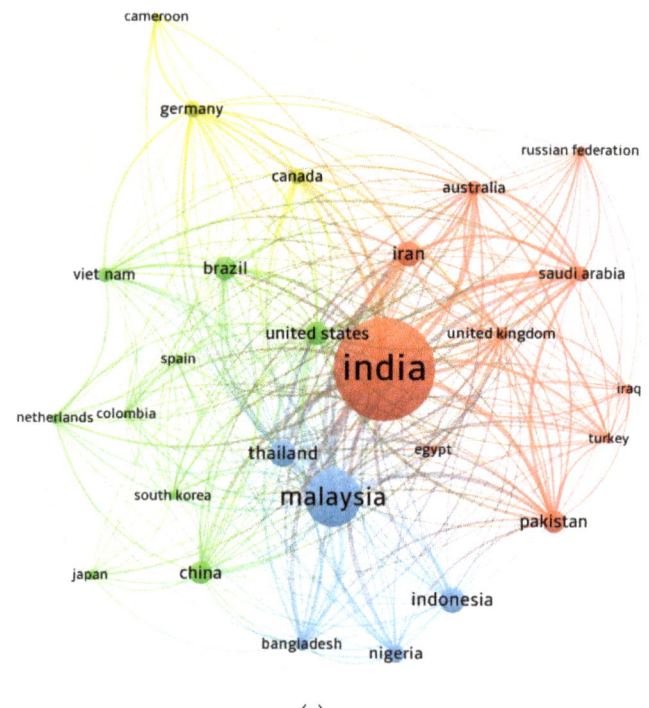

(a)

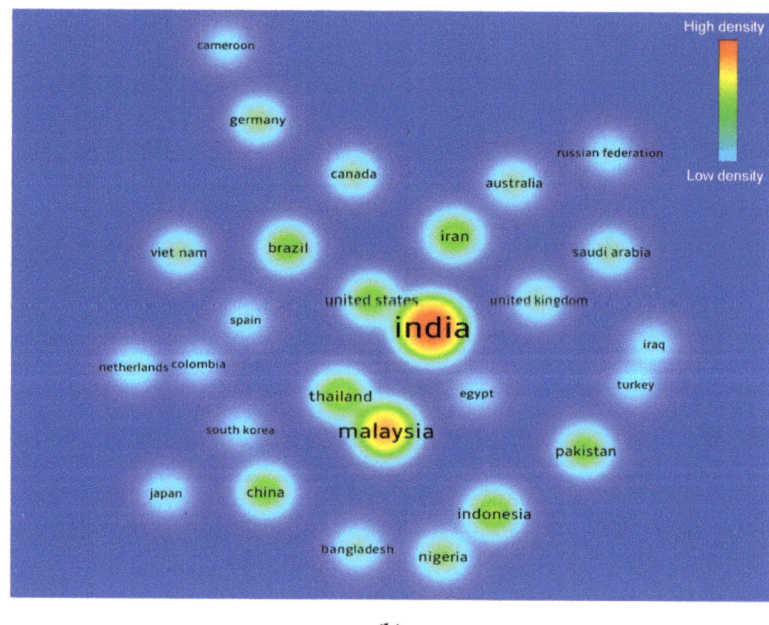

(b)

Figure 10. Scientific visualization countries with at least ten publications in the related research area up to 2021: (**a**) network visualization; (**b**) density visualization.

4. Discussions and Future Perspectives

This study provided a statistical overview and mapping of various aspects of the literature on RHA concrete. Previous manual review studies have limitations in terms of their ability to comprehensively and accurately connect diverse sections of the literature. This study identified sources (journals) that published most articles, most commonly employed keywords in the published papers, articles and authors having most citations, and countries actively involved in the research of RHA concrete. The analysis of keywords identified that RHA had been examined for its possible applications as SCM in conventional concrete, self-compacting concrete, and high-performance concrete due to the presence of high silica content in its chemical composition [114–118]. In addition, the use of RHA is also researched for manufacturing geopolymer concrete [119–121]. RHA provides several advantages when used in concrete. RHA has been investigated for prospective use as a cement or fine aggregate substitute in concrete. The issues associated with manufacturing and the use of cement might be decreased [122]. Also, because natural aggregate extraction consumes a significant amount of energy and results in higher CO_2 emissions [123]. As a consequence, concerns regarding natural resource depletion may be alleviated. Thus, the incorporation of RHA into concrete minimizes the need for cement and fine aggregate, resulting in a more environmentally friendly construction material [124]. By incorporating RHA into construction materials, waste management difficulties can be solved [73]. In addition, the most active and contributing countries in terms of publications were identified from the literature and their connections based on citations. The statistical and graphical representations of the contributing states will assist developing scholars in creating scientific partnerships, establishing joint ventures, and exchanging novel approaches and ideas. Researchers from countries interested in advancing RHA concrete research can collaborate with professionals in the area and benefit from their knowledge.

Most of the RHA applications stated above are still in their development, and more in-depth analyses are necessary before broadening their applicability [71]. Furthermore, in the present practice, the utilization of RHA concrete in full-scale reinforced concrete structures under service and high loading circumstances has not been examined. Additionally, there are currently no clear standards for the preparation, processing, and use of RHA on a larger scale. In the available literature, researchers have solely relied on their intuition to determine the optimal degree of cement and fine aggregate replacement using RHA [105]. Additionally, previous work has not explored the compatibility and long-term durability of RHA concrete. Steel reinforcement corrosion in RHA-blended concrete must be researched in water, chloride, sulphate, and acidic environments over an extended period of time. Also, because information on the life cycle evaluation of RHA concrete is limited and needs to be thoroughly examined. To enhance the strength of concrete, alternative and supplemental additives such as nano-silica and fibers can be added to RHA concrete. Additionally, the high concentration and coarser character of RHA allow for the formation of a porous and less dense matrix of the concrete. Nonetheless, the addition of nano-clay, short fibers, and nano-silica to concrete has demonstrated the ability to increase its density, shock resistance, and tensile stress resistance. As a result, these additives combined with RHA-blended concrete may provide another sustainable material for future construction.

5. Conclusions

The objective of this study was to conduct a scientometric analysis of the available literature on rice husk ash (RHA) concrete in order to assess various measures. The Scopus database was queried for 917 relevant papers, and the results were analyzed using the VOSviewer program. The following findings were drawn from this study:

- An analysis of publication sources containing documents on RHA concrete research exposed that the topmost three sources are "Construction and building materials", "IOP conference series: materials science and engineering", and "Materials today: proceedings", with 110, 48, and 45 papers, respectively, Also, the leading three publication sources in terms of overall citations are "Construction and building materials"

- with 6797, "Cement and concrete composites" with 2268, and "Journal of cleaner production" with 1579.
- A keyword analysis of the subject study field shows that the five most-often appearing keywords are rice husk ash, compressive strength, concretes, fly ash, and cements. The keyword analysis revealed that RHA had been studied primarily as a supplemental cementitious material (SCM) in concrete.
- Author analysis revealed that only 50 writers had published at least five publications on RHA concrete research. The top writers were classified according to their number of publications, citations, and average citations. Nuruddin M.F., with 16, Zain M.F.M., and Mahmud H.B., with 14 each, and Shafiq N., with 13 papers, are the top three authors in terms of overall publications. With 973 citations, Jaturapitakkul C. leads the field, followed by Zain M.F.M. with 738 and Chindaprasirt P. with 668 citations until 2021. In addition, when the average number of citations is compared, the following authors stand out: C. Jaturapitakkul has around 97, P. Chindaprasirt has approximately 84, and D.D. Bui has approximately 82 average citations.
- According to an analysis of papers providing data on RHA concrete, Ganesan K. [105] received 346 citations for their article "Rice husk ash blended cement: Assessment of optimal level of replacement for strength and permeability properties of concrete". G.C. Isaia [106] and D.-Y. Yoo [107] received 329 and 228 citations, respectively, for their publications and were positioned in the best three. Moreover, only 18 publications acquired more than 200 citations in the subject area from 2011 to 2021.
- The leading nations were assessed based on their participation in RHA concrete research, and it was determined that only 27 countries published at least ten papers. India, Malaysia, and Thailand each delivered 293, 133, and 48 papers, respectively. In addition, these nations received the most citations, with Malaysia receiving 3104, India receiving 3098, and Thailand receiving 2049 citations.
- RHA has been investigated for its potential uses as SCM in conventional concrete, self-compacting concrete, and high-performance concrete because of the high silica concentration in its chemical composition. Furthermore, RHA is being investigated for application in the production of geopolymer concrete.
- The application of RHA in the construction sector will result in green construction by reducing cement demand and conserving natural sources when used as a substitute for cement and fine aggregate.
- The majority of the RHA applications are still under investigation, and further analysis is required before widening their effectiveness.

Supplementary Materials: The following supporting information can be downloaded at: https://www.mdpi.com/article/10.3390/ma15103431/s1. Table S1: Data retrieved from the Scopus database and used for the analysis.

Author Contributions: M.N.A.: conceptualization, funding acquisition, resources, project administration, supervision, writing, reviewing, and editing; W.A.: conceptualization, data curation, software, methodology, investigation, validation, writing—original draft; K.K.: methodology, investigation, writing, reviewing, and editing; M.M.S.: resources, visualization, writing, reviewing, and editing. All authors have read and agreed to the published version of the manuscript.

Funding: This work was supported by the Deanship of Scientific Research, Vice Presidency for Graduate Studies and Scientific Research, King Faisal University, Saudi Arabia [Project No. GRANT461]. The APC was funded by the same "Project No. GRANT461".

Institutional Review Board Statement: Not applicable.

Informed Consent Statement: Not applicable.

Data Availability Statement: The data used in this research have been properly cited and reported in the main text.

Acknowledgments: The authors acknowledge the support from the Deanship of Scientific Research, Vice Presidency for Graduate Studies and Scientific Research, King Faisal University, Saudi Arabia [Project No. GRANT461]. The authors extend their appreciation for the financial support that has made this study possible.

Conflicts of Interest: The authors declare no conflict of interest.

References

1. Eijgelaar, E.; Thaper, C.; Peeters, P. Antarctic cruise tourism: The paradoxes of ambassadorship, "last chance tourism" and greenhouse gas emissions. *J. Sustain. Tour.* **2010**, *18*, 337–354. [CrossRef]
2. Yan, H.; Shen, Q.; Fan, L.C.H.; Wang, Y.; Zhang, L. Greenhouse gas emissions in building construction: A case study of One Peking in Hong Kong. *Build. Environ.* **2010**, *45*, 949–955. [CrossRef]
3. Dimoudi, A.; Tompa, C. Energy and environmental indicators related to construction of office buildings. *Resour. Conserv. Recycl.* **2008**, *53*, 86–95. [CrossRef]
4. Khan, M.; Cao, M.; Chaopeng, X.; Ali, M. Experimental and analytical study of hybrid fiber reinforced concrete prepared with basalt fiber under high temperature. *Fire Mater.* **2021**, *46*, 205–226. [CrossRef]
5. Li, L.; Khan, M.; Bai, C.; Shi, K. Uniaxial Tensile Behavior, Flexural Properties, Empirical Calculation and Microstructure of Multi-Scale Fiber Reinforced Cement-Based Material at Elevated Temperature. *Materials* **2021**, *14*, 1827. [CrossRef]
6. Zhou, S.; Xie, L.; Jia, Y.; Wang, C. Review of Cementitious Composites Containing Polyethylene Fibers as Repairing Materials. *Polymers* **2020**, *12*, 2624. [CrossRef]
7. Khan, M.; Cao, M.; Xie, C.; Ali, M. Effectiveness of hybrid steel-basalt fiber reinforced concrete under compression. *Case Stud. Constr. Mater.* **2022**, *16*, e00941. [CrossRef]
8. Khan, M.; Cao, M.; Ai, H.; Hussain, A. Basalt Fibers in Modified Whisker Reinforced Cementitious Composites. *Period. Polytech. Civ. Eng.* **2022**. [CrossRef]
9. Yuan, X.; Tian, Y.; Ahmad, W.; Ahmad, A.; Usanova, K.I.; Mohamed, A.M.; Khallaf, R. Machine Learning Prediction Models to Evaluate the Strength of Recycled Aggregate Concrete. *Materials* **2022**, *15*, 2823. [CrossRef]
10. Ahmad, W.; Ahmad, A.; Ostrowski, K.A.; Aslam, F.; Joyklad, P.; Zajdel, P. Sustainable approach of using sugarcane bagasse ash in cement-based composites: A systematic review. *Case Stud. Constr. Mater.* **2021**, *15*, e00698. [CrossRef]
11. Ahmad, A.; Farooq, F.; Niewiadomski, P.; Ostrowski, K.; Akbar, A.; Aslam, F.; Alyousef, R. Prediction of compressive strength of fly ash based concrete using individual and ensemble algorithm. *Materials* **2021**, *14*, 794. [CrossRef] [PubMed]
12. Amran, M.; Fediuk, R.; Murali, G.; Avudaiappan, S.; Ozbakkaloglu, T.; Vatin, N.; Karelina, M.; Klyuev, S.; Gholampour, A. Fly Ash-Based Eco-Efficient Concretes: A Comprehensive Review of the Short-Term Properties. *Materials* **2021**, *14*, 4264. [CrossRef] [PubMed]
13. Amran, M.; Murali, G.; Fediuk, R.; Vatin, N.; Vasilev, Y.; Abdelgader, H. Palm oil fuel ash-based eco-efficient concrete: A critical review of the short-term properties. *Materials* **2021**, *14*, 332. [CrossRef] [PubMed]
14. Khan, M.I.; Abbas, Y.M.; Fares, G. Review of high and ultrahigh performance cementitious composites incorporating various combinations of fibers and ultrafines. *J. King Saud Univ. -Eng. Sci.* **2017**, *29*, 339–347. [CrossRef]
15. Martuscelli, C.; Soares, C.; Camões, A.; Lima, N. Potential of fungi for concrete repair. *Procedia Manuf.* **2020**, *46*, 180–185. [CrossRef]
16. Irshidat, M.R.; Al-Nuaimi, N. Industrial Waste Utilization of Carbon Dust in Sustainable Cementitious Composites Production. *Materials* **2020**, *13*, 3295. [CrossRef]
17. Nafees, A.; Javed, M.F.; Khan, S.; Nazir, K.; Farooq, F.; Aslam, F.; Musarat, M.A.; Vatin, N.I. Predictive Modeling of Mechanical Properties of Silica Fume-Based Green Concrete Using Artificial Intelligence Approaches: MLPNN, ANFIS, and GEP. *Materials* **2021**, *14*, 7531. [CrossRef]
18. Shahmansouri, A.A.; Bengar, H.A.; Ghanbari, S. Compressive strength prediction of eco-efficient GGBS-based geopolymer concrete using GEP method. *J. Build. Eng.* **2020**, *31*, 101326. [CrossRef]
19. Miller, S.A. Supplementary cementitious materials to mitigate greenhouse gas emissions from concrete: Can there be too much of a good thing? *J. Clean. Prod.* **2018**, *178*, 587–598. [CrossRef]
20. Habert, G.; Miller, S.A.; John, V.M.; Provis, J.L.; Favier, A.; Horvath, A.; Scrivener, K.L. Environmental impacts and decarbonization strategies in the cement and concrete industries. *Nat. Rev. Earth Environ.* **2020**, *1*, 559–573. [CrossRef]
21. Di Filippo, J.; Karpman, J.; DeShazo, J.R. The impacts of policies to reduce CO_2 emissions within the concrete supply chain. *Cem. Concr. Compos.* **2019**, *101*, 67–82. [CrossRef]
22. Lee, C.T.; Hashim, H.; Ho, C.S.; Van Fan, Y.; Klemeš, J.J. Sustaining the low-carbon emission development in Asia and beyond: Sustainable energy, water, transportation and low-carbon emission technology. *J. Clean. Prod.* **2017**, *146*, 1–13. [CrossRef]
23. Li, X.; Qin, D.; Hu, Y.; Ahmad, W.; Ahmad, A.; Aslam, F.; Joyklad, P. A systematic review of waste materials in cement-based composites for construction applications. *J. Build. Eng.* **2021**, *45*, 103447. [CrossRef]
24. Zhang, B.; Ahmad, W.; Ahmad, A.; Aslam, F.; Joyklad, P. A scientometric analysis approach to analyze the present research on recycled aggregate concrete. *J. Build. Eng.* **2021**, *46*, 103679. [CrossRef]

25. Alyousef, R.; Ahmad, W.; Ahmad, A.; Aslam, F.; Joyklad, P.; Alabduljabbar, H. Potential use of recycled plastic and rubber aggregate in cementitious materials for sustainable construction: A review. *J. Clean. Prod.* **2021**, *329*, 129736. [CrossRef]
26. El-Kassas, A.M.; Elsheikh, A.H. A new eco-friendly mechanical technique for production of rice straw fibers for medium density fiberboards manufacturing. *Int. J. Environ. Sci. Technol.* **2021**, *18*, 979–988. [CrossRef]
27. Elsheikh, A.H.; Panchal, H.; Shanmugan, S.; Muthuramalingam, T.; El-Kassas, A.M.; Ramesh, B. Recent progresses in wood-plastic composites: Pre-processing treatments, manufacturing techniques, recyclability and eco-friendly assessment. *Clean. Eng. Technol.* **2022**, *8*, 100450. [CrossRef]
28. Mikulčić, H.; Klemeš, J.J.; Vujanović, M.; Urbaniec, K.; Duić, N. Reducing greenhouse gasses emissions by fostering the deployment of alternative raw materials and energy sources in the cleaner cement manufacturing process. *J. Clean. Prod.* **2016**, *136*, 119–132. [CrossRef]
29. Khan, M.; Ali, M. Improvement in concrete behavior with fly ash, silica-fume and coconut fibres. *Constr. Build. Mater.* **2019**, *203*, 174–187. [CrossRef]
30. Khan, M.; Cao, M.; Hussain, A.; Chu, S.H. Effect of silica-fume content on performance of $CaCO_3$ whisker and basalt fiber at matrix interface in cement-based composites. *Constr. Build. Mater.* **2021**, *300*, 124046. [CrossRef]
31. Khan, M.; Rehman, A.; Ali, M. Efficiency of silica-fume content in plain and natural fiber reinforced concrete for concrete road. *Constr. Build. Mater.* **2020**, *244*, 118382. [CrossRef]
32. Park, S.; Wu, S.; Liu, Z.; Pyo, S. The role of supplementary cementitious materials (SCMs) in ultra high performance concrete (UHPC): A review. *Materials* **2021**, *14*, 1472. [CrossRef] [PubMed]
33. Xu, Q.; Ji, T.; Gao, S.-J.; Yang, Z.; Wu, N. Characteristics and applications of sugar cane bagasse ash waste in cementitious materials. *Materials* **2019**, *12*, 39. [CrossRef] [PubMed]
34. Supino, S.; Malandrino, O.; Testa, M.; Sica, D. Sustainability in the EU cement industry: The Italian and German experiences. *J. Clean. Prod.* **2016**, *112*, 430–442. [CrossRef]
35. Juenger, M.C.G.; Snellings, R.; Bernal, S.A. Supplementary cementitious materials: New sources, characterization, and performance insights. *Cem. Concr. Res.* **2019**, *122*, 257–273. [CrossRef]
36. Lothenbach, B.; Scrivener, K.; Hooton, R.D. Supplementary cementitious materials. *Cem. Concr. Res.* **2011**, *41*, 1244–1256. [CrossRef]
37. Shanmugasundaram, N.; Praveenkumar, S. Influence of supplementary cementitious materials, curing conditions and mixing ratios on fresh and mechanical properties of engineered cementitious composites—A review. *Constr. Build. Mater.* **2021**, *309*, 125038. [CrossRef]
38. Shi, C.; Qu, B.; Provis, J.L. Recent progress in low-carbon binders. *Cem. Concr. Res.* **2019**, *122*, 227–250. [CrossRef]
39. Zheng, Y.; Wang, C.; Zhou, S.; Luo, C. The self-gelation properties of calcined wollastonite powder. *Constr. Build. Mater.* **2021**, *290*, 123061. [CrossRef]
40. Ahmad, W.; Ahmad, A.; Ostrowski, K.A.; Aslam, F.; Joyklad, P. A scientometric review of waste material utilization in concrete for sustainable construction. *Case Stud. Constr. Mater.* **2021**, *15*, e00683. [CrossRef]
41. Tam, V.W.Y.; Tam, C.M. A review on the viable technology for construction waste recycling. *Resour. Conserv. Recycl.* **2006**, *47*, 209–221. [CrossRef]
42. Tang, P.; Chen, W.; Xuan, D.; Zuo, Y.; Poon, C.S. Investigation of cementitious properties of different constituents in municipal solid waste incineration bottom ash as supplementary cementitious materials. *J. Clean. Prod.* **2020**, *258*, 120675. [CrossRef]
43. He, Z.-H.; Zhu, H.-N.; Zhang, M.-Y.; Shi, J.-Y.; Du, S.-G.; Liu, B. Autogenous shrinkage and nano-mechanical properties of UHPC containing waste brick powder derived from construction and demolition waste. *Constr. Build. Mater.* **2021**, *306*, 124869. [CrossRef]
44. Shi, J.; Liu, B.; He, Z.; Liu, Y.; Jiang, J.; Xiong, T.; Shi, J. A green ultra-lightweight chemically foamed concrete for building exterior: A feasibility study. *J. Clean. Prod.* **2021**, *288*, 125085. [CrossRef]
45. Aprianti, E. A huge number of artificial waste material can be supplementary cementitious material (SCM) for concrete production–a review part II. *J. Clean. Prod.* **2017**, *142*, 4178–4194. [CrossRef]
46. Sohal, K.S.; Singh, R. *Sustainable Use of Sugarcane Bagasse Ash in Concrete Production*; Springer: Singapore, 2021; pp. 397–407.
47. Abdulkadir, T.S.; Oyejobi, D.O.; Lawal, A.A. Evaluation of sugarcane bagasse ash as a replacement for cement in concrete works. *Acta Tech. Corviniensis-Bull. Eng.* **2014**, *7*, 71.
48. Ahmed, A.; Hyndman, F.; Kamau, J.; Fitriani, H. *Rice Husk Ash as a Cement Replacement in High Strength Sustainable Concrete*; Trans Tech Publications Ltd.: Bäch, Switzerland, 2020; pp. 90–98.
49. Zareei, S.A.; Ameri, F.; Dorostkar, F.; Ahmadi, M. Rice husk ash as a partial replacement of cement in high strength concrete containing micro silica: Evaluating durability and mechanical properties. *Case Stud. Constr. Mater.* **2017**, *7*, 73–81. [CrossRef]
50. Ihedioha, J.N.; Ukoha, P.O.; Ekere, N.R. Ecological and human health risk assessment of heavy metal contamination in soil of a municipal solid waste dump in Uyo, Nigeria. *Environ. Geochem. Health* **2017**, *39*, 497–515. [CrossRef]
51. Iftikhar, B.; Alih, S.C.; Vafaei, M.; Elkotb, M.A.; Shutaywi, M.; Javed, M.F.; Deebani, W.; Khan, M.I.; Aslam, F. Predictive modeling of compressive strength of sustainable rice husk ash concrete: Ensemble learner optimization and comparison. *J. Clean. Prod.* **2022**, *348*, 131285. [CrossRef]
52. Prasara-A, J.; Gheewala, S.H. Sustainable utilization of rice husk ash from power plants: A review. *J. Clean. Prod.* **2017**, *167*, 1020–1028. [CrossRef]

53. De Sensale, G.R. Strength development of concrete with rice-husk ash. *Cem. Concr. Compos.* **2006**, *28*, 158–160. [CrossRef]
54. He, Z.-H.; Yang, Y.; Yuan, Q.; Shi, J.-Y.; Liu, B.-J.; Liang, C.-F.; Du, S.-G. Recycling hazardous water treatment sludge in cement-based construction materials: Mechanical properties, drying shrinkage, and nano-scale characteristics. *J. Clean. Prod.* **2021**, *290*, 125832. [CrossRef]
55. Sandhu, R.K.; Siddique, R. Influence of rice husk ash (RHA) on the properties of self-compacting concrete: A review. *Constr. Build. Mater.* **2017**, *153*, 751–764. [CrossRef]
56. Siddika, A.; Mamun, M.; Al, A.; Ali, M. Study on concrete with rice husk ash. *Innov. Infrastruct. Solut.* **2018**, *3*, 18. [CrossRef]
57. Hwang, C.-L.; Huynh, T.-P. Effect of alkali-activator and rice husk ash content on strength development of fly ash and residual rice husk ash-based geopolymers. *Constr. Build. Mater.* **2015**, *101*, 1–9. [CrossRef]
58. Koushkbaghi, M.; Kazemi, M.J.; Mosavi, H.; Mohseni, E. Acid resistance and durability properties of steel fiber-reinforced concrete incorporating rice husk ash and recycled aggregate. *Constr. Build. Mater.* **2019**, *202*, 266–275. [CrossRef]
59. Subashi De Silva, G.; Priyamali, M.W.S. Potential use of waste rice husk ash for concrete paving blocks: Strength, durability, and run-off properties. *Int. J. Pavement Eng.* **2020**, *21*, 1–13. [CrossRef]
60. Sutas, J.; Mana, A.; Pitak, L. Effect of rice husk and rice husk ash to properties of bricks. *Procedia Eng.* **2012**, *32*, 1061–1067. [CrossRef]
61. Kanimozhi, K.; Prabunathan, P.; Selvaraj, V.; Alagar, M. Vinyl silane-functionalized rice husk ash-reinforced unsaturated polyester nanocomposites. *RSC Adv.* **2014**, *4*, 18157–18163. [CrossRef]
62. Elahi, M.M.A.; Shearer, C.R.; Reza, A.N.R.; Saha, A.K.; Khan, M.N.N.; Hossain, M.M.; Sarker, P.K. Improving the sulfate attack resistance of concrete by using supplementary cementitious materials (SCMs): A review. *Constr. Build. Mater.* **2021**, *281*, 122628. [CrossRef]
63. El-Sayed, M.A.; El-Samni, T.M. Physical and chemical properties of rice straw ash and its effect on the cement paste produced from different cement types. *J. King Saud Univ. -Eng. Sci.* **2006**, *19*, 21–29. [CrossRef]
64. Meddah, M.S.; Praveenkumar, T.R.; Vijayalakshmi, M.M.; Manigandan, S.; Arunachalam, R. Mechanical and microstructural characterization of rice husk ash and Al_2O_3 nanoparticles modified cement concrete. *Constr. Build. Mater.* **2020**, *255*, 119358. [CrossRef]
65. Channa, S.H.; Mangi, S.A.; Bheel, N.; Soomro, F.A.; Khahro, S.H. Short-term analysis on the combined use of sugarcane bagasse ash and rice husk ash as supplementary cementitious material in concrete production. *Environ. Sci. Pollut. Res.* **2022**, *29*, 3555–3564. [CrossRef] [PubMed]
66. Bonfim, W.B.; de Paula, H.M. Characterization of different biomass ashes as supplementary cementitious material to produce coating mortar. *J. Clean. Prod.* **2021**, *291*, 125869. [CrossRef]
67. Khan, K.; Ullah, M.F.; Shahzada, K.; Amin, M.N.; Bibi, T.; Wahab, N.; Aljaafari, A. Effective use of micro-silica extracted from rice husk ash for the production of high-performance and sustainable cement mortar. *Constr. Build. Mater.* **2020**, *258*, 119589. [CrossRef]
68. De Sensale, G.R. Effect of rice-husk ash on durability of cementitious materials. *Cem. Concr. Compos.* **2010**, *32*, 718–725. [CrossRef]
69. Sujivorakul, C.; Jaturapitakkul, C.; Taotip, A. Utilization of fly ash, rice husk ash, and palm oil fuel ash in glass fiber–reinforced concrete. *J. Mater. Civ. Eng.* **2011**, *23*, 1281–1288. [CrossRef]
70. Amin, M.N.; Hissan, S.; Shahzada, K.; Khan, K.; Bibi, T. Pozzolanic reactivity and the influence of rice husk ash on early-age autogenous shrinkage of concrete. *Front. Mater.* **2019**, *6*, 150. [CrossRef]
71. Siddika, A.; Mamun, M.A.A.; Alyousef, R.; Mohammadhosseini, H. State-of-the-art-review on rice husk ash: A supplementary cementitious material in concrete. *J. King Saud Univ. -Eng. Sci.* **2021**, *33*, 294–307. [CrossRef]
72. Fernandes, I.J.; Calheiro, D.; Kieling, A.G.; Moraes, C.A.M.; Rocha, T.L.A.C.; Brehm, F.A.; Modolo, R.C.E. Characterization of rice husk ash produced using different biomass combustion techniques for energy. *Fuel* **2016**, *165*, 351–359. [CrossRef]
73. Pode, R. Potential applications of rice husk ash waste from rice husk biomass power plant. *Renew. Sustain. Energy Rev.* **2016**, *53*, 1468–1485. [CrossRef]
74. Goyal, S.K.; Jogdand, S.V.; Agrawal, A.K. Energy use pattern in rice milling industries—A critical appraisal. *J. Food Sci. Technol.* **2014**, *51*, 2907–2916. [CrossRef] [PubMed]
75. Rukzon, S.; Chindaprasirt, P. Use of disposed waste ash from landfills to replace Portland cement. *Waste Manag. Res.* **2009**, *27*, 588–594. [CrossRef] [PubMed]
76. Alex, J.; Dhanalakshmi, J.; Ambedkar, B. Experimental investigation on rice husk ash as cement replacement on concrete production. *Constr. Build. Mater.* **2016**, *127*, 353–362. [CrossRef]
77. Ahsan, M.B.; Hossain, Z. Supplemental use of rice husk ash (RHA) as a cementitious material in concrete industry. *Constr. Build. Mater.* **2018**, *178*, 1–9. [CrossRef]
78. Jamil, M.; Kaish, A.; Raman, S.N.; Zain, M.F.M. Pozzolanic contribution of rice husk ash in cementitious system. *Constr. Build. Mater.* **2013**, *47*, 588–593. [CrossRef]
79. Tran, V.-A.; Hwang, C.-L.; Vo, D.-H. Manufacture and Engineering Properties of Cementitious Mortar Incorporating Unground Rice Husk Ash as Fine Aggregate. *J. Mater. Civ. Eng.* **2021**, *33*, 04021258. [CrossRef]
80. Bheel, N.; Meghwar, S.L.; Sohu, S.; Khoso, A.R.; Kumar, A.; Shaikh, Z.H. Experimental study on recycled concrete aggregates with rice husk ash as partial cement replacement. *Civ. Eng. J.* **2018**, *4*, 2305–2314. [CrossRef]

81. Das, S.; Patra, R.K.; Mukharjee, B.B. Feasibility study of utilisation of ferrochrome slag as fine aggregate and rice husk ash as cement replacement for developing sustainable concrete. *Innov. Infrastruct. Solut.* **2021**, *6*, 1–18. [CrossRef]
82. Ramasamy, V.-W. Compressive strength and durability properties of rice husk ash concrete. *KSCE J. Civ. Eng.* **2012**, *16*, 93–102. [CrossRef]
83. Limbachiya, M.; Meddah, M.S.; Ouchagour, Y. Use of recycled concrete aggregate in fly-ash concrete. *Constr. Build. Mater.* **2012**, *27*, 439–449. [CrossRef]
84. Zakka, W.P.; Lim, N.H.A.S.; Khun, M.C. A scientometric review of geopolymer concrete. *J. Clean. Prod.* **2021**, *280*, 124353. [CrossRef]
85. Udomsap, A.D.; Hallinger, P. A bibliometric review of research on sustainable construction, 1994–2018. *J. Clean. Prod.* **2020**, *254*, 120073. [CrossRef]
86. Yang, H.; Liu, L.; Yang, W.; Liu, H.; Ahmad, W.; Ahmad, A.; Aslam, F.; Joyklad, P. A comprehensive overview of geopolymer composites: A bibliometric analysis and literature review. *Case Stud. Constr. Mater.* **2022**, *16*, e00830. [CrossRef]
87. Xu, Y.; Zeng, J.; Chen, W.; Jin, R.; Li, B.; Pan, Z. A holistic review of cement composites reinforced with graphene oxide. *Constr. Build. Mater.* **2018**, *171*, 291–302. [CrossRef]
88. Xiao, X.; Skitmore, M.; Li, H.; Xia, B. Mapping knowledge in the economic areas of green building using scientometric analysis. *Energies* **2019**, *12*, 3011. [CrossRef]
89. Darko, A.; Chan, A.P.; Huo, X.; Owusu-Manu, D.-G. A scientometric analysis and visualization of global green building research. *Build. Environ.* **2019**, *149*, 501–511. [CrossRef]
90. Aghaei Chadegani, A.; Salehi, H.; Yunus, M.; Farhadi, H.; Fooladi, M.; Farhadi, M.; Ale Ebrahim, N. A comparison between two main academic literature collections: Web of Science and Scopus databases. *Asian Soc. Sci.* **2013**, *9*, 18–26. [CrossRef]
91. Afgan, S.; Bing, C. Scientometric review of international research trends on thermal energy storage cement based composites via integration of phase change materials from 1993 to 2020. *Constr. Build. Mater.* **2021**, *278*, 122344. [CrossRef]
92. Bergman, E.M.L. Finding citations to social work literature: The relative benefits of using Web of Science, Scopus, or Google Scholar. *J. Acad. Librariansh.* **2012**, *38*, 370–379. [CrossRef]
93. Meho, L.I. Using Scopus's CiteScore for assessing the quality of computer science conferences. *J. Informetr.* **2019**, *13*, 419–433. [CrossRef]
94. Zuo, J.; Zhao, Z.-Y. Green building research–current status and future agenda: A review. *Renew. Sustain. Energy Rev.* **2014**, *30*, 271–281. [CrossRef]
95. Darko, A.; Zhang, C.; Chan, A.P. Drivers for green building: A review of empirical studies. *Habitat Int.* **2017**, *60*, 34–49. [CrossRef]
96. Ahmad, W.; Khan, M.; Smarzewski, P. Effect of Short Fiber Reinforcements on Fracture Performance of Cement-Based Materials: A Systematic Review Approach. *Materials* **2021**, *14*, 1745. [CrossRef]
97. Markoulli, M.P.; Lee, C.I.; Byington, E.; Felps, W.A. Mapping Human Resource Management: Reviewing the field and charting future directions. *Hum. Resour. Manag. Rev.* **2017**, *27*, 367–396. [CrossRef]
98. Jin, R.; Gao, S.; Cheshmehzangi, A.; Aboagye-Nimo, E. A holistic review of off-site construction literature published between 2008 and 2018. *J. Clean. Prod.* **2018**, *202*, 1202–1219. [CrossRef]
99. Park, J.Y.; Nagy, Z. Comprehensive analysis of the relationship between thermal comfort and building control research-A data-driven literature review. *Renew. Sustain. Energy Rev.* **2018**, *82*, 2664–2679. [CrossRef]
100. Oraee, M.; Hosseini, M.R.; Papadonikolaki, E.; Palliyaguru, R.; Arashpour, M. Collaboration in BIM-based construction networks: A bibliometric-qualitative literature review. *Int. J. Proj. Manag.* **2017**, *35*, 1288–1301. [CrossRef]
101. Van Eck, N.J.; Waltman, L. Software survey: VOSviewer, a computer program for bibliometric mapping. *Scientometrics* **2010**, *84*, 523–538. [CrossRef]
102. Wuni, I.Y.; Shen, G.Q.; Osei-Kyei, R. Scientometric review of global research trends on green buildings in construction journals from 1992 to 2018. *Energy Build.* **2019**, *190*, 69–85. [CrossRef]
103. Su, H.-N.; Lee, P.-C. Mapping knowledge structure by keyword co-occurrence: A first look at journal papers in Technology Foresight. *Scientometrics* **2010**, *85*, 65–79. [CrossRef]
104. Yu, F.; Hayes, B.E. Applying data analytics and visualization to assessing the research impact of the Cancer Cell Biology (CCB) Program at the University of North Carolina at Chapel Hill. *J. eSci. Librariansh.* **2018**, *7*, 4. [CrossRef]
105. Ganesan, K.; Rajagopal, K.; Thangavel, K. Rice husk ash blended cement: Assessment of optimal level of replacement for strength and permeability properties of concrete. *Constr. Build. Mater.* **2008**, *22*, 1675–1683. [CrossRef]
106. Isaia, G.C.; GastaldinI, A.L.G.; Moraes, R. Physical and pozzolanic action of mineral additions on the mechanical strength of high-performance concrete. *Cem. Concr. Compos.* **2003**, *25*, 69–76. [CrossRef]
107. Yoo, D.-Y.; Banthia, N. Mechanical properties of ultra-high-performance fiber-reinforced concrete: A review. *Cem. Concr. Compos.* **2016**, *73*, 267–280. [CrossRef]
108. Bui, D.D.; Hu, J.; Stroeven, P. Particle size effect on the strength of rice husk ash blended gap-graded Portland cement concrete. *Cem. Concr. Compos.* **2005**, *27*, 357–366. [CrossRef]
109. Sata, V.; Jaturapitakkul, C.; Kiattikomol, K. Influence of pozzolan from various by-product materials on mechanical properties of high-strength concrete. *Constr. Build. Mater.* **2007**, *21*, 1589–1598. [CrossRef]
110. Zhang, M.-H.; Malhotra, V.M. High-performance concrete incorporating rice husk ash as a supplementary cementing material. *ACI Mater. J.* **1996**, *93*, 629–636.

111. Nehdi, M.; Duquette, J.; El Damatty, A. Performance of rice husk ash produced using a new technology as a mineral admixture in concrete. *Cem. Concr. Res.* **2003**, *33*, 1203–1210. [CrossRef]
112. Paris, J.M.; Roessler, J.G.; Ferraro, C.C.; DeFord, H.D.; Townsend, T.G. A review of waste products utilized as supplements to Portland cement in concrete. *J. Clean. Prod.* **2016**, *121*, 1–18. [CrossRef]
113. Wongpa, J.; Kiattikomol, K.; Jaturapitakkul, C.; Chindaprasirt, P. Compressive strength, modulus of elasticity, and water permeability of inorganic polymer concrete. *Mater. Des.* **2010**, *31*, 4748–4754. [CrossRef]
114. Khan, M.N.N.; Jamil, M.; Karim, M.R.; Zain, M.F.M.; Kaish, A.A. Utilization of rice husk ash for sustainable construction: A review. *Res. J. Appl. Sci. Eng. Technol.* **2015**, *9*, 1119–1127. [CrossRef]
115. Raisi, E.M.; Amiri, J.V.; Davoodi, M.R. Influence of rice husk ash on the fracture characteristics and brittleness of self-compacting concrete. *Eng. Fract. Mech.* **2018**, *199*, 595–608. [CrossRef]
116. Thiedeitz, M.; Schmidt, W.; Härder, M.; Kränkel, T. Performance of rice husk ash as supplementary cementitious material after production in the field and in the lab. *Materials* **2020**, *13*, 4319. [CrossRef]
117. Cordeiro, G.C.; Toledo Filho, R.D.; de Moraes Rego Fairbairn, E. Use of ultrafine rice husk ash with high-carbon content as pozzolan in high performance concrete. *Mater. Struct.* **2009**, *42*, 983–992. [CrossRef]
118. Huang, H.; Gao, X.; Wang, H.; Ye, H. Influence of rice husk ash on strength and permeability of ultra-high performance concrete. *Constr. Build. Mater.* **2017**, *149*, 621–628. [CrossRef]
119. Kaur, K.; Singh, J.; Kaur, M. Compressive strength of rice husk ash based geopolymer: The effect of alkaline activator. *Constr. Build. Mater.* **2018**, *169*, 188–192. [CrossRef]
120. Pham, T.M. Enhanced properties of high-silica rice husk ash-based geopolymer paste by incorporating basalt fibers. *Constr. Build. Mater.* **2020**, *245*, 118422.
121. Wen, N.; Zhao, Y.; Yu, Z.; Liu, M. A sludge and modified rice husk ash-based geopolymer: Synthesis and characterization analysis. *J. Clean. Prod.* **2019**, *226*, 805–814. [CrossRef]
122. Alghamdi, H. A review of cementitious alternatives within the development of environmental sustainability associated with cement replacement. *Environ. Sci. Pollut. Res.* **2022**, *29*, 1–19. [CrossRef]
123. Gustavsson, L.; Sathre, R. Variability in energy and carbon dioxide balances of wood and concrete building materials. *Build. Environ.* **2006**, *41*, 940–951. [CrossRef]
124. Gursel, A.P.; Maryman, H.; Ostertag, C. A life-cycle approach to environmental, mechanical, and durability properties of "green" concrete mixes with rice husk ash. *J. Clean. Prod.* **2016**, *112*, 823–836. [CrossRef]

Review

Waste Material via Geopolymerization for Heavy-Duty Application: A Review

Marwan Kheimi [1,*], Ikmal Hakem Aziz [2,3,*], Mohd Mustafa Al Bakri Abdullah [2,3,*], Mohammad Almadani [1] and Rafiza Abd Razak [2,4]

1. Department of Civil Engineering, Faculty of Engineering—Rabigh Branch, King Abdulaziz University, Jeddah 21589, Saudi Arabia; malmadani@kau.edu.sa
2. Geopolymer & Green Technology, Center of Excellence (CEGeoGTech), Universiti Malaysia Perlis (UniMAP), Arau 02600, Malaysia; rafizarazak@unimap.edu.my
3. Faculty of Chemical Engineering Technology, Universiti Malaysia Perlis (UniMAP), Arau 02600, Malaysia
4. Faculty of Civil Engineering Technology, Universiti Malaysia Perlis (UniMAP), Arau 02600, Malaysia
* Correspondence: mmkheimi@kau.edu.sa (M.K.); ikmalhakem@unimap.edu.my (I.H.A.); mustafa_albakri@unimap.edu.my (M.M.A.B.A.)

Abstract: Due to the extraordinary properties for heavy-duty applications, there has been a great deal of interest in the utilization of waste material via geopolymerization technology. There are various advantages offered by this geopolymer-based material, such as excellent stability, exceptional impermeability, self-refluxing ability, resistant thermal energy from explosive detonation, and excellent mechanical performance. An overview of the work with the details of key factors affecting the heavy-duty performance of geopolymer-based material such as type of binder, alkali agent dosage, mixing design, and curing condition are reviewed in this paper. Interestingly, the review exhibited that different types of waste material containing a large number of chemical elements had an impact on mechanical performance in military, civil engineering, and road application. Finally, this work suggests some future research directions for the the remarkable of waste material through geopolymerization to be employed in heavy-duty application.

Keywords: geopolymer; waste material; heavy duty application

Citation: Kheimi, M.; Aziz, I.H.; Abdullah, M.M.A.B.; Almadani, M.; Abd Razak, R. Waste Material via Geopolymerization for Heavy-Duty Application: A Review. *Materials* **2022**, *15*, 3205. https://doi.org/10.3390/ma15093205

Academic Editor: Frank Collins

Received: 11 April 2022
Accepted: 27 April 2022
Published: 29 April 2022

Publisher's Note: MDPI stays neutral with regard to jurisdictional claims in published maps and institutional affiliations.

Copyright: © 2022 by the authors. Licensee MDPI, Basel, Switzerland. This article is an open access article distributed under the terms and conditions of the Creative Commons Attribution (CC BY) license (https://creativecommons.org/licenses/by/4.0/).

1. Introduction

In 2000, the world population was over 6 billion people, and it is predicted to grow by 50% in the next half-century, reaching 9 billion in 2050 [1]. Countless products and goods will be delivered via distribution infrastructure to fulfil the requirements and demands of individuals seeking pleasant and convenient lifestyles. As the global economy grows, people began to purchase more items and goods, resulting in an increase in the number of products created and consumed. Solid trash is generated throughout these processes, which is then collected by municipalities and the private waste management industry for recycling or disposal purposes. As society becomes more prosperous, more garbage is produced. Currently, Asia generates roughly one-fourth of the world's solid waste, although this is predicted to increase to one-third by 2050 [2].

To minimize resources depletion, the Seventh Millennium Development Goal (MDG), which focuses on environmental sustainability through capacity building and sound environmental decision making, calls for the integration of sustainable development strategies into country policies. As a result, one of its key proposals is to Reduce, Reuse, and Recycle, or "3R" [3]. The Society of Solid Waste Management Experts in Asia and the Pacific Island (SWAPI), a network of solid waste management experts, was founded in 2005 with the goal of promoting the 3R's for solid waste, namely, waste reduction, reuse, and recycling, as well as improving waste management to achieve a 3R Society to conserve natural resources and preserve our living environment.

Economic activity, resource consumption, and economic growth are all intricately related to waste volumes. Economic expansion in Southeast Asian countries is driving annual urban growth rates of 6–8%, a pattern that is likely to last several decades. The trend in waste generation is expected to accelerate as economic developments rise. Table 1 shows the trends in waste generation. Economic trends, demographic projections, and municipal solid waste (MSW) per capita generation rates are used to estimate them. Table 1 demonstrates that in middle-income countries like Malaysia, Thailand, Indonesia and Philippines, waste generation will grow by about 0.3 kg/capita. The growth is mostly due to the prevalence of paper, plastic, bulk waste, and other multi-material packaging in middle-income countries' waste streams. The waste generation rate in Singapore, a high-income country, is expected to remain relatively steady until falling drastically below its current level.

The waste generation rate in the other nations—Vietnam, Cambodia, Laos, Brunei, and Myanmar—will rise by four to six times the current amount. The density of organic matter and ash residues in waste streams is larger in low-income nations. Additionally, the growing proportion of plastic and paper garbage in the waste stream will contribute to the rising waste volume. In general, the total amount of waste in ASEAN is expected to increase by around 1 million tonnes per day until 2025, compared to existing waste volumes, due to the projected expanding path of economic development [4].

Table 1. The rate of municipal solid waste generation per capita in urban ASEAN by 2025. Reprinted/adapted with permission from [5]. 2009. Ngoc.

Country	Gross National Product Per Capita (USD)		Waste Generation Rate (kg/cap/day)		Predicted Urban Waste Generation	
	1995	2025	Generation Rates (kg/cap/day)	Total Waste (tons/day)	Municipal Solid Waste (kg/cap/day)	Total (tons/day)
High Income						
Singapore	26,730	36,000	1.1	4840	1.1	4840
Middle Income						
Thailand	2740	6700	0.64	15,715	1.5	3673
Indonesia	980	2400	0.76	96,672	1.0	1272
Philippines	1050	2500	0.52	33,477	0.8	5150
Malaysia	3890	9440	0.81	15,663	1.4	2681
Low Income						
Vietnam	240	950	0.61	19,983	1.0	3276
Brunei	260	750	0.66	149,140	0.95	2169
Cambodia	220	700	0.52	3544	1.1	7497
Myanmar	240	580	0.45	12,118	0.85	2289
Laos	350	850	0.55	1379	0.9	2257

Out of about 300 MtCO$_2$e that comes from emerging countries in South and East Asia, the Intergovernmental Panel on Climate Change (IPCC) estimates that landfill methane will reach 1103 MtCO$_2$e and 1218 MtCO$_2$e in 2020, and 2030, respectively [6]. As a result, proper mitigations must be put in place to prevent future greenhouse gas emission (GHG) from entering the atmosphere. In line with the effort, 3R actions that encourage sustainable waste management often helps to reduce GHG emissions that contribute to the global warming issue. Therefore, as a result of the waste reduction approach, less waste materials were dumped into landfills, reducing the waste material's degradation potential and, consequently, lowering GHG emissions. This is especially important when dealing with the municipal solid waste (MSW)'s organic component. Diverting organic waste from landfills can minimize the conversion of organic compounds into harmful methane gas, which has 21 times the global warming potential of carbon dioxide.

However, the disposal of these wastes in these landfills, based on present regulations, does not provide efficient and effective management of these solid wastes, which continue to pose a serious environmental danger. Furthermore, a significant amount of waste is still generated each year, with the little land area available to dispose of it. Hence, finding creative ways to value and transform these solid wastes for varied applications would contribute to the implementation of a circular economy and the attainment of a sustainable environment.

Type of Waste

External factors, such as geographic location, population standard of living, energy sources, and weather, influence waste composition. Quantifying and classifying the various forms of waste created are the most basic stage in waste source management. It is critical to have a system in place for collecting, sorting, and analysing basic waste information, such as the source of wastes, the quantities of waste generated, their composition and characteristic, seasonal variations, and future generation trends. Since municipal, industrial, agricultural, hazardous, and toxic waste, as well was wastewater, require different treatment methods, this is the best way to identify the method to treat waste.

It is possible to find the exact innovation that meets our needs, but we must acknowledge that it will have a significant impact on society and the environment in the future. Researchers are on the lookout for better technology that ensures sustainable development while also protecting our community. As the human population continues to grow, it appears that human needs are increasing as well, resulting in increased demand for food and other necessities. This rising demand also results in waste problem. Agricultural garbage, industrial waste, and domestic waste are polluting the society today, spreading diseases, and destroying nature's beauty. If this waste is not properly disposed over time, we may not be able to provide a clean and hygienic environment for future generations. It is now our responsibility to appropriately dispose of the waste materials. Garbage can be combined with other materials to be used for various purposes in order to add value to it. Waste materials as reinforcement in composites appear to be a superior option, as it also improves polymer characteristics. Table 2 highlights the various forms of solid waste that are found in our environment and therefore can be effectively utilized.

Table 2. Solid wastes and related possible uses are described in detail.

Type of Waste	Sources of Content	Potential Application	References
Hazardous Waste	Trash from galvanising, tannery waste, and metallurgical waste	Cement brick, tiles, boards	[7,8]
Mining Mineral Waste	Overburden waste tailing from the iron, coal wateriest waste, copper, gold, zinc and aluminium industries	Light-weight aggregate fuel, brick, tiles	[9,10]
Agro Waste	Cotton stalks, husk from packed rice and wheat straw, sawmill waste, jute and banana stalks, nut shells, sisal, and vegetable residue	Insulation boards, particle board, wall panel, roofing sheets, fibrous construction panel, fuel binder, acid resistant cement	[11–13]
Industrial Waste	Bauxite red mud, steel slag, construction detritus, coal combustion residues	Bricks, blocks, cement, paint, wood substitutes, tiles, concrete, and ceramic goods	[14–16]
Non-hazardous Waste	Gypsum waste, lime sludge limestone waste, marble production waste,	Cement clinker, super sulphate hydraulic binder, gypsum plaster, fibrous gypsum, boards, bricks and blocks	[17,18]
Municipal Solid Waste	Soft drink bottle, jar for food, cosmetics product	Replacement binder material, supplementary material in concrete, soil stabilization	[19,20]

Municipal solid waste is generated by households, commercial activities, and other sources with activities that are similar to those of households and commercial enterprise, such as waste from offices, hotels, supermarkets, shops, school, and institutions, as well as municipal services like street cleaning and recreational area maintenance. Food waste, paper plastic, rags, metal, and glass are the most common categories of MSW, along with some hazardous household wastes such as light bulbs, batteries, waste pharmaceuticals, and automotive parts.

Industrial waste is a type of trash produced by manufacturing processes such as factories, mill, and mines. It has been yielded since the beginning of the industrial revolution. Most industrial wastes, such as waste fibre from agriculture and logging, are neither harmful nor toxic. From a wide range of industrial processes, the manufacturing industry generates a variety of waste streams. Basic metals, tobacco products, wood and wood products, and paper and paper products are among the most waste-generating industrial sectors in Southeast Asia, particularly in Singapore and Malaysia. In 2000, the Southeast Asian nations contributed to an estimated 19 million tonnes of industrial waste [5]. Meanwhile, in 2010, the Southern American nations passed legislation on "National policy on Solid Wastes" This policy aims to put an end to the disposal of solid waste at open-air dumps, which are placed where waste is simply dumped on the ground [21].

Plastic is a common packaging material, ranging from the well-known disposable plastic carrying bag to the plastic milk bottle. Single-use plastics are a source of concern since they waste a valuable resource when they end-up in landfills. The paper " The New Plastic Economy" [22] intends to inspire businesses and society to move towards a "circular economy" model for plastic by highlighting impediments to global material flows as well as enablers such as digital technologies [23]. Similarly, the increasing popularity of fiber reinforced polymer composites has been aided by the demand for energy and other limited resources. Vehicles (cars, trains, boats, and planes) can be lighter owing to composites, which improve fuel efficiency. Furthermore, the wind turbine requires lightweight turbines blades, and fibre reinforced composites are an obvious solution. Although composites are long-lasting, waste generated during the manufacturing process is a current concern, and as end-of-life approaches, there will be future concern about 'disposing' of massive composites structure. The Composites UK report [24] identifies the recycling alternative for composite materials and compares the environmental impact of various recycling techniques.

Owing to enhanced features such as high specific stiffness, high specific strength, high impact resistance, high abrasion resistance, better corrosion resistance, and higher chemical resistance, polymer matrix composites are widely employed in a variety of applications. They also have low thermal resistance and a high coefficient of thermal expansion. Polymer composites are made up of a polymer matrix with inorganic or organic fillers, which can either be natural or synthetic. Typically, fillers improve the required properties of polymers while also lowering the associated cost. At the time being, due to their improved thermal, mechanical, chemical, and barrier qualities, polymer composites are being used as engineered materials with a variety of applications [25]. Polymer matrix composites are in high demand across a wide range of industries, including aerospace, automobiles, sport, medicine, electronic, civil, communications, energy, construction, industry, marine, military, and various household item applications [26].

Waste material made of geopolymer can be employed to meet the increased demand. Various studies on mixed-based geopolymers are now underway. Previous paper addressed a wide range of recycling waste material to produce advance material as a non-essential application. In contrast, this review will focus on geopolymerization technology in the most often used integrated waste material generation towards heavy duty application. The geopolymerization method with various waste materials can be implemented for the greenhouse gases reduction in the environment. Finally, based on the gaps revealed in the previous literature, additional research opportunities have been proposed.

2. Geopolymerization

Geopolymerized composites are currently being studied as a possible replacement for traditional Portland cement-based construction materials. Initially, geopolymer research was limited to natural raw materials such as kaolin, metakaolin, silica fumes, and calcined clays; however, in recent years, the scope of research has expanded to include industrial waste products such as fly ash [27,28], clay-based slag [29,30], palm oil fuel [31] ash, and so on (shown Figure 1) to make them more economically and environmentally sustainable. The fact that practically precursor materials (both natural and industrial waste by-products) emit far less CO_2 than cement ensures environment sustainability [32]. Considering the high generation (compared to utilization) of industrial waste by-products, disposal concerns, and their harmful/hazardous nature, immobilization/use as a precursor is even environmentally viable. According to current estimates, using geopolymer as a cement substitute in construction products can reduce overall CO_2 emissions by anywhere from 9% to 64% [32,33]. In fact, these precursor materials are massively generated by industries throughout the world, including fly ash amounting up to 780 million tonnes per year [34,35] (75% to 80% of global annual ash production [36]), palm oil fuel ash, 11 million tonnes per year [37], rice husk ash, 20 million tonnes per year [38,39], red mud alumina, 120 million tonnes per year [40], and the tremendous occurrence of clay kaolin deposits in the earth [41,42]. Furthermore, after accounting for the cost of alkaline activators, the price of geopolymer concrete might be as low as 10–30% lower than conventional cement-based concrete due to reduced price in industrial waste by-products and processing of natural precursor compared to cement.

Figure 1. Common type of precursor materials used in geopolymer material.

Geopolymer precursor material must be alumina (Al_2O_3) and silica (SiO_2) component, preferably in reactive amorphous form, in both natural and by-product forms. For the geopolymerization of these aluminosilicate precursors, alkaline activating solutions such as potassium or sodium hydroxide (KOH, NaOH), and potassium or sodium silicate (K_2SiO_3, Na_2SiO_3) are required. The primary phase begins with the dissolution. In an alkaline media, the species interact ionically, followed by the breakage of the covalent bond between silicon, aluminium, and oxygen atoms. Alkali cations such as Na^+, Ca^{2+}, K^+, Li^+, and other charges balance negatively charged ion linked with tetrahedral Al (III). Following that, precursor ions are transported, oriented, and condensed into monomers. Coagulation and gelation are the next steps in the process. Finally, polycondensation of monomers forms rigid 3D networks of silica aluminates [43]. Figure 2 illustrates a conceptual diagram of the several steps of geopolymerization. However, several researchers focus on the variables that could influence the mechanical properties of geopolymer concrete in either a good or negative way. The main disadvantages of geopolymer concrete, as well as the key limitation on geopolymer concrete applications, were found to be the high workability loss rate, short setting time, and the need for heat curing. The type of alkali activator, alkali dosage, fineness of material, and the molar oxide ratios are the most apparent factors that determine the geopolymer properties.

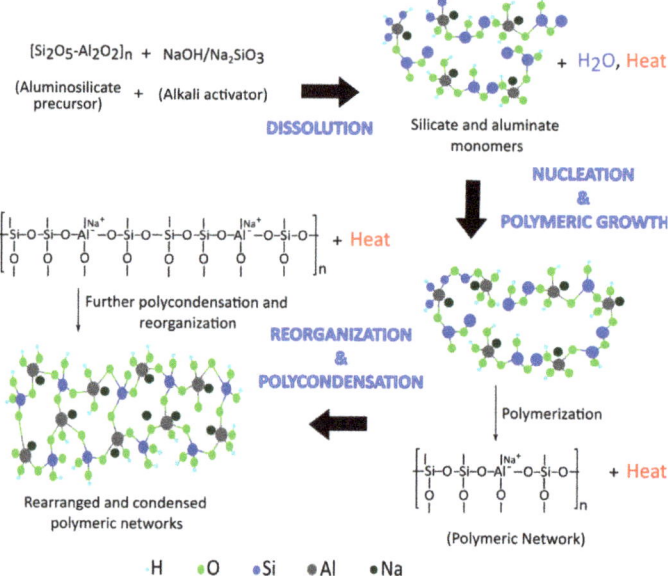

Figure 2. Conceptual process of geopolymerization.

Geopolymer concrete (GPC) is a revolutionary and environmentally friendly concrete that hardens by reacting aluminosilicate waste materials with alkaline activating solutions instead of using cement [44]. GPC allows for a reduction in the requirement for cement production while also providing more outlets for waste materials and industrial by-products. When compared to cement-based concrete, it is projected that using GPC might save up to 43% on energy and lower greenhouse gas emission by 9–80% [45]. This wide range is owing to the complexities of calculating emissions, which depend on a number of factors such as local conditions, transportation, and the mix design itself [19]. In addition, compared to regular concrete, GPC has better durability features, such as chloride resistance [46], high temperature resistance [47,48], freeze-thaw cycles [49], and carbonation resistance [50]. It has been demonstrated that GPC has appropriate compressive and tensile strength [51].

3. Waste-By Products Based Geopolymer

Fly ash is a by-product of the manufacturing process of coal combustion that is split into two classes: class F and class C. The combustion of bituminous coal creates a king of fly ash known as class F fly ash, which has a very low CaO level (FFA). Class C fly ash with high calcium content is also produced using lignite and sub-bituminous coal as new power sources [52]. FFA has a similar composition to natural volcanic ash [53]. Fly ash is a readily available by-product with a microscopic shape of small spherical particles that is commonly utilized as a raw material for manufacturing geopolymer [43,54]. The high free-CaO level of CFA limits its use in the OPC system, and its use in geopolymer preparation has been beyond imagination [55]. The chemical composition of various raw materials is shown in Table 3, FFA and CFA have Si/Al ratios of 1.86–3.09 and 1.82–2.52, respectively. Fly ash has been used in cement and concrete since the early 20th century, and it is often used as a major component. It is better for the environment to use FA instead of cement because it minimizes greenhouse gas emission and construction budgets. FFA has a reasonable price, is readily available, has a nice spherical structure, and the aluminate and amorphous silicate have a high activity. In alkali activator solution, high-strength geopolymers can be easily generated [56].

Due to the existence of amorphous phases, high hardness, and pozzolanic activity, ground granulated blast furnace slag (GGBS) is mostly utilized as a partial alternative of OPC after grinding, depending on the cooling condition [57]. Table 3 shows that GGBS is extremely reactive in geopolymers synthesis, and a satisfactory reaction rate can be achieved at temperatures as low as room temperature. When slag is utilized as a cement alternative, it produces less heat during hydration, which reduces the risk of cracking [58]. GGBS can be utilized in a variety of situations such as to enhance concrete porosity, long-term strength, and resistance to sulphate and alkali silicate reactivity, as well as hydration heat, permeability, and lower water demand [59,60].

The Bayer process, which is employed in industrial aluminium refining, produces red mud (RM) as a by-product. The Bayer method dissolves the soluble component of bauxite with sodium hydroxide at high temperatures and pressures. A small quantity of sodium hydroxide employed in this method will invariably remain in the RM, leading to higher pH value [61]. By eliminating the need for mud drying, using RM in the form of mud saves time and energy. It also reduces the total amount of alkali activator by utilizing high alkalinity red mud, thus lowering the cost of geopolymer manufacturing [62]. The appropriate replacement value of RM for FA-based geopolymers varies depending on NaOH concentration and curing conditions [63]. Furthermore, the geopolymer mixed with red mud has increased strength and durability, according to the research by Liu et al. [64].

Table 3. Chemical composition of various waste by-product geopolymers.

Type of Slag	Chemical Composition (wt %)							
	SiO_2	Al_2O_3	CaO	MgO	Fe_2O_3	K_2O	Na_2O	SO_3
Fly Ash [65]	55.38	28.14	3.45	1.85	3.31	1.39	2.30	0.32
Fly ash [66]	56.00	18.10	7.24	0.93	5.31	1.36	1.21	1.65
Fly ash [67]	65.90	24.00	1.59	0.42	2.87	1.44	0.49	N/A
Fly ash [68]	47.90	25.70	4.11	1.36	14.70	0.67	0.81	0.19
High Calcium Fly Ash [69]	37.30	14.90	17.10	3.72	16.50	1.66	1.74	2.56
High Calcium Fly Ash [70]	34.00	13.50	16.50	3.10	5.00	5.50	1.50	2.80
High Calcium Fly Ash [71]	36.20	19.90	14.20	1.90	11.90	2.40	N/A	3.60
Ground Granulated Blast Furnace Slag [72]	35.34	20.69	31.32	8.11	0.18	0.29	1.36	1.79
Ground Granulated Blast Furnace Slag [73]	18.90	6.43	66.90	1.41	0.74	0.67	N/A	1.97
Ground Granulated Blast Furnace Slag [74]	28.20	9.73	52.69	2.90	0.98	1.22	N/A	1.46

Table 3. Cont.

Type of Slag	Chemical Composition (wt %)						
	SiO_2	Al_2O_3	CaO	MgO	Fe_2O_3	K_2O	Na_2O	SO_3
Ground Granulated Blast Furnace Slag [75]	36.50	9.95	43.38	6.74	0.38	0.35	N/A	N/A
Red Mud [76]	14.40	22.20	2.00	0.17	40.20	0.11	12.70	0.28
Red Mud [77]	16.51	28.05	2.22	0.70	30.32	0.26	8.70	N/A
Red Mud [78]	27.54	30.59	25.48	0.49	4.60	N/A	N/A	1.42
Rice Husk Ash [79]	92.33	0.18	0.63	0.82	0.17	0.15	0.07	N/A
Rice Husk Ash [80]	93.10	0.30	1.50	0.49	0.20	2.30	0.06	N/A
Silica Fume [79]	87.60	0.38	0.57	3.67	0.66	2.36	1.26	N/A
Silica Fume [81]	90.00	1.20	1.00	0.60	2.00	N/A	N/A	0.50
Volcanic Ash [82]	43.32	14.84	8.80	7.70	14.19	1.52	3.04	0.01
High Magnesium Nickel Slag [74]	43.22	4.35	3.45	26.15	10.34	0.18	0.23	0.28

Another waste by-product is rice husk ash (RHA) that is produced from rice husk combustion. RHA, a silica rich agricultural waste, is regarded as a clean alternative for improving the characteristic of geopolymers [83]. The use of RHA in geopolymer concrete can reduce nano-SiO_2 consumption and pollution issues caused by RHA disposal in landfills, particularly in rice-producing countries [84]. RHA has been widely used in self-compacting geopolymer concrete due to its greater reactivity inspired by high silicon concentration and ultra-high specific surface area [85]. Sugarcane bagasse ash is an industrial by-product that has been used as a source of alumina and silicates in volcanic as a product by a number of researchers [86].

The principal by-product of municipal solid waste incineration is bottom ash. Heavy metals are abundant in the bottom ash, which has a small particle size [87]. In recent years, bottom has been increasingly recycled as building binders and concrete [84,88]. Moreover, bottom ash from the burning of municipal sewage sludge is employed in concrete at a concentration of 10–15%, resulting in greater strength than concrete without bottom ash [83]. High silica and alumina concentrations can be found in fly ash, blast furnace slag, red mud, and materials such as rice husk as main biomass ash, making them appropriate as supplementary materials for gelling. Steel slag, volcanic ash, silica fume, waste glass, coal gangue, high-magnesium nickel slag, and other minerals are also often employed. Numerous industrial catalyst residues include enough silicon and aluminium, as well as an amorphous structure that can be used to form synthetic geopolymers, and its compressive strength has been measured to be between 40 to 85 MPa [89]. It is obvious that the raw materials used to discover geopolymer are high in silica and aluminium, and calcium oxide content cannot be neglected.

In conclusion, the raw materials might be an aluminosilicates natural mineral including silicon, aluminium, oxygen, and other possible elements. The right raw materials should have amorphous properties and a high ability to release aluminium easily.

4. Heavy-Duty Applications of Geopolymers

Geopolymer applications can be divided into two groups based on their function: those having varied physical and chemical properties as well as those with physical and mechanical properties. Buildings such as fire prevention buildings, insulation walls, and nuclear power plant can make use of these functional applications for fire prevention, isolation, heat preservation, and adsorption of hazardous ions. Table 4 shows the utilization of waste material-based geopolymer in heavy-duty applications.

Table 4. The potential geopolymer material and possible application are described in detail.

Geopolymer Waste Material	Potential Application	Properties	Ref.
Ground granulated blast furnace slag, Fly ash, granite coarse aggregate	Concrete pavement	50 MPa of compressive strength and 4.72 MPa of flexural strength	[90]
Red mud waste (bauxite residue), slag	Heavy metal removal, composite materials, Adsorbent and coagulant	66 to 86 MPa of Compressive strength	[91,92]
Ferrosilicon slag, alumina waste	Thermal insulation brick	10.9 MPa of compressive strength and 0.59 W/m.k of thermal conductivity	[93]
Metakaolin, bottom ash waste	Thermal insulation brick	47.9 MPa of compressive strength, 1.32 W/m.k of thermal conductivity	[94]
Blast furnace slag, rice husk ash	Acid proof cement	57 MPa of compressive strength	[95]

The stability and safety of a structure will be compromised if it is exposed to rains, ocean, or saline soil over an extended period of time. However, the chemical resistance of geopolymer concrete, particularly sulphate resistance, makes it more suitable for marine building. Geopolymer concrete is comprised of more amorphous phases, smaller porosity, and more mesopore than OPC concrete, and the dense microstructure of geopolymer concrete makes seawater permeation harder [96,97]. When compared to OPC concrete, geopolymer concrete has greater chloride ion erosion resistance and a longer corrosion cracking time, making it an excellent prospect for use as an anti-corrosive coating in the maritime environment [98]. According to Chindaprasirt and Chalee [99], the penetration and corrosion of chloride ion reduced as the molarity of sodium hydroxide increased after the fly ash-based geopolymer was exposed to the tidal zone of the ocean environment for three years after being air-dried in the laboratory for 28 days. Nevertheless, after six years in a salt lake environment, fly ash-based geopolymer concrete is more easily carbonized than OPC concrete, and chloride and sulphate are more easily diffused [100]. However, according to Alzeebaree et al. [101], both carbon fibre and basalt fiber reinforced geopolymer fabric can be employed as the modification material to resist chloride ion erosion. Additionally, fibre reinforced geopolymer concrete allows it to be employed as a structural member instead of ordinary concrete. The permeability of chlorine ion can be reduced by adding OPC to fly ash, whereas the permeability of chlorine ion can be strengthened by introducing metakaolin and nano-SiO_2 [102].

4.1. Geopolymer in Military Application

Geopolymer was used in the heavy duty rigid pavements (turning node, aprons, and taxiways) at a commercial airport Brisbane, Australia [103], as well as the Global Change Institute (GCI) building at the University of Queensland [104]. Pre-stressed geopolymer concrete sleepers have previously been produced by one of Australia's leading concrete sleeper providers and have been successfully used on mainline railway tracks [105]. Considering geopolymer concrete as having better acid resistance and less alkali-silica reaction than typical OPC concrete, it has been recommended for use in the construction of railway sleepers, which are exposed to chemicals and prolonged environmental circumstance [106]. Furthermore, the US Army Corps of Engineer's Waterways Experiment Station (WES) stated that fly ash-based alkali activated aluminosilicate binder can be potentially used to repair deteriorate Army airbase concrete and other special construction demands.

According to previous reports, the US Air Force and Navy, the Royal Air Force (RAF), and the Royal Australian Air Force (RAAF) have all had concrete durability difficulties with their F/A-18 and B-1 parking aprons [107–109]. Military airfield concrete, particularly aprons, has been exposed to severe thermal shocks from jet exhaust and has been discovered saturated with chemical like hydrocarbon fluids (HF); aircraft engine oil, hydraulic fluids, and jet fuel [107]. During engine start-up, the surface temperature of the apron concrete underneath F/A-18 auxiliary power units (APUs) can reach 175 °C in 10–12 min [110]. Figure 3a shows an ancient deep scaling where particles were scraped away from the concrete's wearing surface while Figure 3b depicts an APU in the bottom of an F/A-18 [107]. This type of damage is a source of foreign object debris (FOD), and it is more common in the area where the Auxiliary power unit (APU) exhaust impinges on the concrete. A substantial amount of split engine oil, hydraulic fluid, and vented jet fuel from the aeroplane is also commonly observed in similar areas of pavement where the APU exhaust impinges concrete. It is also worth noting that the jets tend to park in roughly the same spot each time, causing localised damage to some aprons. In the construction of rigid pavements at military airbases, several researchers have proposed substituting OPC with other heat and chemical resistant cementitious materials [111]. The feasibility of geopolymer binder for replacing OPC at military airbases should be examined because it is more resistant to both heat and chemicals and is more durable than OPC. Hence, it can be concluded that geopolymer can be a promising alternative to OPC for repairing apron concrete at military airbases.

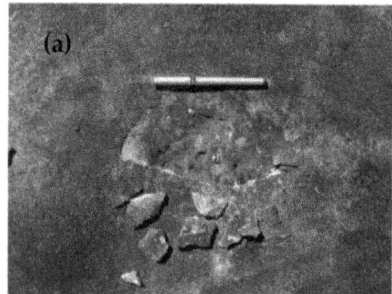 

Figure 3. (**a**) Scaling at the top layer of the military airbase concrete and (**b**) Underbelly of an F/A-18 with the APU in the centre. Reprinted/adapted with permission from [107]. 2018. Shill.

4.2. Geopolymer in Civil-Engineering Application

Structures that are still in the design phase are likely to be subjected to blast and impact loading threats. Due to its high-ultra strength, high ductility, and outstanding toughness, Portland cement-based ultra-high performance concrete (PC-UHPC) has been developed in recent decades to meet the increasing safety requirements of structures to overcome such destructive intensive loadings. Although PC-UHPC has emerged as one of the most promising construction materials for civil and military structures, the extensive use of Portland cement has a negative impact on the environment due to the carbon dioxide emission produced during cement manufacture. Meanwhile, the environment is facing a large increase in the formation of industrial wastes and the consumption of raw material in cement manufacture. As a result, it is required to develop a geopolymer as an alternative binder system that is less expensive and energy-intensive while also being greener in order to reduce or fully replace ordinary Portland cement. However, it has been noted that the higher the geopolymer's strength, the greater the fire resistance. Low-strength and low-density geopolymers are difficult to dehydrate and react to volume changes better in the temperature range of 100 °C to 1000 °C; even after heat exposure, their intensity increases [112]. Figure 4 shows the effect of alkali cation selection on the fly ash-based geopolymer's high temperature exposure strength and durability. Depending on the type of alkali cations utilized, densification of particular substrates and healing of microcracks are useful to increasing strength in different temperature ranges [113]. These findings imply that geopolymers can be tailored to attain stable (and even improved) strength after exposure to a high heat environment. The building's damage caused by fire cannot be ignored. The Windsor Tate Fire in Shanghai, as well as the 9/11 terrorist attack, resulted in massive human and material losses. As a result, refractory materials for building are crucial. Today, continuing to improve the sustainability and ecology of fire-resistant and high-temperature materials is a primary concern. Therefore, the goal of geopolymerization is to turn industrial solid waste into a chemically durable cement binder that is both thermally stable and non-combustible.

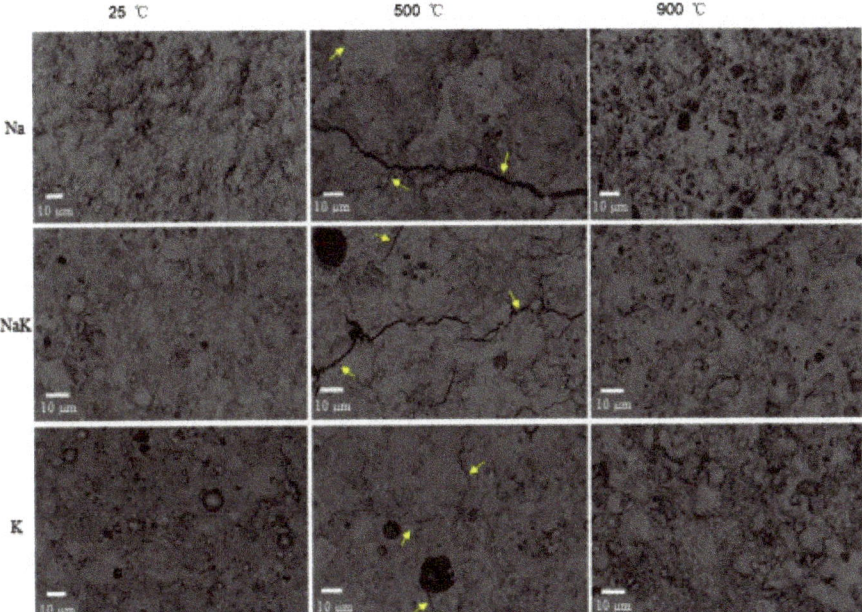

Figure 4. SEM-BSE micrographs depict the formation and healing of micro-cracks in geopolymer at various temperatures (solid arrows showing micro-crack). Reprinted/adapted with permission from [113]. 2018. Lahoti.

4.3. Geopolymer in Road Application

The development of geopolymers in the past and present has centred on the production and use of such materials to replace cement in structural construction. There has not been a lot of research applied in road construction. Several geopolymer research reports for road applications were even at the proof-of-concept level. Tenn et al. [114] investigated the interaction between sodium and potassium-based geopolymer binders and granite or diorite pavement aggregates in order to promote the usage of geopolymer in place of asphalt cement (or bitumen). Camacho-Tauta et al. [115] demonstrated an attempt to improve road fatigue damage resistance by employing a fly ash-based geopolymer as a road base layer. To assess the study material's long-term performance, a full-scale accelerated pavement test was assigned. Compared to a non-treated road base layer, the study found that a geopolymer-treated road base layer could give a reduction in deformability. Waste material-based geopolymer has been implemented as a replacement binder to enhance the properties for road and pavement material as tabulated in Table 5.

Table 5. Research work utilising geopolymers in road applications.

No.	Researcher	Materials	Findings
1	Sukprasert et al. [116]	Fly ash, silty clay, ground granulated blast furnace slag	• Increase packing density • Increased unconfined compressive strength
2	Dave et al. [117]	Ground blast furnace slag, fly ash, silica fume	• Appropriate strength as road repair material • Well durability through ultrasonic pulse velocity test
3	Wongsa et al. [118]	Crumb rubber, river sand, high calcium fly ash	• Average value of thermal conductivity and density • Meet the strength requirement for lightweight concrete
4	Mohammed et al. [119]	Fly ash, crumb rubber	• Reduction in compressive and flexural strength • Higher water absorption

The Netherlands was one of the first countries in Europe to use fly ash and blast furnace slag as a binder for acid-resistant pipe manufacture. Activated alkaline materials used in civil construction subsequently established enterprises in other United Kingdom countries, and eventually extended throughout the continent [120,121]. Conversely, the company with the most building applications is based in Australia. In 2007, the Melbourne-based company, E-cert, developed its own concrete. This company employs a blend of fly ash and ground blast furnace slag that has been alkali activated according to a proprietary dosage and composition. Bridges, highways, and big structures are several of the applications [122].

Figure 5a depicts a section of the Westgate Freeway in Port Melbourne's highway pavement. Since it was a different material, the project had to comply to numerous needs of the local road authority as well as more specialized technical standards in order to gain government clearance. The highway's construction and use were agreed on by a group of multinational construction corporations. Meanwhile, Figure 5b illustrates VicRoad's installation of 55 MPa E-Crete prefabricated panels. Due to the strict inspection in regard to structural concrete, this strength was required [122].

Figure 5. (a) Westgate Freeway paving in Port Melbourne and (b) E-Create Precast Panels. Reprinted/adapted with permission from [122]. 2012. Van Deventer.

5. Conclusions and Suggestions for Future Works

Regardless of the differences in waste material used in geopolymerization, the selection of raw materials depends on the desired application. Waste material-based geopolymer rich in silicon, aluminium, and calcium are easy and highly accessible. This facilitates their disposal waste management rather than building a landfill full of them. This is the main reason behind the low cost of manufacturing geopolymeric composites-based waste material in civil and military application. The utilization of waste materials through geopolymerization should be carried out in future works to mitigate the disposal and environmental issues. Numerous studies have been conducted involving the use of waste-based geopolymers with promising properties for heavy-duty applications such as military, civil-engineering, and road applications.

Based on the identified gaps in this work, future recommendations on waste-material based geopolymer in heavy duty application are listed below:

- Durability works using waste material in advanced application in the civil construction or aerospace fields;
- Establishing standards in order to conduct more advanced tests and research on these waste materials and, as a a result, expand the application as heavy-duty material in civil construction or the aerospace industry;
- Besides construction and airbase application, the geopolymer material can also be implemented as a defence material that consists of lightweight and higher mechanical properties such as bulletproof, Kevlar helmet, and body armour;
- The study on the landfill and waste management cost is crucial in considering the impact of the 3R implementation;
- An alternative activator to hydroxides and silicates that leads to lower environmental impact and can cut the cost of geopolymer production;
- For better understanding and experimental application, standardize dosage and quantify ingredients utilized in the manufacturing of activated alkali component.

Author Contributions: Conceptualization, original draft preparation, writing, review and editing, I.H.A. and M.K.; validation, project administration, supervision, visualization and resources, M.M.A.B.A.; methodology and validation, M.A. and R.A.R. All authors have read and agreed to the published version of the manuscript.

Funding: This research work was funded by Institutional Fund Projects under Grant No. (IFPIP: 434-829-1442). Therefore, the authors gratefully acknowledge technical and financial support from the Ministry of Education and King Abdulaziz University, DSR Jeddah, Saudi Arabia.

Institutional Review Board Statement: Not Applicable.

Informed Consent Statement: Not Applicable.

Data Availability Statement: Not Applicable.

Acknowledgments: The authors gratefully acknowledge the Centre of Excellence Geopolymer and Green Technology (CEGeoGTech) and the Faculty of Chemical Engineering Technology, UniMAP for their expertise and support and the Dean of Scientific Research in King Abdulaziz University, Jeddah, Saudi Arabia.

Conflicts of Interest: The authors declare no conflict of interest.

References

1. Tripathi, A.D.; Mishra, R.; Maurya, K.K.; Singh, R.B.; Wilson, D.W. Estimates for world population and global food availability for global health. In *The Role of Functional Food Security in Global Health*; Elsevier: Amsterdam, The Netherlands, 2019; pp. 3–24.
2. Devadoss, P.M.; Agamuthu, P.; Mehran, S.; Santha, C.; Fauziah, S. Implications of municipal solid waste management on greenhouse gas emissions in Malaysia and the way forward. *Waste Manag.* **2021**, *119*, 135–144. [CrossRef] [PubMed]
3. Pariatamby, A.; Fauziah, S. Sustainable 3R practice in the Asia and Pacific Regions: The challenges and issues. In *Municipal Solid Waste Management in Asia and the Pacific Islands*; Springer: Berlin/Heidelberg, Germany, 2014; pp. 15–40.
4. Inanc, B.; Idris, A.; Terazono, A.; Sakai, S.I. Development of a database of landfills and dump sites in Asian countries. *J. Mater. Cycles Waste Manag.* **2004**, *6*, 97–103. [CrossRef]
5. Ngoc, U.N.; Schnitzer, H. Sustainable solutions for solid waste management in Southeast Asian countries. *Waste Manag.* **2009**, *29*, 1982–1995. [CrossRef] [PubMed]
6. Agamuthu, P. *Landfilling in Developing Countries*; Sage Publications Sage UK: London, UK, 2013; Volume 31, pp. 1–2.

7. Palod, R.; Deo, S.; Ramtekkar, G. Utilization of waste from steel and iron industry as replacement of cement in mortars. *J. Mater. Cycles Waste Manag.* **2019**, *21*, 1361–1375. [CrossRef]
8. Zanelli, C.; Conte, S.; Molinari, C.; Soldati, R.; Dondi, M. Waste recycling in ceramic tiles: A technological outlook. *Resour. Conserv. Recycl.* **2021**, *168*, 105289. [CrossRef]
9. Leiva, C.; Luna-Galiano, Y.; Arenas, C.; Alonso-Fariñas, B.; Fernández-Pereira, C. A porous geopolymer based on aluminum-waste with acoustic properties. *Waste Manag.* **2019**, *95*, 504–512. [CrossRef]
10. Simão, F.V.; Chambart, H.; Vandemeulebroeke, L.; Cappuyns, V. Incorporation of sulphidic mining waste material in ceramic roof tiles and blocks. *J. Geochem. Explor.* **2021**, *225*, 106741. [CrossRef]
11. Priyadarshini, M.; Giri, J.P.; Patnaik, M. Variability in the compressive strength of non-conventional bricks containing agro and industrial waste. *Case Stud. Constr. Mater.* **2021**, *14*, e00506. [CrossRef]
12. Ali, M.; Alabdulkarem, A.; Nuhait, A.; Al-Salem, K.; Iannace, G.; Almuzaiqer, R. Characteristics of agro waste fibers as new thermal insulation and sound absorbing materials: Hybrid of date palm tree leaves and wheat straw fibers. *J. Nat. Fibers* **2021**, *2*, 1–19. [CrossRef]
13. Kazmi, S.M.S.; Munir, M.J.; Patnaikuni, I.; Wu, Y.-F.; Fawad, U. Thermal performance enhancement of eco-friendly bricks incorporating agro-wastes. *Energy Build.* **2018**, *158*, 1117–1129. [CrossRef]
14. Ahmadi, P.F.; Ardeshir, A.; Ramezanianpour, A.M.; Bayat, H. Characteristics of heat insulating clay bricks made from zeolite, waste steel slag and expanded perlite. *Ceram. Int.* **2018**, *44*, 7588–7598. [CrossRef]
15. Lima, M.S.; Thives, L.P. Evaluation of red mud as filler in Brazilian dense graded asphalt mixtures. *Constr. Build. Mater.* **2020**, *260*, 119894. [CrossRef]
16. Ter Teo, P.; Anasyida, A.S.; Kho, C.M.; Nurulakmal, M.S. Recycling of Malaysia's EAF steel slag waste as novel fluxing agent in green ceramic tile production: Sintering mechanism and leaching assessment. *J. Clean. Prod.* **2019**, *241*, 118144. [CrossRef]
17. Hansen, S.; Sadeghian, P. Recycled gypsum powder from waste drywalls combined with fly ash for partial cement replacement in concrete. *J. Clean. Prod.* **2020**, *274*, 122785. [CrossRef]
18. Hamid, N.J.A.; Kadir, A.A.; Hashar, N.N.H.; Pietrusiewicz, P.; Nabiałek, M.; Wnuk, I.; Gucwa, M.; Palutkiewicz, P.; Hashim, A.A.; Sarani, N.A. Influence of Gypsum Waste Utilization on Properties and Leachability of Fired Clay Brick. *Materials* **2021**, *14*, 2800. [CrossRef] [PubMed]
19. Podolsky, Z.; Liu, J.; Dinh, H.; Doh, J.; Guerrieri, M.; Fragomeni, S. State of the art on the application of waste materials in geopolymer concrete. *Case Stud. Constr. Mater.* **2021**, *15*, e00637. [CrossRef]
20. Sharma, R.; Bhardwaj, A. Effect of construction demolition and glass waste on stabilization of clayey soil. In Proceedings of the International Conference on sustainable Waste Management Through Design, Ludhiana, India, 2–3 November 2018; pp. 87–94.
21. de Azevedo, A.R.; Costa, A.M.; Cecchin, D.; Pereira, C.R.; Marvila, M.T.; Adesina, A. Economic potential comparative of reusing different industrial solid wastes in cementitious composites: A case study in Brazil. *Environ. Dev. Sustain.* **2022**, *24*, 5938–5961. [CrossRef]
22. Balwada, J.; Samaiya, S.; Mishra, R.P. Packaging plastic waste management for a circular economy and identifying a better waste collection system using analytical hierarchy process (ahp). *Procedia CIRP* **2021**, *98*, 270–275. [CrossRef]
23. Agenda, I. *The New Plastics Economy Rethinking the Future of Plastics*; The World Economic Forum: Geneva, Switzerland, 2016; p. 36.
24. Job, S.; Leeke, G.; Mativenga, P.; Oliveux, G.; Pickering, S.; Shuaib, N. Composites Recycling—Where Are We Now? 2016. Available online: https://compositesuk.co.uk/system/files/documents/Recycling.pdf (accessed on 26 April 2022).
25. Saxena, D.; Maiti, P. Utilization of ABS from plastic waste through single-step reactive extrusion of LDPE/ABS blends of improved properties. *Polymer* **2021**, *221*, 123626. [CrossRef]
26. Oladele, I.O.; Omotosho, T.F.; Adediran, A.A. Polymer-based composites: An indispensable material for present and future applications. *Int. J. Polym. Sci.* **2020**, *2020*, 8834518. [CrossRef]
27. Aydın, S.; Baradan, B. Mechanical and microstructural properties of heat cured alkali-activated slag mortars. *Mater. Des.* **2012**, *35*, 374–383. [CrossRef]
28. Giasuddin, H.M.; Sanjayan, J.G.; Ranjith, P. Strength of geopolymer cured in saline water in ambient conditions. *Fuel* **2013**, *107*, 34–39. [CrossRef]
29. Ranjbar, N.; Mehrali, M.; Behnia, A.; Alengaram, U.J.; Jumaat, M.Z. Compressive strength and microstructural analysis of fly ash/palm oil fuel ash based geopolymer mortar. *Mater. Des.* **2014**, *59*, 532–539. [CrossRef]
30. Jamil, N.H.; Abdullah, M.; Al Bakri, M.; Che Pa, F.; Hasmaliza, M.; W Ibrahim, W.M.A.; A Aziz, I.H.; Jeż, B.; Nabiałek, M. Phase Transformation of Kaolin-Ground Granulated Blast Furnace Slag from Geopolymerization to Sintering Process. *Magnetochemistry* **2021**, *7*, 32. [CrossRef]
31. He, J.; Jie, Y.; Zhang, J.; Yu, Y.; Zhang, G. Synthesis and characterization of red mud and rice husk ash-based geopolymer composites. *Cem. Concr. Compos.* **2013**, *37*, 108–118. [CrossRef]
32. McLellan, B.C.; Williams, R.P.; Lay, J.; Van Riessen, A.; Corder, G.D. Costs and carbon emissions for geopolymer pastes in comparison to ordinary portland cement. *J. Clean. Prod.* **2011**, *19*, 1080–1090. [CrossRef]
33. Turner, L.K.; Collins, F.G. Carbon dioxide equivalent (CO2-e) emissions: A comparison between geopolymer and OPC cement concrete. *Constr. Build. Mater.* **2013**, *43*, 125–130. [CrossRef]

34. Al Bakri, A.; Abdulkareem, O.A.; Rafiza, A.; Zarina, Y.; Norazian, M.; Kamarudin, H. Review on Processing of low calcium fly ash geopolymer concrete. *Aust. J. Basic Appl. Sci.* **2013**, *7*, 342–349.
35. Duan, P.; Yan, C.; Zhou, W. Influence of partial replacement of fly ash by metakaolin on mechanical properties and microstructure of fly ash geopolymer paste exposed to sulfate attack. *Ceram. Int.* **2016**, *42*, 3504–3517. [CrossRef]
36. Joseph, B.; Mathew, G. Influence of aggregate content on the behavior of fly ash based geopolymer concrete. *Sci. Iran.* **2012**, *19*, 1188–1194. [CrossRef]
37. Vakili, M.; Rafatullah, M.; Ibrahim, M.H.; Salamatinia, B.; Gholami, Z.; Zwain, H.M. A review on composting of oil palm biomass. *Environ. Dev. Sustain.* **2015**, *17*, 691–709. [CrossRef]
38. Sinulingga, K.; Agusnar, H.; Basuki Wirjosentono, Z.M. The effect of mixing rice husk ash and palm oil boiler ash on concrete strength. *Am. J. Phys. Chem.* **2014**, *3*, 9–14.
39. Kamseu, E.; à Moungam, L.B.; Cannio, M.; Billong, N.; Chaysuwan, D.; Melo, U.C.; Leonelli, C. Substitution of sodium silicate with rice husk ash-NaOH solution in metakaolin based geopolymer cement concerning reduction in global warming. *J. Clean. Prod.* **2017**, *142*, 3050–3060. [CrossRef]
40. Ye, N.; Chen, Y.; Yang, J.; Liang, S.; Hu, Y.; Xiao, B.; Huang, Q.; Shi, Y.; Hu, J.; Wu, X. Co-disposal of MSWI fly ash and Bayer red mud using an one-part geopolymeric system. *J. Hazard. Mater.* **2016**, *318*, 70–78. [CrossRef] [PubMed]
41. Liew, Y.-M.; Heah, C.-Y.; Kamarudin, H. Structure and properties of clay-based geopolymer cements: A review. *Prog. Mater. Sci.* **2016**, *83*, 595–629. [CrossRef]
42. Zhang, Z.; Zhu, H.; Zhou, C.; Wang, H. Geopolymer from kaolin in China: An overview. *Appl. Clay Sci.* **2016**, *119*, 31–41. [CrossRef]
43. Azimi, E.A.; Abdullah, M.M.A.B.; Vizureanu, P.; Salleh, M.A.A.M.; Sandu, A.V.; Chaiprapa, J.; Yoriya, S.; Hussin, K.; Aziz, I.H. Strength development and elemental distribution of dolomite/fly ash geopolymer composite under elevated temperature. *Materials* **2020**, *13*, 1015. [CrossRef]
44. Davidovits, J. Geopolymer cement. A Review. *Geopolymer Inst. Tech. Pap.* **2013**, *21*, 1–11.
45. Shobeiri, V.; Bennett, B.; Xie, T.; Visintin, P. A comprehensive assessment of the global warming potential of geopolymer concrete. *J. Clean. Prod.* **2021**, *297*, 126669. [CrossRef]
46. Gunasekara, C.; Law, D.; Bhuiyan, S.; Setunge, S.; Ward, L. Chloride induced corrosion in different fly ash based geopolymer concretes. *Constr. Build. Mater.* **2019**, *200*, 502–513. [CrossRef]
47. Luhar, S.; Chaudhary, S.; Luhar, I. Thermal resistance of fly ash based rubberized geopolymer concrete. *J. Build. Eng.* **2018**, *19*, 420–428. [CrossRef]
48. Aziz, I.H.; Abdullah, M.M.A.B.; Salleh, M.M.; Yoriya, S.; Chaiprapa, J.; Rojviriya, C.; Li, L.Y. Microstructure and porosity evolution of alkali activated slag at various heating temperatures. *J. Mater. Res. Technol.* **2020**, *9*, 15894–15907. [CrossRef]
49. Yuan, Y.; Zhao, R.; Li, R.; Wang, Y.; Cheng, Z.; Li, F.; Ma, Z.J. Frost resistance of fiber-reinforced blended slag and Class F fly ash-based geopolymer concrete under the coupling effect of freeze-thaw cycling and axial compressive loading. *Constr. Build. Mater.* **2020**, *250*, 118831. [CrossRef]
50. Li, N.; Farzadnia, N.; Shi, C. Microstructural changes in alkali-activated slag mortars induced by accelerated carbonation. *Cem. Concr. Res.* **2017**, *100*, 214–226. [CrossRef]
51. Ibrahim, M.; Johari, M.A.M.; Rahman, M.K.; Maslehuddin, M.; Mohamed, H.D. Enhancing the engineering properties and microstructure of room temperature cured alkali activated natural pozzolan based concrete utilizing nanosilica. *Constr. Build. Mater.* **2018**, *189*, 352–365. [CrossRef]
52. Guo, X.; Shi, H.; Wei, X. Pore properties, inner chemical environment, and microstructure of nano-modified CFA-WBP (class C fly ash-waste brick powder) based geopolymers. *Cem. Concr. Compos.* **2017**, *79*, 53–61. [CrossRef]
53. Temuujin, J.; Surenjav, E.; Ruescher, C.H.; Vahlbruch, J. Processing and uses of fly ash addressing radioactivity (critical review). *Chemosphere* **2019**, *216*, 866–882. [CrossRef]
54. Rashad, A.M. A brief on high-volume Class F fly ash as cement replacement—A guide for Civil Engineer. *Int. J. Sustain. Built Environ.* **2015**, *4*, 278–306. [CrossRef]
55. Wongsa, A.; Kunthawatwong, R.; Naenudon, S.; Sata, V.; Chindaprasirt, P. Natural fiber reinforced high calcium fly ash geopolymer mortar. *Constr. Build. Mater.* **2020**, *241*, 118143. [CrossRef]
56. Gupta, R.; Bhardwaj, P.; Mishra, D.; Prasad, M.; Amritphale, S. Formulation of mechanochemically evolved fly ash based hybrid inorganic–organic geopolymers with multilevel characterization. *J. Inorg. Organomet. Polym. Mater.* **2017**, *27*, 385–398. [CrossRef]
57. Aziz, I.H.; Abdullah, M.M.A.B.; Heah, C.-Y.; Liew, Y.-M. Behaviour changes of ground granulated blast furnace slag geopolymers at high temperature. *Adv. Cem. Res.* **2020**, *32*, 465–475. [CrossRef]
58. Salvador, R.P.; Rambo, D.A.; Bueno, R.M.; Silva, K.T.; de Figueiredo, A.D. On the use of blast-furnace slag in sprayed concrete applications. *Constr. Build. Mater.* **2019**, *218*, 543–555. [CrossRef]
59. Amran, Y.M.; Alyousef, R.; Alabduljabbar, H.; El-Zeadani, M. Clean production and properties of geopolymer concrete; A review. *J. Clean. Prod.* **2020**, *251*, 119679. [CrossRef]
60. Aziz, I.H.; Abdullah, M.M.A.B.; Salleh, M.A.A.M.; Ming, L.Y.; Li, L.Y.; Sandu, A.V.; Vizureanu, P.; Nemes, O.; Mahdi, S.N. Recent Developments in Steelmaking Industry and Potential Alkali Activated Based Steel Waste: A Comprehensive Review. *Materials* **2022**, *15*, 1948. [CrossRef]

61. Nie, Q.; Hu, W.; Huang, B.; Shu, X.; He, Q. Synergistic utilization of red mud for flue-gas desulfurization and fly ash-based geopolymer preparation. *J. Hazard. Mater.* **2019**, *369*, 503–511. [CrossRef]
62. Yang, Z.; Mocadlo, R.; Zhao, M.; Sisson, R.D., Jr.; Tao, M.; Liang, J. Preparation of a geopolymer from red mud slurry and class F fly ash and its behavior at elevated temperatures. *Constr. Build. Mater.* **2019**, *221*, 308–317. [CrossRef]
63. Yeddula, B.S.R.; Karthiyaini, S. Experimental investigations and GEP modelling of compressive strength of ferrosialate based geopolymer mortars. *Constr. Build. Mater.* **2020**, *236*, 117602. [CrossRef]
64. Liu, Y.; Qin, Z.; Chen, B. Experimental research on magnesium phosphate cements modified by red mud. *Constr. Build. Mater.* **2020**, *231*, 117131. [CrossRef]
65. Gholampour, A.; Ozbakkaloglu, T.; Ng, C.-T. Ambient-and oven-cured geopolymer concretes under active confinement. *Constr. Build. Mater.* **2019**, *228*, 116722. [CrossRef]
66. Guo, X.; Yang, J. Intrinsic properties and micro-crack characteristics of ultra-high toughness fly ash/steel slag based geopolymer. *Constr. Build. Mater.* **2020**, *230*, 116965. [CrossRef]
67. Hu, Y.; Tang, Z.; Li, W.; Li, Y.; Tam, V.W. Physical-mechanical properties of fly ash/GGBFS geopolymer composites with recycled aggregates. *Constr. Build. Mater.* **2019**, *226*, 139–151. [CrossRef]
68. Khan, I.; Xu, T.; Castel, A.; Gilbert, R.I.; Babaee, M. Risk of early age cracking in geopolymer concrete due to restrained shrinkage. *Constr. Build. Mater.* **2019**, *229*, 116840. [CrossRef]
69. Mohammed, B.S.; Haruna, S.; Wahab, M.; Liew, M.; Haruna, A. Mechanical and microstructural properties of high calcium fly ash one-part geopolymer cement made with granular activator. *Heliyon* **2019**, *5*, e02255. [CrossRef] [PubMed]
70. De Rossi, A.; Ribeiro, M.; Labrincha, J.; Novais, R.; Hotza, D.; Moreira, R. Effect of the particle size range of construction and demolition waste on the fresh and hardened-state properties of fly ash-based geopolymer mortars with total replacement of sand. *Process Saf. Environ. Prot.* **2019**, *129*, 130–137. [CrossRef]
71. Nuaklong, P.; Wongsa, A.; Sata, V.; Boonserm, K.; Sanjayan, J.; Chindaprasirt, P. Properties of high-calcium and low-calcium fly ash combination geopolymer mortar containing recycled aggregate. *Heliyon* **2019**, *5*, e02513. [CrossRef]
72. Yu, J.; Chen, Y.; Chen, G.; Wang, L. Experimental study of the feasibility of using anhydrous sodium metasilicate as a geopolymer activator for soil stabilization. *Eng. Geol.* **2020**, *264*, 105316. [CrossRef]
73. Alrefaei, Y.; Wang, Y.-S.; Dai, J.-G. The effectiveness of different superplasticizers in ambient cured one-part alkali activated pastes. *Cem. Concr. Compos.* **2019**, *97*, 166–174. [CrossRef]
74. Bouaissi, A.; Li, L.-y.; Abdullah, M.M.A.B.; Bui, Q.-B. Mechanical properties and microstructure analysis of FA-GGBS-HMNS based geopolymer concrete. *Constr. Build. Mater.* **2019**, *210*, 198–209. [CrossRef]
75. Alonso, M.; Gascó, C.; Morales, M.M.; Suárez-Navarro, J.; Zamorano, M.; Puertas, F. Olive biomass ash as an alternative activator in geopolymer formation: A study of strength, radiology and leaching behaviour. *Cem. Concr. Compos.* **2019**, *104*, 103384. [CrossRef]
76. Li, Z.; Zhang, J.; Li, S.; Gao, Y.; Liu, C.; Qi, Y. Effect of different gypsums on the workability and mechanical properties of red mud-slag based grouting materials. *J. Clean. Prod.* **2020**, *245*, 118759. [CrossRef]
77. Yang, T.; Wang, Y.; Sheng, L.; He, C.; Sun, W.; He, Q. Enhancing Cd (II) sorption by red mud with heat treatment: Performance and mechanisms of sorption. *J. Environ. Manag.* **2020**, *255*, 109866. [CrossRef] [PubMed]
78. Shi, W.; Ren, H.; Huang, X.; Li, M.; Tang, Y.; Guo, F. Low cost red mud modified graphitic carbon nitride for the removal of organic pollutants in wastewater by the synergistic effect of adsorption and photocatalysis. *Sep. Purif. Technol.* **2020**, *237*, 116477. [CrossRef]
79. Villaquirán-Caicedo, M.A. Studying different silica sources for preparation of alternative waterglass used in preparation of binary geopolymer binders from metakaolin/boiler slag. *Constr. Build. Mater.* **2019**, *227*, 116621. [CrossRef]
80. Liang, G.; Zhu, H.; Zhang, Z.; Wu, Q.; Du, J. Investigation of the waterproof property of alkali-activated metakaolin geopolymer added with rice husk ash. *J. Clean. Prod.* **2019**, *230*, 603–612. [CrossRef]
81. Cheah, C.B.; Tan, L.E.; Ramli, M. The engineering properties and microstructure of sodium carbonate activated fly ash/slag blended mortars with silica fume. *Compos. Part B Eng.* **2019**, *160*, 558–572. [CrossRef]
82. Baenla, J.; Mbah, J.B.; Ndjock, I.D.L.; Elimbi, A. Partial replacement of low reactive volcanic ash by cassava peel ash in the synthesis of volcanic ash based geopolymer. *Constr. Build. Mater.* **2019**, *227*, 116689. [CrossRef]
83. Tosti, L.; van Zomeren, A.; Pels, J.R.; Comans, R.N. Technical and environmental performance of lower carbon footprint cement mortars containing biomass fly ash as a secondary cementitious material. *Resour. Conserv. Recycl.* **2018**, *134*, 25–33. [CrossRef]
84. Nuaklong, P.; Jongvivatsakul, P.; Pothisiri, T.; Sata, V.; Chindaprasirt, P. Influence of rice husk ash on mechanical properties and fire resistance of recycled aggregate high-calcium fly ash geopolymer concrete. *J. Clean. Prod.* **2020**, *252*, 119797. [CrossRef]
85. Raisi, E.M.; Amiri, J.V.; Davoodi, M.R. Mechanical performance of self-compacting concrete incorporating rice husk ash. *Constr. Build. Mater.* **2018**, *177*, 148–157. [CrossRef]
86. Yadav, A.L.; Sairam, V.; Muruganandam, L.; Srinivasan, K. An overview of the influences of mechanical and chemical processing on sugarcane bagasse ash characterisation as a supplementary cementitious material. *J. Clean. Prod.* **2020**, *245*, 118854. [CrossRef]
87. Maldonado-Alameda, A.; Giro-Paloma, J.; Svobodova-Sedlackova, A.; Formosa, J.; Chimenos, J. Municipal solid waste incineration bottom ash as alkali-activated cement precursor depending on particle size. *J. Clean. Prod.* **2020**, *242*, 118443. [CrossRef]
88. Nagrockienė, D.; Daugėla, A. Investigation into the properties of concrete modified with biomass combustion fly ash. *Constr. Build. Mater.* **2018**, *174*, 369–375. [CrossRef]

89. Lee, Y.; Kang, S. Influence of Blended Activator on Microstructure, Crystal Phase and Physical Properties of Spent Catalyst Slag-Based Geopolymer. *J. Nanosci. Nanotechnol.* **2016**, *16*, 11313–11318. [CrossRef]
90. Marathe, S.; Shetty, T.S.; Mithun, B.; Ranjith, A. Strength and durability studies on air cured alkali activated pavement quality concrete mixes incorporating recycled aggregates. *Case Stud. Constr. Mater.* **2021**, *15*, e00732. [CrossRef]
91. Lemougna, P.N.; Wang, K.T.; Tang, Q.; Cui, X.M. Study on the development of inorganic polymers from red mud and slag system: Application in mortar and lightweight materials. *Constr. Build. Mater.* **2017**, *156*, 486–495. [CrossRef]
92. Ye, N.; Yang, J.; Ke, X.; Zhu, J.; Li, Y.; Xiang, C.; Wang, H.; Li, L.; Xiao, B. Synthesis and characterization of geopolymer from Bayer red mud with thermal pretreatment. *J. Am. Ceram. Soc.* **2014**, *97*, 1652–1660. [CrossRef]
93. Ahmed, M.M.; El-Naggar, K.; Tarek, D.; Ragab, A.; Sameh, H.; Zeyad, A.M.; Tayeh, B.A.; Maafa, I.M.; Yousef, A. Fabrication of thermal insulation geopolymer bricks using ferrosilicon slag and alumina waste. *Case Stud. Constr. Mater.* **2021**, *15*, e00737. [CrossRef]
94. Júnior, L.U.T.; Taborda-Barraza, M.; Cheriaf, M.; Gleize, P.J.; Rocha, J.C. Effect of bottom ash waste on the rheology and durability of alkali activation pastes. *Case Stud. Constr. Mater.* **2022**, *16*, e00790.
95. Alomayri, T.; Adesina, A.; Das, S. Influence of amorphous raw rice husk ash as precursor and curing condition on the performance of alkali activated concrete. *Case Stud. Constr. Mater.* **2021**, *15*, e00777. [CrossRef]
96. Hassan, A.; Arif, M.; Shariq, M. Influence of microstructure of geopolymer concrete on its mechanical properties—A review. *Adv. Sustain. Constr. Mater. Geotech. Eng.* **2020**, *35*, 119–129.
97. Shahedan, N.F.; Abdullah, M.M.A.B.; Mahmed, N.; Kusbiantoro, A.; Tammas-Williams, S.; Li, L.-Y.; Aziz, I.H.; Vizureanu, P.; Wysłocki, J.J.; Błoch, K. Properties of a New Insulation Material Glass Bubble in Geo-Polymer Concrete. *Materials* **2021**, *14*, 809. [CrossRef] [PubMed]
98. Reddy, D.V.; Edouard, J.-B.; Sobhan, K. Durability of fly ash–based geopolymer structural concrete in the marine environment. *J. Mater. Civ. Eng.* **2013**, *25*, 781–787. [CrossRef]
99. Chindaprasirt, P.; Chalee, W. Effect of sodium hydroxide concentration on chloride penetration and steel corrosion of fly ash-based geopolymer concrete under marine site. *Constr. Build. Mater.* **2014**, *63*, 303–310. [CrossRef]
100. Pasupathy, K.; Berndt, M.; Sanjayan, J.; Rajeev, P.; Cheema, D.S. Durability of low-calcium fly ash based geopolymer concrete culvert in a saline environment. *Cem. Concr. Res.* **2017**, *100*, 297–310. [CrossRef]
101. Alzeebaree, R.; Çevik, A.; Mohammedameen, A.; Niş, A.; Gülşan, M.E. Mechanical performance of FRP-confined geopolymer concrete under seawater attack. *Adv. Struct. Eng.* **2020**, *23*, 1055–1073. [CrossRef]
102. Nuaklong, P.; Sata, V.; Chindaprasirt, P. Properties of metakaolin-high calcium fly ash geopolymer concrete containing recycled aggregate from crushed concrete specimens. *Constr. Build. Mater.* **2018**, *161*, 365–373. [CrossRef]
103. Glasby, T.; Day, J.; Genrich, R.; Aldred, J. EFC geopolymer concrete aircraft pavements at Brisbane West Wellcamp Airport. *Concrete 2015*, *2015*, 1–9.
104. Ali, A.M.; Sanjayan, J.; Guerrieri, M. Performance of geopolymer high strength concrete wall panels and cylinders when exposed to a hydrocarbon fire. *Constr. Build. Mater.* **2017**, *137*, 195–207.
105. Ferdous, W.; Manalo, A. Failures of mainline railway sleepers and suggested remedies—Review of current practice. *Eng. Fail. Anal.* **2014**, *44*, 17–35. [CrossRef]
106. Mirza, O.; Shill, S.K.; Johnston, J. Performance of precast prestressed steel-concrete composite panels under static loadings to replace the timber transoms for railway bridge. *Structures.* **2019**, *19*, 30–40. [CrossRef]
107. Shill, S.K.; Al-Deen, S.; Ashraf, M. Concrete durability issues due to temperature effects and aviation oil spillage at military airbase–A comprehensive review. *Constr. Build. Mater.* **2018**, *160*, 240–251. [CrossRef]
108. Shill, S.K.; Al-Deen, S.; Ashraf, M. *Thermal and Chemical Degradation of Portland Cement Concrete in the Military Airbase*; EasyChair, 2019; pp. 2314–2516. Available online: https://easychair.org/publications/preprint/b24Q (accessed on 10 April 2022).
109. Shill, S.K.; Al-Deen, S.; Ashraf, M. Saponification and scaling in ordinary concrete exposed to hydrocarbon fluids and high temperature at military airbases. *Constr. Build. Mater.* **2019**, *215*, 765–776. [CrossRef]
110. Al-Deen, S.; Duanne, B.W.; Shill, S.K.; Ashraf, M. Durability Issues of Military Airfield Rigid Pavements due to Combined Influence of Chemical Oil Spills and Repeated Thermal Shocks from Jet Fighter Exhaust. In Proceedings of the Biennial National Conference of the Concrete Institute of Australia & International Congress on Durability of Concrete, Combined Conference, Adelaide, Australia, 2017; Deakin University: Melbourne, Australia, 2017; pp. 1–9.
111. Shill, S.K.; Al-Deen, S.; Ashraf, M.; Hutchison, W. Resistance of fly ash based geopolymer mortar to both chemicals and high thermal cycles simultaneously. *Constr. Build. Mater.* **2020**, *239*, 117886. [CrossRef]
112. Rickard, W.D.; Kealley, C.S.; Van Riessen, A. Thermally induced microstructural changes in fly ash geopolymers: Experimental results and proposed model. *J. Am. Ceram. Soc.* **2015**, *98*, 929–939. [CrossRef]
113. Lahoti, M.; Wong, K.K.; Tan, K.H.; Yang, E.-H. Effect of alkali cation type on strength endurance of fly ash geopolymers subject to high temperature exposure. *Mater. Des.* **2018**, *154*, 8–19. [CrossRef]
114. Tenn, N.; Allou, F.; Petit, C.; Absi, J.; Rossignol, S. Formulation of new materials based on geopolymer binders and different road aggregates. *Ceram. Int.* **2015**, *41*, 5812–5820. [CrossRef]
115. Camacho-Tauta, J.; Reyes-Ortiz, O.; da Fonseca, A.V.; Rios, S.; Cruz, N.; Rodrigues, C. Full-scale evaluation in a fatigue track of a base course treated with geopolymers. *Procedia Eng.* **2016**, *143*, 18–25. [CrossRef]

116. Sukprasert, S.; Hoy, M.; Horpibulsuk, S.; Arulrajah, A.; Rashid, A.S.A.; Nazir, R. Fly ash based geopolymer stabilisation of silty clay/blast furnace slag for subgrade applications. *Road Mater. Pavement Des.* **2021**, *22*, 357–371. [CrossRef]
117. Dave, N.; Sahu, V.; Misra, A.K. Development of geopolymer cement concrete for highway infrastructure applications. *J. Eng. Des. Technol.* **2020**, *18*, 1321–1333. [CrossRef]
118. Wongsa, A.; Sata, V.; Nematollahi, B.; Sanjayan, J.; Chindaprasirt, P. Mechanical and thermal properties of lightweight geopolymer mortar incorporating crumb rubber. *J. Clean. Prod.* **2018**, *195*, 1069–1080. [CrossRef]
119. Mohammed, B.S.; Liew, M.S.; Alaloul, W.S.; Al-Fakih, A.; Ibrahim, W.; Adamu, M. Development of rubberized geopolymer interlocking bricks. *Case Stud. Constr. Mater.* **2018**, *8*, 401–408. [CrossRef]
120. de Oliveira, L.B.; de Azevedo, A.R.; Marvila, M.T.; Pereira, E.C.; Fediuk, R.; Vieira, C.M.F. Durability of geopolymers with industrial waste. *Case Stud. Constr. Mater.* **2022**, *16*, e00839. [CrossRef]
121. Provis, J.L.; Bernal, S.A. Geopolymers and related alkali-activated materials. *Annu. Rev. Mater. Res.* **2014**, *44*, 299–327. [CrossRef]
122. Van Deventer, J.S.; Provis, J.L.; Duxson, P. Technical and commercial progress in the adoption of geopolymer cement. *Miner. Eng.* **2012**, *29*, 89–104. [CrossRef]

Article

Evaluation of Mechanical and Microstructural Properties and Global Warming Potential of Green Concrete with Wheat Straw Ash and Silica Fume

Kaffayatullah Khan [1,2], Muhammad Ishfaq [3], Muhammad Nasir Amin [1,2,*], Khan Shahzada [3], Nauman Wahab [4] and Muhammad Iftikhar Faraz [2,5]

1. Department of Civil and Environmental Engineering, College of Engineering, King Faisal University, Al-Ahsa 31982, Saudi Arabia; kkhan@kfu.edu.sa
2. Al Bilad Bank Scholarly Chair for Food Security in Saudi Arabia, The Deanship of Scientific Research, The Vice Presidency for Graduate Studies and Scientific Research, King Faisal University, Al-Ahsa 31982, Saudi Arabia; mfaraz@kfu.edu.sa
3. Department of Civil Engineering, University of Engineering and Technology, Peshawar 25120, Pakistan; mishfaq948@gmail.com (M.I.); khanshahzada@uetpeshawar.edu.pk (K.S.)
4. Department of Civil and Environmental Engineering, University of Rome "La Sapienza", Via Eudossiana 18, 00184 Rome, Italy; nauman.wahab@uniroma1.it
5. Department of Mechanical Engineering, College of Engineering, King Faisal University, Al-Ahsa 31982, Saudi Arabia
* Correspondence: mgadir@kfu.edu.sa; Tel.: +966-13-589-5431; Fax: +966-13-581-7068

Abstract: Cement and concrete are among the major contributors to CO_2 emissions in modern society. Researchers have been investigating the possibility of replacing cement with industrial waste in concrete production to reduce its environmental impact. Therefore, the focus of this paper is on the effective use of wheat straw ash (WSA) together with silica fume (SF) as a cement substitute to produce high-performance and sustainable concrete. Different binary and ternary mixes containing WSA and SF were investigated for their mechanical and microstructural properties and global warming potential (GWP). The current results indicated that the binary and ternary mixes containing, respectively, 20% WSA (WSA20) and 33% WSA together with 7% SF (WSA33SF7) exhibited higher strengths than that of control mix and other binary and ternary mixes. The comparative lower apparent porosity and water absorption values of WSA20 and WSA33SF7 among all mixes also validated the findings of their higher strength results. Moreover, SEM–EDS and FTIR analyses has revealed the presence of dense and compact microstructure, which are mostly caused by formation of high-density calcium silicate hydrate (C-S-H) and calcium hydroxide (C-H) phases in both blends. FTIR and TGA analyses also revealed a reduction in the portlandite phase in these mixes, causing densification of microstructures and pores. Additionally, N_2 adsorption isotherm analysis demonstrates that the pore structure of these mixes has been densified as evidenced by a reduction in intruded volume and a rise in BET surface area. Furthermore, both mixes had lower CO_2-eq intensity per MPa as compared to control, which indicates their significant impact on producing green concretes through their reduced GWPs. Thus, this research shows that WSA alone or its blend with SF can be considered as a source of revenue for the concrete industry for developing high-performance and sustainable concretes.

Keywords: wheat straw ash; global warming potential; compressive strength; water absorption; microstructure and pore structure

Citation: Khan, K.; Ishfaq, M.; Amin, M.N.; Shahzada, K.; Wahab, N.; Faraz, M.I. Evaluation of Mechanical and Microstructural Properties and Global Warming Potential of Green Concrete with Wheat Straw Ash and Silica Fume. *Materials* 2022, 15, 3177. https://doi.org/10.3390/ma15093177

Academic Editor: Dumitru Doru Burduhos Nergis

Received: 4 March 2022
Accepted: 18 April 2022
Published: 27 April 2022

Publisher's Note: MDPI stays neutral with regard to jurisdictional claims in published maps and institutional affiliations.

Copyright: © 2022 by the authors. Licensee MDPI, Basel, Switzerland. This article is an open access article distributed under the terms and conditions of the Creative Commons Attribution (CC BY) license (https://creativecommons.org/licenses/by/4.0/).

1. Introduction

Concrete is a heterogeneous material formulated from natural and synthetic components, combined with water, and forms a versatile, durable, and essential building material. Worldwide, concrete production has significantly increased due to urbanization and urban development, reaching 25 billion tons per year [1]. Subsequently, the yearly production of cement has also soared to 4.1 billion tons globally [2]. Besides its usefulness, cement and concrete production has resulted in massive carbon emissions as well. Apart from the carbon emissions, the usage of raw materials in cement production is also responsible for the depletion of natural resources and its associated carbon footprint [3]. Hence, the annual manufactured carbon emissions from the cement industry are more than 5% of the total global anthropogenic emissions [4]. This vast amount of CO_2 generation is also a leading factor in creating the issues related to global warming and climate change.

These issues can be controlled by following the sustainability approach in the construction sector. One such approach is to replace cement with supplementary cementitious materials (SCMs) at the construction site or during cement manufacturing. Another approach is to conserve natural resources by replacing aggregates with alternate ecofriendly materials. Hence, the search for supplementary cementitious materials has resulted in the usage of industrial byproducts such as volcanic ash [5,6], silica fume (SF) [7,8], fly ash [9,10], electric arc furnace slag [11,12], tire ash [13,14], copper slag [15,16], etc., and agricultural byproducts such as sugarcane bagasse ash [17,18], rice husk ash [19,20], woodwaste ash [21,22], wheat straw ash (WSA) [23,24], and others. The utilization of these agroindustrial wastes has been proved to be promising both as supplementary cementitious material to replace cement and/or fine aggregates [24]. Furthermore, the usage of these byproducts offers many advantages such as improved mechanical and durability properties, reduction in waste generation, reduction in cost by replacing cement/aggregates, and most importantly, the decrease in carbon emissions [25,26]. Therefore, Luhar et al. [4] reviewed the usage of agriculture waste in concrete. They concluded that agrowaste could be used as an effective supplementary cementitious material in concrete, resulting in the development of green concrete. Rattanachu et al. [27] observed that 20% replacement of OPC with finely ground rice husk ash can significantly improve the compressive strength of concrete. In addition, the utilization of SCMs also improves the durability of concrete [2]. In another study, Rashad [3] replaced fine aggregate with metakaolin at 10%, 20%, 30%, 40%, and 50% by weight. The experimental study revealed increased splitting tensile strength, compressive strength, and abrasion resistance of metakaolin mixed concrete at 40% replacement.

In developing countries, agrowaste is one of the significant issues which arise from different food crops such as sugarcane, rice, and wheat. Wheat is one of the most important cereal crops and a major food source for around 2.5 billion people globally [28]. The worldwide wheat production was estimated to be 750 million tons from 2016 to 2017 [29]. Pakistan ranked high in the wheat-producing countries around the globe. In the year 2017–2018, Pakistan produced about 26.6 million tons of wheat. In the Gulf region, the Kingdom of Saudi Arabia is known as a major wheat-producing country. The annual wheat production of the Kingdom of Saudi Arabia is estimated at 700,000 tons [30]. Pan and Sano [31] stated that one kg of wheat grain yields around 1.3 kg wheat straw. Primarily, the wheat straw was utilized to feed cattle. However, open field burning is also practiced in some cases, which causes air pollution (smog) and health issues such as respiratory diseases in that region.

Burned wheat straw produces ash, which is pozzolanic in nature. However, the pozzolanic efficiency of the WSA mainly depends on its source; therefore, its composition varies with regions primarily due to soil properties and climatic conditions [32]. In addition, the burning temperature, exposure time, and particle sizes also play a vital role in its pozzolanic behavior. Biricik et al. [33] reported that wheat straw burned for 5 h at 570 and 670 °C resulted in high-quality WSA. However, WSA produced at a heating temperature of 670 °C showed superior pozzolanic properties. Similarly, an increase in silica content with an increase in burning temperature was also recorded by Amin et al. [29]. Memon et al. [34] investigated the effects of various burning temperatures (500, 600, 700, and 800 °C) on the pozzolanic efficacy of WSA. They observed the transition of amorphous silica into crystalline form with an increasing temperature beyond 600 °C. Therefore, several researchers observed favorable temperatures, which ranged between 570 and 670 °C for an improved WSA production.

Research studies have highlighted the efficacy of the WSA both as filling and pozzolanic material in concrete and mortar. The addition of WSA in mortar increased the compressive and flexural strength [35]. Moreover, the improvement in the mechanical properties of cement mortar was attributed to its filling ability. Furthermore, the addition of WSA in cement mortar or concrete also improves its durability. Qudoos et al. [36] noted that extensively ground WSA exhibited higher compressive strength at all ages than control specimen with 20% cement replacement. Similarly, Amin et al. [29] also observed that 15–20% cement replacement with WSA showed significant enhancement in the strength and ductility of concrete samples at 91 days. WSA as cement or sand replacement decreased the water adsorption and increased the resistance against acid and sulfate attacks. Al-Akhras [37] used WSA as cement replacement up to 15% by weight and concluded that concrete containing WSA showed better resistance to freeze and thaw damage than control specimens. In addition to its standalone performance as a pozzolanic material, the behavior of WSA was also evaluated as a binary mix with other pozzolanic materials such as metakaolin, fly ash, bentonite clay, millet husk ash, and others [3,26,32,38]. The usage of bentonite clay along with WSA was effective in consuming the free lime. Moreover, it improved the resistance of the cementitious matrix against acid attack [38]. The effectiveness of WSA and millet husk ash (MHA) combined was evaluated by Bheel et al. [39]. The results showed improvement in the flexural, tensile, and compressive strength of the specimens containing 15% MHA and 30% WSA combined.

The binary and ternary blends of various agroindustrial wastes have increased its pozzolanic activity. Among many, SF is widely used component of ternary blends to enhance mechanical as well as durability performance of concrete. Generally, SF is obtained as a byproduct during the production of silicon and ferrosilicon alloys at a very high temperature around 2000 °C in an electric furnace arc [40]. Oxygen is eliminated by heating highly pure quartz with coke or coal. An ultrafine powder is obtained, having high porosity and specific surface area in which the silica ranged between 85% and 95% [41]. SF is most commonly used in concrete as a dry, densified form consisting of agglomerates of size from 10 microns to several millimeters. These agglomerates may only partially disintegrate during normal concrete mixing [42]. In order to disperse SF effectively, sonification techniques [43] or the Holland method [44] of mixing SF concrete in a laboratory mixer are necessary for improving the microstructure and pore size of the materials. Murthi et al. [45] presented the effects of OPC, bagasse ash, and nanosilica on the fresh and hardened properties of high-performance concrete. The experimental investigation showed that the addition of nanosilica significantly enhanced the early age strength; however, it reduced the setting time. SF is a highly siliceous material, and its addition to the cement matrix increases the formation of calcium silicate hydrate (C-S-H) gel, resulting in a denser microstructure of the cementitious system. The cement replacement with 10% SF and 20% WSA in lightweight concrete improved its density and strength for structural applications [46]. The microstructure investigation of cementitious matrix containing a ternary blend of OPC, SF, and WSA indicated better mechanical and durability performance than the control specimen [47]. Based on previous research, it has been highlighted that the usefulness of the SCMs increases with the addition of SF. SF, being one of the most reactive siliceous materials, could accelerate the pozzolanic activity inside the cementitious system when used in combination with agro-industrial ashes. Despite several studies, it has been noted that the existing literature is limited to the use of WSA as an individual SCM with low replacement levels.

The purpose of this study is to investigate the novel use of WSA and its blend with SF as a cement substitute to produce environmentally sustainable concrete that does not compromise on its mechanical properties. Therefore, the impact of WSA and its blend with SF, as a high-volume replacement of cement, on the mechanical, durability, and microstructural characteristics of concrete were investigated. Additionally, X-ray fluorescence (XRF) and X-ray diffraction (XRD) methods were used to measure the physical and chemical properties of cement, WSA, and SF. Concrete specimens were prepared, namely, control (100% cement), three binary with only WSA (C/WSA: 90/10, 80/20, and 70/30), and three ternary containing WSA along with SF (C/WSA/SF: 70/25/5, 60/33/7, and 50/40/10). In ternary mixes, a large amount of cement was substituted (up to 50%). Mechanical properties, such as the compressive and split tensile strengths with aging (7, 28, and 91 days), and water absorption (WA) and apparent porosity (AP) after aging for 91 days were evaluated for hardened concrete samples. Additionally, the effect of WSA and its blend with SF on the microstructure and pore structure of the cement paste matrix was studied by employing scanning electron microscopy/energy-dispersive X-ray spectroscopy (SEM–EDS), Fourier transform infrared (FTIR) spectroscopy, thermogravimetric analysis (TGA), and nitrogen (N_2) adsorption isotherm analysis. At the end, the global warming potential (GWP) of all the concrete mixes was calculated in kg CO_2-equivalent per unit concrete and kg CO_2-equivalent per unit concrete/MPa by using the green concrete lifecycle assessment (LCA) tool.

2. Materials and Methods

2.1. Materials

The main binders used in this study were conventional OPC, consistent with ASTM C150, and commercially available SF and WSA (processed in the laboratory). The specified physical and chemical properties of these materials are listed in Table 1.

Table 1. Physical and chemical properties of cement, WSA, and SF.

	OPC	WSA	SF
	Physical properties		
Specific gravity (kg/m^3)	3.20	2.21	2.23
Blain fineness (m^2/g)	0.344	-	21.5
	Chemical properties (oxides, % by weight)		
SiO_2	21.3	65.1	93.3
Al_2O_3	5.56	9.10	-
Fe_2O_3	3.24	2.67	0.58
$SiO_2 + Al_2O_3 + Fe_2O_3$ *	-	76.87	-
CaO	63.4	5.90	1.82
MgO	0.93	1.11	0.28
Na_2O	0.13	0.40	0.19
K_2O	0.62	10.5	0.88
SO_3	2.25	1.13	-
LOI **	2.41	4.03	2.25
	Mineral composition (%) ***		
C_2S	47.9	-	-
C_3S	24.7	-	-
C_3A	8.76	-	-
C_4AF	9.86	-	-

* ASTM C618-15; ** LOI = loss on ignition; *** data from local cement manufacturer.

Besides binder materials, aggregates from local sources were obtained for using as fillers in concrete. Table 2 shows the sieve analysis results for fine and coarse aggregates. Blending percentages for coarse aggregates are 50% (20 mm down) and 50% (10 mm down). In accordance with ASTM C33, fine aggregate possessed specific gravity of 2.60 and water absorption of 1.03%, while its fineness modulus was 2.54 (Table 2). Specifically, the coarse aggregates had a specific gravity of 2.65 and water absorption of 0.82%, respectively.

Table 2. Sieve analysis of aggregates (ASTM C136).

Sieve #	Sieve Size (mm)	Weight Retained (%)	Cumulative Passing (%)	Cumulative Retained (%)
Coarse aggregate (CA)				
1 inch	25	0	100	0
3/4 inch	19	0	100	0
1/2 inch	12.5	50.5	50	50
3/8 inch	9.5	26.1	23	77
No. 4	4.75	23.4	0	100
Fine aggregate (FA)				
3/8 inch	9.5	0	100	0
No. 4	4.75	0	100	0
No. 8	2.36	4.54	95.46	4.54
No. 16	1.18	16.1	79.33	20.67
No. 30	0.600	30.9	48.40	51.60
No. 50	0.300	26.4	22.02	77.98
No. 100	0.150	21.3	0.67	99.33
Pan	-	0.67	-	-
Fineness Modulus of FA (**FM**) = (0 + 4.54 + 20.67 + 51.6 + 77.98 + 99.33)/100 = **2.54**				

2.1.1. Burning and Grinding of Wheat Straw Ash

The chemical properties of WSA are largely dependent on the source (wheat straw), the organic composition, and the sintering temperature [20]. Further, climatic conditions the chemical composition of WSA is influenced by both climatic conditions and geographical location [48]. The WSA obtained was burned at control temperature at 550 for 4 and 8 h and finally at 800 °C for 30 min in a kiln. Each of these samples exposed to different elevated temperatures were subjected to analysis using XRF, XRD, and FTIR to determine their chemical and mineralogical properties. Accordingly, XRD and FTIR results revealed that the maximum amount of amorphous silica was achieved for the sample when burned at 550 °C for 4 h (Figure 1). Based on chemical composition, FTIR, and XRD analyses, the WSA obtained after burning at 550 °C for 4 h was ground to obtain a fine powder. After cooling under normal air, the WSA was ground in a rotary mill for 12 h at 15 rpm. Using this ground WSA, binary concrete specimens containing only WSA and ternary concrete specimens having blends of WSA with SF were prepared in order to evaluate its effectiveness as a partial substitute of cement. The physical and chemical properties of WSA are listed in Table 1.

2.1.2. Concrete Mixture Proportions

In addition to CC, three binary and three ternary concrete mixture proportions were designed. The CC having 100% cement, whereas the binary concrete mixtures were designed by partial substitution of cement with 10%, 20%, and 30% WSA, and are respectively identified as WSA10, WSA20, and WSA30. The ternary mixtures incorporating blends of WSA and SF were designed by substituting high amounts of cement ranging from 30% to 50% as 25% WSA + 5% SF (WSA25SF5), 33% WSA + 7% SF (WSA33SF7), and 40% WSA + 10% SF (WSA40SF10). Table 3 shows the corresponding ingredients of each concrete mixture and the related details of test specimens.

Adding SF to ternary concrete mixtures was primarily to increase the percent of cement replaced by WSA without compromising the mechanical or durability characteristics. Thus, to improve the mechanical strength of binary concrete as compared to those of CC, SF is added at a rate of 5%, 7%, and 10% to concrete mixtures containing a high percentage of WSA as 25%, 33%, and 40%, respectively. The water-to-binder ratio was maintained at 0.35 for all the concrete mixtures. In accordance with the constant water-to-binder ratio and binder content (457 kg/m^3), the water content for the concrete mixtures was maintained at 160 kg/m^3. Consequently, to achieve the target workability corresponding to slump values of 120 ± 30 mm, additional dosages (wt.%) of the naphthalene-based water-reducing admixture were employed for each concrete mixture, as listed in Table 3.

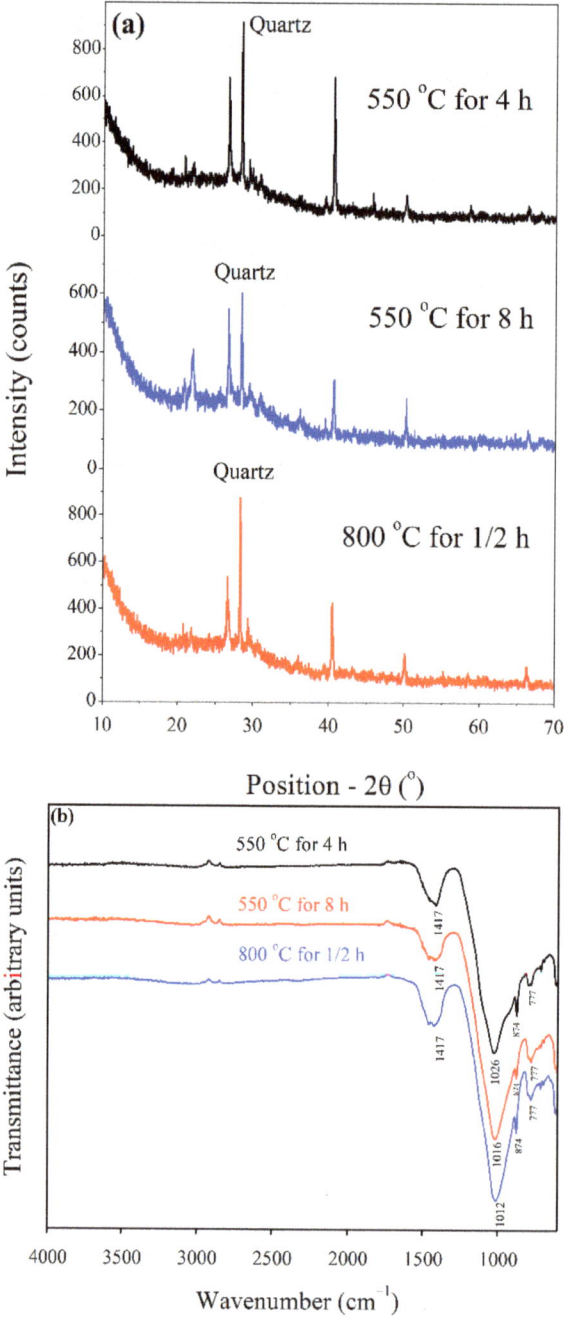

Figure 1. (a) XRD pattern and (b) FTIR of WSA after heat treatment at 550 °C for 4 and 8 h, and at 800 °C for 30 min.

Table 3. Mixture proportions for control, binary, and ternary concretes (w/b = 0.35; a/b = 3.37; s/a = 0.40).

Mix ID (Total # of Specimens for Each = 18 and Cured for 7, 28, 91 Days)	W (kg/m³)	Binder, b (kg/m³)			Aggregates, a (kg/m³)			Superplasticizer (% of b)	Slump (mm)
		C	WSA	SF	FA (S)	CA 10 mm	CA 20 mm		
Control concrete (CC)		457	-	-	617			1	
Concrete containing 10% WSA (WSA10)		411	46	-				1	
Concrete containing 20% WSA (WSA20)		366	91	-				1	
Concrete containing 30% WSA (WSA30)	160	320	137	-		462	462	1.1	120 ± 30
Concrete containing 25% WSA and 5% SF (WSA25SF5)		320	114	23				1.2	
Concrete containing 33% WSA and 7% SF (WSA33SF7)		274	151	32				1.3	
Concrete containing 40% WSA and 10% SF (WSA40SF10)		229	182	46				1.4	

2.2. Methods

2.2.1. Concrete Mixing and Preparation of Concrete Specimens

Concrete ingredients were mixed using a rotating pan mixer driven by power, following ASTM C192 guidelines. As soon as the required slump was achieved, cylindrical specimens measuring 100 mm in diameter and 200 mm in height were poured with fresh concrete according to ASTM C39. A total of 18 specimens were cast for each concrete mixture to determine the evolution of compressive and splitting tensile strengths after 7, 28, and 91 days of aging (3 identical specimens at each age). The cylindrical molds were covered with plastic sheets after fabrication and kept under standard laboratory temperature of $20 \pm 1\ °C$ and relative humidity of $60 \pm 5\%$ for 24 h. Concrete specimens were demolded 24 h after casting and moist cured in a curing tub until being tested at 7, 28, and 91 days. The upper and lower surfaces of concrete specimens were leveled by an end-surface grinder after reaching the desired curing period.

2.2.2. Compressive and Splitting Tensile Strength Tests

As per ASTM C39, compression strength tests were conducted on cylindrical specimens using a 200-ton-capacity universal testing machine (UTM, SHIMADZU, Kyoto, Japan) assembly at 0.2 MPa/s loading rate. Additionally, the split tensile test was performed in accordance with ASTM C496 by using a 200-ton UTM at a loading rate of 1 MPa/min.

2.2.3. Water Absorption

In addition to compressive and splitting tensile tests, tests for WA were also performed on hardened concrete samples of all the concrete mixtures according to ASTM C948. To perform WA tests, 100 mm diameter and 50 mm thick concrete samples were cut from 91-day moist-cured concrete specimens. Samples of cut concrete were soaked in water at 21 °C for 24 h and then periodically weighed until the saturated-surface dry weight (SSD) stabilized. If less than 0.5% difference in weight occurred between successive SSD weight measurements, the weight was considered stable. The letter "B" refers to the most recent yielded weight of the samples. Afterward, the mass of the specimens suspended in water was measured to the nearest 0.01 g and labeled as "A". Concrete samples were oven-dried at 100–110 °C, and their weights were taken every 24 h. Upon achieving a weight loss of <0.5% of the last measured weight, the sample was cooled inside a vacuum desiccator at room temperature and weighed, which is referred to as "C". To calculate WA and AP, the following equations were used:

$$\text{Water Absorption (\%)} = (B - C)/C \times 100 \qquad (1)$$

$$\text{Apparent Porosity (\%)} = (B - C)/(B - A) \times 100 \qquad (2)$$

2.2.4. Casting and Curing of Paste Samples for Microstructure and Pore Structure Analysis

Paste samples for the control (100% cement) and other mixes with various amounts of WSA (binary) and blends of WSA with SF (ternary), as partial substitutes for cement (weight percentage), were prepared. Using a Hobart mixer (Hobart, IN, USA), a paste of standard consistency was obtained for all mixes. The freshly mixed cement paste was poured into small plastic containers (20 mm diameter and 50 mm height) immediately after mixing. The plastic containers were subsequently sealed and capped prior to curing. Following 91 days in the cured state, the specimens were dried with a solvent exchange method to inhibit the hydration process. Finally, the flaky slices and powder specimens were then washed with isopropanol for 15 min. In order to remove the isopropanol present in the paste samples, the samples were dried in an oven at 40 °C for 30 min. The samples were then sealed in plastic bags for storage before testing.

Scanning Electron Microscopy/Energy-Dispersive X-ray Spectroscopy (SEM–EDS) Analysis

The SEM–EDS analysis of all concrete mixes was performed on hardened fragments in the form of slices of cement paste using a JSM-IT100 scanning electron microscope. Isopropanol was used to dry the fragments of hardened paste using the solvent exchange method. After that, all specimens were examined for changes in morphology and composition.

Nitrogen Adsorption Isotherm Technique

In addition to the SEM–EDS study, N_2 sorptiometry was conducted on all the tested concrete mixes. The surface area and pores of the powdered specimen (weighing approximately 0.3 g) obtained from the 91-day cured hardened paste were measured by N_2 sorption analysis (NOVA2200e, Quanta chrome, Boynton Beach, FL, USA) at 273 K. First, the samples were degassed to remove airborne contaminants that had absorbed during curing. Afterward, N_2 adsorption was performed at ambient temperatures and controlled pressure on the specimens.

Fourier Transform Infrared (FTIR) Analysis

All the mixes studied were additionally characterized using FTIR using a PerkinElmer Spectrum Two FTIR spectrometer to determine their individual phases. The powdered samples of all of the studied mixes were dried and examined under an infrared light source, and the IR spectra were taken in the wavenumber range 400 to 4000 cm^{-1}.

Thermogravimetric Analysis (TGA) of Cement Pastes

To perform TGA of all the concrete mixes, the powdered specimens obtained from 91-day cured pastes were placed in ceramic vessels fitted with thermogravimetric analyzers. The specimens were heated inside the thermal gravimetric analyzer to a temperature of 20 to 1000 °C at a rate of 10 °C/min, using N_2 as a medium under static conditions. Moreover, alumina powder was used as a reference for thermal stability at elevated temperatures. Finally, a comparison plot was developed using the built-in software to show the loss of weight of various paste specimens in various temperature ranges.

2.2.5. Methodology to Calculate Global Warming Potential (kg CO_2-eq) of Concrete Mixes

In this study, the GWP of all concrete mixes was calculated as equivalent to kg CO_2 (CO_2-eq) emissions using the green concrete LCA tool. This tool is typically designed to calculate the environmental impact of concrete, its constituent materials (including cement, aggregates, admixtures, and SCMs), and consumption of fuels and water. The environmental carbon footprint effect of all the seven concrete mixes used in this study was evaluated [49]. The study focuses on the major processes associated with raw materials extraction and production. A summary of the assumptions used for the different production technologies, geographic locations, distances, modes of transport, and material types is given in Table 4. Table 5 shows the percentages of power sources in the local electricity grid mix [50].

Table 4. Assumptions used in LCA calculations.

User Input Data:	Type of Material	
Ordinary Portland cement (OPC)	ASTM Type I	
SCMs	Silica fume (SF), Wheat straw ash (WSA)	
Admixture	Superplasticizer	
Electricity grid mix for:	**Location**	
Cement supplier	Cherat, Pak	
Fine aggregates supplier	Larunspur, Pak	
Coarse aggregates supplier	Taxila, Pak	
Gypsum supplier	Dera Ismael Khan, Pak	
SF supplier	Karachi, Pak	
Wheat straw ash collection	Punjab, Pak	
Transportation details for:	**Mode**	**Distance (km)**
Cement raw materials to cement plant	Truck Class 8b	5
Gypsum to cement plant	Truck Class 8b	300
Cement to concrete plant	Truck Class 8b	200
Fine aggregates to concrete plant	Truck Class 8b	200
Coarse aggregates to concrete plant	Truck Class 8b	100
Admixture to concrete plant	Truck Class 2	1000
WSA to concrete plant	Truck Class 8b	200
SF to concrete plant	Truck Class 8b	1200
Technology options for:	**Type of technology selected**	
Cement raw materials prehomogenization	Dry, raw storing, preblending	
Cement raw materials grinding	Dry, raw grinding, tube mill	
Cement raw meal blending/homogenization	Dry, raw meal blending, storage	
Clinker pyroprocessing	Preheater/precalciner kiln with US average kiln fuel mix	
Clinker cooling	Reciprocating grate cooler (modern)	
Cement finish milling/grinding/blending	Tube mill	
Cement PM control technology	ESP	
Conveying within the cement plant	Screw pump	20 m between process stations
Concrete batching plant loading/mixing	Mixer loading (central mix)	
Concrete batching plant PM control	Fabric filter	

Table 5. Electricity grid mix percentage for Pakistan [50].

User Input Data:	National Grid (%)
Coal	12
Natural gas	29
Fuel oil	20
Pet coke	-
Nuclear	2.5
Hydropower	34
Biomass	0.5
Geothermal	-
Solar	1
Wind	1

3. Results and Discussion

3.1. Compressive and Tensile Strength Evolution of Binary (C/WSA) and Ternary (C/WSA/SF) Concretes

A comparison of compressive strength development between CC and those containing WSA alone (binary mixes) and WSA jointly with SF (ternary mixes) is presented in Figure 2a. As listed in Table 3, a relatively low cement substitution rate in binary concrete mixes was set as 10%, 20%, and 30%, whereas in ternary concrete mixes a slightly higher cement replacement was used as 30%, 40%, and 50% owing to highly reactive SF. The purpose of adding different percentage of SF (5%, 7%, and 10%) in combination with different percentages of WSA (25%, 33%, and 40%) was to explore the optimum cement replacement without affecting the strength and durability properties of concrete as compared to control and binary concrete mixes. In addition to technical benefits in terms of mechanical properties, other aspects of blended concrete having WSA with SF in lowering the GWP were also evaluated and compared to those of control and binary concretes. To avoid the effects of external factors, the specimens of all the concrete mixtures studied were water cured under uniform temperature conditions of 20 °C until the age of testing.

As shown in Figure 2a, the binary concrete WSA20 demonstrated highest compressive strength evolution among binary mixes at all testing ages, including the CC. However, a reduction in compressive strength was observed for other binary mixes (WSA10 and WSA30) as compared to CC, regardless of aging. From the current results, it can be seen that the rate and the reduction in compressive strength was higher in WSA30 as compared to that of WSA10. The low strength of WSA10 than that of CC is due to its lower degree of pozzolanic activity, whereas the significantly low compressive strength of WSA30 is due to addition of high amount of WSA, which consequently affected the pozzolanic activity in a significant manner. These results suggested addition of 20% WSA as an optimum amount without compromising the compressive strength of concrete.

To achieve high sustainability in terms of lesser GWP, efforts were made to regain the reduction in the compressive strength of binary mixes containing high percentages of WSA by adding different percentages of SF (5%, 7%, and 10%). The test results show significant improvement in strength of ternary mix with an equal percentage of cement substitution (WSA25SF5) to that of the corresponding binary mix having WSA alone (WSA30) at all ages. Owing to addition of 5% SF, the improvement in strength that occurred remained slightly lower than that of CC. However, an encouraging response was noticed for ternary mix (WSA33SF7) containing a slightly high percentages of SF (7%) in presence of high percentage of WSA (33%), where the compressive strength was higher than that of CC at all ages. These results demonstrated fast early-age hydration and better packing and filling abilities due to slightly increased amount of very fine SF along with the later-age pozzolanic reaction of high-volume WSA. With a further increase in cement substitution with 10% SF in presence of 40% WSA (WSA40SF10), a slight reduction in strength was observed as compared to CC at all ages. This is because a high percentage of cement substitution (50%) affected both the early hydration and later-age pozzolanic reaction due to the production of less calcium hydroxide (C-H). However, the strength performance of this ternary mix (WSA40SF10) is commendable, despite of its high cement substitution, as it either shows higher compressive strength than other binary mixes (WSA10 and WSA30) or comparable to ternary mix WSA25SF5, despite of its low cement substitution. These results are, once again, attributed to the addition of a high percentage of very fine SF (10%) due to its better packing and filling abilities at early ages (7 days). Among all the mixes, the highest compressive strength at 7 days was demonstrated by the binary mix WSA20, while at 28 and 91 days by the ternary mix WSA33SF7.

In contrast to compressive strength, Figure 2b demonstrated a higher splitting tensile strength (STS) development for all concretes tested (binary and ternary) than that of CC. However, among binary and ternary mixes, their trends of STS development with respect to cement substitutions remained similar to that of compressive strength development. Moreover, similar to the compressive strength, the evolution of STS for WSA20 and WSA33SF7 was significantly higher as compared to CC and all other binary and ternary mixes. The only exception was at 7 days when WSA33SF7 exhibited higher STS than only CC. It is worth mentioning that the WSA20 concrete exhibited highest STS among all the concretes tested at all ages.

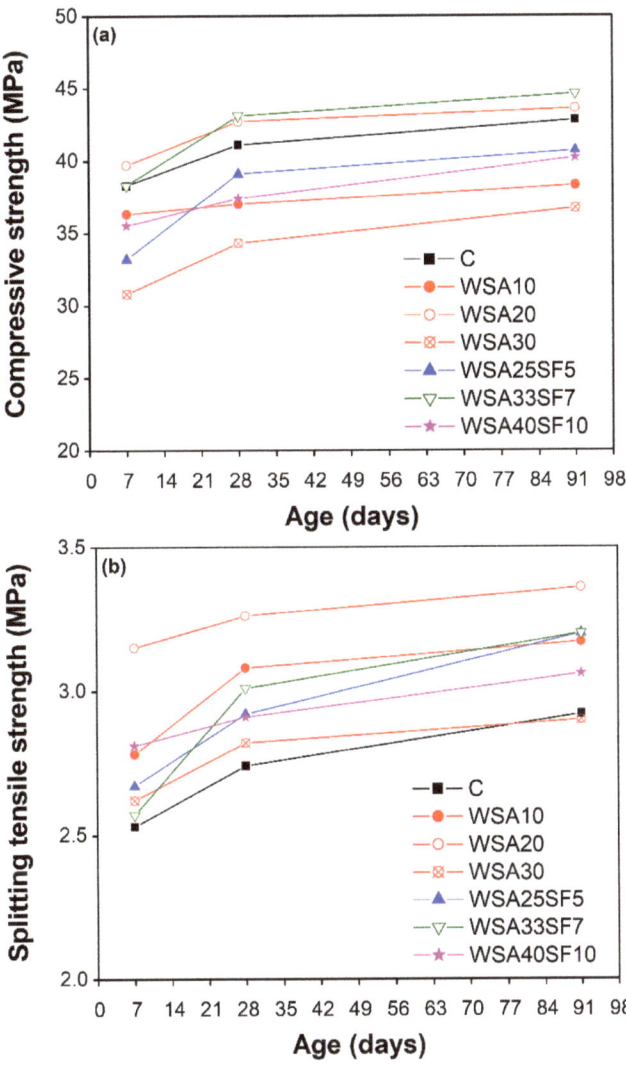

Figure 2. Comparison of strengths of control concrete and concrete having different percentages of WSA alone and blends of WSA with SF: (**a**) compressive strength and (**b**) splitting tensile strength.

To summarize, the current findings demonstrated a decrease in compressive strength owing to a low (WSA10) or high (WSA30) cement replacement with WSA. Contrarily, a significant increase in both compressive and splitting tensile strength was obtained for 20% cement replacement with WSA (WSA20). Similar to binary concretes, the compressive strength of ternary concrete mixes decreased owing to a low (WSA25SF5) or high (WSA40SF10) cement replacements. However, a significant increase in both compressive and splitting strengths was observed for ternary concrete blend having 33% WSA jointly with 7% SF (WSA33SF7).

3.2. Comparison of the Water Absorption and Apparent Porosity of Control Concrete and Binary and Ternary Concrete Mixtures

The comparison of the other important properties (WA and AP) of the binary and ternary concrete mixtures with that of the CC was also performed in addition to the strength characteristics. This is because the durability of the hardened concretes can be indirectly assessed based on the WA and AP values. The different trends of changes in the WA and AP according to various amounts of cement substitution with WSA alone (binary concretes) and WSA with SF (ternary concretes) are depicted in Figures 3 and 4. It can be seen in Figure 3 that a lower WA compared to CC was exhibited by all binary and ternary concretes regardless of the amount of cement substitution, which is attributed to their lower AP values as compared to CC (Figure 4). A decrease in the WA of binary concrete mixtures with increasing percent substitution of cement with WSA was observed up to a certain replacement level (20%). Thus, a slightly higher WA was exhibited by binary concrete having 30% WSA when compared with other binary concretes having 10% or 20% WSA (Figure 3). These binary concretes also demonstrated a similar trend of AP (Figure 4). A slightly higher value of AP exhibited by concrete containing a high amount of WSA (30%) is probably due to a relatively slower rate of the pozzolanic reaction owing to replacement of high percentage of cement with WSA. On the contrary, ternary concretes having high cement substitutions with blends of WSA and SF demonstrated a remarkable decrease in WA and AP values. The results demonstrated that, despite similar cement substitutions (30%), the ternary mix WSA25SF5 exhibited slightly lower WA and AP to that of corresponding binary mix WSA30. However, contrary to this, the WA and AP of this ternary mix were slightly higher than the other binary mixes with 10% and 20% WSA. Increasing the cement replacement from 30% to 40% in ternary concrete (WSA33SF7) resulted in further decrease in WA and AP values. The results demonstrated that the ternary concrete WSA33SF7 with an even higher percent replacement of cement exhibited lower WA and AP than the ternary concrete WSA25SF5 and all binary concretes having relatively lower cement substitutions (10%, 20%, or 30%). This was because SF particles of very fine size leads to significant pore refinement in ternary concrete mixtures compared to WSA. The ternary concrete having a very high cement substitution of 50% (WS40SF10) yielded slightly higher WA and AP values when compared to those of WSA33SF7 with relatively low cement substitution (40%). This was due to potentially slower rates of pore refinement and pozzolanic reaction in WSA40SF10 concrete owing to high cement substitutions.

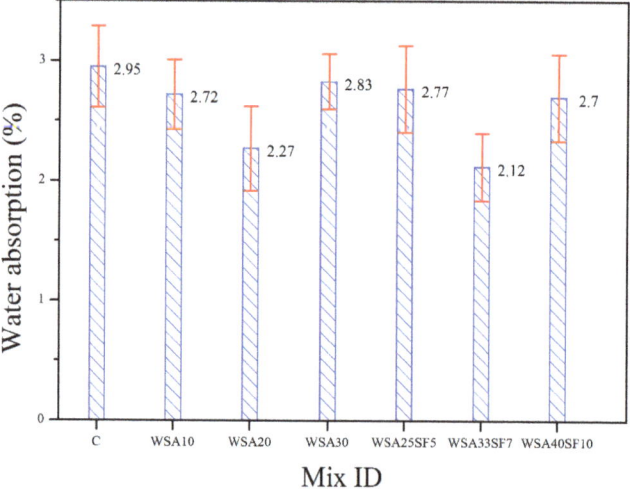

Figure 3. Comparison of the results for water absorption between the control and other concretes (binary and ternary mixes) after 91 days of standard curing.

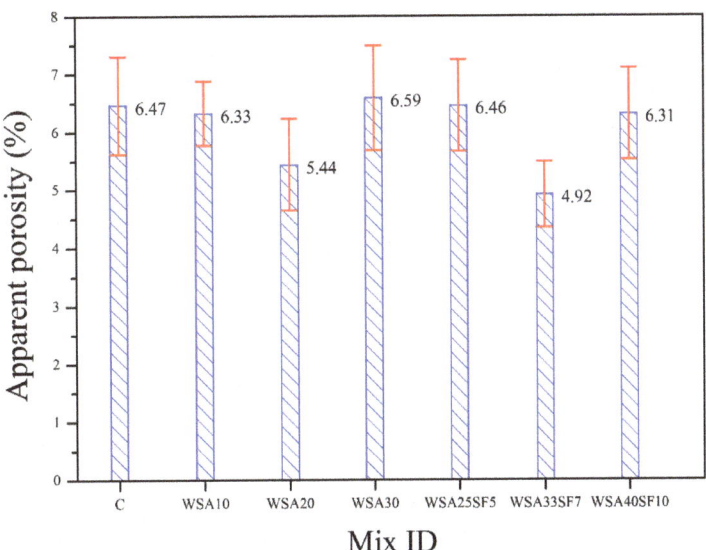

Figure 4. Comparison of the results for apparent porosity between control and other concretes (binary and ternary mixes) after 91 days of standard curing.

3.3. Evaluation of Compressive and Tensile Strength Correlation of Concrete by Prediction Models

Regardless of the type of mixture proportions, curing conditions, aging, or binder type and its content, the values of the experimental STS of concrete can be correlated with their corresponding compressive strength, mainly due to the existing consistency between their general trends of development with aging. In addition to the existing codes such as ACI 318 [51], ACI 363 [52], and CEB-FIP model code 1990 [53], various correlations between compressive and tensile strength were developed by researchers depending upon their specific experimental data of the curing and testing conditions, geometry of specimen, and the types of concrete [54–60]. Nevertheless, a consistent equation in general form $[f_{sp} = a \, (f'_c)^b]$ is used for this correlation by all researchers, including the existing model codes. This equation presents f_{sp} as the unknown STS of concrete to be predicted (MPa), whereas the compressive strength obtained directly from the experiments is represented by f'_c (MPa). In the equation, the parameters a and b are the constants that consider the dissimilarity of increasing rate between both mechanical properties. Based on different test results by researchers, the value of b varies, for example, as 0.67, 0.50, and 0.71 by the CEB-FIP model code [53], ACI 318 [51], and Kim et al. [57], respectively. The reason for dissimilarity in the b values arises since both ACI 318 and Kim et al. used specified and mean compressive strength, respectively, while the CEB-FIP model code used compressive strength associated with the specific characteristic compressive strength. Moreover, experimental results of 28 days were used in developing most of these correlations using Type-I cement for normal concrete subjected to standard moist curing at 20 °C. Despite consideration of the effect of various influencing factors (different binder types, curing, and aging) on the rate of both properties by some researchers [57], the effects of some other influencing factors such as the type of concrete using SCMs, geometry of specimen, and seasonal variations were overlooked. Having considered this important factor, it is desirable to evaluate suitability of existing correlations to predict the STS of concretes produced in this study containing various percentage of WSA alone (WSA10, WSA20, WSA30) and blends of WSA with SF (WSA25SF5, WSA33SF7, WSA40SF10).

Figure 5 depicts various existing correlations between compressive and tensile strengths of concrete. To evaluate the STS based on the experimental compressive strength, the current results of various binary and ternary concrete specimens along with CC were drawn with respect to 3, 7, and 28 days of aging and compared to existing models [51–60]. The experimental results of compressive strength were used to estimate the values of STS for the prediction models. With the exception of Noguchi-Tomosawa [58] and JSCE-2012 design codes [59], and De Larrard and Malier [60], all other existing models significantly overestimated STS, as shown in Figure 5. For the CC and WSA33SF7 at any known value of the compressive strength, a close match of the STS with that of the experimental compressive–tensile strength was predicted, regardless of aging, when using the JSCE-2012 model [$f_{sp} = 0.23\ (f'_c)^{2/3}$]. The Noguchi–Tomosawa model [$f_{sp} = 0.291\ (f'_c)^{0.637}$], on the other hand, resulted in a close match for the binary (WSA10, WSA20, WSA30) and other ternary (WSA25SF5, WSA40SF10) concretes. Similarly, a reasonably good estimate of the STS for any value of the compressive strength for these mixes was also predicted by the proposed model of De Larrard and Malier [$f_{sp} = 0.6 + 0.06\ (f'_c)$].

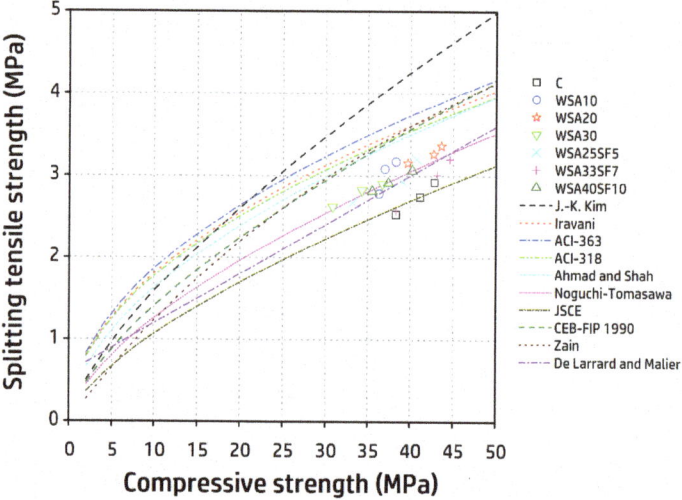

Figure 5. Comparison of the correlation between experimental compressive and tensile strengths for different concrete mixtures with existing prediction models.

From these findings, it is revealed that the correlation between the compressive and splitting tensile strengths of concrete is not significantly influenced by the type of binder and aging. Kim et al. [57] also noted the same independency with respect to the cement type, aging, and curing temperature with no effect on the correlation of compressive and tensile strengths of concrete. The models proposed by either De Larrard and Malier [60] or Noguchi-Tomosawa [58], hence, are considered safe in estimating the STS of all the concrete mixtures studied with the exception of the CC and WSA33SF7 concrete. The JSCE model [59], however, accurately predicted STS of the CC and WSA33SF7 concrete. Furthermore, the JSCE model, with slight underestimation, satisfactorily predicted the STS of all the concrete tested. The underestimation of the STS with respect to the current experimental values is considered safe because it is used as the criterion of crack control.

3.4. SEM–EDS Analysis of Control (C), Binary (C/WSA), and Ternary (C/WSA/SF) Cementitious Pastes

Figure 6 shows the results of SEM–EDS analysis for different mixes. As shown in this figure, the effects of WSA and SF on the microstructure of cementitious paste were examined through EDS analyses that were performed on SEM micrographs. The purpose of the EDS analysis with SEM was to study the crystal structure changes in C-H and C-S-H phases of paste matrix. Based on the computation from EDS analyses, a comparison of Ca/Si ratios among different mixes is presented in Table 6.

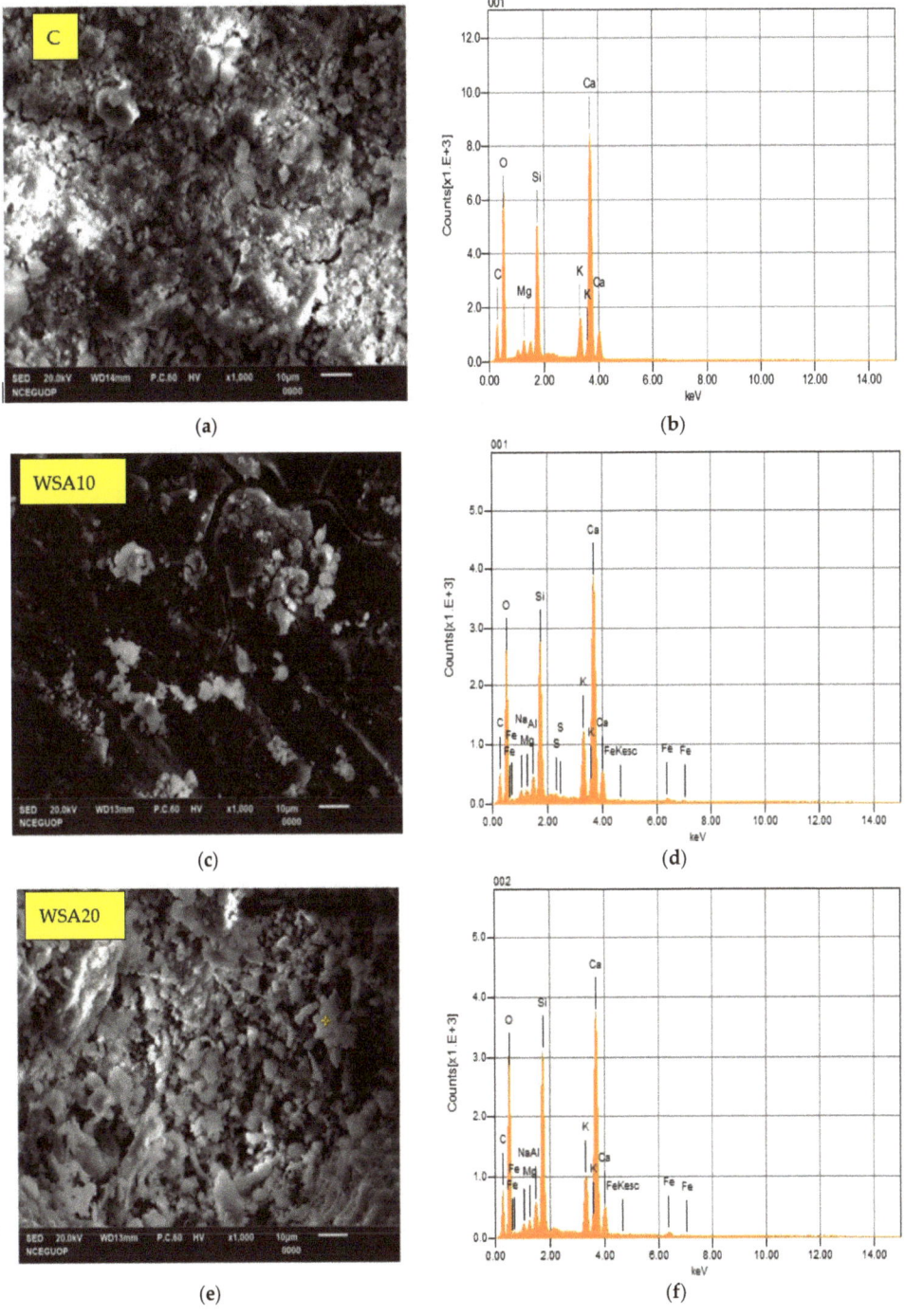

Figure 6. Cont.

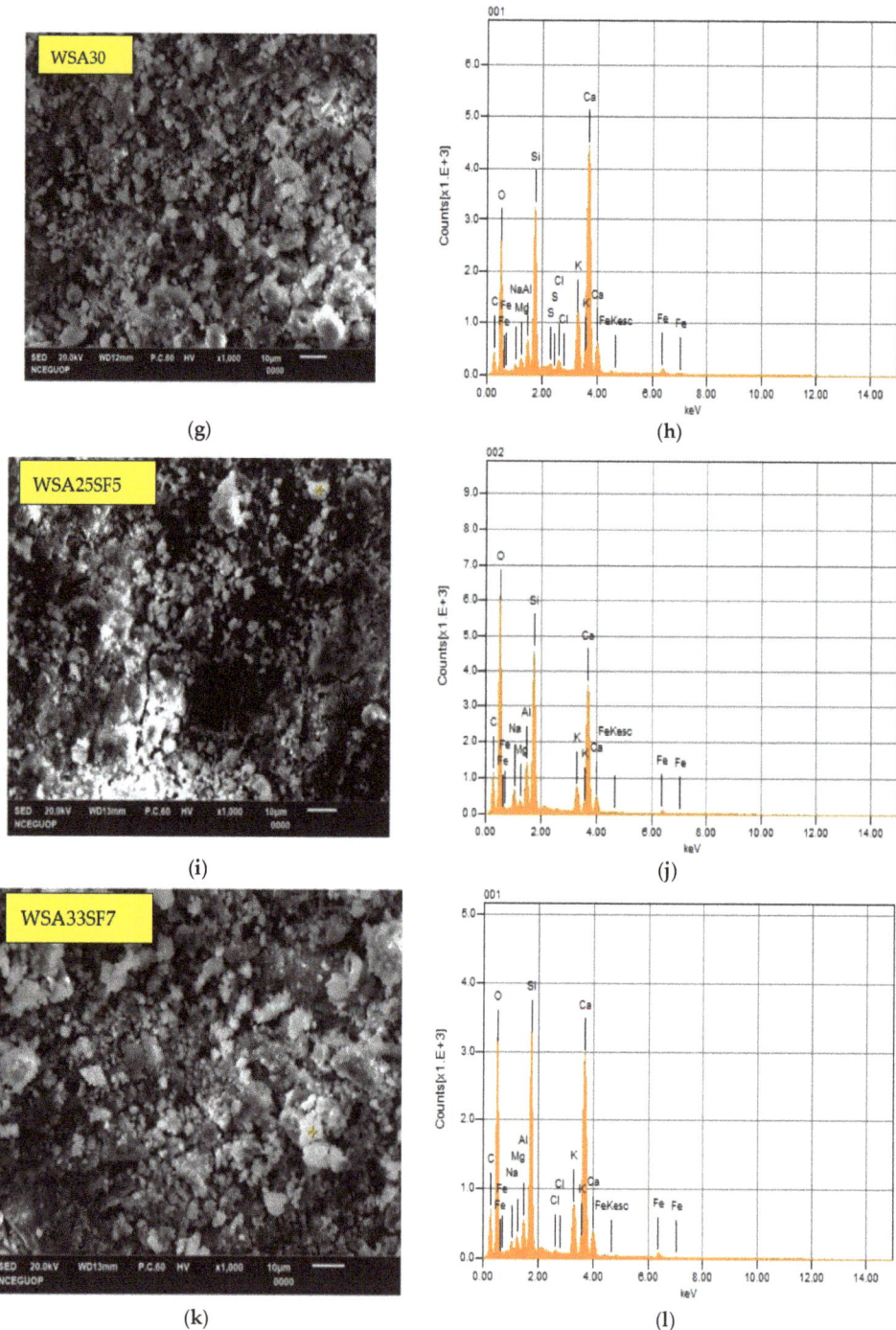

Figure 6. Cont.

(m) (n)

Figure 6. SEM–EDS spectrum of control, binary (C/WSA), and ternary (C/WSA/SF) paste samples (**a,b**) control, (**c,d**) WSA10, (**e,f**) WSA20, (**g,h**) WSA30, (**i,j**) WSA25SF5, (**k,l**) WSA33SF7, (**m,n**) WSA40SF10.

Table 6. Ca/Si atomic ratio from EDS analysis of control and mixes containing different percentages of WSA alone and WSA with SF.

Mix ID	Ca/Si Atomic Ratio	
	C-S-H Phase	C-H Phase
CC	1.93	3.3
WSA10	1.63	3.1
WSA20	1.42	2.65
WSA30	1.6	2.94
WSA25SF5	1.05	2.16
WSA33SF7	0.91	2.1
WSA40SF10	1.24	2.3

According to the findings of past researchers [61–63], the Ca/Si ratio for the C-S-H phase in paste sample ranges between 0.5 to 2.0, while 2.0 or higher for C-H phases. Among all mixes, both C-H and C-S-H phases of the control mix exhibited a highest Ca/Si ratio, at 3.30 and 1.93, respectively. As illustrated in Table 6, the lower Ca/Si ratio values of both the binary and ternary mixes is due to the substitution of cement with WSA and SF that led to decreased porosity of the paste matrix by forming high-density C-H and C-S-H phases in these mixes. Among binary mixes, WSA20 exhibited lowest Ca/Si ratio at 2.65 and 1.42 for C-H and C-S-H, respectively. Therefore, current results suggest a 20% replacement of cement with WSA without compromising the mechanical and microstructural performance of concrete. Moreover, all the ternary mixes showed even lower Ca/Si ratio as compared to binary mixes. A significant decrease in Ca/Si ratio in ternary mixes is attributed mainly to the fine, amorphous, and highly reactive nature of SF, which is used in ternary mixes in the presence of WSA. The addition of SF along with WSA results in the formation of additional C-S-H phases by utilizing C-H phases in the paste matrix. Furthermore, the addition of SF further enhances both the density of C-S-H and C-H phases and the compactness of cement paste. Among ternary mixes, the lowest Ca/Si ratio was observed for WSA33SF7, which indicates its improved microstructural properties due to the formation of high-density C-H (2.1) and C-S-H (0.91) phases. It is expected that such formation of high-density C-S-H and C-H phases would result in densification and refinement of the microstructure, leading to enhanced performance in practical engineering applications.

3.5. Fourier Transform Infrared (FTIR) Analysis of Cement Pastes

As illustrated in Figure 7, all the paste samples show appearance of IR bands at the same location; however, their intensities differ. This is attributed to the formation of hydration products such as C-S-H and C-H [64]. The peaks from 900 to 1100 cm^{-1} are associated with vibrations of Si-O bands in C-S-H phase [65]. The IR bands show a higher relative intensity of the Si-O band in paste samples containing 20% WSA (WSA20). Further, the samples of ternary blends having WSA and SF showed better results than those of control and binary mixes. This shift in the Si-O band is associated with polymerization of silica. A slight shift (960 cm^{-1}) was detected in WSA20. On the other hand, the ternary concrete samples with different percentages of SF (5%, 7%, and 10%) showed a broader and significant shift (980 cm^{-1}). A shift toward a higher wavenumber in the spectrum of binary and ternary pastes suggests formation of a high amount of C-S-H gels. A large amount of C-S-H gels are produced from the nucleation sites provided by the fine particles of SF, as also observed earlier through SEM–EDS analysis. The development of high compressive strength in these mixes can be linked to the formation of a large amount of C-S-H gels. Moreover, the peaks at 720, 875, and 1415 cm^{-1} are associated with calcite formed as a result of carbonation [66].

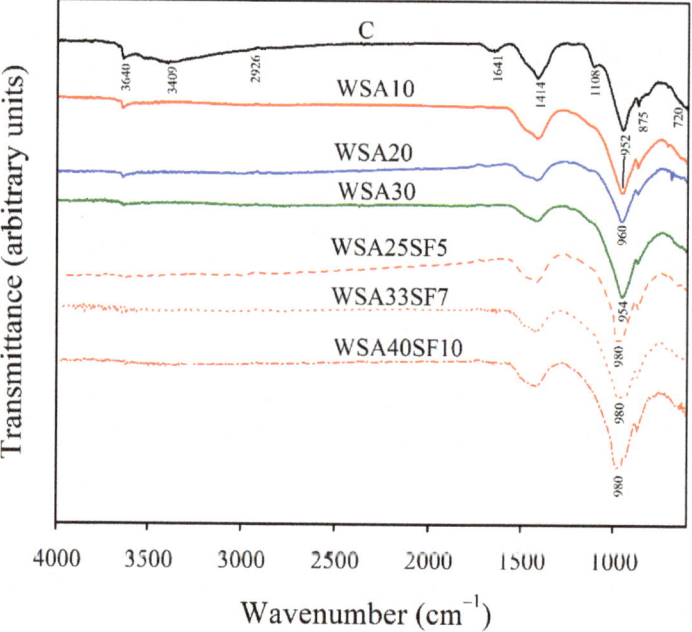

Figure 7. FTIR spectra of control, binary (C/WSA), and ternary (C/WSA/SF) paste samples after 91 days of curing.

In all the paste samples, the peak at 3645 cm^{-1} indicates the presence of free OH groups, which suggests the presence of the portlandite phase. The control sample exhibits a wider and more visible peak as compared to all other binary and ternary samples. However, this portlandite peak was reduced in all the binary mixes, which indicates the extent of portlandite consumption caused by the presence of amorphous silica in WSA. Interestingly, in ternary mixes containing WSA and SF, the peak remained very small or almost disappeared, indicating a high pozzolanic reactivity, which consequently results in greater consumption of C-H. This leads to formation of more C-S-H gels in these mixes [67]. As discussed in the preceding section, similar evidence of this large amount of C-S-H gels in these mixes was also noticed in SEM–EDS analyses.

3.6. Nitrogen Adsorption Results (Surface Area and Pore Structure of Cement Pastes)

Figure 8 shows the comparison of cumulative nitrogen intrusion volume with respect to pore width for different paste samples after curing for 91 days. The results demonstrate lesser pore volume for ternary mixes having WSA together with SF as compared to control as well as binary mixes containing WSA. The relatively lesser pore volume of ternary mixes indicates formation of dense pore structure due to accelerated pozzolanic reactivity caused by the highly reactive SF. The least amount of nitrogen intrusion volume (0.043 cm^3/g) was observed for a ternary mix with 40% cement substitution (WSA33SF7) followed by WSA25SF5 (0.045 cm^3/g), WSA40SF10 (0.046 cm^3/g), WSA10 (0.048 cm^3/g), WSA20 (0.049 cm^3/g), control (0.050 cm^3/g), and WSA30 (0.051 cm^3/g). The binary mixes with lower percent of WSA (WSA10 and WSA20) exhibited lesser nitrogen intrusion volume as compared to control, however, the maximum nitrogen intrusion volume was observed for WSA30. This is most obviously due to reduced pozzolanic reactivity when a high percent of cement replaced with WSA, which consequently had led to increased porosity and vascularity in the paste matrix.

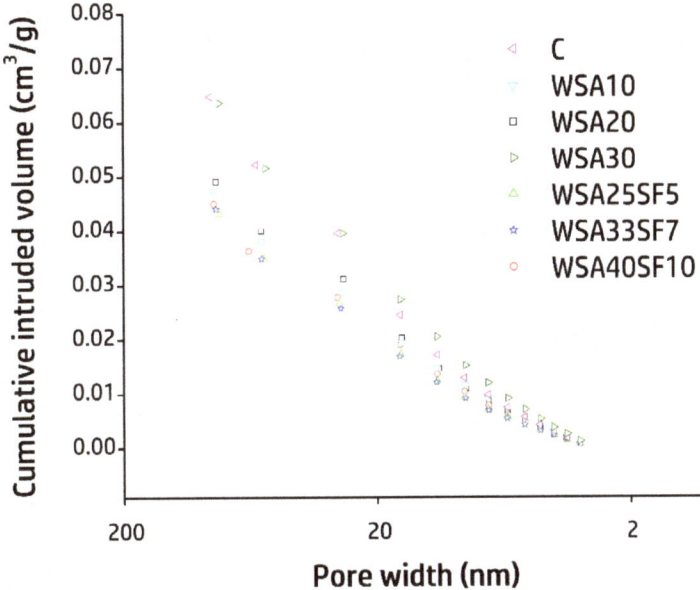

Figure 8. Comparison of the cumulative nitrogen intrusion volumes versus pore widths for control, binary (C/WSA), and ternary (C/WSA/SF) paste samples after 91 days of curing.

A comparison of Brunauer–Emmett–Teller (BET) surface areas between control and other binary and ternary mixes is presented in Figure 9. All ternary mixes exhibited higher BET surface areas as compared to those of binary and CC. Similar to the least amount of nitrogen intruded volume, WSA33SF7 demonstrated a larger BET surface area (17.6 m^2/g) as compared to all other mixes. An increased BET surface area of the ternary mix indicates its improved and denser microstructure of C-S-H gels [68,69]. As described in the preceding section, SEM analysis with EDS demonstrated a dense cementitious matrix with few pores. In general, BET surface area increased with increasing percent of WSA and SF in ternary mixes, except for WSA40SF10 (15.7 m^2/g), which showed a slight reduction that could be due to a high amount of cement substitution leading to a large amount of unreacted WSA in the mixture [70]. Similar to ternary mixes, binary mixes, especially those containing relatively low percentage of WSA such as WSA10 (12.8 m^2/g) or WSA20 (13.0 m^2/g), also showed slightly larger BET surface areas than that of control (12.7 m^2/g), which ultimately had led to their improved microstructure of C-S-H gel. However, contrary to this, BET surface area decreased with further increase in percent of WSA (more than 20%) as was observed for WSA30 (12.0 m^2/g). This clearly indicates the presence of unreacted WSA that had caused the hydration products to jam the pores and form a porous paste matrix. These results suggest a restricted use of WSA in binary mixes by not more than 20% when used as a sole substitute of cement.

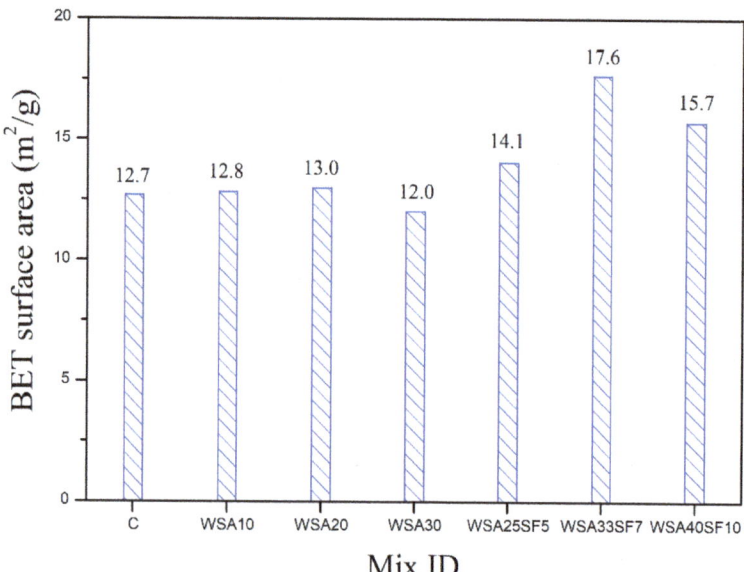

Figure 9. Comparison of BET surface area between control, binary (C/WSA) and ternary (C/WSA/SF) paste samples after 91 days of curing.

3.7. Thermogravimetric Analysis (TGA) of Cement Pastes

The thermal decomposition of 91-day cured cement paste samples of control, binary, and ternary mixtures was performed by TGA to evaluate the effect of WSA and SF on the amount of C-H (%), which occurred due to the weight loss between 400 and 500 °C [71,72]. Based on the TGA results, a comparison of weight loss (%) with respect to temperature is presented between control, binary, and ternary mixes (Figure 10). Consequently, using the TGA results, the amount of C-H for different mixes and their normalized C-H (%) values with respect to OPC was calculated, as listed in Table 7. The normalized C-H values for each mix were calculated by dividing their C-H values with their respective OPC (%) content.

From Table 7, it can be seen that the maximum C-H content exists in control sample and a gradual decrease in C-H phase occurred in binary mixes with increasing percent substitution of cement with WSA. The gradual decrease in C-H content in binary mixes is obviously due to their decreasing cement content and partly because of the pozzolanic reactivity caused by WSA due to the formation of hydration products as a result of C-H consumption. However, as compared to binary mixes, a significant decrease in C-H phase occurred in ternary mixes, which was due to the simultaneous effects of high reactivity of SF along with pozzolanic reaction caused by WSA. This would ultimately lead to significantly high densification and compactness of C-S-H gel microstructure for ternary mixes as compared to binary and control mixes.

Similar to C-H, it can be further seen from Table 7 that the normalized C-H values of all the binary and ternary mixes are lower as compared to control. However, in comparison to ternary, all the binary mixes exhibited relatively higher normalized C-H values, which demonstrated their lesser pozzolanic reactivity. In fact, the same was also evidenced earlier through XRD, SEM–EDS, and FTIR analyses. Moreover, the reason for relatively lower normalized C-H in ternary mixes, despite their large volume substitution of cement, is mainly due to the high reactivity of SF jointly with pozzolanic WSA, which ultimately causes a seeding effect to produce more C-S-H gel. These important findings with clear scientific proofs may justify the effectiveness of using both WSA and SF jointly as a high-volume replacement of cement for the production of a strong, durable, and sustainable concrete.

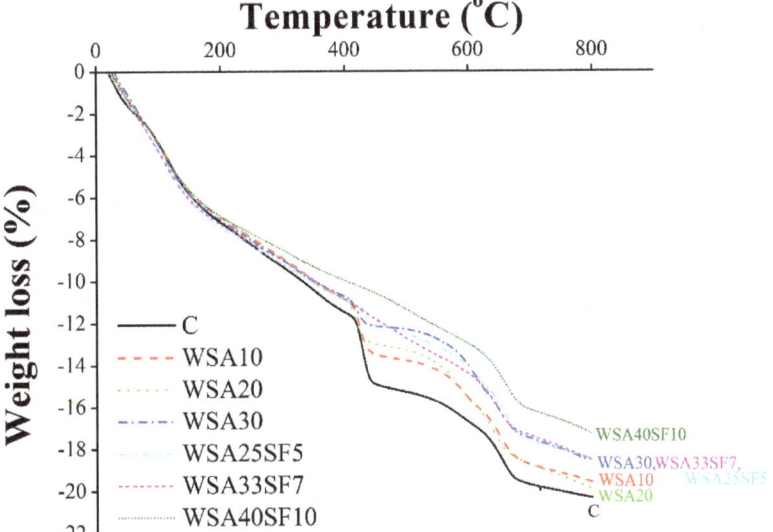

Figure 10. Comparison of thermogravimetric analysis of 91-day cured paste samples among control, binary (C/WSA), and ternary (C/WSA/SF) paste samples after 91 days of curing.

Table 7. Actual amount of calcium hydroxide (C-H) and normalized with respect to the percent content of cement in each mix (C-H/OPC) after 91 days of cement hydration.

Mix ID	C-H (%)	C-H/OPC (%)
CC	18.5	18.5
WSA10	13.87	15.4
WSA20	11.28	14.1
WSA30	8.85	12.6
WSA25SF5	7.47	10.7
WSA33SF7	7.26	12.1
WSA40SF10	6.44	12.9

3.8. Global Warming Potential

3.8.1. Comparison of CO_2-eq for Unit Concrete Production among Control, Binary, and Ternary Concrete Mixes

Figure 11 shows the comparison of estimated GWP (kg CO_2-eq per unit volume concrete) between control and other binary as well as ternary concrete mixtures. In this figure, the distribution of CO_2-eq for each concrete mixture is illustrated by its ingredients and major production processes. The value of CO_2-eq for all concrete mixtures was calculated by using the green LCA tool according to the data listed in Tables 4 and 5. Subsequently, the total emission for each concrete mixture was calculated by adding its direct and supply chain emissions associated with the quarrying, production, and transportation processes that occur within a systems' boundary.

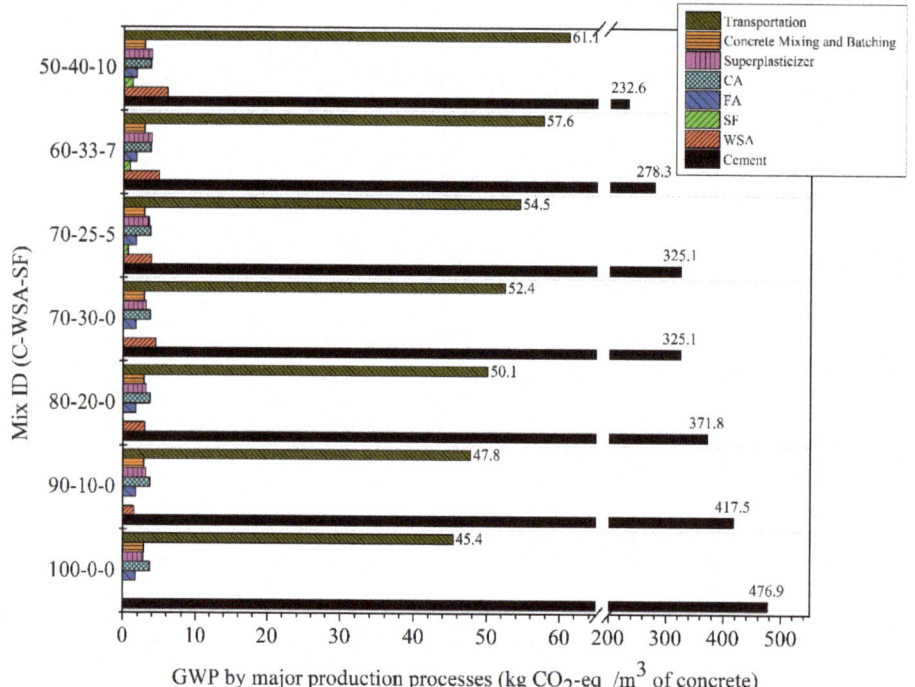

Figure 11. Comparison of GWP distribution by concrete ingredient and phase between control, binary (C/WSA), and ternary (C/WSA/SF) concrete mixtures.

It can be seen in Figure 11 that among all the different ingredients used in unit concrete production, cement is responsible for maximum CO_2 emission for all the mixtures. Following cement production, transportation of the raw materials and their products represents the second-highest source of CO_2 emissions, which varies between 8.5% and 19.5%. Most importantly, the CO_2 emissions associated with noncementitious materials are very low and remained almost same for all concrete mixes. According to the calculations, the SCMs (WSA and SF) constituted only 0.3% to 1.4% of total calculated CO_2 emissions. Similarly, concrete batching and mixing account for very small percentages, between 0.4% and 0.5%.

According to the comparison, the CC exhibited the highest GWP at 533 kg CO_2-eq/m^3 among all mixes (Figure 11). On the other hand, the ternary concrete mix containing WSA together with SF (WSA40SF10) as 50% cement substitution possessed lowest GWP at 313 kg CO_2-eq/m^3. Generally, by decreasing the amount of Portland cement in concrete mixes and increasing the amount of SCMs, the carbon footprint for cement production decreased from 89% for CC to 44% for WSA40SF10.

3.8.2. Comparison of Normalized CO_2-eq for Unit Concrete Production per Unit Strength (MPa) among Control, Binary, and Ternary Concrete Mixes

Figure 12 shows the comparison of normalized values of GWPs as CO_2-eq/m^3/MPa among all the mixes, including control, binary, and ternary (Mix #1 to 7). As illustrated in Figure 12, the normalized values of CO_2-eq/m^3/MPa for all the mixes were drawn with respect to their compressive strength and aging (7, 28, and 91 days). The CO_2-eq intensity of different mixes is used as a measure to evaluate their important impact on compressive strength of concrete and associated GWP per unit concrete volume and strength [73].

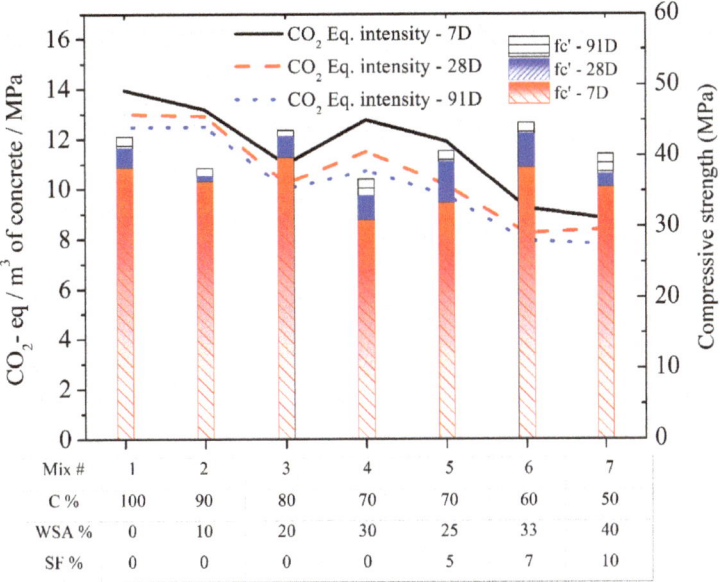

Figure 12. Comparison of normalized GWPs as kg CO_2-eq with respect to strength among control, binary (C/WSA), and ternary (C/WSA/SF) concrete mixtures.

It can be seen in Figure 12 that each individual mix, in general, showed a gradual decreasing CO_2-eq/m^3/MPa trend due to increasing compressive strength with aging. Moreover, it can be further noted that the amount of CO_2-eq/m^3/MPa decreases with decreasing quantity of cement in all the binary and ternary mixtures. The only exception is the binary mix WSA30, where it slightly increased as compared to WSA20. This is due to drop of strength in WSA30 because of the high amount of cement substitution (30%) in this mix as compared to WSA20. A sharp decrease in CO_2-eq/m^3/MPa in a binary (WSA20) and ternary mix (WSA33SF7) is due to their high compressive strengths as compared to other mixes.

The current results dictate the possibility of lowering CO_2-eq intensities by replacing cement with only one type of SCM (WSA) or blends of SCMs (WSA with SF). According to the current findings, the intensities of CO_2-eq improved significantly in ternary blends as compared to binary, without compromising any strength potentials. For instance, despite of almost similar compressive strength of binary WSA20 (43.6 MPa) and ternary WSA33SF7 (44.6) mixes at 91 days, CO_2-eq intensity decreased to 7.9 kg/m^3/MPa in ternary mix (mix #6 having 60% cement) from 10 kg/m^3/MPa in binary mix (mix #3 having 80% cement). Furthermore, the addition of higher amounts of WSA (40%) and SF (10%) in ternary mix (mix #7 having 50% cement) causes a further reduction in CO_2-eq intensity as 7.8 kg/m^3/MPa as compared to 12.5 kg/m^3/MPa in binary mix (mix #2 having 90% cement). This might be due to slightly higher 91-day strength of ternary mix (40.2 MPa) as compared to binary mix (38.3 MPa). These findings are equally applicable for these mixes at other ages as well, such as 7 and 28 days. Consequently, these results suggest that the intensities of CO_2-eq can be reduced without compromising the strength of concrete by optimizing concrete mixes with appropriate amount of cement replacement and selection of a suitable type of SCMs with their correctly chosen or tested blend percentages.

4. Conclusions

This study investigated the use of high-volume WSA and its blend with SF as a partial substitute of cement for the development of high-performance and sustainable concrete. Besides control (100% cement), several other concrete mixtures were prepared by partially substituting cement with only WSA as binary system (10%, 20%, and 30%) and WSA together with SF as ternary system (25%/5%, 33%/7%, and 40%/10%). Subsequently, the influence of adding WSA and WSA with SF on the hardened mechanical properties (compressive and tensile strengths, WA, and AP) was assessed and compared to those of CC. Finally, paste samples were prepared for all mixes to examine their microstructures and pore structures using SEM–EDS, FTIR, TGA, and N_2 adsorption techniques to scientifically understand the impact of adding WSA and SF on the resulting paste matrix. At the end, GWP as kg CO_2-eq per unit volume of concrete and kg CO_2-eq per unit volume of concrete/MPa were also calculated using the LCA tool and compared among different mixes used in this study.

The main findings of this study are summarized as follows:

The current findings demonstrated a decrease in strength of binary concrete corresponding to their relatively low (WSA10) and high (WSA30) cement substitution rates, while a significant increase in strength was observed for moderate substitution rate of cement (20%) for binary concrete WSA20. In a very similar manner, the ternary mixes also showed a decreasing trend of strengths both at low (WSA25SF5) and high (WSA40SF10) blends of cement substitution, and the significant increase in strength was obtained for the moderate ternary blend of 33% WSA and 7% SF (WSA33SF7). Moreover, these increased strengths of WSA20 and WSA33SF7 were validated by their relatively lower apparent porosity and water absorption values among all mixes.

A correlation between experimental compressive and tensile strengths showed close agreement to models proposed by Noguchi–Tomosawa and De Larrard and Malier. Based on the results of current findings, it is recommended to use these models to properly estimate the tensile strength of tested concretes containing WSA alone or WSA jointly with SF, except the CC and WSA33SF7, as the JSCE model demonstrated a close agreement for these two concrete mixes. In addition, the JSCE model safely predicts the tensile strength of other binary and ternary concrete mixes with their slightly underestimated values.

Analysis of SEM–EDS data reveals that the incorporation of WSA as 20% replacement of cement (WSA20) leads to the densification of the paste matrix by decreasing the Ca/Si ratio in both the C-S-H and C-H phases. Furthermore, adding a 7% SF jointly with 33% WSA in the ternary mix (WSA33SF7) resulted in the lowest Ca/Si ratio among all the concrete mixtures tested. These findings suggest better refinement of microstructure for these mixes, which ultimately would lead to improvement of their engineering performance.

A visible shift in Si-O band was observed through FTIR analysis for almost all the binary and ternary mixes. Comparatively, the shift was more pronounced in all ternary mixes containing WSA together with SF, which clearly demonstrates the presence of high levels of C-S-H gels. Moreover, the portlandite peaks (3641–3644 cm^{-1}) were also significantly smaller in all ternary mixes as compared to binary mixes. This, consequently, suggests an improved pozzolanic reactivity and the formation of more C-S-H gels that ultimately leads to a more refined and denser microstructure.

Similar to FTIR analysis, a densification of the paste matrix and refinement of the pore structure also suggested by the results of N_2 adsorption tests was due to the decrease in intruded pore volume and an increase in BET surface area, especially for mixes WSA20, WSA25SF5, and WSA33SF7. However, other mixes such as those containing a large amount of WSA (WSA30 and WSA40SF10) showed a smaller surface area and more intruded pore volume. This could be possibly due to the presence of unreacted WSA in these mixes that might have caused high porosity and vascularity in their paste matrices.

As a matter of further validation of these findings, TGA results also showed a reduction in the portlandite phase of binary and ternary mixes. This occurred especially in those binary mixes that contained high doses of WSA (WSA20 and WSA30), partly because of a lesser amount of cement in these mixes and due to the pozzolanic reactivity of amorphous silica present in WSA. A further reduction in the proportion of portlandite phase occurred in ternary mixes due to very fine and amorphous silica present in both WSA and SF, which would ultimately consume the C-H phase to generate additional C-S-H phases and lead to densification of the paste matrix.

The GWP per unit volume of concrete (CO_2-eq/m^3) mixes decreased with decreasing amount of Portland cement simultaneously with an increase in amounts of SCMs. The highest GWP of 533 was generated by the control mixture (100% OPC) while the least GWP, only 313, was produced by the ternary concrete mixture (WSA40SF10) having 40% WSA jointly with 10% SF and only 50% OPC.

Regardless of aging, all the binary and ternary concrete mixes containing SCMs (WSA or WSA with SF) exhibited lower CO_2-eq intensities as compared to CC, with the only exception of a binary mix having 10% WSA that showed almost identical CO_2-eq intensities to that of control at later ages of 28 and 91 days. Furthermore, as compared to binary, ternary mixes having WSA together with SF showed good potential for further reducing the normalized CO_2-eq intensities/MPa at all ages (7, 28, and 91 days). The ternary mixes containing highest percentages of SCMs, at 40% (WSA33SF7) and 50% (WSA40SF10), resulted in lower CO_2-eq intensity/MPa as compared to that of CC, regardless of aging. Consequently, it may be safe concluding that the efforts for using larger amounts of regionally available SCMs can have an important positive effect on producing green concretes together with reduced GWP without compromising any strength potentials.

Author Contributions: Conceptualization, K.K., M.I., M.N.A. and K.S.; Methodology, K.K., M.I. and N.W. Validation, M.N.A. and K.K.; Formal analysis, M.N.A. and K.K.; Investigation, M.N.A.; Resources, M.N.A. and K.S.; Data curation, M.I.; Writing—original draft preparation, M.N.A., K.K., M.I. and N.W.; Writing—review and editing, M.N.A.; Visualization, M.I.F. and K.S.; Supervision, M.N.A., K.S. and K.K.; Project administration, M.N.A.; Funding acquisition, M.N.A., M.I.F. and K.K. All authors have read and agreed to the published version of the manuscript.

Funding: This work was supported by the Al Bilad Bank Scholarly Chair for Food Security in Saudi Arabia, the Deanship of Scientific Research, Vice Presidency for Graduate Studies and Scientific Research, King Faisal University, Saudi Arabia [Grant No. CHAIR21]. The APC was funded by the same "Grant No. CHAIR21".

Institutional Review Board Statement: Not applicable.

Informed Consent Statement: Not applicable.

Data Availability Statement: All the data utilized in current research are available upon a reasonable request from the corresponding author.

Acknowledgments: The authors acknowledge the Al Bilad Bank Scholarly Chair for Food Security in Saudi Arabia, the Deanship of Scientific Research, Vice Presidency for Graduate Studies and Scientific Research, King Faisal University, Saudi Arabia, for the financial support (Grant No. CHAIR21).

Conflicts of Interest: The authors declare no conflict of interest. The founding sponsors had no role in the design of the study; in the collection, analyses, or interpretation of data; in the writing of the manuscript; nor in the decision to publish the results.

References

1. Zareei, S.A.; Ameri, F.; Bahrami, N.; Shoaei, P.; Moosaei, H.R.; Salemi, N. Performance of sustainable high strength concrete with basic oxygen steel-making (BOS) slag and nano-silica. *J. Build. Eng.* 2019, *25*, 100791. [CrossRef]
2. Koushkbaghi, M.; Kazemi, M.J.; Mosavi, H.; Mohseni, E. Acid resistance and durability properties of steel fiber-reinforced concrete incorporating rice husk ash and recycled aggregate. *Constr. Build. Mater.* 2019, *202*, 266–275. [CrossRef]
3. Rashad, A.M. A preliminary study on the effect of fine aggregate replacement with metakaolin on strength and abrasion resistance of concrete. *Constr. Build. Mater.* 2013, *44*, 487–495. [CrossRef]
4. Luhar, S.; Cheng, T.-W.; Luhar, I. Incorporation of natural waste from agricultural and aquacultural farming as supplementary materials with green concrete: A review. *Compos. Part B Eng.* 2019, *175*, 107076. [CrossRef]
5. Khan, K.; Amin, M.N.; Saleem, M.U.; Qureshi, H.J.; AlFaiad, M.; Qadir, M.G. Effect of Fineness of Basaltic Volcanic Ash on Pozzolanic Reactivity, ASR Expansion and Drying Shrinkage of Blended Cement Mortars. *Materials* 2019, *12*, 2603. [CrossRef] [PubMed]
6. Celik, K.; Hay, R.; Hargis, C.W.; Moon, J. Effect of volcanic ash pozzolan or limestone replacement on hydration of Portland cement. *Constr. Build. Mater.* 2018, *197*, 803–812. [CrossRef]

7. Mohan, A.; Mini, K.M. Strength and durability studies of SCC incorporating silica fume and ultra fine GGBS. *Constr. Build. Mater.* 2018, *171*, 919–928. [CrossRef]
8. Jeong, Y.; Kang, S.-H.; Kim, M.O.; Moon, J. Acceleration of cement hydration from supplementary cementitious materials: Performance comparison between silica fume and hydrophobic silica. *Cem. Concr. Compos.* 2020, *112*, 103688. [CrossRef]
9. Booya, E.; Gorospe, K.; Das, S.; Loh, P. The influence of utilizing slag in lieu of fly ash on the performance of engineered cementitious composites. *Constr. Build. Mater.* 2020, *256*, 119412. [CrossRef]
10. Mahmud, S.; Manzur, T.; Samrose, S.; Torsha, T. Significance of properly proportioned fly ash based blended cement for sustainable concrete structures of tannery industry. *Structures* 2021, *29*, 1898–1910. [CrossRef]
11. Amin, M.N.; Khan, K.; Saleem, M.U.; Khurram, N.; Niazi, M.U.K. Influence of Mechanically Activated Electric Arc Furnace Slag on Compressive Strength of Mortars Incorporating Curing Moisture and Temperature Effects. *Sustainability* 2017, *9*, 1178. [CrossRef]
12. Yoon, H.; Seo, J.; Kim, S.; Lee, H.; Park, S. Hydration of calcium sulfoaluminate cement blended with blast-furnace slag. *Constr. Build. Mater.* 2020, *268*, 121214. [CrossRef]
13. Thakare, A.A.; Siddique, S.; Sarode, S.N.; Deewan, R.; Gupta, V.; Gupta, S.; Chaudhary, S. A study on rheological properties of rubber fiber dosed self-compacting mortar. *Constr. Build. Mater.* 2020, *262*, 120745. [CrossRef]
14. Tian, L.; Qiu, L.-C.; Liu, Y. Fabrication of integrally hydrophobic self-compacting rubberized mortar with excellent waterproof ability, corrosion resistance and stable mechanical properties. *Constr. Build. Mater.* 2021, *304*, 124684. [CrossRef]
15. Wang, D.; Wang, Q.; Huang, Z. Reuse of copper slag as a supplementary cementitious material: Reactivity and safety. *Resour. Conserv. Recycl.* 2020, *162*, 105037. [CrossRef]
16. Gopalakrishnan, R.; Nithiyanantham, S. Microstructural, mechanical, and electrical properties of copper slag admixtured cement mortar. *J. Build. Eng.* 2020, *31*, 101375. [CrossRef]
17. Zhang, P.; Liao, W.; Kumar, A.; Zhang, Q.; Ma, H. Characterization of sugarcane bagasse ash as a potential supplementary cementitious material: Comparison with coal combustion fly ash. *J. Clean. Prod.* 2020, *277*, 123834. [CrossRef]
18. Yadav, A.L.; Sairam, V.; Muruganandam, L.; Srinivasan, K. An overview of the influences of mechanical and chemical processing on sugarcane bagasse ash characterisation as a supplementary cementitious material. *J. Clean. Prod.* 2020, *245*, 8854. [CrossRef]
19. Rumman, R.; Bari, M.; Manzur, T.; Kamal, M.; Noor, M. A Durable Concrete Mix Design Approach using Combined Aggregate Gradation Bands and Rice Husk Ash Based Blended Cement. *J. Build. Eng.* 2020, *30*, 101303. [CrossRef]
20. Khan, K.; Ullah, M.F.; Shahzada, K.; Amin, M.N.; Bibi, T.; Wahab, N.; Aljaafari, A. Effective use of micro-silica extracted from rice husk ash for the production of high-performance and sustainable cement mortar. *Constr. Build. Mater.* 2020, *258*, 119589. [CrossRef]
21. Tamanna, K.; Raman, S.N.; Jamil, M.; Hamid, R. Utilization of wood waste ash in construction technology: A review. *Constr. Build. Mater.* 2019, *237*, 117654. [CrossRef]
22. Caldas, L.R.; Da Gloria, M.Y.R.; Pittau, F.; Andreola, V.M.; Habert, G.; Filho, R.D.T. Environmental impact assessment of wood bio-concretes: Evaluation of the influence of different supplementary cementitious materials. *Constr. Build. Mater.* 2020, *268*, 121146. [CrossRef]
23. Šupić, S.; Malešev, M.; Radonjanin, V.; Bulatović, V.; Milović, T. Reactivity and Pozzolanic Properties of Biomass Ashes Generated by Wheat and Soybean Straw Combustion. *Materials* 2021, *14*, 1004. [CrossRef]
24. Thomas, B.S.; Yang, J.; Mo, K.H.; Abdalla, J.A.; Hawileh, R.A.; Ariyachandra, E. Biomass ashes from agricultural wastes as supplementary cementitious materials or aggregate replacement in cement/geopolymer concrete: A comprehensive review. *J. Build. Eng.* 2021, *40*, 102332. [CrossRef]
25. Charitha, V.; Athira, V.; Jittin, V.; Bahurudeen, A.; Nanthagopalan, P. Use of different agro-waste ashes in concrete for effective upcycling of locally available resources. *Constr. Build. Mater.* 2021, *285*, 122851. [CrossRef]
26. Bheel, N.; Ali, M.O.A.; Kirgiz, M.S.; Galdino, A.G.D.S.; Kumar, A. Fresh and mechanical properties of concrete made of binary substitution of millet husk ash and wheat straw ash for cement and fine aggregate. *J. Mater. Res. Technol.* 2021, *13*, 872–893. [CrossRef]
27. Rattanachu, P.; Toolkasikorn, P.; Tangchirapat, W.; Chindaprasirt, P.; Jaturapitakkul, C. Performance of recycled aggregate concrete with rice husk ash as cement binder. *Cem. Concr. Compos.* 2020, *108*, 103533. [CrossRef]
28. Memon, S.; Javed, U.; Haris, M.; Khushnood, R.; Kim, J. Incorporation of Wheat Straw Ash as Partial Sand Replacement for Production of Eco-Friendly Concrete. *Materials* 2021, *14*, 2078. [CrossRef] [PubMed]
29. Amin, M.N.; Murtaza, T.; Shahzada, K.; Khan, K.; Adil, M. Pozzolanic Potential and Mechanical Performance of Wheat Straw Ash Incorporated Sustainable Concrete. *Sustainability* 2019, *11*, 519. [CrossRef]
30. Sadik, M.W.; El Shaer, H.M.; Yakot, H.M. Recycling of Agriculture and Animal Farm Wastes into Compost Using Compost Activator in Saudi Arabia. *J. Environ. Sci. Int.* 2010, *5*, 397–403.
31. Pan, X.; Sano, Y. Fractionation of wheat straw by atmospheric acetic acid process. *Bioresour. Technol.* 2005, *96*, 1256–1263. [CrossRef]
32. Ahmad, J.; Tufail, R.; Aslam, F.; Mosavi, A.; Alyousef, R.; Javed, M.F.; Zaid, O.; Niazi, M.K. A Step towards Sustainable Self-Compacting Concrete by Using Partial Substitution of Wheat Straw Ash and Bentonite Clay Instead of Cement. *Sustainability* 2021, *13*, 824. [CrossRef]

33. Biricik, H.; Aköz, F.; Berktay, I.; Tulgar, A.N. Study of pozzolanic properties of wheat straw ash. *Cem. Concr. Res.* **1999**, *29*, 637–643. [CrossRef]
34. Memon, S.A.; Wahid, I.; Khan, M.K.; Tanoli, M.A.; Bimaganbetova, M. Environmentally Friendly Utilization of Wheat Straw Ash in Cement-Based Composites. *Sustainability* **2018**, *10*, 1322. [CrossRef]
35. Al-Akhras, N.M.; Abu-Alfoul, B.A. Effect of wheat straw ash on mechanical properties of autoclaved mortar. *Cem. Concr. Res.* **2002**, *32*, 859–863. [CrossRef]
36. Qudoos, A.; Kim, H.G.; Atta-Ur-Rehman, A.-U.; Ryou, J.-S. Effect of mechanical processing on the pozzolanic efficiency and the microstructure development of wheat straw ash blended cement composites. *Constr. Build. Mater.* **2018**, *193*, 481–490. [CrossRef]
37. Al-Akhras, N.M. Durability of wheat straw ash concrete exposed to freeze–thaw damage. *Proc. Inst. Civ. Eng.-Constr. Mater.* **2011**, *164*, 79–86. [CrossRef]
38. Khushnood, R.A.; Rizwan, S.A.; Memon, S.A.; Tulliani, J.-M.; Ferro, G.A. Experimental Investigation on Use of Wheat Straw Ash and Bentonite in Self-Compacting Cementitious System. *Adv. Mater. Sci. Eng.* **2014**, *2014*, 1–11. [CrossRef]
39. Bheel, N.; Ibrahim, M.H.W.; Adesina, A.; Kennedy, C.; Shar, I.A. Mechanical performance of concrete incorporating wheat straw ash as partial replacement of cement. *J. Build. Pathol. Rehabil.* **2020**, *6*, 1–7. [CrossRef]
40. Mehta, A.; Kar, D.; Ashish, K. Silica fume and waste glass in cement concrete production: A review. *J. Build. Eng.* **2019**, *15*, 100888. [CrossRef]
41. Sinyoung, S.; Kunchariyakun, K.; Asavapisit, S.; MacKenzie, K. Synthesis of belite cement from nano-silica extracted from two rice husk ashes. *J. Environ. Manag.* **2017**, *190*, 53–60. [CrossRef] [PubMed]
42. Diamond, S.; Sahu, S.; Thaulow, N. Reaction products of densified silica fume agglomerates in concrete. *Cem. Concr. Res.* **2004**, *34*, 1625–1632. [CrossRef]
43. Rodriguez, E.D.; Bernal, S.A.; Provis, J.L.; Payá, J.; Monzó, J.M.; Borrachero, M.V. Structure of Portland Cement Pastes Blended with Sonicated Silica Fume. *J. Mater. Civ. Eng.* **2012**, *24*, 1295–1304. [CrossRef]
44. Holland, T.C. Federal Highway Administration Report FHWA-IF-05-016. In *Silica Fume User's Manual*; Federal Highway Administration: Washington, DC, USA, 2005; p. 193.
45. Murthi, P.; Poongodi, K.; Awoyera, P.O.; Gobinath, R.; Saravanan, R. Enhancing the Strength Properties of High-Performance Concrete Using Ternary Blended Cement: OPC, Nano-Silica, Bagasse Ash. *Silicon* **2019**, *12*, 1949–1956. [CrossRef]
46. Sadrmomtazi, A.; Sobhani, J.; Mirgozar, M.A.; Najimi, M. Properties of multi-strength grade EPS concrete containing silica fume and rice husk ash. *Constr. Build. Mater.* **2012**, *35*, 211–219. [CrossRef]
47. Xu, W.T.; Lo, T.Y.; Wang, W.L.; Ouyang, D.; Wang, P.G.; Xing, F. Pozzolanic Reactivity of Silica Fume and Ground Rice Husk Ash as Reactive Silica in a Cementitious System: A Comparative Study. *Materials* **2016**, *9*, 146. [CrossRef] [PubMed]
48. Jamil, M.; Khan, M.; Karim, M.; Kaish, A.B.M.A.; Zain, M. Physical and chemical contributions of Rice Husk Ash on the properties of mortar. *Constr. Build. Mater.* **2016**, *128*, 185–198. [CrossRef]
49. Gursel, A.P. Life-Cycle Assessment of Concrete: Decision-Support Tool and Case Study Application, University of California, Berkeley. 2014. Available online: https://escholarship.org/uc/item/5q24d64s (accessed on 1 December 2021).
50. Azam, A.; Rafiq, M.; Shafique, M.; Ateeq, M.; Yuan, J. Causality Relationship Between Electricity Supply and Economic Growth: Evidence from Pakistan. *Energies* **2020**, *13*, 837. [CrossRef]
51. *ACI 318-08: Building Code Requirements for Structural Concrete and Commentary*; American Concrete Institute: Farmington Hills, MI, USA, 2008.
52. ACI Committee 363. *State of the Art Report on High-Strength Concrete*; American Concrete Institute: Farmington Hills, MI, USA, 1992.
53. Telford, T. *CEB-FIP MODEL CODE 1990, DESIGN CODE*; Comité Euro-International du Béton: Laussane, Switzerland, 1993.
54. Zain, M.; Mahmud, H.; Ilham, A.; Faizal, M. Prediction of splitting tensile strength of high-performance concrete. *Cem. Concr. Res.* **2002**, *32*, 1251–1258. [CrossRef]
55. Iravani, S. Mechanical properties of high-performance concrete. *ACI Mater. J.* **1996**, *93*, 416–426.
56. Shah, S.P.; Ahmad, S.H. Structural Properties of High Strength Concrete and its Implications for Precast Prestressed Concrete. *PCI J.* **1985**, *30*, 92–119. [CrossRef]
57. Kim, J.-K.; Han, S.H.; Song, Y.C. Effect of temperature and aging on the mechanical properties of concrete: Part I. Experimental results. *Cem. Concr. Res.* **2002**, *32*, 1087–1094. [CrossRef]
58. Tomosawa, F.; Noguchi, T.; Tamura, M. The Way Concrete Recycling Should Be. *J. Adv. Concr. Technol.* **2005**, *3*, 3–16. [CrossRef]
59. Design, L. *Standard Specifications for Concrete Structures*; Japan Society for Civil Engineers: Tokyo, Japan, 2013; pp. 34–35. (In Japanese)
60. De Larrard, F.; Malier, Y. Engineering Properties of Very High Performance Concretes. *Mater. Struct.* **2018**, *20*, 85–114. [CrossRef]
61. Gmira, A.; Pellenq, R.-M.; Rannou, I.; Duclaux, L.; Clinard, C.; Cacciaguerra, T.; Lequeux, N.; Van Damme, H. A Structural Study of Dehydration/Rehydration of Tobermorite, a Model Cement Compound. *Stud. Surf. Sci. Catal.* **2002**, *144*, 601–608. [CrossRef]
62. Li, J.; Yu, Q.; Huang, H.; Yin, S. Effects of Ca/Si Ratio, Aluminum and Magnesium on the Carbonation Behavior of Calcium Silicate Hydrate. *Materials* **2019**, *12*, 1268. [CrossRef] [PubMed]
63. Kunther, W.; Lothenbach, B.; Skibsted, J. Influence of the Ca/Si ratio of the C–S–H phase on the interaction with sulfate ions and its impact on the ettringite crystallization pressure. *Cem. Concr. Res.* **2015**, *69*, 37–49. [CrossRef]

64. Yu, P.; Kirkpatrick, R.J.; Poe, B.; McMillan, P.F.; Cong, X. Structure of Calcium Silicate Hydrate (C-S-H): Near-, Mid-, and Far-Infrared Spectroscopy. *J. Am. Ceram. Soc.* **2004**, *82*, 742–748. [CrossRef]
65. Fernández-Carrasco, L.; Torrens-Martín, D.; Morales, L.; Martínez-Ramírez, L.M.A.S. Infrared Spectroscopy in the Analysis of Building and Construction Materials. *Eng. Technol.* **2012**, *1*, 369–382. [CrossRef]
66. Hughes, T.L.; Methven, C.M.; Jones, T.G.; Pelham, S.E.; Fletcher, P.; Hall, C. Determining cement composition by Fourier transform infrared spectroscopy. *Adv. Cem. Based Mater.* **1995**, *2*, 91–104. [CrossRef]
67. Kupwade-Patil, K.; Chin, S.H.; Johnston, M.L.; Maragh, J.; Masic, A.; Büyüköztürk, O. Particle Size Effect of Volcanic Ash towards Developing Engineered Portland Cements. *J. Mater. Civ. Eng.* **2018**, *30*, 04018190. [CrossRef]
68. Bose, B.; Davis, C.R.; Erk, K.A. Microstructural refinement of cement paste internally cured by polyacrylamide composite hydrogel particles containing silica fume and nanosilica. *Cem. Concr. Res.* **2021**, *143*, 106400. [CrossRef]
69. Bodor, E.; Skalny, J.; Brunauer, S.; Hagymassy, J.; Yudenfreund, M. Pore structures of hydrated calcium silicates and portland cements by nitrogen adsorption. *J. Colloid Interface Sci.* **1970**, *34*, 560–570. [CrossRef]
70. Kupwade-Patil, K.; Al-Aibani, A.F.; Abdulsalam, M.F.; Mao, C.; Bumajdad, A.; Palkovic, S.D.; Büyüköztürk, O. Microstructure of cement paste with natural pozzolanic volcanic ash and Portland cement at different stages of curing. *Constr. Build. Mater.* **2016**, *113*, 423–441. [CrossRef]
71. Helmi, M.; Hall, M.; Stevens, L.; Rigby, S. Effects of high-pressure/temperature curing on reactive powder concrete microstructure formation. *Constr. Build. Mater.* **2016**, *105*, 554–562. [CrossRef]
72. Scrivener, K.; Snellings, R.; Lothenbach, B. *A Practical Guide to Microstructural Analysis of Cementitious Materials*; CRC Press: Boca Raton, FL, USA, 2016.
73. Loser, R.; Münch, B.; Lura, P. A volumetric technique for measuring the coefficient of thermal expansion of hardening cement paste and mortar. *Cem. Concr. Res.* **2010**, *40*, 1138–1147. [CrossRef]

Article

The Influence of Sintering Temperature on the Pore Structure of an Alkali-Activated Kaolin-Based Geopolymer Ceramic

Mohd Izrul Izwan Ramli [1,2], Mohd Arif Anuar Mohd Salleh [1,2,*], Mohd Mustafa Al Bakri Abdullah [1,2], Ikmal Hakem Aziz [1,2], Tan Chi Ying [2], Noor Fifinatasha Shahedan [2], Winfried Kockelmann [3], Anna Fedrigo [3], Andrei Victor Sandu [4,5,6,*], Petrica Vizureanu [4,7,*], Jitrin Chaiprapa [8] and Dumitru Doru Burduhos Nergis [4,*]

1. Faculty of Chemical Engineering Technology, Universiti Malaysia Perlis (UniMAP), Perlis 02600, Malaysia; izrulizwan@unimap.edu.my (M.I.I.R.); mustafa_albakri@unimap.edu.my (M.M.A.B.A.); ikmalhakem@unimap.edu.my (I.H.A.)
2. Geopolymer & Green Technology, Center of Excellence (CEGeoGTech), Universiti Malaysia Perlis (UniMAP), Perlis 02600, Malaysia; chiying95@outlook.com (T.C.Y.); fifinatasha@unimap.edu.my (N.F.S.)
3. STFC, Rutherford Appleton Laboratory, ISIS Facility, Harwell OX11 0QX, UK; winfried.kockelmann@stfc.ac.uk (W.K.); anna.fedrigo@stfc.ac.uk (A.F.)
4. Faculty of Materials Science and Engineering, Gheorghe Asachi Technical University of Iasi, D. Mangeron 41, 700050 Iasi, Romania
5. Romanian Inventors Forum, St. P. Movila 3, 700089 Iasi, Romania
6. National Institute for Research and Development in Environmental Protection INCDPM, Splaiul Independentei 294, 060031 Bucuresti, Romania
7. Technical Sciences Academy of Romania, Dacia Blvd 26, 030167 Bucharest, Romania
8. Synchrotron Light Research Institute, Muang, Nakhon Ratchasima 30000, Thailand; jitrin@slri.or.th
* Correspondence: arifanuar@unimap.edu.my (M.A.A.M.S.); sav@tuiasi.ro (A.V.S.); peviz@tuiasi.ro (P.V.); doru.burduhos@tuiasi.ro (D.D.B.N.)

Abstract: Geopolymer materials are used as construction materials due to their lower carbon dioxide (CO_2) emissions compared with conventional cementitious materials. An example of a geopolymer material is alkali-activated kaolin, which is a viable alternative for producing high-strength ceramics. Producing high-performing kaolin ceramics using the conventional method requires a high processing temperature (over 1200 °C). However, properties such as pore size and distribution are affected at high sintering temperatures. Therefore, knowledge regarding the sintering process and related pore structures on alkali-activated kaolin geopolymer ceramic is crucial for optimizing the properties of the aforementioned materials. Pore size was analyzed using neutron tomography, while pore distribution was observed using synchrotron micro-XRF. This study elucidated the pore structure of alkali-activated kaolin at various sintering temperatures. The experiments showed the presence of open pores and closed pores in alkali-activated kaolin geopolymer ceramic samples. The distributions of the main elements within the geopolymer ceramic edifice were found with Si and Al maps, allowing for the identification of the kaolin geopolymer. The results also confirmed that increasing the sintering temperature to 1100 °C resulted in the alkali-activated kaolin geopolymer ceramic samples having large pores, with an average size of ~80 µm^3 and a layered porosity distribution.

Keywords: geopolymer; pore; tomography imaging; sintering

1. Introduction

Geopolymer is an inorganic compound material used in construction as a sealant and heat-resistant material [1]. It is a three-dimensional (3D) aluminosilicate structure that is activated using suitable precursor raw materials. Kaolin is an inorganic material that has been identified as geopolymer-compatible with excellent performance. Wang et al. [2] reported that the kaolin structure is significantly influenced by the calcination temperature. A change in the aluminium species influences the structural changes of geopolymer after being heated to 900 °C. The calcium aluminosilicate framework fills the pores between akermanite crystals after being heated up to 1200 °C.

Apart from the geopolymerization component, sintering plays a vital role in producing geopolymer ceramic. Sintering is defined as a thermally activated adhesion process, which increases the contact between particles and their respective coalescence. Sintering closes some of the open pores, decreasing the water absorption rate and increasing pore strength. The dense heated geopolymer has a glassy phase, making it a ceramic. Traditionally, ceramic vitrification begins at 900 °C, marked by the melting of several solid phases that bind present solid particles, enhancing bonding strength [3,4]. The solid reaction product usually consists of an open-pore volume fraction that was reported to be ~<1–40% [5]. After the sintering process, gas adsorption–desorption and mercury intrusion porosimetry are standard methods used to investigate pore structures [6]. Pores ranging from 1 μm to 0.5 mm are also commonly investigated using SEM and nitrogen adsorption. However, these measurements suffer from several drawbacks, rendering them unsuitable for observing cementitious materials. Both are destructive and can potentially alter pore structures. Therefore, advanced techniques such as tomography using neutron sources have been explored to understand the sintering process' effect on the pore structure of kaolin-based geopolymer. It has been demonstrated that neutron tomography imaging is a suitable characterization method for pore structures. An understanding of the pore structure after the sintering process can be applied for tailoring the resulting materials' properties. Also, it has been established that the nondestructive testing (NDT) of high-resolution 3D tomography is beneficial as it elucidates qualitative and quantitative pore formations [7]. The utilization of this tomography technique to investigate porosity and pore size distribution is advantageous and effective. Moreover, extensive quantitative research has been conducted on the pores of ceramic materials such as alumina ceramic using X-ray computed tomography, per Lo. et al. [8]. Nickerson et al. also studied the porosities formed in ceramics and their permeability using X-ray computed tomography [9].

In this study, tomography imaging with a neutron source was used to elucidate the effect of sintering on the pore structure of kaolin-based geopolymer. Neutron attenuation coefficients resulted in different image contrasts relative to those generated by conventional X-ray tomography, producing high-resolution images suitable for determining correlations between pore size, density, and absorption performance. Correlations were linked to the elemental distribution obtained using micro-X-ray fluorescence at a synchrotron source. This work successfully characterized and investigated the pore structure of kaolin-based geopolymer.

2. Experimental Section

Materials, Sample Preparation, and Characterization

A precursor of kaolin (supplied by Associated Kaolin Industries Sdn. Bhd., Petaling Jaya, Malaysia) was used for the synthesis of geopolymer. The NaOH was in pellet form with 97% purity, and the Na_2SiO_3 consisted of 9.4% Na_2O, 30.1% SiO_2, and 60.5% H_2O, with $SiO_2/Na_2O = 3.2$. The other characteristics were: specific gravity at 20 °C = 1.4 kg/cm^3 and viscosity = 0.4 Pa s. To form the geopolymer samples, the kaolin was activated with alkaline activator solution, namely, sodium hydroxide (NaOH) and sodium silicate (Na_2SiO_3) solution, at ambient temperature. The NaOH clear solution was mixed with sodium silicate solution and cooled to ambient temperature one day before mixing [10]. The solid–liquid and Na_2SiO_3/NaOH were fixed at 1.0 (NaOH molarity 8 M) and 1.5, respectively, on the basis of previous research on the optimum design of kaolin geopolymer [11]. The kaolin materials were mixed with an alkaline activator solution for 5 min; then, the homogenized mixture was poured into a mold. Then, after curing for 14 days, the kaolin-based geopolymer was sintered at 900, 1000, and 1100 °C for 2 h at a heating rate of 10 °C/min in an electrically heated furnace. The details of sample preparation are illustrated in Figure 1.

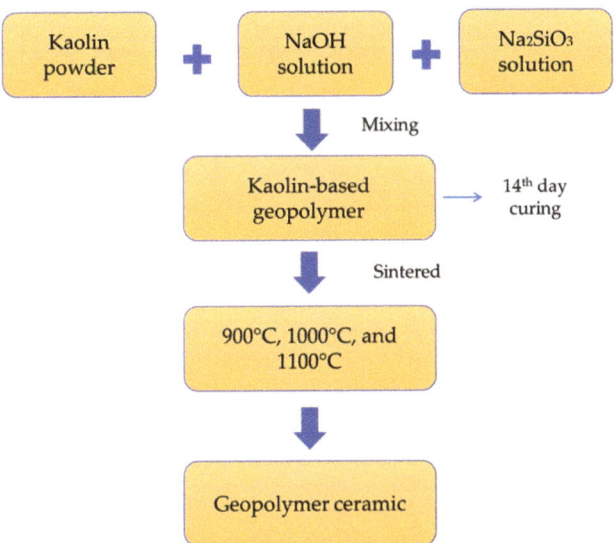

Figure 1. The process of creating kaolin-based geopolymer ceramic.

The unsintered and sintered samples of pore microstructures were imaged using the JSM-6460LA Scanning Electron Microscope (JEOL, Peabody, MA, USA) equipped with secondary electron detectors. The voltage and working distance were fixed at 10 kV and 10 mm, respectively. The surface area and pore volume were measured using Brunauer–Emmet–Teller (BET) equipment (TriStar 3000, Micromeritics Instrument Corporation, GA, USA). The adsorbed quantity correlated with the particles' total surface areas and pore volume in the unsintered and sintered samples. The samples' thicknesses were 0.5–1 mm. For contrast variation measurements, the samples were placed horizontally in a sample holder and the solvent was added dropwise to the center of the disc.

Neutron images of the samples were acquired at the IMAT beamline, ISIS neutron source, Rutherford Appleton Laboratory, United Kingdom [12]. The IMAT tomography camera was equipped with a 2048 × 2048 pixel Andor Zyla sCMOS 4.2 PLUS. The camera pixel size was 29 μm. The samples were inserted into an aluminum tube that was fixed on the rotating platform and placed at a distance, L, of 10 m from the beam aperture and a distance, d, of 25 mm from the neutron screen. The diameter (D) of the beam aperture was 40 mm, resulting in an L/D ratio of 250. We collected 868 projections, with an exposure time for each projection of 30 s and a total scan time of approximately 6 h/tomogram. Several open-beam and dark images were collected for flat fielding before and after each tomography scan. The images were analyzed using ImageJ and the Octopus reconstruction package (XRE, Ghent, Belgium). The unsintered and sintered geopolymer samples' elemental distributions were determined using synchrotron μ-XRF at BL6b beamline at the Synchrotron Light Research Institute (SLRI) in Bangkok, Thailand. A polycapillary lens was used to initiate a micro-X-ray beam with a size of 30 × 30 μm on the samples, with continuous synchrotron radiation. The X-ray energy range used was 2–12 keV without the monochromator feature. The detection limit at the sub parts per million concentration level can be obtained at larger than 100 nm, with sensitivities approaching the attogram (10−18 g) level [13]. The experiments were conducted in a helium (He) gas atmosphere with 30 s of exposure at each point. The data were obtained and analyzed using PyMca [14].

The samples were fabricated in powder form for phase analyses. The XRD analysis was performed using an XRD-6000 Shimadzu X-ray diffractometer (Cu Kα radiation (λ = 1.5418 A)). The operating parameters were 40 kV, 35 mA, at 2θ of 10–80°, at a 1°/min scan rate. The XRD patterns were then analyzed using X'pert HighScore Plus. The density was calculated, and water absorption tests were conducted per ASTM C642-13 (ASTM C642-13, Standard Test Method for Density, Absorption, and Voids in Hardened Concrete, ASTM International, United States (2013)). The weight of the samples after and before the samples were immersed in water was recorded, and the percentages of water absorption for the samples after sintering at 900 and 1100 °C were determined.

3. Results and Discussions

3.1. Density and Water Absorption Analysis

In order to examine the pore structure in kaolin-based geopolymer ceramic, the density and water absorption of kaolin-based geopolymer samples were investigated. The densities of the unsintered and sintered kaolin at 900 and 1100 °C after 3 days are shown in Figure 2. The densities of the unsintered and sintered samples at all temperatures decreased as time increased. The unsintered samples had the highest density of 1610 kg/cm^3, while the samples sintered at 1100 °C had the lowest density of 1203 kg/cm^3. Therefore, we speculate that the formation of large pores created in the kaolin at 1100 °C resulted in the lowest density, while sintering at 900 °C resulted in the formation of small pores in the kaolin-based geopolymer samples. In addition, the unsintered samples contained small and open pores, while the sintered samples had large and closed pores, which translated into a high material density. The existence of these larger pores was likely due to the growth of sintered necks, which was reflected in the phase evolution. The details of phase crystallization are discussed in Section 3.6.

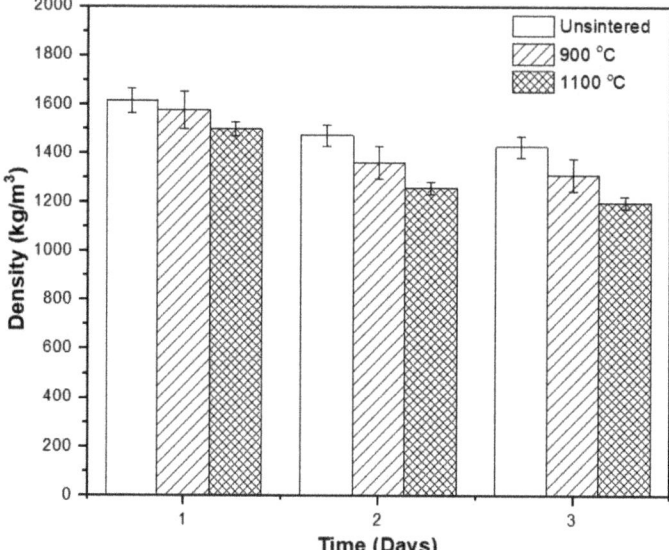

Figure 2. The density of kaolin geopolymer over 3 days for unsintered and sintered samples at 900 and 1100 °C.

Figure 3 shows the percentage water absorption of the kaolin-based geopolymer ceramic samples after sintering at 900 and 1100 °C after 3 days. After 3 days, the highest value percentage water absorption occurred with sintering at 1100 °C. The percentage water absorption continuously increased with sintering temperature and time. This was in accordance with results published by Faris et al. [15]. The higher sintering temperature resulted in larger pores due to water removal, and the increased pore size increased the water absorption capacity of the kaolin-based geopolymer samples. The high volume of open pores in the samples may have contributed to the high water absorption due to a high surface area, which was reported by Zulkifli et al. [11].

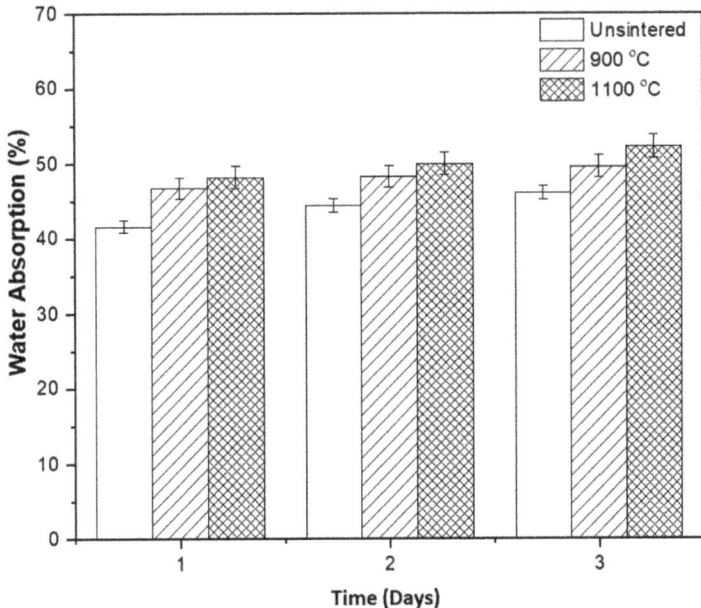

Figure 3. The water absorption of kaolin geopolymer over 3 days for unsintered and sintered samples at 900 and 1100 °C.

3.2. Pore Structure Analysis

The Brunauer–Emmett–Teller (BET) method was used to determine the surface area of the unsintered and sintered kaolin-based geopolymers. The specific surface area and pore volume of unsintered and sintered geopolymers are depicted in Figure 4. Smaller particles resulted in larger surface areas. This is because sintered kaolin-based geopolymer has a larger surface area due to the removal of volatiles and impurities from the sample's surface. Sutama et al. [16] stated that the formation of pores on the sample may lower the compressive strength.

The unsintered kaolin-based geopolymer had the lowest surface area (2.3 m^2/g) and pore volume (0.01 cm^3/g). After sintering at 900 °C, the kaolin-based geopolymer's surface area (up to 245 m^2/g) and pore volume (up to 0.025 cm^3/g) increased relative to those of the unsintered kaolin-based geopolymer. Then, after sintering at 1100 °C, the surface area increased to 270 m^2/g and pore volume increased to 0.04 cm^3/g. The kaolin-based geopolymer was assumed to consist of mesopores in a small quantity, resulting in a higher surface area after sintering at high temperatures.

3.3. Microstructure Analysis

An SEM revealed the morphological features of kaolin-based geopolymer ceramic samples at sintering temperatures of (a) unsintered, (b) 900, and (c) 1100 °C, as shown in Figure 5. The unsintered kaolin showed the presence of well-defined clay platelets and an incomplete reaction of kaolin, as shown in Figure 5a. After sintering at 900 and 1100 °C, the images clearly showed the presence of pores and cracks in all of the heated kaolin-based geopolymer ceramic samples. The pores formed a network, which resulted in increased internal porosity. The kaolin-based geopolymer surface became glassy and glossy when sintered at 900 °C (Figure 5b). This microstructure change was attributed to moisture hydration and phase transformation, as reported by Dudek et al. [17]. It can also be seen in Figure 5b that the kaolin-based geopolymer ceramic samples sintered at 900 °C had a higher porosity, alongside cracks and voids.

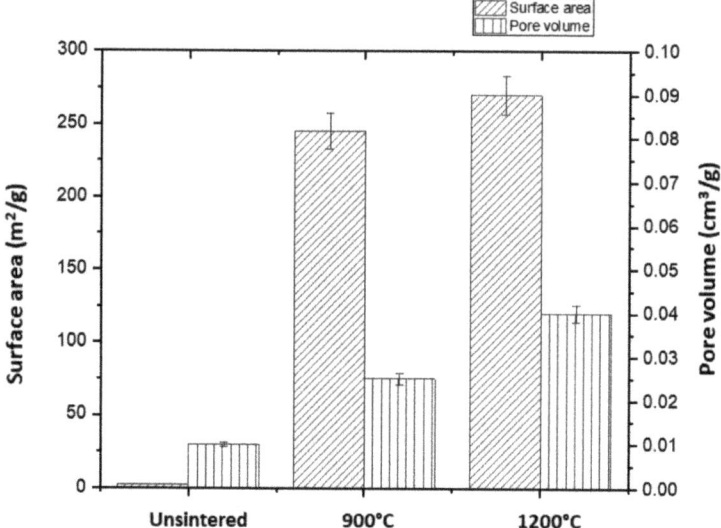

Figure 4. Surface area and pore volume of kaolin geopolymer samples versus sintering temperature.

Increasing the sintering temperature up to 1100 °C increased the number of large pores. The pore size distribution of kaolin-based geopolymer was ~50 μm for the unsintered samples. After sintering at 1100 °C, the pore size increased to 80 μm, similar to the findings from the tomography analysis. The pore sizes in kaolin directly affect its mechanical strength. The SEM images also showed significant cracks due to moisture evaporation and shrinkage during the sintering process. The loosely grained structure of kaolinite can also cause cracks, and the presence of voids at the interface of loosening grains can result in increased total porosity.

3.4. Neutron Tomography Imaging Analysis

Segmentation was carried out in a small area to analyze porosity data in the kaolin-based geopolymer samples quantitatively, and the 3D reconstruction images are shown in Figure 6. The kaolin-based geopolymer samples' widths, lengths, and thicknesses, shown in Figure 6a–c, were 2900, 1740, and 1740 μm, respectively. The white color indicates the solid kaolin-based geopolymer, while blue indicates the air (pore) space. The total number of pores for this region was estimated to be 197, and after sintering at 900 and 1100 °C, the total number of pores decreased to 182 and 125, respectively. Neutron tomography made imaging very small pores at high resolutions possible, and the results are shown in Figure 6d–f. In the case of the unsintered kaolin, the pore size was ~50 μm^3, and when sintered at 900 and 1100 °C, the pore size increased to 68 and 82 μm^3, respectively. Figure 6g displays pore numbers and sizes. These sizes are in agreement with those measured in the SEM images shown in Figure 4.

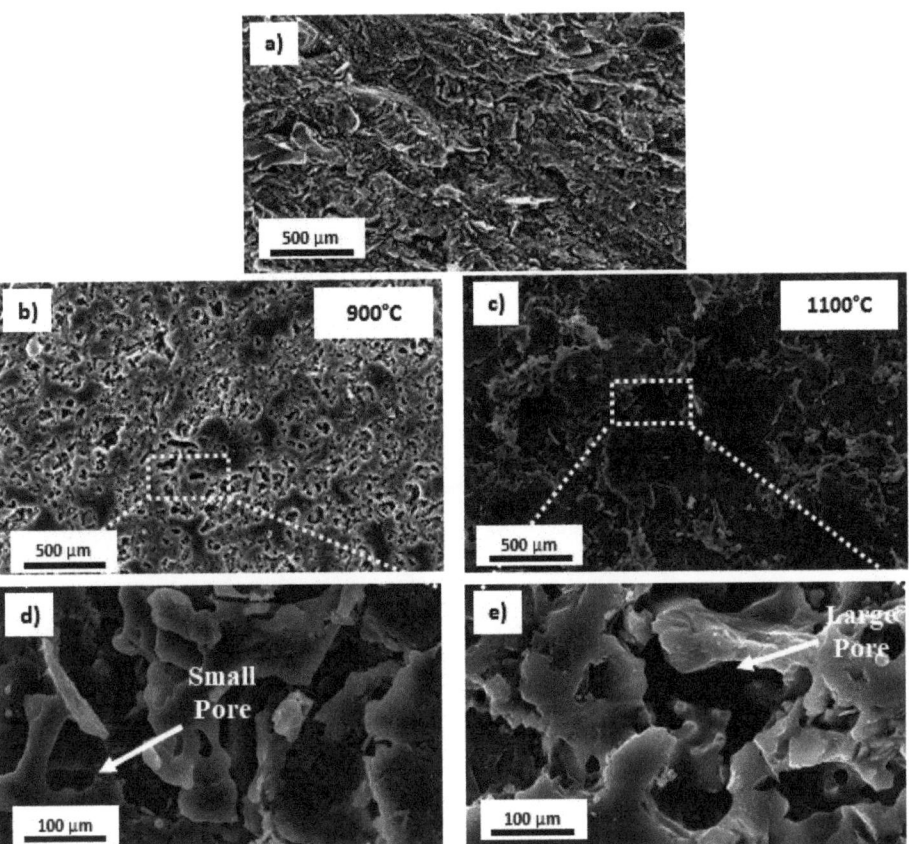

Figure 5. SEM micrograph of (**a**) unsintered, (**b,d**) sintered at 900 °C, and (**c,e**) sintered at 1100 °C kaolin-based geopolymer.

When sintered, the small pores merged to become large(r) pores due to moisture hydration after sintering. Our images show the isolated closed pores in the 3D volume, and it was, in fact, a network of fully connected open pores in 3D. Interestingly, after sintering, the pore distribution of the kaolin-based geopolymer became layered, as shown in Figure 6b,c. The layer distance between porosities was estimated to be ~120-130 μm when sintered at 900 and 1100 °C because the kaolin-based geopolymer exhibited low reactivity with the alkaline silicate solution.

A layered structure was caused by the sintering of the kaolin-based geopolymer at a higher temperature. The layered structure was indicated by the transformation of pore appearance, as shown in Figure 7. The pore transformation was attributed to the larger surface area causing necking reactions between particles (Figure 7b). During sintering, atoms diffuse from an area of higher chemical potential to an area of lower chemical potential. Small pores then merge to form larger pores. The layered grain structure represented the disorganized kaolinite structure (grey color) that was due to dehydroxylation. The dehydroxylation of kaolin resulted in the destruction of the crystalline structure and the transformation of the mullite phase, as confirmed by an XRD analysis. These findings are consistent with ElDeeb et al. [18], who posited that the hydroxylation of clay sheets occurs with high-temperature sintering.

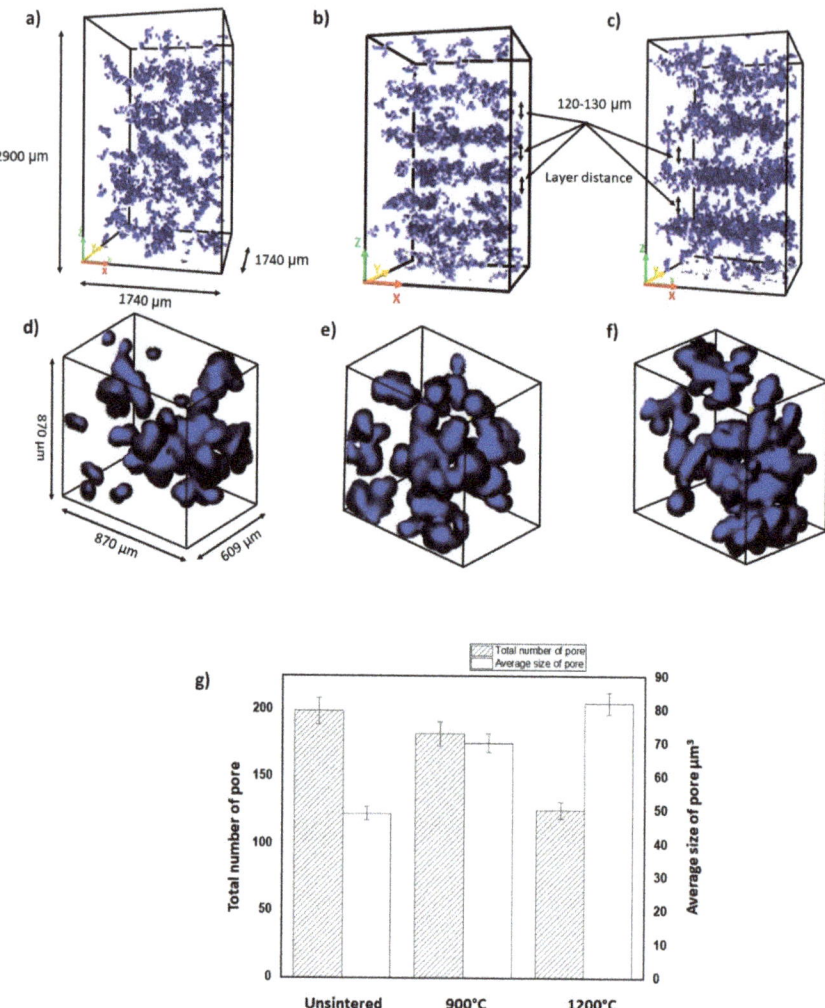

Figure 6. Tomography imaging of (**a**) unsintered and sintered geopolymer at (**b**) 900 and (**c**) 1100 °C. (**d–f**) Tomography imaging with zoom and higher resolution and (**g**) total pore numbers and average pore sizes.

3.5. Elemental Distribution Analysis

The kaolin-based geopolymer ceramic samples were further characterized using micro-XRF mapping to understand their elemental distribution vis-à-vis sintering temperatures. Figure 8 illustrates the localized micro-XRF mapping of the kaolin-based geopolymer ceramic samples that were (Figure 8a) unsintered or heated to 900 (Figure 8b) or 1100 °C (Figure 8c), signifying where the (main) elements Si and Al were critically located within the samples. The distributions of the main elements within the geopolymer ceramic edifice were confirmed using synchrotron micro-XRF. The distribution of Si combined with the Al map allowed for the identification of the kaolin-based geopolymer ceramic backbone (kaolinite). The red, green, and blue spots represent the high, medium, and low intensities, respectively, for each distribution element at the integrated area.

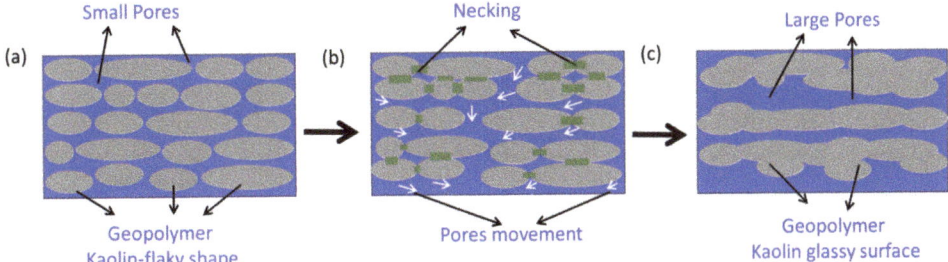

Figure 7. Sintering mechanism of pore transformation in various environments: (**a**) unsintered, (**b**) 900 °C, and (**c**) 1100 °C.

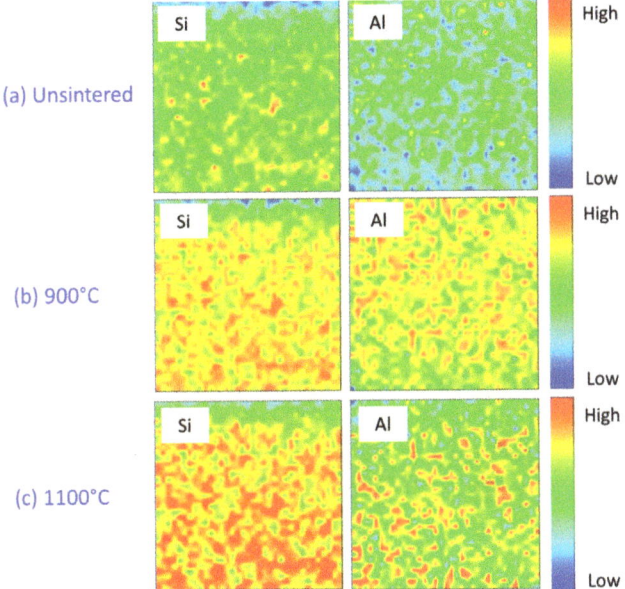

Figure 8. Micro-XRF elemental distribution maps of Si and Al in kaolin geopolymer ceramic at various sintering temperatures.

The various sintering temperatures resulted in significant changes in the Si and Al element distributions, edging the material towards phase transformation. A high concentration of Si (Figure 8a) represented the kaolinite grain. Upon obtaining the pore microstructure of the kaolin-based geopolymer ceramic (Figure 8b), the Si and Al regions showed higher intensities, reflecting the presence of the minerals quartz and nepheline, as depicted in Figure 8 and the next section. At 1100 °C, Si and Al were of higher intensities in a localized area, reflecting the formation of mullite. This Si–Al-rich crystalline mineral contributed to the pores' microstructure appearance, as shown in Figure 5b,c.

3.6. Mineral Phase Transformation

Figure 9 shows an XRD diffractogram of the kaolin-based geopolymer ceramic when (a) unsintered or heated to (b) 900 or (c) 1100 °C. The unsintered kaolin-based geopolymer showed the presence of crystalline phases such as kaolinite, quartz, and tridymite. A geopolymerization reaction was initiated by the dissolution of aluminosilicate materials in an alkali activator (combination of NaOH and Na_2SiO_3 solutions). Next, the products of dissolution underwent nucleation growth and polymerization processes before hardening at the polycondensation stage. There have been several findings obtained with a similar method for producing kaolin-based geopolymer at room temperatures [19,20]. Additionally, kaolinite was traced as a major mineral in spectra of kaolin-based geopolymer samples [21]. Owing to the lower activity of pure kaolin, a number of distinctive kaolinite peaks remained in the diffractogram of the kaolin-based geopolymer [22]. However, these kaolinite peaks decreased at high sintering temperatures, as shown in Figure 9b.

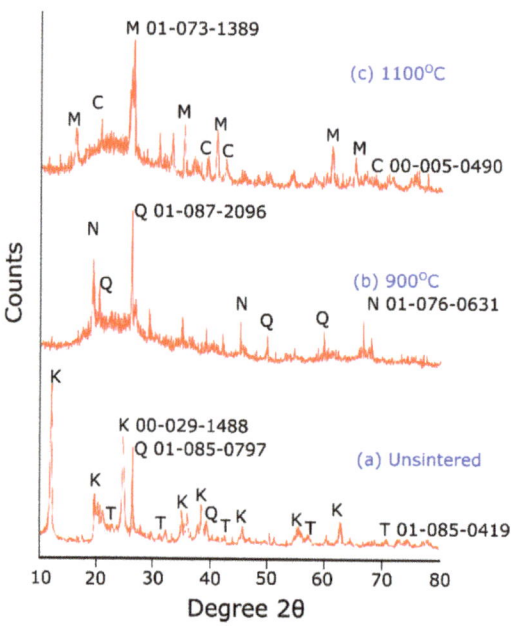

Figure 9. Phase transformation of kaolin geopolymer when (**a**) unsintered, (**b**) sintered at 900 °C, and (**c**) sintered at 1100 °C. M, mullite; C, cristobalite; Q, quartz; K, kaolin; N, nepheline; T, tridymite.

Sintering temperatures up to 1100 °C introduced the formation of the mullite phase (Figure 8c). The mullite phase is present in this sintering region, manifesting superior thermochemical stabilities [23,24]. Furthermore, the appearance of cristobalite was due to unreacted quartz (SiO_2) after the decomposition of kaolinite at 900 °C [25]. The liberation of SiO_2 corresponded to the kaolinite–mullite transformation, which yields to Al–Si spinel phase [26]. This was corroborated with the elemental distribution analysis obtained using micro-XRF, as the sintered kaolin geopolymer ceramic samples showed a high intensity at the Si and Al regions at 1100 °C (Figure 7c). The transformation of mullite is described by the chemical reaction in Equation (1) [27,28]:

$$2Si_3Al_4O_{12}\text{Al-Si spinel} \rightarrow 23Al_2O_3 \cdot 2SiO_2 \text{ mullite} + 5SiO_2 \tag{1}$$

4. Conclusions

This manuscript summarizes the effects of sintering temperature on the pore structure of an alkali-activated kaolin-based geopolymer ceramic. Sintering temperatures significantly affected the size and number of pores in the kaolin-based geopolymer. The material's density and water absorption confirmed the presence of pores after the sintering process. Microstructural analyses showed that sintering at 1100 °C resulted in large pore sizes relative to the material's unsintered counterpart. Tomography imaging also confirmed the presence of a layered pore structure after sintering. The pore size at 900 °C was 50 µm^3, and after sintering at 900 and 1100 °C, the pore size increased to 68 and 82 µm^3, respectively.

Author Contributions: Conceptualization and methodology and writing, M.I.I.R. and D.D.B.N.; supervision and resources, M.A.A.M.S.; methodology and formal analysis and investigation, I.H.A., T.C.Y., W.K., A.F. and N.F.S.; interpretation and review and editing, M.M.A.B.A., A.V.S., P.V. and J.C. All authors have read and agreed to the published version of the manuscript.

Funding: The neutron tomography studies of the geopolymer ceramic used for reinforcement materials in a solder alloy for a robust electric/electronic solder joint were financed by Ministry of Education Malaysia under reference no: JPT.S (BPKI)2000/016/018/019(29). This work was supported by the TUIASI's internal grants program (GI_PUBLICATIONS/2021), financed by the Romanian Government. This work also supported by Ministry of Higher Education Malaysia regarding the use of the ISIS Neutron and Muon Source and funded by the UK Department of Business, Energy and Industrial Strategy (BEIS) and Malaysia and delivered by the British Council.

Institutional Review Board Statement: Not applicable.

Informed Consent Statement: Not applicable.

Data Availability Statement: Not applicable.

Acknowledgments: The authors gratefully acknowledged the Ministry of Higher Education Malaysia regarding the use of the ISIS Neutron and Muon Source. A grant was funded by the UK Department of Business, Energy and Industrial Strategy (BEIS) and Malaysia and delivered by the British Council.

Conflicts of Interest: The authors declare no conflict of interest.

References

1. Wen, J.; Zhou, Y.; Ye, X. Study on Structure and Properties of Kaolin Composites-based Geopolymers. *Chem. Eng. Trans.* **2018**, *66*, 463–468.
2. Wang, M.; Jia, D.; He, P.; Zhou, Y. Influence of calcination temperature of kaolin on the structure and properties of final geopolymer. *Mater. Lett.* **2010**, *64*, 2551–2554. [CrossRef]
3. Lemougna, P.N.; Yliniemi, J.; Ismailov, A.; Levanen, E.; Tanskanen, P.; Kinnunen, P.; Roning, J.; Illikainen, M. Recycling lithium mine tailings in the production of low temperature (700–900 C) ceramics: Effect of ladle slag and sodium compounds on the processing and final properties. *Constr. Build. Mater.* **2019**, *221*, 332–344. [CrossRef]
4. Aziz, I.H.; Al Bakri Abdullah, M.M.; Yong, H.C.; Ming, L.Y.; Hussin, K.; Surleva, A.; Azimi, E.A. Manufacturing parameters influencing fire resistance of geopolymers: A review. *Proc. Inst. Mech. Eng. Part L J. Mater. Des. Appl.* **2019**, *233*, 721–733. [CrossRef]
5. Maitland, C.; Buckley, C.; O'connor, B.; Butler, P.; Hart, R. Characterization of the pore structure of metakaolin-derived geopolymers by neutron scattering and electron microscopy. *J. Appl. Crystallogr.* **2011**, *44*, 697–707. [CrossRef]
6. Li, J.; Mailhiot, S.; Sreenivasan, H.; Kantola, A.M.; Illikainen, M.; Adesanya, E.; Kriskova, L.; Telkki, V.-V.; Kinnunen, P. Curing process and pore structure of metakaolin-based geopolymers: Liquid-state 1H NMR investigation. *Cem. Concr. Res.* **2021**, *143*, 106394. [CrossRef]
7. Jan, D.; De Kock, T.; Fronteau, G.; Derluyn, H.; Vontobel, P.; Dierick, M.; Hoorebeke, L.; Jacobs, P.; Cnudde, V. Neutron radiography and X-ray computed tomography for quantifying weathering and water uptake processes inside porous limestone used as building material. *Mater. Charact.* **2014**, *88*, 86–99. [CrossRef]
8. Lo, C.; Sano, T.; Hogan, J.D. Microstructural and mechanical characterization of variability in porous advanced ceramics using X-ray computed tomography and digital image correlation. *Mater. Charact.* **2019**, *158*, 109929. [CrossRef]
9. Nickerson, S.; Shu, Y.; Zhong, D.; Könke, C.; Tandia, A. Permeability of Porous Ceramics by X-ray CT Image Analysis. *Acta Mater.* **2019**, *172*, 121–130. [CrossRef]
10. Jamil, N.H.; Abdullah, M.M.A.B.; Pa, F.C.; Mohamad, H.; Ibrahim, W.M.A.W.; Chaiprapa, J. Influences of SiO2, Al2O3, CaO and MgO in phase transformation of sintered kaolin-ground granulated blast furnace slag geopolymer. *J. Mater. Res. Technol.* **2020**, *9*, 14922–14932. [CrossRef]
11. Zulkifli, N.N.I.; Abdullah, M.M.A.B.; Przybył, A.; Pietrusiewicz, P.; Salleh, M.A.A.M.; Aziz, I.H.; Kwiatkowski, D.; Gacek, M.; Gucwa, M.; Chaiprapa, J. Influence of Sintering Temperature of Kaolin, Slag, and Fly Ash Geopolymers on the Microstructure, Phase Analysis, and Electrical Conductivity. *Materials* **2021**, *14*, 2213. [CrossRef]
12. Minniti, T.; Watanabe, K.; Burca, G.; Pooley, D.E.; Kockelmann, W. Characterization of the new neutron imaging and materials science facility IMAT. *Nucl. Instrum. Methods Phys. Res. Sect. A Accel. Spectrometers Detect. Assoc. Equip.* **2018**, *888*, 184–195. [CrossRef]
13. Aziz, I.H.; Abdullah, M.M.A.B.; Salleh, M.M.; Yoriya, S.; Chaiprapa, J.; Rojviriya, C.; Li, L.Y. Microstructure and porosity evolution of alkali activated slag at various heating temperatures. *J. Mater. Res. Technol.* **2020**, *9*, 15894–15907. [CrossRef]
14. Solé, V.; Papillon, E.; Cotte, M.; Walter, P.; Susini, J. A multiplatform code for the analysis of energy-dispersive X-ray fluorescence spectra. *Spectrochim. Acta B* **2007**, *62*, 63–68. [CrossRef]

15. Faris, M.A.; Abdullah, M.M.A.B.; Muniandy, R.; Abu Hashim, M.F.; Błoch, K.; Jeż, B.; Garus, S.; Palutkiewicz, P.; Mohd Mortar, N.A.; Ghazali, M.F. Comparison of Hook and Straight Steel Fibers Addition on Malaysian Fly Ash-Based Geopolymer Concrete on the Slump, Density, Water Absorption and Mechanical Properties. *Materials* **2021**, *14*, 1310. [CrossRef]
16. Sutama, A.; Saggaff, S.; Saloma, S.; Hanafiah, H. Properties And Microstructural Characteristics Of Lightweight Geopolymer Concrete With Fly Ash And Kaolin. *Int. J. Sci. Technol. Res.* **2019**, *8*, 57–64.
17. Dudek, M.; Sitarz, M. Analysis of changes in the microstructure of geopolymer mortar after exposure to high temperatures. *Materials* **2020**, *13*, 4263. [CrossRef]
18. ElDeeb, A.; Brichkin, V.; Kurtenkov, R.; Bormotov, I. Extraction of alumina from kaolin by a combination of pyro-and hydro-metallurgical processes. *Appl. Clay Sci.* **2019**, *172*, 146–154. [CrossRef]
19. Matalkah, F.; Aqel, R.; Ababneh, A. Enhancement of the mechanical properties of kaolin geopolymer using sodium hydroxide and calcium oxide. *Procedia Manuf.* **2020**, *44*, 164–171. [CrossRef]
20. Yunsheng, Z.; Wei, S.; Zongjin, L. Composition design and microstructural characterization of calcined kaolin-based geopolymer cement. *Appl. Clay Sci.* **2010**, *47*, 271–275. [CrossRef]
21. Heah, C.; Kamarudin, H.; Mustafa Al Bakri, A.; Bnhussain, M.; Luqman, M.; Khairul Nizar, I.; Ruzaidi, C.; Liew, Y. Kaolin-based geopolymers with various NaOH concentrations. *Int. J. Miner. Metall. Mater.* **2013**, *20*, 313–322. [CrossRef]
22. Slaty, F.; Khoury, H.; Wastiels, J.; Rahier, H. Characterization of alkali activated kaolinitic clay. *Appl. Clay Sci.* **2013**, *75*, 120–125. [CrossRef]
23. Djangang, C.N.; Kamseu, E.; Ndikontar, M.K.; Nana, G.L.L.; Soro, J.; Melo, U.C.; Elimbi, A.; Blanchart, P.; Njopwouo, D. Sintering behaviour of porous ceramic kaolin–corundum composites: Phase evolution and densification. *Mater. Sci. Eng. A* **2011**, *528*, 8311–8318. [CrossRef]
24. Abtew, M.A.; Boussu, F.; Bruniaux, P.; Loghin, C.; Cristian, I.; Chen, Y.; Wang, L. Ballistic impact performance and surface failure mechanisms of two-dimensional and three-dimensional woven p-aramid multi-layer fabrics for lightweight women ballistic vest applications. *J. Ind. Text.* **2021**, *50*, 1351–1383. [CrossRef]
25. Burduhos Nergis, D.D.; Vizureanu, P.; Sandu, A.V.; Burduhos Nergis, D.P.; Bejinariu, C. XRD and TG-DTA Study of New Phosphate-Based Geopolymers with Coal Ash or Metakaolin as Aluminosilicate Source and Mine Tailings Addition. *Materials* **2022**, *15*, 202. [CrossRef] [PubMed]
26. Hui-Teng, N.; Cheng-Yong, H.; Abdullah, M.M.A.B.; Yong-Sing, N.; Bayuaji, R. Study of fly ash geopolymer and fly ash/slag geopolymer in term of physical and mechanical properties. *Eur. J. Mater. Sci. Eng.* **2020**, *5*, 187–198. [CrossRef]
27. Zhou, H.; Qiao, X.; Yu, J. Influences of quartz and muscovite on the formation of mullite from kaolinite. *Appl. Clay Sci.* **2013**, *80*, 176–181. [CrossRef]
28. Chen, Y.-F.; Wang, M.-C.; Hon, M.-H. Phase transformation and growth of mullite in kaolin ceramics. *J. Eur. Ceram. Soc.* **2004**, *24*, 2389–2397. [CrossRef]

Article

XRD and TG-DTA Study of New Phosphate-Based Geopolymers with Coal Ash or Metakaolin as Aluminosilicate Source and Mine Tailings Addition

Dumitru Doru Burduhos Nergis [1], Petrica Vizureanu [1,2,*], Andrei Victor Sandu [1,3], Diana Petronela Burduhos Nergis [1] and Costica Bejinariu [1,*]

1. Faculty of Materials Science and Engineering, Gheorghe Asachi Technical University of Iasi, 700050 Iasi, Romania; doru.burduhos@tuiasi.ro (D.D.B.N.); sav@tuiasi.ro (A.V.S.); burduhosndiana@yahoo.com (D.P.B.N.)
2. Materials Science and Engineering Section, Technical Sciences Academy of Romania, Dacia Blvd 26, 030167 Bucharest, Romania
3. Romanian Inventors Forum, St. P. Movila 3, 700089 Iasi, Romania
* Correspondence: peviz2002@yahoo.com (P.V.); costica.bejinariu@tuiasi.ro (C.B.)

Abstract: Coal ash-based geopolymers with mine tailings addition activated with phosphate acid were synthesized for the first time at room temperature. In addition, three types of aluminosilicate sources were used as single raw materials or in a 1/1 wt. ratio to obtain five types of geopolymers activated with H_3PO_4. The thermal behaviour of the obtained geopolymers was studied between room temperature and 600 °C by Thermogravimetry-Differential Thermal Analysis (TG-DTA) and the phase composition after 28 days of curing at room temperature was analysed by X-ray diffraction (XRD). During heating, the acid-activated geopolymers exhibited similar behaviour to alkali-activated geopolymers. All of the samples showed endothermic peaks up to 300 °C due to water evaporation, while the samples with mine tailings showed two significant exothermic peaks above 400 °C due to oxidation reactions. The phase analysis confirmed the dissolution of the aluminosilicate sources in the presence of H_3PO_4 by significant changes in the XRD patterns of the raw materials and by the broadening of the peaks because of typically amorphous silicophosphate (Si–P), aluminophosphate (Al–P) or silico-alumino-phosphate (Si–Al–P) formation. The phases resulted from geopolymerisation are berlinite ($AlPO_4$), brushite ($CaHPO_4 \cdot 2H_2O$), anhydrite ($CaSO_4$) or ettringite as AFt and AFm phases.

Keywords: phosphate-based geopolymers; thermal behaviour; thermogravimetry-differential thermal analysis; phase analysis

1. Introduction

Globally there is a continuing concern for the research and development of green materials for civil engineering, in particular for the replacement of those based on Ordinary Portland Cement (OPC) [1,2]. Thus, it is essential to improve both the conventional technologies for the exploitation of natural resources and the technologies for obtaining concrete based on OPC. The purpose of these changes is to convert existing cement plants into facilities suitable for the manufacture of green concrete, such as geopolymers [3]. Geopolymers are eco-friendly materials created through the geopolymerisation chemical reaction which occurs after mixing an aluminosilicate source with an aqueous solution. This multiple-stage reaction consists of (1) dissolution of the aluminosilicate source in acidic medium, (2) Si–O–Al network and gel formation and (3) formation of a geopolymer structure [4].

A comprehensive review of alkali-activated geopolymers was conducted by Almutairi et al. [5], according to their study, multiple aluminosilicate wastes, such as red mud, ground granulated blast slag, rice husk ash, fly ash and glass powder can be used as raw materials for the alkali-activated cement. Moreover, the resulting material will exhibit high chemical stability and an 80% reduction in CO_2 emissions compared with their main competitor (OPC-based materials). In another study [6], another aluminosilicate, volcanic pumice dust, was mixed with OPC or cement kiln dust to obtain high-performance geopolymers. In terms of compressive strength, the optimum mixture was that of volcanic pumice dust and cement kiln dust, while the porosity and water absorption at a curing age of 28 days were almost the same for all mixtures. In another study, Arpajirakul S. et al. [7] evaluated the effect of urea-Ca^{2+} addition on soft soil stabilisation by microbially-induced calcite preparation technique. According to their study, the compressive strength can be increased up to 18.50% at an optimum urea addition rate of 7.5 mmol/h.

The interest in geopolymer development has been boosted by their versatility, association with a multitude of aluminosilicate sources, production parameters and potential activators [8]. However, the multitude of production process parameters as well as the variety of sources of raw materials, slows down the market transition and industrial development of these materials. Another barrier in their industrial development is the high price of activation solutions, especially sodium silicate [9]. Therefore, recent studies in the field have been focused on the discovery and development of new activation solutions which are cheap and have a low environmental impact. Thus, phosphoric acid has become the main competitor of alkaline activators used to date in geopolymers [10,11]. However, the replacement of alkaline solutions with phosphoric acid has led to a major change in the formation mechanism of geopolymers; therefore, the obtained materials may have chemical, structural, thermal and mechanical characteristics which are different from those of alkaline-activated geopolymers. Accordingly, to reach the industrial development potential of phosphoric acid-based geopolymers, it is necessary to study all of their characteristics.

Wang Y.S. et al. [12] successfully synthesized phosphate-based geopolymers using silica fume mixed with metakaolin as an aluminosilicate source and monoaluminium phosphate for the activator solution. According to their study, at an Al/P ratio of one, the optimum compressive strength was obtained due to the formation of $SiO_2 \cdot Al_2O_3 \cdot P_2O_5 \cdot nH_2O$ and $AlH_3(PO_4)_2 \cdot 3H_2O$. Moreover, these main reaction products of geopolymerisation will be converted to berlinite when exposed to high temperatures. In another study [13], the authors evaluated the thermal behaviour and water resistance of geopolymers obtained from a mixture of metakaolin with phosphoric acid (10 M) as an activator. According to their study, the geopolymers with a Si/P ratio of 0.82, cured at 60 °C for 24 hours and aged in air for 28 days will exhibit a 50% higher compressive strength than those aged in water. The phenomenon was correlated with the hydrolysis of Si–O–P bonds during ageing. However, even in these curing conditions, the obtained geopolymers showed comparable mechanical properties with Ordinary Portland Cement (OPC) materials; therefore, these materials can be used in civil engineering applications.

Bai C. et al. [14] synthesized foams with a homogeneous microstructure by mixing metakaolin with phosphoric acid (85 wt.%) and water at a molar ratio of $H_3PO_4/Al_2O_3 = 1.8$, $SiO_2/Al_2O_3 = 2.4$ and $H_2O/H_3PO_4 = 6.7$, respectively. The geopolymer foam produced had a total open porosity as high as 76.8 vol%, and compressive strength of 0.64 MPa. In addition, when exposed to high temperature a 6.4% shrinkage and close to 90% weight loss was observed. The shrinkage was associated with the mesopores decreasing and densification due to dehydration of the structure, yet they concluded that the obtained material is suitable to replace the conventional highly porous materials in industrial applications.

Zribi M. et al. [15] evaluated the structure of phosphate-based geopolymers by combining four different techniques (magic angle spinning nuclear magnetic resonance (MAS-NMR), Fourier transform infrared spectroscopy (FTIR), X-ray diffraction powder (XRD) and scanning electron microscopy (SEM)). According to their study, the material obtained by mixing metakaolin with phosphoric acid at an Al/P ratio of one and cured at 60 °C for 24 h exhibited an amorphous structure composed of an aluminium phosphate geopolymeric network dispersed in a base created from Si–O–Si, Si–O–Al and Si–O–P units.

According to Djobo J.N.Y. et al. [16], the phosphate geopolymers are obtained due to the reaction of Al^{3+} ions with the proton H^+ and H_2PO_4—species resulting from the deprotonation of commercial H_3PO_4. Moreover, when the aluminosilicate source is rich in different types of metal compounds besides Al^{3+}, the obtaining reaction involves Fe^{2+}/Fe^{3+}, Ca^{2+} and Mg^{2+} ions dissolution. The resulting ions will react with the phosphate species in the following order: $Ca^{2+} = Mg^{2+} > Al^{3+}$, Fe^{2+}/Fe^{3+}, resulting in calcium phosphate, magnesium phosphate aluminophosphate, silico-aluminophosphate, silicophosphate and iron phosphate phases.

One of the reasons why the interest in the development of geopolymers activated with acid solutions based on potassium is increasing, is the superior compressive strength of the resulting materials, compared to those which are alkaline-activated [17]. Moreover, the economic aspects also tip the scales in favour of the acid-activated ones, this aspect is due to the fact that the sodium hydroxide solutions do not present a sufficiently high geopolymerization potential when used alone. To develop suitable mechanical properties for civil engineering applications, the alkaline activator most often consists of a mixture of sodium hydroxide solution and sodium silicate. The sodium silicate is in a larger amount than the sodium hydroxide solution. This aspect greatly influences the price of the final product due to the high purchase price of sodium silicate. Given that the industrial development of a product is limited by its price, obtaining a cheap alternative to geopolymers and their development is essential for the transition from conventional materials (OPC-based concrete) to sustainable materials (geopolymer concrete).

So far it has been observed that in the case of potassium-based geopolymers, the compressive strength developed over time is higher than that of sodium-based geopolymers.

The effect of the Si–Al ratio on the mechanical and structural characteristics of geopolymers has been reported in a multitude of studies [18,19]. According to these studies, it was observed that the best properties are obtained for Si/Al between 1.5 and 1.9. It has also been observed that another chemical ratio with a significant influence on geopolymers is that between Na/Al [20]. Its optimal value is close to one, however, a decrease in the ratio leads to the production of a structure with high porosity, and its increase results in an improvement of the compressive strength. In the case of acid-activated geopolymers, the chemical ratio of primary interest is that between P and Al. According to previous studies [12,21], it influences the mechanical properties of geopolymers in a similar way to the ratio of Na to Al, but to our knowledge, no study has evaluated the thermal behaviour on ambient cured geopolymers activated with phosphoric acid in different P/Al ratios.

Multiple previous publications have focused on the effects of phosphoric acid activation on the mechanical characteristics of geopolymers. Moreover, most of the studies use metakaolin as a raw material because it has a simple chemical composition compared with other precursors. Therefore, there is a lack of information on the thermal behaviour and phase transition of phosphate-based geopolymers, especially on those which use coal ash or other by-products as aluminosilicate sources. This study aims to evaluate the influence of curing parameters and phosphate acid concentration on the thermal behaviour and phase transition of coal ash-based geopolymers with mine tailings content. As presented in [22,23], the minerals containing sulfides, such as pyrite, pyrrhotite and arsenopyrite can be oxidized when mixed with water or oxygen. Therefore, harmful metals can be released into the environment. Another advantage of using mine tailings in geopolymer development is related to their capacity for immobilizing harmful species during the hardening process. During the formation of ettringite, different elements can be replaced with others that have a similar radius and oxidation state. Accordingly, multiple metals, the harmful one included, will be encapsulated into the structure of the geopolymer, as follows: Ca^{2+} will be replaced by Mg^{2+}, Co^{2+} and Zn^{2+}; Al^{3+} will be replaced by Cr^{3+}, Sr^{3+} and Fe^{3+}; and SO_4^{2-} will be replaced by oxyanions of Cr and As.

Accordingly, this study investigates the effect of room temperature hardening on phase transition during geopolymerisation in five types of geopolymers obtained by mixing three types of raw materials (coal ash, metakaolin and mine tailings). Moreover, the thermal behaviour of these materials was analysed up to 600 °C.

2. Materials and Methods

The obtained geopolymer was manufactured by mixing the raw material with a commercially available acid solution of o-phosphate (H_3PO_4) with 85 wt.% solid content. The solid to liquid ratio was optimized to assure an Al/P ratio of 1, for both types of geopolymers.

2.1. Materials

2.1.1. Coal Ash (CA)

In this study, local coal ash from CET II Holboca, Iasi, Romania was collected and processed. To ensure experimental repeatability, the collected powder was firstly dried in a chamber with a static atmosphere, i.e., without ventilation, to avoid fine particle removal, and was secondly sifted to remove the particles with a diameter above 100 μm. The drying method is common and is presented in the literature [20], while the sifting method has been presented in a previous study [4]. The coal ash used in the study had a particle size distribution of 3.4 μm (d50) and a specific surface area of 1.6 m^2/g as determined by a Coulter LS 200 laser scattering particle size distribution analyser (Beckman Coulter Inc., Pasadena, CA, USA). In addition, the bulk density of coal ash was 2.16 ± 0.01 g/cm^3 evaluated with a Le Chatelier densimeter (Recherches & Realisations Remy, Montauban, France).

According to XRF analysis, the coal ash used in this study belongs to class F fly ashes. From a microstructural point of view (Figure 1), the aluminosilicate waste collected is a mixture of fly ash and bottom particles, as both spherical and irregular porous particles can be seen.

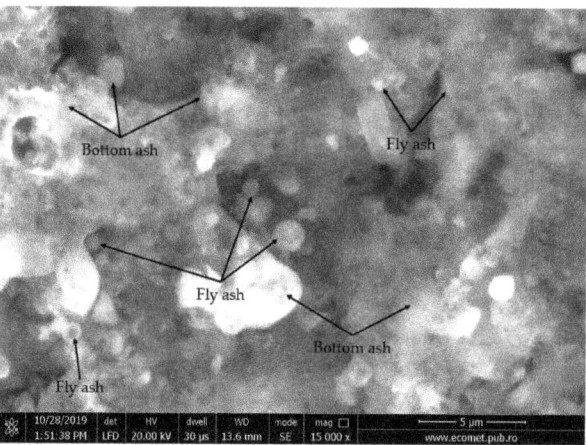

Figure 1. Coal ash morphology showing bottom and fly ash particles.

Coal ash is a mineral residue resulting from coal combustion in thermal power plants, it has small particles in the range of 0.01 to 300 μm. The chemical composition of coal is critical since it influences the ultimate properties of geopolymers. Silicon dioxide, aluminium oxide, iron oxide, and calcium oxide are the chemical compounds with the highest concentrations in their composition. Its chemical composition, on the other hand, changes depending on the type of coal and the furnace operation.

2.1.2. Metakaolin (MK)

The metakaolin used in this research was produced by the calcination of commercially available kaolin clay [12] at low temperatures (heated up to 700 °C at a rate of 10 °C/min and maintained for 30 min.). As a result, the starting material was converted into an aluminosilicate source with strong pozzolanic activity. According to the XRF analysis, the metakaolin contains a high concentration of silicon and aluminium oxides (Table 1). The metakaolin used in the study had a particle size distribution of 9.2 µm (d50), a specific surface area of 16.8 m^2/g and a bulk density of 0.22 ± 0.01 g/cm^3.

Table 1. Oxide composition of raw materials, coal ash (CA), metakaolin (MK) and mine tailings (MT).

Sample	Oxide	SiO$_2$	Al$_2$O$_3$	Fe$_x$O$_y$	CaO	K$_2$O	MgO	TiO$_2$	CuO	Na$_2$O	P$_2$O$_5$	SO$_3$	Oth.*	L.O.I.**
CA	%, wt.	46.1	27.6	9.8	6.2	2.3	1.9	1.3	0.0	0.6	0.4	-	0.3	3.5
MK	%, wt.	52.1	42.5	1.2	0.7	0.5	0.2	0.9	0.0	-	0.2	-	0.4	1.3
MT	%, wt.	16.2	2.6	38.9	0.4	0.6	-	0.2	0.5	-	0.3	11.4	0.9	28.1

Oth.*—oxides in a concentration lower than 0.1% (traces of S, Cl, Cr, Zr, Ni, Sr, Zn and Cu). L.O.I.**—Loss on ignition.

2.1.3. Mine Tailings (MT)

In this study mine tailings from barite mine were collected and subjected to calcination to remove water, organic matter and to improve reactivity. The parameters of the calcination process were the same as those used for the kaolin calcination. However, according to the oxide chemical composition analysis, the MT used in this study exhibit a high content of Fe oxides and a much lower content of SiO$_2$ and Al$_2$O$_3$ than CA or MK. The mine tailings used in the study had a particle size distribution of 46 µm (d50), a specific surface area of 0.33 m^2/g and a bulk density of 2.83 ± 0.01 g/cm^3.

2.1.4. Activator Solution

As an activator, an acid solution of o-phosphate (H$_3$PO$_4$) with 85 wt.% was diluted in distilled water to obtain a corresponding activation solution for an H$_3$PO$_4$/Al$_2$O$_3$ ratio of 1:1. According to previous studies [12,13], this ratio exhibits the best performance in terms of the compressive strength of obtained geopolymers.

Using the materials presented above, five types of geopolymers were synthetized. Table 2 shows the experimental mixes used in this study. The samples were designed to observe the influence of 50% by the mass addition of each raw material onto another. Accordingly, geopolymers with 100% of the solid component of coal ash (CA–geo) and metakaolin (MK–geo) were obtained, while because of the low geopolymerisation potential of mine tailings, a geopolymer with 100% mine tailing in the solid component could not be synthesized (the sample could not be cured at room temperature). In order to evaluate the addition of each component, three types of blended geopolymers have been designed, one from a mixture of metakaolin and coal ash (CAMK), one from a mixture of mine tailings and coal ash (MTCA) and another one with mine tailings and metakaolin (MTMK). Accordingly, these samples were obtained from a mixture of two of the raw materials at a 1 to 1 weight ratio.

Table 2. Mix design and parameters.

Sample Code	Coal Ash, wt.%	Metakaolin, wt.%	Mine Tailings, wt.%	Al/P Molar Ratio	Curing, °C
CA-geo	100	-	-	1	22 ± 2
MK-geo	-	100	-	1	22 ± 2
CAMK	50	50	-	1	22 ± 2
MTCA	50	-	50	1	22 ± 2
MTMK	-	50	50	1	22 ± 2

All the mixtures have been activated with a phosphorous solution at a solid/liquid ratio of 0.9 (to assure workability). Moreover, to assure better homogeneity, the geopolymers with multiple aluminosilicate sources were mixed in a dry state for 2 min, and for 5 min after the liquid addition, according to the procedure presented in [18,19]. Accordingly, the mixtures were poured into 20 × 20 × 20 mm^3 moulds, covered with a thin layer of plastic (to avoid moisture loss) and cured at room temperature (22 ± 2 °C) for 28 days before testing. The schematic representation of samples obtained is presented in Figure 2.

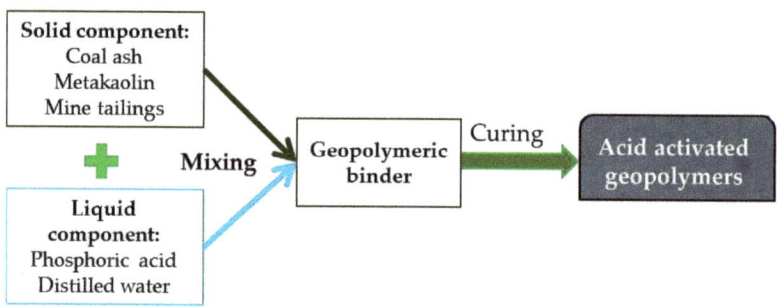

Figure 2. Process flow diagram of the obtained samples.

2.2. Methods

The thermal behaviour of the acid-activated geopolymers obtained in this study was evaluated through differential thermal analysis (DTA) combined with thermogravimetric analysis (TGA). Moreover, X-ray diffraction (XRD) was involved to analyse the phase transition following the geopolymerisation reaction.

2.2.1. X-ray Diffraction

The mineralogical composition of the acid-activated geopolymers was analysed by XRD using X'Pert Pro MPD equipment (Malvern Panalytical Ltd., Eindhoven, The Netherlands). The CuKα radiation was recorded in the range of 5° and 60° 2θ using a single channel detector. The radiation was collected at a step size of 0.013°, while the copper X-ray tube was operated at a 40 mA current intensity and a 45 kV voltage.

The XRD analysis was performed on samples in a powder form, which were grounded after the curing period.

2.2.2. Simultaneous Thermal Analysis

The transformations of phases and weight evolution was evaluated in the 25–600 °C temperature range using STA PT–1600 equipment (Linseis, Selb, Germany). The heating was performed in a static air atmosphere on samples lighter than 50 mg at a rate of 10 °C/min.

3. Results and Discussion

3.1. Phase Transition Analysis

The XRD patterns were evaluated using Highscore software v5.1, while the identification of the peaks was conducted using the PANalytical 2021 database. The peak search was realized considering only those with a minimum significance of 5.00, and a peak base with Gonio of 2.00, through the minimum second derivative method.

The XRD analysis of the CA samples (Figure 3) confirms the presence of multiple phases, such as quartz (96-900-9667), calcite (96-900-0967), anorthite (96-900-0362) and hematite-proto (96-900-2163). The detected hematite-proto contains Fe, H and O. Moreover, the XRD pattern of the metakaolin showed a high content of a typical amorphous structure. Compared to the CA pattern, the one specific to MK showed multiple peaks with high intensity. The most significant peak is positioned close to 25° (2θ) and corresponds to kaolinite (96-900-9231), the second peak, positioned close to 12.5° (2θ) corresponds to the same phase, while the peak close to 26.6° (2θ) corresponds to quartz (96-900-9667).

Close to 35° (2θ) multiple peaks confirm the presence of muscovite (96-901-4938), while chloritoid (96-900-5444) was detected with the most significant peak close to 20° (2θ). The detected kaolinite contains Al, Si and O, the detected quartz contains Si and O, the detected muscovite contains K, Al, Si and O and the detected chloritoid contains Al, Fe, Mg, Si and O. As can be seen from Figure 3, in the case of MK, almost all of the identified peaks confirm the presence of kaolinite, quartz, muscovite and chloritoid.

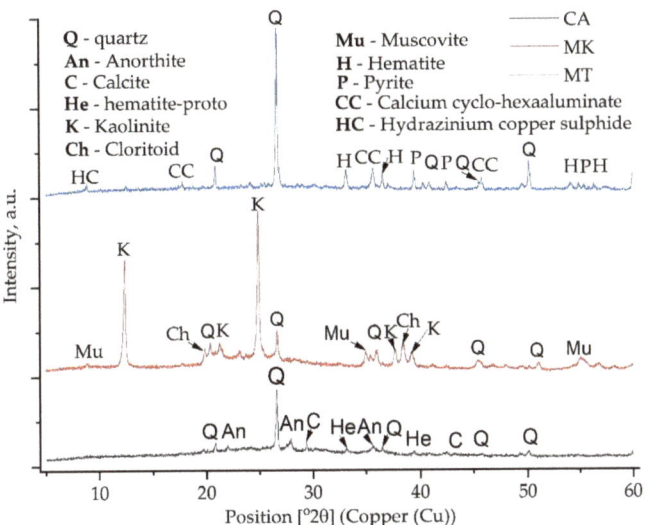

Figure 3. The XRD pattern of the raw materials used in this study.

In the case of MT, the XRD pattern confirmed the presence of multiple phases rich in Fe or Si. Accordingly, quartz (96-901-5023), hematite (96-901-5965), pyrite (96-900-0595), calcium cyclo-hexaaluminate (96-100-0040) and hydrazinium copper sulfide (96-430-7636) were detected.

Due to the reaction between the activator and the aluminosilicate source, the dissolution of aluminates and silicates occurred, resulting in disordered and amorphous silicophosphate (Si–P), aluminophosphate (Al–P) or silico-alumino-phosphate (Si–Al–P) gels. However, the differences between the XRD spectra of the raw materials and the spectra of the acid-activated geopolymers are low because the Si–P, Al–P and Si–Al–P phases are typically amorphous (Figure 4). The existence of quartz and kaolinite at the same position was evident in previous studies [24,25]. Moreover, it can be observed that, due to the geopolymerisation reaction, multiple peaks disappeared, such as the peak around 22° (2θ) and the one around 29.5° (2θ). In addition, the intensity of the peaks decreased significantly. However, one new peak can be observed around 40.5° (2θ).

The calcite disappearance from the CA after the reaction with the phosphoric acid, i.e., the peak around 28° (2θ), disappeared due to geopolymerisation. This could be attributed to brushite formation or to the following chemical reaction (Equation (1)):

$$3Ca^{2+} + 2PO_4^{3-} + H_2O = 3CaO - P_2O_5 - H_2O \text{ (C-P-H gel)} \qquad (1)$$

In a phosphate acid environment, the Al oxides will be dissolute after the Ca compounds; therefore, the C–P–H gel will be formed before Al-P [10]. Therefore, by comparing the samples with MK against those without MK, it can be stated that the samples with higher Ca content will exhibit a lower setting time, due to the solubility differences between the divalent metals and the trivalent one.

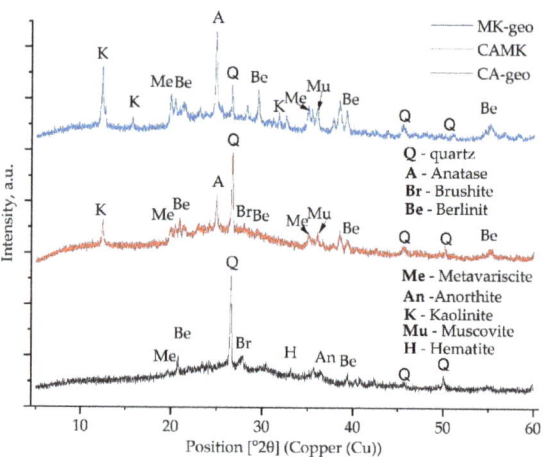

Figure 4. The XRD pattern of the acid-activated geopolymers without mine tailings.

The broad peak at 27° (2θ) indicates the formation of the berlinite phase (AlPO$_4$) which has a similar XRD pattern to quartz and will confer high mechanical properties to the final geopolymer [10,26]. Moreover, by comparing the XRD pattern of the MK with that of MK–geo, it can be observed that all of the peaks between at 26° (2θ) and 35° (2θ) appeared due to the reaction between the acid and the raw material. Furthermore, the intensity of all the peaks specific to kaolinite have been reduced.

In addition, as observed in [27,28], multiple Al–O–P phases, such as phosphotridymite, phosphocristobalite, aluminium phosphate or aluminium phosphate hydroxide appear, but they overlap with the patterns of other phases. However, a hydrated form of aluminium phosphate confirmed as metavariscite was detected [29].

In the samples with mine tailings (Figure 5), the characteristic diffraction of anhydrite (CaSO$_4$) disappeared, which indicates that the structures of CaSO$_4$ were dissolute due to the phosphoric acid presence. Furthermore, the resulting Ca contributes to the formation of brushite or amorphous calcium phosphate [30] and as crystalline ettringite in Aft and Afm monosulfate [31,32]. In addition, the peak around 12° (2θ), which corresponds to hydrazinium copper sulfide, increased significantly in the MTCA geopolymer.

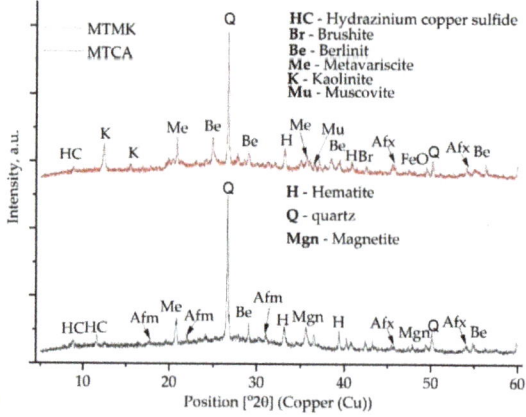

Figure 5. The XRD pattern of the acid-activated geopolymers with mine tailings.

3.2. Thermal Behaviour Analysis

The TG-DTA analysis was used in this study to analyse the thermal behaviour of five types of geopolymers. Accordingly, it was observed that the addition of another aluminosilicate source influenced the metakaolin or coal ash-based geopolymers. By evaluating the heat flux and the mass evolution during the heating of the sample, the volatile compounds were eliminated, while the transition of different phases were observed. Moreover, by overlapping the TGA with the DTA curve, the mass loss or gain at specific temperatures could be correlated with the heat flux, to confirm the presence of a specific phase and its amount.

The samples analysed in this study showed multiple peaks on the DTA curves, which have been correlated with the water evaporation and oxidation processes. As can be seen in Figure 6, the samples without mine tailings exhibited similar behaviour, except for the samples CAMK which showed an endothermic peak around 220–260 °C. Moreover, the samples with mine tailings exhibited important oxidation reactions above 400 °C.

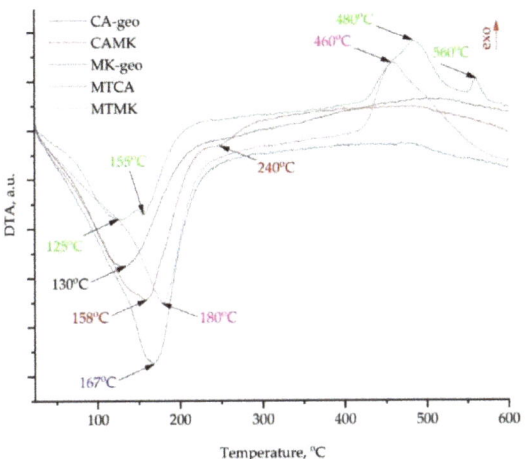

Figure 6. DTA plots of the studied geopolymers.

Moreover, as can be seen in Figure 6, the acid-activated geopolymers exhibit similar behaviour to the alkali-activated ones. Accordingly, the DTA curve of the coal ash-based geopolymer (CA–geo) shows multiple peaks, while the most significant one is the endothermic peak around 131 °C. This peak is correlated to the removal of the water molecules, which can exist as a free or chemical bond with the components from the geopolymers' structure. Therefore, the use of phosphoric acid as an activator will lead to the development of a porous structure with zeolitic channels, which keep the water at temperatures much higher than evaporation. Firstly, the hygroscopic water is removed until 120 °C, this exists in the structure of the geopolymers due to their hygroscopicity. Secondly, up to 300 °C, the physically strong bond molecules of water are removed as follows: (i) up to 200 °C the crystallization water bounded in the structure during the crystal formation from the aqueous solution is created by mixing the aluminosilicate source with the activator; (ii) during heating between the 180 and 300 °C temperature range, the molecules from intracrystalline type or network type hydrogels are removed. In approximately the same temperature range, zeolitic water will be removed from the channels. The behaviour is similar also for MK-geo and CAMK, the peak temperature being changed in accordance with the amount of water and the size of the sample. Moreover, the metakaolin-based sample showed the largest peak as it has a higher amount of gel pores and zeolitic channels compared with the coal ash-based ones. Accordingly, the minimum value of the first peak was moved to higher temperatures, close to 167 °C. The blended (mix of two aluminosilicate sources, coal ash and metakaolin) geopolymer, CAMK sample, showed a broader peak into the water removal temperature range. This change could be correlated with the influence of the thermal behaviour on water removal at a low temperature from large pores specific to coal ash-based geopolymers overlapping next to water removal from small pores specific to metakaolin-based geopolymers.

By comparing the broadening of the first peak, it can be stated that the metakaolin-based geopolymers contain a higher quantity of water in small pores compared with those based on coal ash. These results are fully in line with the pore size distribution in geopolymers evaluated by NMR in previous studies, where the authors from [33] discovered a large amount of pores around 2.5 nm in metakaolin-based geopolymers, while in [34], the experiments on coal ash geopolymers showed that the first peak, on the relative pore size distribution, was positioned at higher relaxation time, i.e., the pores have a higher diameter.

As the samples are heated up above 300 °C, the chemically bound water will be removed. Accordingly, the fluctuations from the DTA curves correspond to the decomposition of acids, basics and neutral groups, which are formed between a metal and the OH groups. Considering the chemical composition of the raw materials, those can be Fe(III), Ti(IV), Si(IV), Na, K, P, Mg(II), Ca(II) and Al(III).

In the case of the blended geopolymer, the Na and Ca addition from the coal ash significantly affected the amount of water retained in Ca- or Al–silicate–hydrate channels and pores. This phenomenon can be due to the high concentration of Al brought into the system by the metakaolin. In other words, by mixing these two raw materials, the addition of these three elements (Na, Ca and Al) will have a significant impact on the condensation of C–A–S–H, C–S–H and N–A–S–H and, consequently, on the three-dimensional aluminosilicate network [35]. Accordingly, in the case of the CAMK sample, the water evaporation reaction from sodium–aluminosilicate hydrogel (N–A–S–H gel) is much more dominant in the hydrogels range (close to 185 °C), while the endothermic peak related to water removal from C–S–H and C–A–S–H structures can be observed around 240 °C. Moreover, in the case of acid-activated geopolymer, structures such as –Si–O–Al–O–P– gel will lose the water from the network in the same temperature range [10].

Above this temperature, no significant peak can be identified in the case of coal ash- or metakaolin-based geopolymers. However, when mine tailings are involved in the mixture, the thermal behaviour is significantly changed. Consequently, the separation of the endothermic peaks in the range of water evaporation is much clearer. Accordingly, the MTCA curve shows two minimum points, the first being close to 125 °C, and the second being around 155 °C. However, in the case of the metakaolin-mine tailings blended geopolymer, the DTA curve shows only one peak in this temperature range, which has the minimum point at 180 °C. Therefore, the MK presence contributed to a zeolite-like structure formation. Another significant difference between the samples with and without mine tailings is the appearance of the endothermic peaks above 400 °C.

By comparing the DSC curves of the studied samples with those of alkali-activated geopolymers presented in [36], it can be observed that in the 20–300 °C temperature range, the thermal behaviour is almost the same. Both types of materials exhibit water evaporation from large pores and zeolitic channels, followed by its removal from hydrogels.

The first exothermic peak with the maximum point around 460 °C for the MTMK sample, and around 480 °C for the MKCA sample, is the conversion of magnetite to hematite [37]. Considering the provenience of the mine tailings (dams), in the same temperature range, the exothermic reactions can be associated with humic acid disintegration [38]. Moreover, by heating sulfide ores such as pyrite in air atmospheres, the following chemical reaction can occur:

$$4FeS_2(s) + 11O_2(g) \rightarrow 2Fe_2O_3(s) + 8SO_2(g) \tag{2}$$

The second exothermic peak appears due to the recrystallization of precipitated sulphate apatite at a lower temperature as a result of the transition of sulphate ions [39].

The weight loss during heating is related to the evaporation of water from the highly porous structure, which includes C–S–H and C–A–S–H formations that retain the activator in a liquid state even after multiple days of ageing. At higher temperatures, the weight loss is due to the decomposition of portlandite and other phases. Accordingly, the DTA curves exhibit endothermic peaks at corresponding temperatures, except for the samples with mine tailings addition which show high exothermic peaks in the range of 420–520 °C and 560–590 °C.

According to the TGA plots (Figure 7), up to 300 °C, the mass loss of the CA–geo sample was 12.7 wt.%, the mass loss of the CAMK sample was 16.2 wt.%, the mass loss of MK–geo was 18.7 wt.%, the mass loss of MTCA was 10.5 wt.% and the mass loss of MTMK was 14.4 wt.%. The mass change behaviour had the same trend as the first peak from the DTA curves, i.e., the samples with high water content showed a broader peak. Moreover, by correlating the type of raw material with the mass loss, it can be stated that the raw materials influence the water retention in the following order: metakaolin < coal ash < mine tailings. Therefore, this confirms that the samples with metakaolin have lower pores and channels which retain water at high temperatures.

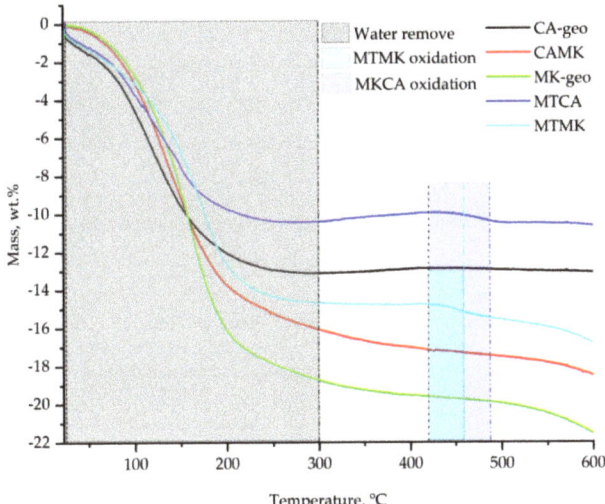

Figure 7. TGA plots of the studied geopolymers.

In the 300–600 °C temperature range, the mass loss was very low for the CA–geo and MTCA samples which showed an extra decrease of 0.2 wt.%. However, in the same temperature range, the MTMK sample showed an extra mass loss of 2.1 wt.%, CAMK showed a loss of 2.3 wt.% and MK–geo showed a loss of 2.8 wt.%. Starting around 420 °C, with a maximum peak around 460 °C for the MTMK sample and 480 °C for MKCA, an exothermic reaction occurred. The weight gain corresponding to this oxidizing reaction was close to 0.5 wt.% for MTMK and 0 wt.% for MTCA. Accordingly, due to oxidation, only a small amount of oxygen remained in the sample, while other chemical elements evacuated.

4. Conclusions

This study examined the thermal behaviour and mineralogical composition of acid-activated geopolymers. Coal ash, metakaolin, mine tailings and different mixtures of these three aluminosilicate sources were used to synthesize silico-aluminophosphate geopolymers. Based on the experimental results reported, the following conclusions are drawn:

- In the 20–300 °C temperature range, the geopolymers obtained with H_3PO_4 acid exhibited similar thermal behaviour to those activated with a mix of NaOH and Na_2SiO_3,

- In the 400–600 °C temperature range, the geopolymers with mine tailings addition exhibited exothermic reactions, while those without mine tailings addition did not show significant phase transition,
- Up to 600 °C, the total mass loss of the Ca–geo was 12.9 wt.%, 21.5 wt.% for the MK and 18.5 wt.% for CAMK. The MT addition decreased the mass loss at 10.5 wt.% when mixed with CA and 14.4 wt.% when mixed with MK,
- The XRD analysis confirmed the formation of the ettringite phase in the geopolymers with MT addition and berlinite, brushite or metavariscite in those based on coal ash or metakaolin.

Author Contributions: Conceptualization, writing original draft and investigation, D.D.B.N.; writing original draft, formal analysis, C.B.; project administration and scientific supervision, P.V.; data curation, validation and writing—reviewing and editing, A.V.S.; methodology, resources, formal analysis and investigation, D.P.B.N. All authors have read and agreed to the published version of the manuscript.

Funding: This work was supported by the publications grant of the Gheorghe Asachi Technical University of Iași—TUIASI—Romania, project number GI/P4/2021.

Institutional Review Board Statement: Not applicable.

Informed Consent Statement: Not applicable.

Data Availability Statement: The study did not report any data.

Conflicts of Interest: The authors declare no conflict of interest.

References

1. Habert, G.; Ouellet-Plamondon, C. Recent update on the environmental impact of geopolymers. *RILEM Tech. Lett.* **2016**, *1*, 17. [CrossRef]
2. Krishna, R.S.; Mishra, J.; Zribi, M.; Adeniyi, F.; Saha, S.; Baklouti, S.; Shaikh, F.U.A.; Gökçe, H.S. A review on developments of environmentally friendly geopolymer technology. *Materialia* **2021**, *20*, 101212. [CrossRef]
3. Yahya, Z.; Abdullah, M.M.A.B.; Mohd Ramli, N.; Burduhos-Nergis, D.D.; Abd Razak, R. Influence of Kaolin in Fly Ash Based Geopolymer Concrete: Destructive and Non-Destructive Testing. *IOP Conf. Ser. Mater. Sci. Eng.* **2018**, *374*, 012068. [CrossRef]
4. Burduhos Nergis, D.D.; Vizureanu, P.; Corbu, O. Synthesis and characteristics of local fly ash based geopolymers mixed with natural aggregates. *Rev. Chim.* **2019**, *70*, 1262–1267. [CrossRef]
5. Almutairi, A.L.; Tayeh, B.A.; Adesina, A.; Isleem, H.F.; Zeyad, A.M. Potential applications of geopolymer concrete in construction: A review. *Case Stud. Constr. Mater.* **2021**, *15*, e00733. [CrossRef]
6. Zeyad, A.M.; Magbool, H.M.; Tayeh, B.A.; Garcez de Azevedo, A.R.; Abutaleb, A.; Hussain, Q. Production of geopolymer concrete by utilizing volcanic pumice dust. *Case Stud. Constr. Mater.* **2022**, *16*, e00802. [CrossRef]
7. Arpajirakul, S.; Pungrasmi, W.; Likitlersuang, S. Efficiency of microbially-induced calcite precipitation in natural clays for ground improvement. *Constr. Build. Mater.* **2021**, *282*, 122722. [CrossRef]
8. Zhang, X.; Bai, C.; Qiao, Y.; Wang, X.; Jia, D.; Li, H.; Colombo, P. Porous geopolymer composites: A review. *Compos. Part A Appl. Sci. Manuf.* **2021**, *150*, 106629. [CrossRef]
9. Roy, S.; Adhikari, G.R.; Gupta, R.N. Use of gold mill tailings in making bricks: A feasibility study. *Waste Manag. Res.* **2007**, *25*, 475–482. [CrossRef] [PubMed]
10. Guo, H.; Zhang, B.; Deng, L.; Yuan, P.; Li, M.; Wang, Q. Preparation of high-performance silico-aluminophosphate geopolymers using fly ash and metakaolin as raw materials. *Appl. Clay Sci.* **2021**, *204*, 106019. [CrossRef]
11. Louati, S.; Baklouti, S.; Samet, B. Acid based geopolymerization kinetics: Effect of clay particle size. *Appl. Clay Sci.* **2016**, *132–133*, 571–578. [CrossRef]
12. Wang, Y.S.; Dai, J.G.; Ding, Z.; Xu, W.T. Phosphate-based geopolymer: Formation mechanism and thermal stability. *Mater. Lett.* **2017**, *190*, 209–212. [CrossRef]
13. Bewa, C.N.; Tchakouté, H.K.; Fotio, D.; Rüscher, C.H.; Kamseu, E.; Leonelli, C. Water resistance and thermal behavior of metakaolin-phosphate-based geopolymer cements. *J. Asian Ceram. Soc.* **2018**, *6*, 271–283. [CrossRef]
14. Bai, C.; Conte, A.; Colombo, P. Open-cell phosphate-based geopolymer foams by frothing. *Mater. Lett.* **2017**, *188*, 379–382. [CrossRef]
15. Zribi, M.; Samet, B.; Baklouti, S. Structural Characterization of Phosphate-Based Geopolymer. *Lect. Notes Mech. Eng.* **2020**, 36–42. [CrossRef]
16. Djobo, J.N.Y.; Yanou Nkwaju, R. Preparation of acid aluminum phosphate solutions for metakaolin phosphate geopolymer binder. *RSC Adv.* **2021**, *11*, 32258–32268. [CrossRef]

17. Xu, W.; Dai, J.G.; Ding, Z.; Wang, Y. Polyphosphate-modified calcium aluminate cement under normal and elevated temperatures: Phase evolution, microstructure, and mechanical properties. *Ceram. Int.* **2017**, *43*, 15525–15536. [CrossRef]
18. Zhang, F.; Zhang, L.; Liu, M.; Mu, C.; Liang, Y.N.; Hu, X. Role of alkali cation in compressive strength of metakaolin based geopolymers. *Ceram. Int.* **2017**, *43*, 3811–3817. [CrossRef]
19. Norkhairunnisa, M.; Fariz, M.N.M. Geopolymer: A review on physical properties of inorganic aluminosilicate coating materials. *Mater. Sci. Forum* **2015**, *803*, 367–373. [CrossRef]
20. Siyal, A.A.; Azizli, K.A.; Man, Z.; Ullah, H. Effects of Parameters on the Setting Time of Fly Ash Based Geopolymers Using Taguchi Method. *Procedia Eng.* **2016**, *148*, 302–307. [CrossRef]
21. Louati, S.; Baklouti, S.; Samet, B. Geopolymers Based on Phosphoric Acid and Illito-Kaolinitic Clay. *Adv. Mater. Sci. Eng.* **2016**, *2016*, 2359759. [CrossRef]
22. Kiventerä, J.; Piekkari, K.; Isteri, V.; Ohenoja, K.; Tanskanen, P.; Illikainen, M. Solidification/stabilization of gold mine tailings using calcium sulfoaluminate-belite cement. *J. Clean. Prod.* **2019**, *239*, 118008. [CrossRef]
23. Lindsay, M.B.J.; Moncur, M.C.; Bain, J.G.; Jambor, J.L.; Ptacek, C.J.; Blowes, D.W. Geochemical and mineralogical aspects of sulfide mine tailings. *Appl. Geochem.* **2015**, *57*, 157–177. [CrossRef]
24. Sargent, P. The development of alkali-activated mixtures for soil stabilisation. In *Handbook of Alkali-Activated Cements, Mortars and Concretes*; Woodhead Publishing: Sawston, UK, 2015; pp. 555–604. [CrossRef]
25. Daou, I.; Lecomte-Nana, G.L.; Tessier-Doyen, N.; Peyratout, C.; Gonon, M.F.; Guinebretiere, R. Probing the Dehydroxylation of Kaolinite and Halloysite by In Situ High Temperature X-ray Diffraction. *Minerals* **2020**, *10*, 480. [CrossRef]
26. Tchakouté, H.K.; Rüscher, C.H.; Kamseu, E.; Andreola, F.; Leonelli, C. Influence of the molar concentration of phosphoric acid solution on the properties of metakaolin-phosphate-based geopolymer cements. *Appl. Clay Sci.* **2017**, *147*, 184–194. [CrossRef]
27. Celerier, H.; Jouin, J.; Mathivet, V.; Tessier-Doyen, N.; Rossignol, S. Composition and properties of phosphoric acid-based geopolymers. *J. Non. Cryst. Solids* **2018**, *493*, 94–98. [CrossRef]
28. Celerier, H.; Jouin, J.; Tessier-Doyen, N.; Rossignol, S. Influence of various metakaolin raw materials on the water and fire resistance of geopolymers prepared in phosphoric acid. *J. Non. Cryst. Solids* **2018**, *500*, 493–501. [CrossRef]
29. Zhang, B.; Guo, H.; Deng, L.; Fan, W.; Yu, T.; Wang, Q. Undehydrated kaolinite as materials for the preparation of geopolymer through phosphoric acid-activation. *Appl. Clay Sci.* **2020**, *199*, 105887. [CrossRef]
30. Hamdi, N.; Ben Messaoud, I.; Srasra, E. Production of geopolymer binders using clay minerals and industrial wastes. *Comptes Rendus Chim.* **2019**, *22*, 220–226. [CrossRef]
31. Ambroise, J.; Péra, J. Immobilization of calcium sulfate contained in demolition waste. *J. Hazard. Mater.* **2008**, *151*, 840–846. [CrossRef] [PubMed]
32. Christensen, A.N.; Jensen, T.R.; Hanson, J.C. Formation of ettringite, $Ca_6Al_2(SO_4)_3(OH)_{12} \cdot 26H_2O$, AFt, and monosulfate, $Ca_4Al_2O_6(SO_4) \cdot 14H_2O$, AFm-14, in hydrothermal hydration of Portland cement and of calcium aluminum oxide—Calcium sulfate dihydrate mixtures studied by in situ synchrotron X-ray powder diffraction. *J. Solid State Chem.* **2004**, *177*, 1944–1951. [CrossRef]
33. Li, J.; Mailhiot, S.; Sreenivasan, H.; Kantola, A.M.; Illikainen, M.; Adesanya, E.; Kriskova, L.; Telkki, V.V.; Kinnunen, P. Curing process and pore structure of metakaolin-based geopolymers: Liquid-state 1H NMR investigation. *Cem. Concr. Res.* **2021**, *143*, 106394. [CrossRef]
34. Burduhos Nergis, D.D.; Vizureanu, P.; Ardelean, I.; Sandu, A.V.; Corbu, O.C.; Matei, E. Revealing the influence of microparticles on geopolymers' synthesis and porosity. *Materials* **2020**, *13*, 3211. [CrossRef]
35. Kalombe, R.M.; Ojumu, V.T.; Eze, C.P.; Nyale, S.M.; Kevern, J.; Petrik, L.F. Fly Ash-Based Geopolymer Building Materials for Green and Sustainable Development. *Materials* **2020**, *13*, 5699. [CrossRef]
36. Nergis, D.D.B.; Al Bakri Abdullah, M.M.; Sandu, A.V.; Vizureanu, P. XRD and TG-DTA study of new alkali activated materials based on fly ash with sand and glass powder. *Materials* **2020**, *13*, 343. [CrossRef] [PubMed]
37. Sandeep Kumar, T.K.; Viswanathan, N.N.; Ahmed, H.; Dahlin, A.; Andersson, C.; Bjorkman, B. Investigation of Magnetite Oxidation Kinetics at the Particle Scale. *Metall. Mater. Trans. B Process Metall. Mater. Process. Sci.* **2019**, *50*, 150–161. [CrossRef]
38. Rustschev, D.; Atanasov, O. Thermal and Group Analysis of Peat. *J. Therm. Anal.* **1983**, *27*, 439–442. [CrossRef]
39. Tõnsuaadu, K.; Gross, K.A.; Pluduma, L.; Veiderma, M. A review on the thermal stability of calcium apatites. *J. Therm. Anal. Calorim.* **2012**, *110*, 647–659. [CrossRef]

Article

Concrete Compressive Strength by Means of Ultrasonic Pulse Velocity and Moduli of Elasticity

Bogdan Bolborea [1,2,*], Cornelia Baera [2,3,4], Sorin Dan [1], Aurelian Gruin [2], Dumitru-Doru Burduhos-Nergis [5,*] and Vasilica Vasile [6]

1. Civil Engineering Faculty, Politehnica University of Timisoara, 300223 Timisoara, Romania; sorin.dan@upt.ro
2. NIRD URBAN-INCERC Timişoara Branch, 300223 Timisoara, Romania; cornelia.baera@incd.ro (C.B.); aurelian.gruin@incd.ro (A.G.)
3. Research Center in Engineering and Management, Politehnica University of Timisoara, 300191 Timisoara, Romania
4. Civil Engineering Faculty, IOSUD-UTCN Doctoral School, Technical University of Cluj-Napoca, 400020 Cluj-Napoca, Romania
5. Faculty of Materials Science and Engineering, Gheorghe Asachi Technical University, 700050 Iasi, Romania
6. NIRD URBAN-INCERC Bucharest Branch, 021652 Bucharest, Romania; valivasile67@yahoo.com
* Correspondence: bogdan.bolborea@student.upt.ro (B.B.); doru.burduhos@tuiasi.ro (D.-D.B.-N.); Tel.: +40-766-678-029 (B.B.)

Citation: Bolborea, B.; Baera, C.; Dan, S.; Gruin, A.; Burduhos-Nergis, D.-D.; Vasile, V. Concrete Compressive Strength by Means of Ultrasonic Pulse Velocity and Moduli of Elasticity. *Materials* **2021**, *14*, 7018. https://doi.org/10.3390/ma14227018

Academic Editor: Jie Hu

Received: 3 November 2021
Accepted: 17 November 2021
Published: 19 November 2021

Publisher's Note: MDPI stays neutral with regard to jurisdictional claims in published maps and institutional affiliations.

Copyright: © 2021 by the authors. Licensee MDPI, Basel, Switzerland. This article is an open access article distributed under the terms and conditions of the Creative Commons Attribution (CC BY) license (https://creativecommons.org/licenses/by/4.0/).

Abstract: Developing non-destructive methods (NDT) that can deliver faster and more accurate results is an objective pursued by many researchers. The purpose of this paper is to present a new approach in predicting the concrete compressive strength through means of ultrasonic testing for non-destructive determination of the dynamic and static modulus of elasticity. For this study, the dynamic Poisson's coefficient was assigned values provided by technical literature. Using ultra-sonic pulse velocity (UPV) the apparent density and the dynamic modulus of elasticity were determined. The viability of the theoretical approach proposed by Salman, used for the air-dry density determination (predicted density), was experimentally confirmed (measured density). The calculated accuracy of the Salman method ranged between 98 and 99% for all the four groups of specimens used in the study. Furthermore, the static modulus of elasticity was deducted through a linear relationship between the two moduli of elasticity. Finally, the concrete compressive strength was mathematically determined by using the previously mentioned parameters. The accuracy of the proposed method for concrete compressive strength assessment ranged between 92 and 94%. The precision was established with respect to the destructive testing of concrete cores. For this research, the experimental part was performed on concrete cores extracted from different elements of different structures and divided into four distinct groups. The high rate of accuracy in predicting the concrete compressive strength, provided by this study, exceeds 90% with respect to the reference, and makes this method suitable for further investigations related to both the optimization of the procedure and = the domain of applicability (in terms of structural aspects and concrete mix design, environmental conditions, etc.).

Keywords: concrete; compressive strength; NDT; ultrasonic pulse velocity; modulus of elasticity

1. Introduction

Non-destructive testing methods (NDT) are essential tools in estimating concrete properties (mechanical or physical). A comprehensive analysis of the mechanical properties is useful in the process of structural optimization, as well as in terms of budget efficiency.

In the case of Reinforced Concrete (RC) structures, one of the key properties is the compressive strength. An investigation from this point of view can provide an overview of the structural integrity of a building. Such an analysis helps civil engineers in optimizing the process of structural intervention by deepening the understanding of how the building works from the structural point of view and also considering the concrete mix design and the associated physical, mechanical, and durability characteristics. Therefore, the interventions can be targeted on those elements that have a deficient behavior, which can induce negative effects into the structure [1–7].

Traditionally, the concrete compressive strength is determined through destructive testing (DT) which is considered the most reliable testing, and thus been referred to as the reference method. In DT, there are identified three possible situations:

- The samples are prepared and tested in the laboratory;
- The samples are collected on the construction site during the concrete casting, followed by curing and testing in the laboratory;
- The concrete cores are extracted directly from the structure (specific elements) and, after specific processing and conditioning, they are tested in the laboratory.

In the first situation, this testing is performed in order to evaluate or calibrate different mixing sequences of concrete mixture [8,9]. The samples collected on the construction site during the concrete casting are usually considered in cases of new buildings, to assess and confirm the concrete class [10] or to check compressive strength at intermediate specific terms (identity tests at 1 day, 2 days, 7 days, etc.). In the case of existing buildings, DT is performed by extracting concrete cores from certain concrete elements, such as columns, beams, slabs, diaphragms, etc., and then subjecting them to a compressive load until failure [11]. The preparation of the cores is made with respect to specific regulations and procedures in order to ensure the necessary testing accuracy [12].

According to [13,14] the analysis procedure depends on the amount of available information regarding the existing structure to be evaluated: information about the construction geometry, elements detailing, and material type determine the appropriate Knowledge Level (KL) of the structure under study. There are considered to be three KLs, defined as follows: KL1—limited, KL2—normal, and KL3—full, and to each of them is assigned a Confidence Factor (CF): (CF_{KL1} = 1.35; CF_{KL2} = 1.20; and CF_{KL3} = 1.00). The confidence factor is used as a correction factor for incomplete knowledge and level of uncertainty [11]. To reach a superior level of confidence (KL3 level, for instance), a large number of cores must be extracted from the structure, which can cause several inconveniences such as they can be time- and resource-consuming and also affect the structure itself due to specimens' extraction. Furthermore, the compressive strength may vary within the same element, due to the specific heterogeneity of concrete [15].

NDT represents a possible, viable alternative, mainly in terms of cost efficiency and also as they are fast in delivering results. However, NDT techniques measure indicators that are sensitive to a specific concrete property. For example, the ultrasonic pulse velocity and rebound hammer are sensitive to mechanical properties such as the compressive strength and porosity of concrete [16]. Another major problem pointed out by Angst [17] is the fact that the relation between mechanical properties and measured indicators is not constant. This is attributed to several causes, strongly connected to concrete physical characteristics (its specific heterogeneity, the porosity, water content, aggregate maximum dimension, etc.) and also to element exposure, measured data accuracy, and limited number of measurements.

Over the years, several NDT methods have been developed with the main purpose of estimating, as correctly as possible, the mechanical properties of materials and elements. A short overview of the NDT methods used on concrete structures is presented thus:

- One of the first and most used NDTs is the visual inspection (VI) [18]. It focuses on identifying visible pathologies in concrete such as cracks, voids, and spalling and trying to understand what caused them. It is a subjective and limited method, as internal characteristics of the structure cannot be determined, but it is also an important first step for further evaluation, providing useful information for establishing optimum and adequate methodologies (DT or NDT) for further investigation.
- The rebound hammer (RHS) determines the surface hardness. The main advantages are the fact that is a simple to use method with low cost and energy. The method tests

the concrete strength on a depth of 2–3 cm, this being the reason why it should be combined with other methods that tests the concrete elements in depth [19].
- The radiographic testing (RT) consists of a radioactive isotope source that transmits photons continuously through the concrete element, photons which are developed on a radiation sensitive film. It is mainly used for visualizing interior features of an element. There are many types of radiography that each have a specific application. In order to detect voids, cracks, or other interior defects, gamma-ray radiography was found to be useful [20]. This method is rarely used due to safety concerns.
- The carbonatation testing (CT) is made by spraying an exposed surface of the concrete element with a solution containing 1% phenolphthalein. The calcium hydroxide reacts with the solution resulting in a pink color, while the carbonated area will remain uncolored [21].
- Infrared thermography (IT) is used in order to determine the internal voids, cracks, or delamination by measuring the time delay before the temperature changes [22].
- The electromagnetic and radar testing (E&RT) are the most used NDT methods for the identification of reinforcements position, diameter, and distance from the surface [23].
- The drilling resistance method (DRM) presumes estimating the concrete compressive strength by counting the time required to drill a certain distance in the concrete element with a constant force and rotation speed. Serkan et al. [24] proved the accuracy of the method and presented good results when it was combined with the rebound hammer testing.
- Ultrasonic pulse velocity (UPV) is an NDT which has been extensively investigated for decades [25]. The method consists of measuring the transit time of an ultrasonic pulse from a transmitter to a receiver, knowing the distance between the two transducers. A short amount of time is needed for the ultrasonic pulse to pass through an element result in a high velocity, meaning a compact and homogenous material. This is an indication of the element's strength.

The most used NDT methods, for estimating concrete compressive strength, are the Schmidt rebound hammer, the ultrasonic pulse velocity, and the sonic rebound (SonReb) which consists of a combination of the first two methods. Făcăoaru et al. [26] developed and described the procedure which consists of applying some correctional factors based on cement type and dosage, granulometry and type of aggregates, and concrete age. SonReb has a high degree of efficiency and is still used worldwide, successfully, in estimating concrete compressive strength. Still, one of the most important disadvantages of the method is its requirements of mix design information regarding the evaluated concrete; in the case of older structures this information is not always available, which may lead to unprecise results. The viability of the method should also be verified on various types of concrete mix design developed with different additions, waste, or by-products (mineral, rubber, plastic, glass, etc.), which gained large diversity in recent years due to environmental protection requirements and Circular Economy implementation [2–4].

Researchers tried to develop various relationships between the measured indicators and the mechanical properties of concrete, by using different techniques, such as response surface (RS) [27–34], data fusion (DF) [35–37], and artificial neural networks (ANN) [38–43]. The empirical relationships developed over the years have a different structure: linear (LN) [40,41], polynomial (PL) [44,45], and power (PW) [46,47]. Sbartai et al. [34] report a satisfactory level in predicting concrete properties based on ultrasonic pulse velocity, ground penetration radar (GPR), electrical resistivity measurements, and data interpretation through the means of the response surface. However, when the data is interpreted with the help of ANN, the results have a higher rate of predictability. Asteris et al. [40] developed and optimized an ANN that considers the ultrasonic pulse velocity and Schmidt rebound hammer as the input values needed in order to estimate the concrete compressive strength. Based on the statistical parameters employed to evaluate the performance, the developed ANN model proved to have high efficiency in estimating the compressive strength, both when applied on its own database and also applied on other databases of different researchers. Khademi et al. [41] compared different techniques used to predict the 28 days compressive strength of concrete. In their mentioned study, a multiple linear regression (MLR), an artificial neural network, and adaptive neuro-fuzzy inference system models (ANFIS) were implemented with the purpose of finding the most accurate method of estimating concrete compressive strength. It was concluded that both ANN and ANFIS models can predict the concrete compressive strength more accurately than MLR, which proved to be unreliable. This is due to the fact that these models consider the non-linear correlation between the variables used as input data. It was also concluded that the accuracy of prediction is influenced by the number of input variables.

Breysse concluded [16] that a universal law between NDT and concrete compressive strength does not exist, despite the fact that many authors tried to find one.

This paper aims to present a methodology in estimating the concrete compressive strength by using ultrasonic pulse velocity as the only on-site testing method and a series of mathematical relations connecting the UPV with the moduli of elasticity (dynamic and static) and finally with the compressive strength.

2. Materials and Methods

2.1. The Destructive Method (DT)

This method is considered to deliver the most reliable results regarding the concrete compressive strength and, in this study, all results are reported to this method, considered to be the reference one. Consequently, the precision rate of the proposed method is also established with respect to the DT, as reference base of evaluation. DT consists of extracting concrete cores from the existing elements, cores which are then subjected to a series of laboratory processing and conditioning after which they are subjected to compression load until failure. The resulting compressive bearing capacity (f_{car}) is corrected by a series of coefficients described in Equation (1) provided by Romanian Norm NP 137 [12], thus resulting in the equivalent concrete compressive strength (f_{is}).

$$f_{is} = a \cdot b \cdot c \cdot e \cdot g \cdot d \cdot f_{car} \quad (1)$$

where: f_{is}—equivalent concrete compressive strength (MPa); a—coefficient that takes into account the influence of the core diameter; b—coefficient that takes into account the height/diameter ratio; c—coefficient that takes into account the influence of the degraded layer; e—coefficient that takes into account the nature of the leveling layer; g—coefficient that takes into account the humidity of the concrete core; d—coefficient that takes into account the position and diameter of the reinforcement bars; and f_{car}—resulted compressive bearing capacity (MPa).

As mentioned in the previous paragraph, destructive testing inflicts damage on the tested element; therefore, the number of cores must be maintained to a minimum in order to preserve the structural integrity of the element. For this reason, it is possible that the obtained values, calculated on an insufficient number of specimens, namely extracted cores from a designated element or assembly, do not reflect the overall value of the compressive strength of the targeted element. Additionally, in some cases, the extraction of the concrete core itself can prove to be difficult or even impossible to perform due to technological conditions such as the position of the designated element in the structure, the possibility to fix the drilling machine in order to extract the concrete core, etc.

2.2. Ultrasonic Pulse Velocity (UPV)

The ultrasonic pulse velocity was used as the on-site NDT testing method. In accordance with the theory of sound propagation in solids, the velocity of the ultrasonic signal depends on the density and elastic modulus of the material subjected to testing [48].

A calibration between compressive strength and ultrasonic pulse velocity for each concrete sample assures enough dependability for the two indicators [49]. Naik et al. [50]

presented a full review of the method. ultrasonic pulse velocity can be determined with Equation (2) presented by Romanian Norm NP 137 [12].

$$V_L = L/T \tag{2}$$

where: V_L—ultrasonic pulse velocity (km/s); L—path length in concrete (mm); and T—transit time (µs).

2.3. Theoretical Considerations

Modulus of elasticity of concrete (E) is a property of concrete that estimates the potential deformation of a structural element under service conditions [51]. The factors influencing this property are the dosage of cement, concrete age and class, the binder characteristics, and proportions.

The static modulus of elasticity (E_s) is a fundamental parameter that is defined by the stress–strain diagram under static loads [51] and it is generally estimated based on design code, not on direct measurements.

The dynamic modulus of elasticity (E_d), in comparison to E_s, is defined by the ratio of stress–strain under vibratory conditions [52]. The most common techniques for determining E_d are resonance frequency or UPV [53], but a study conducted by Luo and Bungey [54] presented a new approach by using surface waves in order to determine E_d. For this study, E_d was determined accordingly to Romanian Guide GE 039 [55] via UPV using Equation (3).

$$E_d = \frac{(1+\Theta_d)\cdot(1-2\cdot\Theta_d)}{1-\Theta_d}\cdot\frac{\gamma}{g}\cdot V_L^2 \tag{3}$$

where: E_d—dynamic modulus of elasticity (MPa); Θ_d—dynamic Poisson's ratio; γ—air dry density (kg/m^3); g—gravitational acceleration (m/s^2); and V_L—ultrasonic pulse velocity (km/s).

Romanian Guide GE 039 [55] presents a mathematical expression, Equation (4), for the determination of the dynamic Poisson's ratio, but for this study, the dynamic modulus of elasticity was assumed the value presented by the technical literature [55], namely $\Theta_d = 0.25$ (for concrete preserved in the air).

$$\Theta_d = \frac{(2\cdot n\cdot l)^2}{V_L^2} \tag{4}$$

where: n—fundamental resonant frequency (cycles/sec); l—length of specimen (m); and V_L—ultrasonic pulse velocity (km/s).

Thereby, when considering the $\Theta_d = 0.25$ the values of the function depending on the dynamic Poisson's ratio becomes:

$$f(\Theta_d) = \frac{(1+\Theta_d)\cdot(1-2\cdot\Theta_d)}{1-\Theta_d} = 0.83 \tag{5}$$

Inserting Equation (5) in Equation (3) results the dynamic modulus of elasticity has the following expression:

$$E_d = 0.83\cdot\frac{\gamma}{g}\cdot V_L^2 \tag{6}$$

Regarding the air-dry density Salman [56] and Panzera et al. [57] conducted studies to find a linear correlation between air-dry density (γ) and UPV. In this study, Equation (7), presented by Salman [56] was used to determine the air-dry density of concrete.

$$\gamma = 114.8\cdot V_L + 1813 \tag{7}$$

where: γ—air-dry density (kg/m^3) and V_L—ultrasonic pulse velocity (km/s).

In order to establish the accuracy of the proposed equation, the air-dry density of concrete was experimentally determined. The samples were weighed and measured with the purpose of determining the apparent volume. Comparing the mean values of air-dry density obtained experimentally (γ_e) with the mean values of the predicted ones using Equation (7) (γ_t), it was shown it reached a precision rate of 98%.

Furthermore, the theoretical air-dry density was used in this study as it was proven to be efficient, thus the method remained completely non-destructive and depended only on UPV.

Romanian Guide GE 039 [55] stipulates that the ratio between E_s and E_d ranges, in general, between 0.85–0.95. For this study, the correlation between the two moduli of elasticity was determined by experimentally. Therefore, each modulus of elasticity (E_s and E_d) was calculated individually and then a direct link between them was established. E_d was determined via UPV (Equation (6)) and E_s was determined via DT (Equation (8)).

For determining the static modulus of elasticity, with the air-dry density determined with Equation (7) and compressive strength obtained destructively (f_{is}) determined with Equation (1), using the mathematical relationship presented by Noguchi et al. [51] (Equation (8)), a static modulus of elasticity could be determined.

$$E_s = 2.1 \cdot 10^5 \cdot \left(\frac{\gamma}{2.3}\right)^{1.5} \cdot (f_c/200)^{1/2} \qquad (8)$$

where: E_s—static modulus of elasticity (MPa); $f_c = f_{is}$—concrete strength (MPa); and $\gamma = \gamma_t$—concrete air-dry density determined via UPV (kg/m^3).

The dynamic modulus of elasticity was mathematically calculated with Equation (6), using the ultrasonic pulse velocity.

Comparing the values of the two moduli of elasticity, determined for each specimen separately, it was established a direct and linear link between them described in Equation (9).

$$E_s = 0.75 \cdot E_d \qquad (9)$$

Using Equation (9), the static modulus of elasticity can now be determined only from the ultrasonic pulse velocity measurements and using Equation (7) the air-dry density can be obtained through the same measurements. Therefore, in Equation (8) the only unknown parameter remains concrete compressive strength (f_c). Extracting that parameter and rewriting Equation (8) results in a relationship (Equation (10)) where the compressive strength value depends only on parameters that can be determined via UPV.

$$f_c = (E_s^2 \cdot 200) \cdot [2.1 \cdot 10^5 \cdot (\gamma/2.3)^{1.5}]^2 \qquad (10)$$

2.4. Experimental Procedure

The study was conducted on 90 concrete cores with a diameter of 74 and 94 mm, extracted from different elements of different structures (Figure 1). The elements include the raft foundation, columns, beams, and reinforced concrete walls. After the extractions of the cores, the processing was conducted in accordance with Romanian Norm NP 137 [12]. The specimens were cut at both ends with a wet diamond disk and then dry air stored in laboratory conditions T: (21 ± 3) °C and RH: (50 ± 5)%, in accordance to Romanian Norm NP 137 [12]. The core specimens were cured for five days before weighting and UPV testing. This conditioning was performed to avoid that the humidity resulting from the wet cutting would affect the UPV data. Figure 1 presents the concrete sample after specific cutting and conditioning and before the destructive testing.

Figure 1. Concrete core specimens.

The as-received state density was determined with respect to SR EN 12390-7 [58] by specimens air-dry curing and then their measuring (diameter and height, for volume calculation), followed by their weighting. Then, the specimens were tested via UPV with a Tico Proceq device equipped with 54 kHz transducers (Figure 2). The coupling agent for the transducers was Vaseline. The last step was testing destructively the specimens, in compression. This procedure was conducted with respect to SR EN 12390-3 [59] which stipulates the curing and the testing conditions: the recommended dimensions of the concrete cores, the air-dry exposure, the preparation, and positioning, etc., as well as the loading rate. The destructive testing was conducted with a hydraulic press at a loading rate of 0.6 MPa/s.

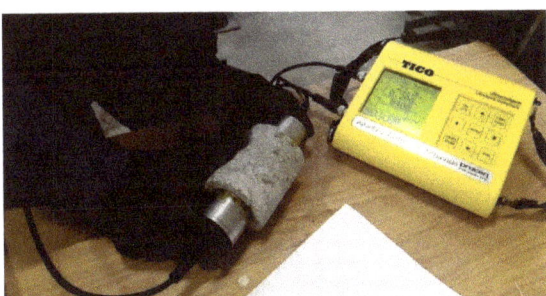

Figure 2. UPV testing.

Figure 3 presents the flowchart of the proposed method, consisting of the sequence of the major considered steps, in terms of experimental testing (black curve contour) and the corresponding parameters (light blue contour) determined by using the previously collected data.

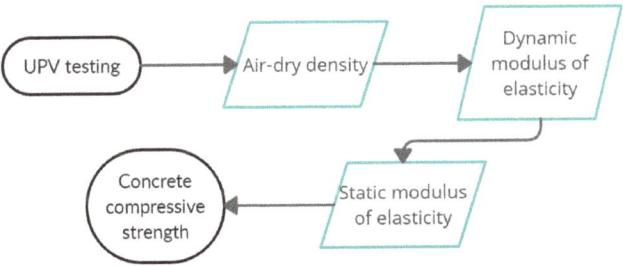

Figure 3. Flowchart of the presented method.

3. Results and Discussions
3.1. Proposed Method Compared to DT

For a more accessible interpretation of the results, the core specimens were divided into four groups. The considered division criterion is the value of compressive strength (f_{is}) determined via the Destructive Method, as follows:

- Group 1 [15–20) MPa: f_{is} ranges from 15 to 20 Mpa;
- Group 2 [20–25) Mpa: f_{is} ranges from 20 to 25 Mpa;
- Group 3 [25–30) Mpa: f_{is} ranges from 25 to 30 Mpa;
- Group 4 $\geq$ 30 MPa: f_{is} exceeds 30 MPa.

The theoretical methods for assessing the concrete air-dry density and compressive strength, for the four considered groups of specimens, were statistically evaluated with respect to the experimental and reference procedures, in terms of coefficient of variation (CoV), and also the accuracy (A_c).

CoV is defined by Everitt [60] by the means of Equation (11), as the ratio between the standard deviation (σ) and the mean value (μ) of the group of specimens where applied.

$$CoV = \frac{\sigma}{\mu} \qquad (11)$$

where: CoV—coefficient of variation (%); σ—standard deviation; and μ—mean value.

The air-dry density and compressive strength (measured and predicted) were also analyzed in terms of accuracy, defined in accordance with ISO 5725-1 [61] as the ratio between the predicted value (result of the proposed method) and the "true" value, provided by the reference method. Accuracy was calculated by the use of Equation (12).

$$A_c = \frac{\text{Predicted value}}{\text{Measured value}} \cdot 100 \qquad (12)$$

where: A_c—accuracy (%).

Table 1 presents the air-dry density values for each of the four specimen groups, determined by using both methods: the experimental method, (comprises specimens' measurement and weighing) and the Salman theoretical method (based on UPV individual values) [56]. The accuracy was calculated by using as input data the mean values recorded for each of the four groups of core specimens.

Table 1. Air-dry density.

	Density (kg/m³)								
	Measured Density (γ_e)				Predicted Density (γ_t)				Accuracy (%)
	Min	Mean	Max	CoV (%)	Min	Mean	Max	CoV (%)	
Group 1 [15–20) MPa	2149	2237	2319	1.7	2215	2236	2254	0.4	99
Group 2 [20–25) MPa	2210	2300	2384	2.1	2241	2263	2286	0.5	98
Group 3 [25–30) MPa	2221	2317	2411	2.7	2259	2286	2347	0.8	98
Group 4 $\geq$ 30 MPa	2256	2312	2365	1.3	2281	2308	2322	0.6	99

Figures 4 and 5 present a graphical representation of the mean values and accuracy of measured and predicted density. With an accuracy ranging between 98% and 99%, the theoretical Salman method [56] for determining the density via UPV testing proves to be a viable approach. This conclusion is also supported by the CoV values, ranging from 1.3% to 2.7% for the reference method (experimental measurements) and presenting a more compact range of smaller CoV values, from 0.4% to 0.8%, for the theoretical, Salman approach.

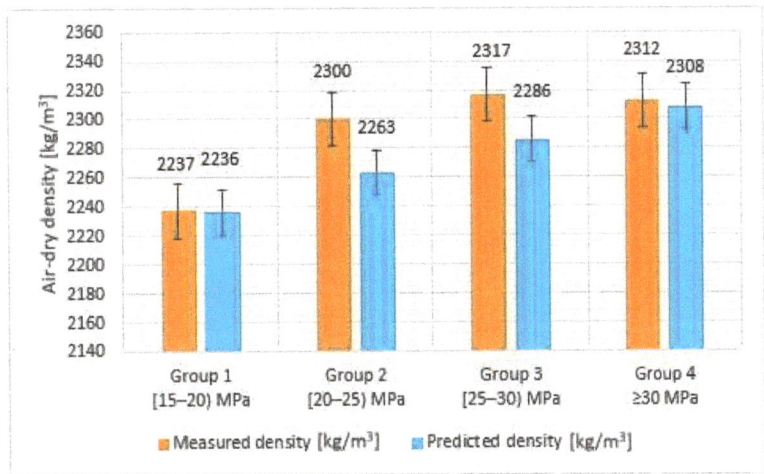

Figure 4. Air-dry density, mean values.

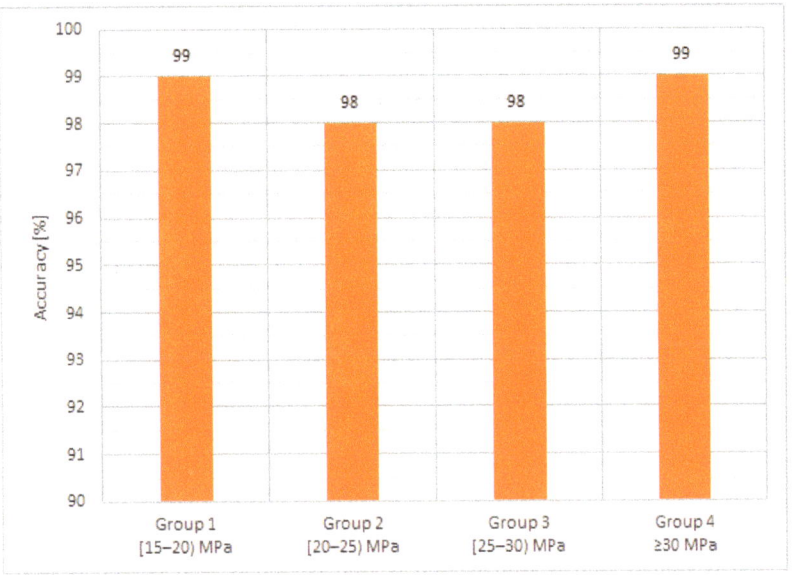

Figure 5. Air-dry density, the accuracy of the NDT method with respect to the reference.

Table 2 presents the results of compressive strength for each of the four core groups, determined by both methods: the proposed method (UPV testing and interpretation via moduli of elasticity) and the reference, destructive testing.

Table 2. Compressive strength.

	Compressive Strength (MPa)								Accuracy (%)
	DT Compressive Strength (f_{is})				NDT Compressive Strength (f_c)				
	Min	Mean	Max	CoV (%)	Min	Mean	Max	CoV (%)	
Group 1 [15–20) MPa	15.8	17.9	20.0	6.8	15.1	18.4	21.6	8.9	94
Group 2 [20–25) MPa	18.3	22.6	25.0	7.1	19.1	23.2	27.9	9.6	93
Group 3 [25–30) MPa	25.1	27.4	29.6	4.7	22.4	27.4	30.7	7.7	94
Group 4 ≥30 MPa	30.2	33.0	38.9	6.3	26.9	33.3	37.0	10	94

In terms of compressive strength, which is the main focus of the study, a graphical representation of the mean values and accuracy for each of the four groups is presented in Figures 6 and 7. The specimens sorting into four distinct groups function of the compressive strength, as previously specified, was considered proper for a better understanding of the phenomenon and to facilitate the data processing and the scattering of results. The wide range of values for the DT compressive strength (f_{is}) can be noted, with a minimum of 15.8 MPa and a maximum of 38.9 MPa. Additionally, the corresponding values for the NDT testing (f_c) range from 15.1 MPa to 37.0 MPa (Table 2).

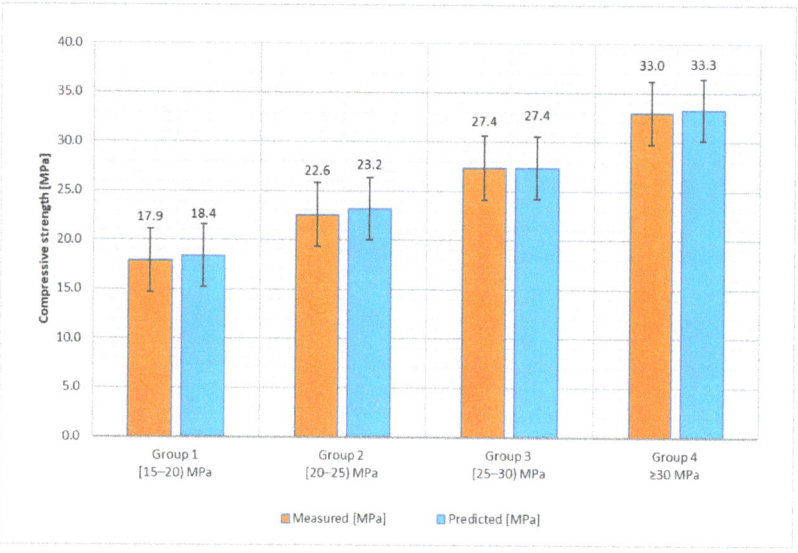

Figure 6. Compressive strength, mean values.

The accuracy of the proposed method for compressive strength determination via UPV and moduli of elasticity was also calculated using Equation (12), with respect to the mean values for each of the four groups of core specimens, presented in Table 2. The obtained results proved to be satisfactory, as the variation is very small with respect to the DT, regardless of the wide range of values. This conclusion is also supported by the CoV values (Table 2), ranging from 4.7 to 6.8% for the reference method and offering a similar compact range of values from 7.7 to 10% for the proposed method.

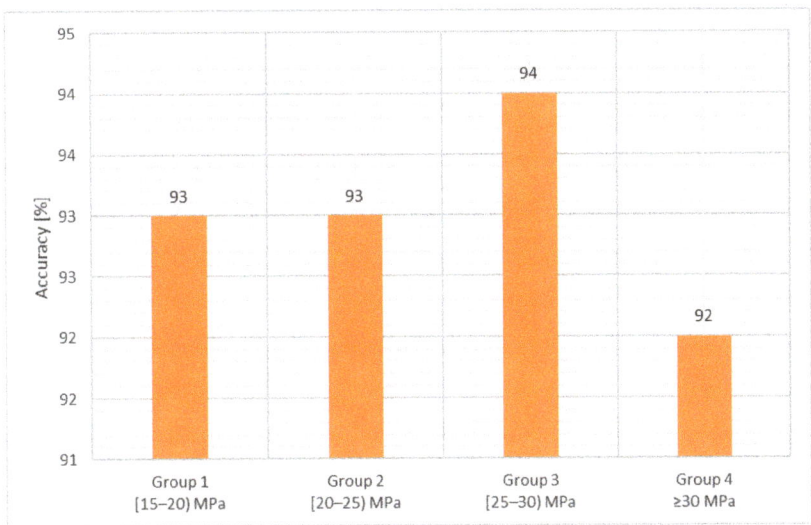

Figure 7. Compressive strength, the accuracy of the proposed method related to the reference.

Furthermore, when analyzing each group, it can be noticed that for groups 1, 2, and 4, the theoretical compressive strength, evaluated via the proposed method, tends to be a little overestimated with respect to the reference (2.8, 2.7 and 0.9%, respectively), while in the third group there is a clear match of the values.

Figure 7 graphically presents the accuracy evaluation, in terms of mean values. It can be noticed that the values are ranging from 93 to 94%.

As presented in Figure 7, the precision of this method in terms of Ac reaches 93% for the first two groups, 94% for the third group, and 92% for the fourth group. Considering all the values, the mean value of the precision is up to 93%.

Figure 8 presents a graphical representation of the correlation between experimentally determined compressive strength (f_{is}) and predicted compressive strength (f_c) is presented.

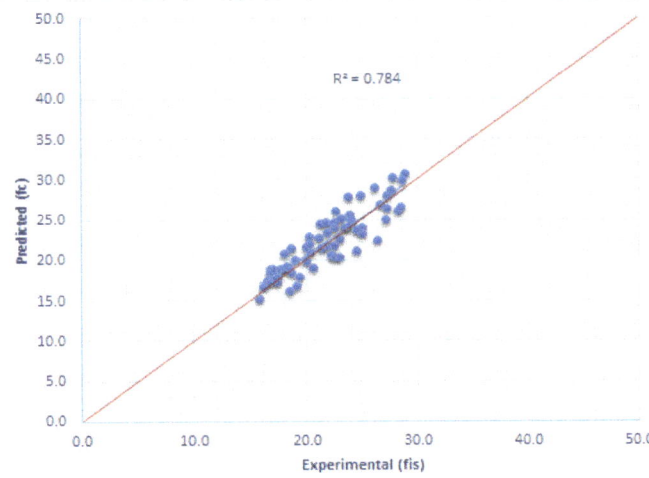

Figure 8. Correlation between experimental and calculated compressive strength.

A good correlation is achieved between the two sets of values. The coefficient of determination r², which is a statistical indicator of the quality of the theoretical model, is in this case equal to 0.78.

3.2. Proposed Method and SonReb Method Compared to Destructive Method

For a better validation of the proposed method, the obtained results are compared to both the SonReb method and the DT. For the SonReb method, six structural elements (columns) were investigated on-site by using ultrasonic pulse velocity and Rebound Hammer Schmidt. Each element was tested in three sections, with five UPV and nine RHS measurements/section. In Table 3, the resulting mean values of the UPV, RHS, and compressive strength from each method are presented.

Table 3. The results obtained from each presented method.

Element	Mean UPV (km/s)	Mean RHS (div)	Mean Concrete Compressive Strength		
			Proposed Method	SonReb Method	Destructive Method
Column 1	4.324	40	26.1	25.7	26.7
Column 2	4.142	37	28.6	21.8	27.7
Column 3	3.985	38	24.7	19.7	22.5
Column 4	3.716	41	18.9	17.4	17.7
Column 5	3.632	35	17.3	11.4	17.4
Column 6	3.758	35	19.7	12.9	20.0

A graphical representation of the accuracy is presented in Figure 9, namely a comparative analysis between the proposed method vs. SonReb method, both of them evaluated with respect to the reference, namely, the destructive method.

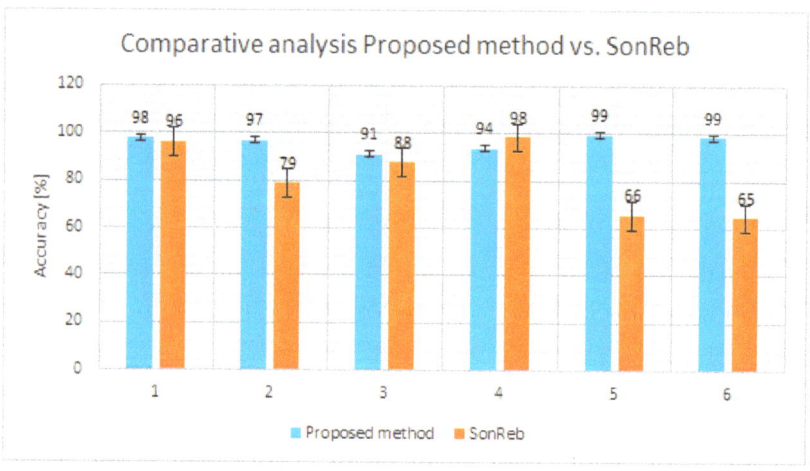

Figure 9. Graphical representation of the accuracy evaluation of each method, proposed and SonReb method, with respect to the reference method (destructive testing).

For the six elements investigated through both NDT methods, the precision rate in the case of the proposed method reaches up to 96%, while in the case of the SonReb method the precision reaches up to 82%.

4. Conclusions

The aim of this paper is to present the results obtained by combining the on-site measurements of UPV and theoretical interpretation using a set of equations developed by different researchers linking the values of ultrasonic pulse velocity to the dynamic modulus of elasticity, static modulus of elasticity, and finally concrete compressive strength.

Estimating concrete compressive strength through this method delivered results with high accuracy, which, in this case, ranged between 84 and 100%. It can be noticed that the high level of accuracy remains the same regardless of the range value of compressive strength, which in this study is between 15.8 and 38.9 MPa. Additionally, the coefficient of variation (CoV) shows reduced values, ranging along compact intervals, both for the dry-air density evaluation (from 0.4 to 0.8%, for the theoretical, Salman approach) and also for the compressive strength evaluation (from 7.7 to 10%, for proposed method). Further investigations will also consider the methodology and statistical approach proposed by Breysse et al. [62].

As this method relies only on UPV measurements, the on-site surface preparation and testing process must be performed with a high level of precision; otherwise, the results will have a higher level of uncertainty.

In contrast to the SonReb method, the proposed one has the advantage, so far, that there is no requirement of information about the classical concrete mix design such as cement type and dosage, granulometry, and nature of aggregates. This information is often difficult to obtain, especially in the case of buildings where the concrete mix was produced on-site with no known recipe. In the SonReb method, not knowing these parameters correct can lead to errors up to $\pm30\%$. In this case, although the concrete mix design was known, hence all the coefficients were correctly assumed, the accuracy of the SonReb method had a lower value than the proposed method on each analyzed concrete column when compared to DT.

The current results are clearly encouraging, offering new research perspective for method optimization and further confirmation to prove its viability, especially in terms of concrete mix design diversity, which has experienced an exponential growth in the last decades. The single and multiple additions in concrete compositions, waste, or by-products generated by the industry, may induce concrete hardened state changes which lead to complex investigation in terms of overall behavior, NDT included. A preliminary approach on this area represents the on-going study of the current research.

Author Contributions: Conceptualization, B.B. and A.G.; methodology, B.B. and V.V.; validation, B.B. and V.V.; formal analysis, C.B. and D.-D.B.-N.; investigation A.G.; resources, B.B. and S.D.; data curation, D.-D.B.-N. and S.D.; writing—original draft, B.B. and C.B.; writing—review and editing, V.V. and D.-D.B.-N.; visualization, B.B. and V.V.; supervision, C.B., A.G. and S.D. All authors have read and agreed to the published version of the manuscript.

Funding: This research received no external funding.

Institutional Review Board Statement: Not applicable.

Informed Consent Statement: Not applicable.

Acknowledgments: This paper is supported by: Programme: Research for sustainable and ecological integrated solutions for space development and safety of the built environment, with advanced potential for open innovation—"ECOSMARTCONS", Programme code: PN 19 33 04 02: "Sustainable solutions for ensuring the population health and safety within the concept of open innovation and environmental preservation", financed by the Romanian Government. Project "Entrepreneurial competences and excellence research in doctoral and postdoctoral programs—ANTREDOC", project co-funded by the European Social Fund financing agreement no. 56437/24.07.2019. The authors would like to thank Michael Grantham of Sandberg LLP, University of Leeds and Queen's University Belfast, Institute of Concrete Technology and Editorial Board of the Journal "Case Studies in Construction Materials", UK for his support and contribution in technical editing, language editing and proofreading.

Conflicts of Interest: The authors declare no conflict of interest.

References

1. Chalangaran, N.; Farzampour, A.; Paslar, N. Nano Silica and Metakaolin Effects on the Behavior of Concrete Containing Rubber Crumbs. *CivilEng* **2020**, *1*, 264–278. [CrossRef]
2. Chalangaran, N.; Farzampour, A.; Paslar, N.; Fatemi, H. Experimental investigation of sound transmission loss in concrete containing recycled rubber crumbs. *Adv. Concr. Constr.* **2021**, *11*, 447–454.
3. Baeră, C.; Szilagyi, H.; Lăzărescu, A. Developing Fibre Engineered Cementitious Materials with Self-Healing abilities (SH-FECM) by using polyvinyl alcohol fibres and supplementary powder addition. In Proceedings of the Fib Symposium, Krakow, Poland, 27–29 May 2019.
4. Baeră, C.; Vasile, V.; Matei, C.; Gruin, A.; Szilagyi, H.; Perianu, I.A. Developments fo Green Cementitious Materials by Using the Abrasive Waterjet Garnet Wastes: Preliminary Studies. *Adv. Mat. Res.* **2021**, *1164*, 87–96.
5. D'Ambrisi, A.; De Stefano, M.; Tanganelli, M.; Viti, S. The Effect of Common Irregularities on the Seismic Performance of Existing RC Framed Buildings. In *Seismic Behaviour and Design of Irregular and Complex Civil Structures*; Springer: Dordrecht, The Netherlands, 2013; pp. 47–58.
6. De Stefano, M.; Tanganelli, M.; Viti, S. Torsional effects due to concrete strength variability in existing buildings. *Earthq. Struct.* **2015**, *8*, 379–399. [CrossRef]
7. Viti, S.; Tanganelli, M.; De Stefano, M. The concrete strength variability as source of irregularity for RC existing buildings. *Geotech. Geol. Eng.* **2016**, *40*, 149–158.
8. Chandrappa, A.K.; Biligiri, K.P. Influence of mix parameters on pore properties and modulus of pervious concrete: An application of ultrasonic pulse velocity. *Mater. Struct.* **2016**, *49*, 5255–5271. [CrossRef]
9. Cuibuş, A.; Kiss, Z.; Gorea, M. Influence of mineral additions on the physical-mechanical properties of concretes. *Rev. Rom. Mater.* **2014**, *44*, 225–235.
10. Aryal, R.; Mishra, A.K. In-Situ compressive strength assessment of concrete in under-construction residential buildings at Gaindakot municipality. *Mater. Today Proc.* **2020**. [CrossRef]
11. Pucinotti, R. Assessment of In Situ characteristic concrete strength. *Constr. Build. Mater.* **2013**, *44*, 63–73. [CrossRef]
12. iTeh. NP 137. In *Normative for In Situ Evaluation of the Concrete Compressive Strength of the Existing Constructions*; iTeh: Kamen, Germany, 2014.
13. European Union. EN 1998-3. In *Eurocode 8—Design of Structures for Earthquake Resistance—Part 3: Assessment and Retrofitting of Buildings*; European Union: Brussels, Belgium, 2005.
14. Ministry of Development, Public Works and Administration of Romania. *P100-3. Design Code—Part 3: Provisions for Seismic Assessment of Existing Buildings*; M.Of., p I, nr. 1003 bis/13.12.2019; Ministry of Development, Public Works and Administration of Romania: Bucharest, Romania, 2019.
15. Masi, A.; Chiauzzi, L. An experimental study on the within-member variability of in situ concrete strength in RC building structures. *Constr. Build. Mater.* **2013**, *47*, 951–961. [CrossRef]
16. Breysse, D. Nondestructive evaluation of concrete strength: An historical review and a new perspective by combining NDT methods. *Constr. Build. Mater.* **2012**, *33*, 139–163. [CrossRef]
17. Angst, U.M.; Polder, R. Spatial variability of chloride in concrete within homogeneously exposed areas. *Cem. Concr. Res.* **2014**, *56*, 40–51. [CrossRef]
18. Zambon, I.; Vidovic, A.; Strauss, A.; Matos, J. Condition Prediction of Existing Concrete Bridges as a Combination of Visual Inspection and Analytical Models of Deterioration. *Appl. Sci.* **2019**, *9*, 148–157. [CrossRef]
19. Brencich, A.; Cassini, G.; Pera, D.; Riotto, G. Calibration and Reliability of the Rebound (Schmidt) Hammer Test. *JACE* **2013**, *1*, 66–78. [CrossRef]
20. Garney, G. *Defects Found through Non-Destructive Testing Methods of Fiber Reinforced Polymeric Composites*; California State University: Long Beach, CA, USA, 2006.
21. Valls, S.; Vazquez, E. Accelerated carbonatation of sewage sludge–cement–sand mortars and its environmental impact. *Cem. Concr. Res.* **2001**, *31*, 1271–1276. [CrossRef]
22. Vavilov, V.P.; Burleigh, D.D. Review of pulsed thermal NDT: Physical principles, theory and data processing. *NDT E Int.* **2015**, *73*, 28–52. [CrossRef]
23. Dobriec, L.; Jasinski, R.; Mazur, W. Accuracy of Eddy-Current and Radar Methods Used in Reinforcement Detection. *Materials* **2019**, *12*, 1168–1192.
24. Serkan, K.; Muhammet, A.; Oguz, G. Estimation of In-situ concrete strength using drilling resistance. *MATEC Web Conf. Concr. Solut.* **2019**, *289*, 06001.
25. Shariati, M.; Ramli-Sulong, N.H.; Aragnejad, K.H.M.M.; Shafigh, P.; Sinaei, H. Assessing the strength of reinforced concrete structures through Ultrasonic Pulse Velocity and Schmidt Rebound Hammer tests. *Sci. Res. Essays.* **2011**, *6*, 213–220.
26. Făcăoaru, I. *Non-Destructive Testing of Concrete in Romania. Symposium on Non-Destructive Testing of Concrete and Timber*; Thomas Telford Publishing: London, UK, 1969; pp. 39–49.
27. Haque, M.; Ray, S.; Mita, A.F.; Bhattacharjee, S.; Bin Shams, M.J. Prediction and Optimization of the Fresh and Hardened Properties of Concrete Containing Rice Hush Ash and Glass Fiber Using Response Surface Methodology. *Case Stud. Constr. Mater.* **2021**, *14*, e00505.

28. Shahmansouri, A.A.; Nematzadeh, M.; Behnood, A. Mechanical properties of GGBFS-based geopolymer concrete incorporating natural zeolite and silica fume with an optimum design using response surface method. *J. Build. Eng.* **2021**, *36*, 102138. [CrossRef]
29. Zhang, Q.; Feng, X.; Chen, X.; Lu, K. Mix design for recycled aggregate pervious concrete based on response surface methodology. *Constr. Build. Mater.* **2020**, *259*, 119776. [CrossRef]
30. Sultana, N.; Hossain, S.M.Z.; Alam, M.S.; Hashish, M.M.A.; Islam, M.S. An experimental investigation and modeling approach of response surface methodology coupled with crow search algorithm for optimizing the properties of jute fiber reinforced concrete. *Constr. Build. Mater.* **2020**, *243*, 118216. [CrossRef]
31. Awolusi, T.F.; Oke, O.L.; Akinkurolere, O.O.; Sojobi, A.O. Application of response surface methodology: Predicting and optimizing the properties of concrete containing steel fibre extracted from waste tires with limestone powder as filler. *Case Stud. Constr. Mater.* **2019**, *10*, e00212. [CrossRef]
32. Sun, Y.; Yu, R.; Shui, Z.; Wang, X.; Qian, D.; Rao, B.; Huang, J.; He, Y. Understanding the porous aggregates carrier effect on reducing autogenous shrinkage of Ultra-High Performance Concrete (UHPC) based on response surface method. *Constr. Build. Mater.* **2019**, *222*, 130–141. [CrossRef]
33. Rooholamini, H.; Hassani, A.; Aliha, M.R.M. Evaluating the effect of macro-synthetic fibre on the mechanical properties of roller-compacted concrete pavement using response surface methodology. *Constr. Build. Mater.* **2018**, *159*, 517–529. [CrossRef]
34. Sbartai, Z.M.; Laurens, S.; Elachachi, S.M.; Payan, C. Concrete properties evaluation by statistical fusion of NDT techniques. *Constr. Build. Mater.* **2012**, *37*, 943–950. [CrossRef]
35. Castanedo, F. A Review of Data Fusion Techniques. *Sci. World J.* **2013**, *2013*, 704504. [CrossRef]
36. Bloch, I. Information Combination Operators for Data Fusion: A Comparative Review with Classification. *IEEE Trans. Syst. Man.* **1996**, *26*, 52–67. [CrossRef]
37. Pollard, E. Evaluation de Situations Dynamiques Multicibles par Fusion de Donnees Spatio-Temporelles. Ph.D. Thesis, Grenoble University, Grenoble, France, 2010.
38. Yaman, M.A.; Elaty, A.; Taman, M. Predicting the ingredients of self-compacting concrete using artificial neural network. *Alex. Eng. J.* **2017**, *56*, 523–532. [CrossRef]
39. Concha, N.; Oreta, A.W. An improved prediction model for bond strength of deformed bars in RC using UPC test and Artificial Neural Network. *Int. J. Geomate* **2020**, *18*, 179–184. [CrossRef]
40. Asteris, P.G.; Mokos, V.G. Concrete compressive strength using artificial neural networks. *Neural. Comput. Appl.* **2020**, *32*, 11807–11826. [CrossRef]
41. Khademi, F.; Akbari, M.; Jamal, S.M.; Nikoo, M. Multiple linear regression, artificial neural network, and fuzzy logic prediction of 28 days compressive strength of concrete. *Front. Struct. Civ.* **2017**, *11*, 90–99. [CrossRef]
42. Ramyar, K.; Kol, P. Destructive and non-destructive test methods for estimating the strength of concrete. *Cem. Concr. World.* **1996**, *2*, 46–54.
43. Erdal, M. Prediction of the Compressive Strength of Vacuum Processed Concretes Using Artificial Neural Network and Regression Techniques. *Sci. Res. Essays.* **2009**, *4*, 1057–1065.
44. Cristofaro, M.T.; Viti, S.; Tanganelli, M. New predictive models to evaluate concrete compressive strength using the SonReb method. *J. Build. Eng.* **2020**, *27*, 100962. [CrossRef]
45. Samarin, A.; Dhir, R. *Determination of In Situ Concrete Strength Rapidly and Confidently by Nondestructive Testing*; ACI Symposium Publication: Detroit, MI, USA, 1987; Volume 82, pp. 77–94.
46. Faella, G.; Guadagnuolo, M.; Donadio, A.; Ferri, L. Calibrazione sperimentale del metodo SonReb per costruzioni della Provincia di Caserta degli anni '60 '80. In Proceedings of the 14th Anidis Conference, Bari, Italia, 18–22 September 2011.
47. Cristofaro, M.T.; D'Ambrisi, A.; De Stefano, M. Nuovi modelli previsionali per la stima della resistenza a compressione del calcestruzzo con il metodo Sonreb. In Proceedings of the Atti del XIII Convegno Nazionale L'Ingegneria Sismica, Bologna, Italia, 28 June–2 July 2009.
48. Hobbs, B.; Tchoketch, K.M. Non-destructive testing techniques for the forensic engineering investigation of reinforced concrete buildings. *Forensic Sci. Int.* **2007**, *167*, 167–172. [CrossRef]
49. Popovics, S.; Popovics, J. A critique of the ultrasonic pulse velocity method for testing concrete. *NDT E Int.* **1997**, *30*, 260.
50. Naik, T.R.; Malhotra, V.M.; Popovics, J.S. The ultrasonic pulse velocity Method. In *Handbook on Nondestructive Testing of Concrete*, 2nd ed.; CRC Press: Boca Raton, FL, USA, 2004.
51. Noguchi, T.; Nemati, K.M. Relationship between compressive strength and modulus of elasticity of High-Strength Concrete. *J. Struct. Constr. Eng.* **1995**, *60*, 1–10.
52. Lu, X.; Sun, Q.; Feng, W.; Tian, J. Evaluation of dynamic modulus of elasticity of concrete using impact-echo method. *Constr. Build. Mater.* **2013**, *47*, 231–239. [CrossRef]
53. Gutierres, P.A.; Canovas, M.F. The modulus of elasticity of high-performance concrete. *Mater. Struct.* **1995**, *28*, 559–568. [CrossRef]
54. Luo, Q.; Bungey, J. Using compression wave ultrasonic transducers to measure the velocity of surface waves and hence determine dynamic modulus of elasticity for concrete. *Constr. Build. Mater.* **1996**, *4*, 237–242.
55. GE 039. *Guide to the In-Site and Laboratory Determination of the Static and Dynamic Moduli of Elasticity of Concrete*; O.M.L.P.T.L. no. 1224/06.09.2001; Monitorul Oficial: Romania, 2001. Available online: https://www.monitoruloficial.ro/article--Publishing_House--55.html (accessed on 3 November 2021).

56. Salman, G.A. Density and Ultrasonic Pulse Velocity Investigation of Self-Compacting Carbon Fiber-Reinforced Concrete. *J. Eng. Technol.* **2018**, *36*, 88–99.
57. Panzera, T.H.; Rubio, J.C.; Bowen, C.R.; Vasconcelos, W.L.; Strecker, K. Correlation between Structure and Pulse Velocity of Cementations Composites. *Adv. Cem. Res.* **2009**, *20*, 101–108. [CrossRef]
58. ASRO. SR EN 12390-7. In *Testing Hardened Concrete—Part 7: Density of Hardened Concrete*; ASRO: Bucharest, Romania, 2019.
59. ASRO. SR EN 12390-3. In *Testing Hardened Concrete—Part 3: Compressive Strength of Test Specimens*; ASRO: Bucharest, Romania, 2019.
60. Everitt, B. *The Cambridge Dictionary of Statistics*; Cambridge University Press: Cambridge, UK; New York, NY, USA, 1998.
61. International Organization for Standardization. ISO 5725-1. In *Accuracy (Trueness and Precision) of Measurement Methods and Results—Part 1: General Principles and Definitions*; International Organization for Standardization: Geneva, Switzerland, 1994.
62. Breysse, D.; Balayssac, J.-P.; Biondi, S.; Corbett, D.; Goncalves, A.; Grantham, M.; Luprano, V.A.M.; Masi, A.; Monteiro, A.V.; Sbartai, Z.M. Recommendation of RILEM TC249-ISC on non destructive in situ strength assessment of concrete. *Mater. Struct.* **2019**, *52*, 71. [CrossRef]

Article

Behavior of Alkali-Activated Fly Ash through Underwater Placement

Zarina Yahya [1,2,*], Mohd Mustafa Al Bakri Abdullah [1,3], Long-yuan Li [4], Dumitru Doru Burduhos Nergis [5], Muhammad Aiman Asyraf Zainal Hakimi [2], Andrei Victor Sandu [5,6,*], Petrica Vizureanu [5,*] and Rafiza Abd Razak [1,2]

[1] Centre of Excellence Geopolymer and Green Technology (CEGeoGTech), Universiti Malaysia Perlis (UniMAP), Perlis 01000, Malaysia; mustafa_albakri@unimap.edu.my (M.M.A.B.A.); rafizarazak@unimap.edu.my (R.A.R.)
[2] Faculty of Civil Engineering Technology, Universiti Malaysia Perlis (UniMAP), Perlis 01000, Malaysia; aimanzasyraf@gmail.com
[3] Faculty of Chemical Engineering Technology, Universiti Malaysia Perlis (UniMAP), Perlis 01000, Malaysia
[4] School of Marine Science and Engineering, University of Plymouth, Plymouth PL4 8AA, UK; long-yuan.li@plymouth.ac.uk
[5] Faculty of Materials Science and Engineering, Gheorghe Asachi Technical University of Iași, Boulevard D. Mangeron, No. 51, 700050 Iasi, Romania; dumitru-doru.burduhos-nergis@academic.tuiasi.ro
[6] Romanian Inventors Forum, Str. P. Movila 3, 700089 Iasi, Romania
* Correspondence: zarinayahya@unimap.edu.my (Z.Y.); sav@tuiasi.ro (A.V.S.); peviz@tuiasi.ro (P.V.)

Abstract: Underwater concrete is a cohesive self-consolidated concrete used for concreting underwater structures such as bridge piers. Conventional concrete used anti-washout admixture (AWA) to form a high-viscosity underwater concrete to minimise the dispersion of concrete material into the surrounding water. The reduction of quality for conventional concrete is mainly due to the washing out of cement and fine particles upon casting in the water. This research focused on the detailed investigations into the setting time, washout effect, compressive strength, and chemical composition analysis of alkali-activated fly ash (AAFA) paste through underwater placement in seawater and freshwater. Class C fly ash as source materials, sodium silicate, and sodium hydroxide solution as alkaline activator were used for this study. Specimens produced through underwater placement in seawater showed impressive performance with strength 71.10 MPa on 28 days. According to the Standard of the Japan Society of Civil Engineers (JSCE), the strength of specimens for underwater placement must not be lower than 80% of the specimen's strength prepared in dry conditions. As result, the AAFA specimens only showed 12.11% reduction in strength compared to the specimen prepared in dry conditions, thus proving that AAFA paste has high potential to be applied in seawater and freshwater applications.

Keywords: alkali-activated; underwater placement; class C fly ash; seawater; fresh water

1. Introduction

The construction of structures involving concrete underwater placement usually require additional considerations due to its unique circumstances. Typically, the effective placement of conventional concrete mixture underwater depends on two main factors: the mix design of concrete itself and placement method during concreting [1,2]. For the mix design of concrete, additions of anti-washout admixtures (AWA) and viscosity-modifying admixtures (VMAs) are necessary for conventional concrete to minimise the washout effect and the ability to self-consolidate during the underwater placement [3–5]. Concrete resistance against washout can also be improved using mineral admixture with high fineness. The most used mineral admixture includes silica fume, ground granulated blast furnace slag (GGBS) and fly ash (FA) [2,6]. Heniegal [7] confirmed that the inclusion of fly ash and silica fume with the addition of limestone or bentonite powder improved the flowing properties of conventional concrete and minimised the washout effect.

Past researchers had also investigated the use of seawater as replacement for water in ordinary Portland cement (OPC) concrete. The justification for using seawater in concrete is for offshore structures where it involves underwater concreting. According to Wang et al. [8] it is possible to mix seawater with cement where the early strength increased due to the existence of Cl^- and Na^+ ions. Specimens with low water/cement (w/c) ratio showed more significant early strength development. For 28 days strength, it also showed an increment about 10% compared to specimen produced with fresh water but it induced corrosion on the rebars [9].

Meanwhile for the placement method, the concreting process can be made using Tremie pipe for mass concrete. This method required steel pipe with a hopper attached to the upper end and injectable plug on the bottom of pipe. The pipe is immersed in water and when the pipe is full of concrete, its bottom is opened for the concreting process. Using this technique, a hydro crane is required to lift the Tremie pipe after finishing concrete placement and the bottom of the pipe need to be kept in the fresh concrete during the process to avoid washout effect [2]. For a small concrete placement underwater, the skip and toggle bag methods are most suitable [1]. The concrete is filled up into different sized buckets, where the top covers are sealed to prevent water infiltration during the lowering process of the concrete placement. The bottom door of the bucket is slowly opened during concreting to allow free flowing of the concrete.

Moreover, for underwater concrete mostly refer to the Standard by Japan Society of Civil Engineers (JSCE). This standard stated that the w/c ratio should be in range 0.50 to 0.55 when placing reinforced concrete in seawater and fresh water [10]. The w/c ratio can be increased up to 0.60 and 0.65 when concreting for non-reinforced concrete. For the strength of hardened concrete through underwater placement, JSCE standards required the compressive strength of specimens attain a minimum of 80% strength with respect to specimen cast in dry conditions [2,10].

The manufacturing of ordinary Portland cement (OPC) consumes a lot of natural resources, is energy intensive and contributed to carbon dioxide (CO_2) emission to the atmosphere [11]. It was estimated that 7% of CO_2 emission comes from the OPC industry [12], which is about 1.35 billion tons per annum. This is a serious environmental concern, and research endeavours involve finding a suitable alternative binder to replace OPC in concrete. The literature [13] refers to the work of Davidovits that found geopolymer in 1978 which also known as amorphous alkali aluminosilicate, and are sometimes referred to as inorganic polymers, geocements, or alkali-activated cements. This new alternative binder is produced by activating source materials with alkaline activators, and its classification is dictated by the content of silica, aluminium, and calcium. If the source materials are made up of mainly silica and aluminium (Class F fly ash, metakaolin, or some natural pozzolan), its final product is the sodium aluminosilicate hydrate (N-A-S-H) backbone of the geopolymer [14]. If the source materials are made up of calcium, aluminium, and silica (Class C fly ash and slag), then the main product after hardening is calcium silicate hydrate (C-S-H) or calcium alumino silicate hydrate (C-A-S-H), which also can be described as alkali-activated materials (AAM) [15].

Fly ash is an industrial waste material that is ubiquitous due to the increasing demand for energy, which is met by increasing coal-fired power plant's usage. The world coal production is expected to rise between 2006 and 2030 by almost 60%, with volumes output to 7011 Mtce by 2030 [16–19]. The management of fly ash disposal is always concerned by environmentalist since only 20–30% of the generated fly ash is reused whereas the rest was disposed either in landfills or ponds [17,18]. Therefore, the use of fly ash as aluminosilicates sources in AAM production is a waste-to-health approach that could also mitigate environmental concerns.

The parameters that influence the properties of AAM have been intensively investigated [19–25], and AAM are known to be resistant against aggressive ions, freeze-thaw resistance, have high early and long-term strength, and excellent fire resistance [13,26–30]. The main issue of using conventional concrete for underwater structure is its resistance to washout. Concrete resistance to washout depends on the content of fine fraction in the binder, water cement ratio, and cement content. Concrete resistance to washout depends on the content of fine fractions in the binder, water cement ratio, and cement. Previous studies are mainly focused on underwater concrete placement using OPC as a binder and addition of special admixture for construction offshore structures such as bridge piers, but studies involving the application of AAM for underwater concreting remain scarce. The current study investigated the performance of alkali-activated fly ash (AAFA) paste through underwater placement in seawater and freshwater (river water and lake water). The compressive strength, changes in pH, X-ray fluorescence (XRF) and Field Emission Scanning Electron Microscope coupled with Energy Dispersive X-ray spectroscopy (FESEM-EDS) are analysed, respectively.

2. Materials and Methods

2.1. Materials

In this study, fly ash was used as source materials for AAM which is supplied by Cement Industries of Malaysia Berhad (CIMA), Perlis, Malaysia. Noted that the major elements in fly ash are silica (SiO_2), alumina (Al_2O_3), ferum (Fe_2O_3) and calcium (CaO). According to the American Society for Testing Materials (ASTM C618), the ash containing more than 70 wt.% of SiO_2, Al_2O_3, Fe_2O_3, and low CaO is considered to be Class F; while that with total of SiO_2, Al_2O_3, and Fe_2O_3 ranging within 50–70 wt.% defined Class C. Due to the relatively high calcium content (22.30%), the fly ash used in this experiment is classified as Class C according to the ASTM C618 [31].

Waterglass or sodium silicate solution was supplied by South Pacific Chemical Industries Sdn. Bhd. (SPCI), Malaysia. The waterglass consists of 30.1% SiO_2, 9.4% Na_2O and 60.5% H_2O (modulus $SiO_2/Na_2O = 3.2$). Its specific gravity and viscosity are 1.4 g/cm^3 and 0.4 Pas, respectively.

Sodium hydroxide (NaOH) powder brand Formosoda-P from Taipei, Taiwan with 99% purity was used. The desired concentration of NaOH solution was prepared 24 h before experiments by diluting NaOH powder with distilled water. The activator solution was prepared by mixing waterglass and NaOH solution at a ratio of 2.5.

2.2. Collection of Water Samples

In this study, the seawater, river water, and lake water samples are collected around Perlis, Malaysia. The water collected was left in the laboratory to allow the impurities to precipitate at the container's bottom. Later, the water was transferred to a plastic tank via infiltration and the pH value for each type of water was recorded.

2.3. Specimens Preparation

The concentration of NaOH solution is fixed at 12 M [32], the ratio of waterglass-to-NaOH and ratio fly ash-to-alkaline activator fixed at 2.0 and 2.5 respectively [22]. The details of mix design are summarized in Table 1. The fly ash and alkaline activator were mixed and stirred for 5 min using a mechanical mixer. Then the fresh AAFA paste poured into the 50 mm × 50 mm × 50 mm [33] moulds that already placed in a container with seawater, river water, and lake water as shown in Figure 1. The AAFA specimens were left in the container for 3, 7, and 28 days, respectively. The pH level and temperature of the water before and after the placement of AAFA paste were recorded. For the control specimens, the AAFA paste was prepared in dry conditions (without underwater placement).

Table 1. Mix design for AAFA paste.

Parameter	Indicator
Ratio fly ash/alkaline activator	2
Ratio waterglass/NaOH	2.5
Mass of fly ash (wt.%)	66.7
Mass of NaOH solution (wt.%)	9.6
Mass of sodium silicate solution (wt.%)	23.8

Figure 1. AAFA paste poured into mould in seawater.

2.4. Testing and Analysis Methods

The setting time of AAFA paste through underwater placement was measured using the Vicat test [33]. The test was conducted at room temperature using the Vicat apparatus, where the mould was placed beneath the water level. The initial setting and final setting time of the AAFA paste were recorded.

The specimen's compressive strength was determined based on ASTM C109 [34] using Instron 5582 Mechanical Tester (Instron, Massachusetts, United States America). A minimum of three specimens was tested for each mix design, and the average results recorded. Total of all 108 specimens were produced for this testing. The AAFA were tested on 3rd days, 7th days, and 28th days for both the control specimens and specimens cast underwater.

The chemical composition of the AAFA paste after going through underwater placement and dry condition is determined using X-ray fluorescence (XRF). XRF was conducted using the PA Nanalytic PW 4030, MiniPAL 4 (Malvern Panalytical, Malvern, United Kingdom) X-ray fluorescence spectrometer. After 28 days placement in various water types, the specimens were crushed to a powder form for analyses.

A JSM-7001F (JEOL, Tokyo, Japan) model of Field Emission Scanning Electron equipped with energy dispersive spectroscopy (EDS) was used to image the AAFA's morphology and determine its elemental composition after underwater placement. The specimens were cut into small pieces and coated with carbon using Auto Fine Coater (JEOL, Tokyo, Japan). The images were observed with accelerating voltage of 15 kV for all specimens.

3. Results and Discussion

3.1. pH Value and Temperature of Water

The water's pH value and temperature before and after underwater placement of the AAFA are illustrated in Figures 2 and 3. The seawater's original pH is 7.5, while the river water and lake water pH value are 7.4. All types of water recorded an increment in value when the AAFA paste was placed into the tanks. The seawater's pH value recorded the lowest increment of 0.6, while the highest increment was 1.8 from the lake water. From the underwater placement of AAFA paste, there is washout effect with the increment in the surrounding water's pH value. However, visible changes in the pH are evident, probably due to the smaller tank (275 mm length × 160 mm width × 160 mm depth) used during underwater placement of the AAFA specimens.

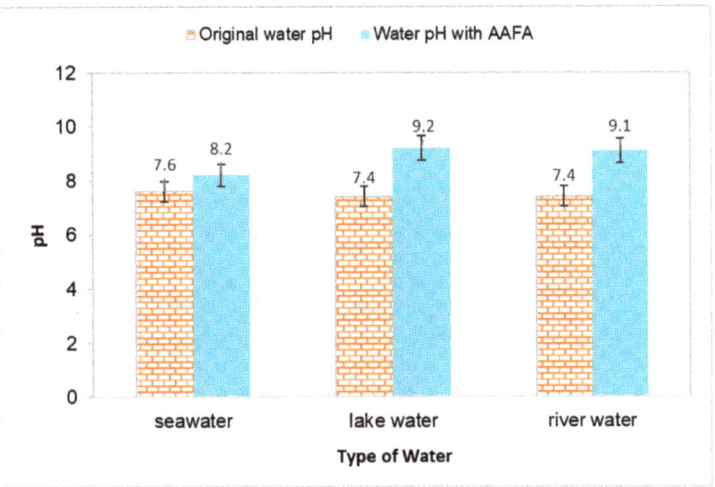

Figure 2. Effect on pH value for different type of water due to placement of the AAFA paste.

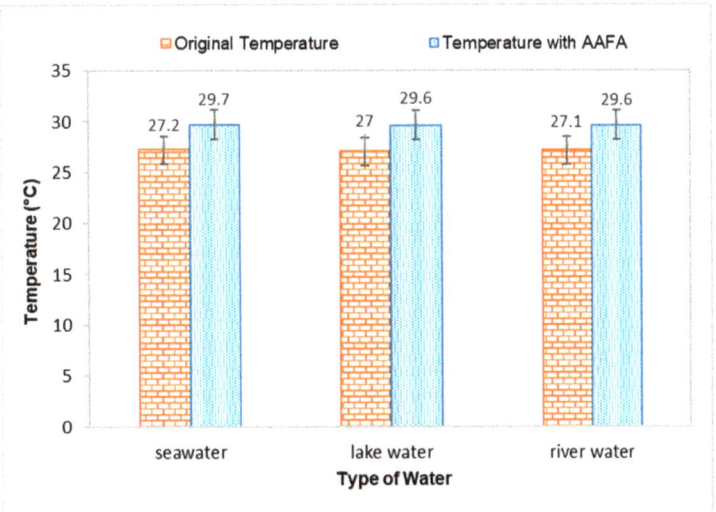

Figure 3. Effect on temperature when casting the AAFA in water.

The temperature for all types of water increased when the AAFA paste was placed into the tanks, proving that the AAFA reaction is exothermic, as heat is released during the hardening process, which increased the temperature of the water. Temperature increment in the seawater and river water was 2.5 °C, while lake water had a temperature increment of 2.6 °C. Previous researchers reported that the reaction between the source materials and alkaline activator is an exothermic reaction when the AAFA is cured at high temperature [33–37]. However, it was observed in this study that even though the AAFA paste was placed in water, heat release can still be detected.

3.2. Setting Time

The setting time of cement or binder occurred when it loses its plasticity and slowly formed into hard rock type material. The initial setting time can be defined as the time taken by the paste to start stiffening; whereas the final setting time is when the paste begins to harden and able to carry some loads. The initial and final setting times for the AAFA paste through underwater placement in various water types are shown in Table 2. For the initial setting time, the AAFA specimens placed in seawater were recorded at the fastest time of 26 min, while the river water specimens recorded with the longest time of 30 min. For final setting time, the specimens in seawater and lake water recorded the same value of 35 min and for river water it recorded 37 min. The final setting time of Class C fly ash in dry condition (room temperature) is usually between 1–2 h, depending on the content of calcium (CaO) [38]. For the current study, the initial and final setting time of the AAFA specimens casted in dry condition reported 31 min and 40 min, respectively. This finding in agreement with past research where it was found that Class C fly ash recorded initial and final setting times of 32.15 min and 60.00 min for specimens casted in dry condition [39]. Source materials rich in Ca content have quick setting time in dry condition due to the higher dissolution rate of Ca^{2+} compared to Si^{4+} and Al^{3+}. The reaction product of the source materials rich in calcium is expected to form of Ca-rich phases that will develop the fundamental skeleton of the AAFA network. According to previous research, reaction products such as calcium silicate hydrate (C-S-H), calcium aluminate silicate hydrate (C-A-S-H) and sodium calcium aluminate silicate hydrate (N, C-A-S-H) are expected to be present in Ca-rich phase [40].

Table 2. Setting time of AAFA cast underwater.

Types of Water	Setting Times (minutes)	
	Initial	Final
Dry Condition	31 ± 0.5	40 ± 0.5
Seawater	26 ± 0.5	35 ± 0.5
River Water	30 ± 0.5	37 ± 0.5
Lake Water	28 ± 0.5	35 ± 0.5

The quick setting time for underwater placement of the AAFA in seawater is due to Cl^- ions which react with cations in the AAFA paste such as Na^+ and Ca^{2+}. The formation of calcium chloride ($CaCl_2$) is widely known as an accelerator for early strength development as well as a minimiser for the setting time [41]. Additionally, during underwater placement of the AAFA specimens, the existence of water helps to improve the properties of AAFA. According to Duxson et al. [42], water helps accelerate Si and Al dissolution process from the source materials by providing discontinuous gel nanopores to the paste, hence improved its performance. For practical application, it is suggested to use the retarding admixture to control the setting time, as it can delay the setting time and keep the AAFA concrete workable throughout the placing process.

3.3. Compressive Strength

The compressive strength of the AAFA specimens through underwater placement was evaluated on 3rd, 7th, and 28th days. All specimens displayed increment in strength with respect to aging days as per Figure 4. The AAFA specimens cast in seawater displayed the highest compressive strength for all the aging days; for example, on 3rd days, the AAFA specimens reported a strength of 36.8 MPa. Meanwhile, the lowest compressive strength was found in the specimens cast in river water with strength 34.6 MPa on 3rd days. For control specimens (dry condition), the compressive strength on 3rd days is 79.1 MPa.

The specimens cast in seawater recorded a decrease in their compressive strength by 54% compared to the specimens cast in dry condition for 3rd days of testing.

The compressive strength of the AAFA specimens cast in seawater was reported to be 46.0MPa on the 7th days, whereas its dry counterpart has a reported strength of 79.9 MPa. This translates to a 20% strength increment in the AAFA specimens cast in seawater from day 3 to day 7. However, the strength increment from day 7 to day 28 was even higher, which is 55%. In contrast, the 28 day's strength of the specimens cast in dry condition was found to be 80.9 MPa, indicating that it has a lower strength increment. According to Kumar et al. [43], the increment of strength with respect to time can be attributed to calcium silicate hydrate (C-S-H) formation. Further discussion about the reaction product will be provided in Section 3.4.

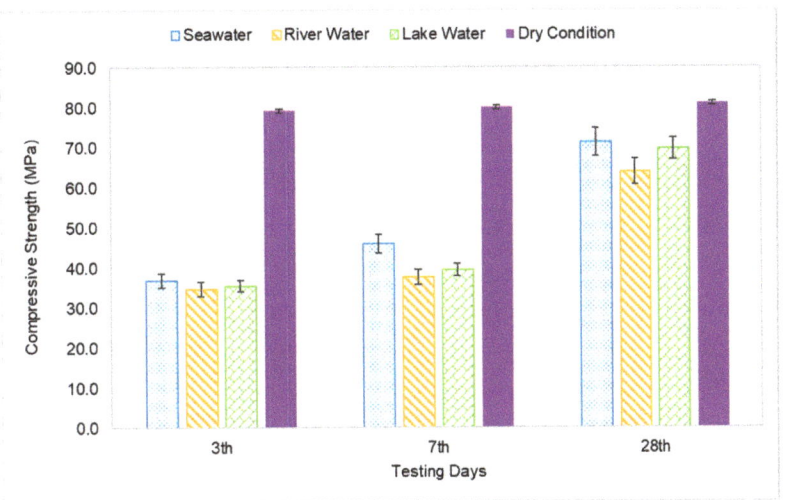

Figure 4. Compressive strength of AAFA paste cast in water.

The AAFA cast in dry condition exhibited almost complete strength development within 3 days, as the strength increment from 3rd days to 28th days testing was only 2%. For the AAFA cast in water, the strength slowly increased from 3rd days to 28th days for all water types. In the case of the 28-day strength, the AAFA specimens cast in seawater recorded a strength decrease by 12% relative to the specimen cast in dry condition, implying that the AAFA can be used for constructing a structure in water due to its impressive strength. Normally, conventional concrete (ordinary Portland cement) requires anti-washout admixture (AWA) and high range water reducer admixture (HRWR) before it can be used as binder materials in construction, especially for underwater structures. However, using AAFA only requires materials rich in silica and alumina, as well as alkaline activator. Additionally, the raw materials used in this case (silica and alumina sources) are waste materials, which falls in line with green technology promotion.

3.4. Chemical Composition Analysis

The chemical composition of control and AAFA specimens placed in different types of water is presented in Table 3. All the AAFA paste showed an increment in SiO_2 content due to the reaction of the fly ash with waterglass (Na_2SiO_3). Meanwhile the content of Al_2O_3 showed a reduction in AAFA paste relative to the raw fly ash. This is due to the participation of Al_2O_3 in setting time of the AAFA via acceleration of the condensation of the AAFA product formation [44,45]. The content of Fe_2O_3 also increased in the AAFA paste, especially those cast in dry condition, hence contributing to the maximum compressive strength. It was suggested that Fe^{3+} contributed to the formation of AAFA network due to the similar charge and ionic radius with Al^{3+} [46,47]. However, the increment of most chemical composition between different specimens is almost similar, which is related to the compressive strength. The compressive strength of the AAFA depends on a few factors such as the formation of reaction products, distribution of Si-Al ratio, calcium content, and the surface reaction between the unreacted Si-Al particles [48,49].

Table 3. Comparison of chemical composition for all specimens.

Composition	Fly Ash	AAFA Paste (wt. %)			
		Dry Condition	Seawater	River Water	Lake Water
SiO_2	30.80	34.30	34.60	34.20	34.00
Al_2O_3	13.10	10.60	10.80	10.70	10.70
CaO	22.30	21.50	22.00	21.20	21.20
Fe_2O_3	22.99	24.75	24.38	23.64	23.47
MgO	4.00	3.10	3.10	3.00	3.20
TiO_2	0.89	0.94	0.93	0.88	0.88
K_2O	1.60	1.43	1.42	1.30	1.33
SO_3	2.67	2.02	1.20	0.93	0.91
MnO	0.21	0.22	0.22	0.21	0.20
Si/Al ratio	2.35	3.24	3.20	3.20	3.18
Ca/Si ratio	0.72	0.63	0.64	0.62	0.62
Fe/Si ratio	0.75	0.72	0.70	0.69	0.69
Strength (MPa)	-	80.9	71.1	63.7	69.5

The molar ratio of Si/Al, Ca/Si, and Fe/Si of raw fly ash and the AAFA paste was calculated based on the result from XRF. For the ratio of Si/Al, the compressive strength increased when the Si/Al ratio increased due to the formation of Si-O-Si bonds. The maximum ratio of Si/Al is contributed by the AAFA specimens cast in dry condition. Nevertheless, the ratios between the specimens do not differ much.

The Ca/Si ratio for the source materials rich in Ca content is also linked with the compressive strength of the AAFA. The Ca/Si is responsible for the formation of C-S-H, and according to Timakul et al. [50], the compressive strength increased alongside the Ca/Si ratio. However, in this study, the ratio of Ca/Si is almost similar (~0.64–0.63) between the AAFA specimens cast in dry condition and seawater, which resulted in less of a difference in terms of compressive strength. For the AAFA cast in river water and lake water, the ratio of Ca/Si is similar (Ca/Si~0.62). Additionally, the formation of C-S-H, as AAFA reaction product and/or as OPC hydration product is entirely different. For the formation of C-S-H as hydration of OPC, the ratio of Ca/Si is in the range of 1.2 to 2.3, which is much higher relative to the AAFA [51,52]. The ratio of Fe/Si also plays essential role in forming the reaction product of the AAFA. The specimens' compressive strength increased when the ratio of Fe/Si increased due to the formation of ferro-sialate-siloxo and/or ferro-sialate-disiloxo poly. The XRF result indicated that iron oxide is involved in the forming of the AAFA network and contributed to the AAFA's strength.

3.5. Microstructure and Elemental Composition of Reaction Products

The morphology image of fly ash shown spheres particles shapes with smooth surface and various sizes of particles as in Figure 5. Figures 6–9 show the microstructure and EDS of the specimens at three selected spots (represented by the spectrum numbers) in the matrix. Elements such as Si, Na, Fe, Al, Ca, and O were identified in the AAFA matrix for each specimen. The selected spot for each specimen is often different, which means that the EDS elemental composition is incomparable between each specimen.

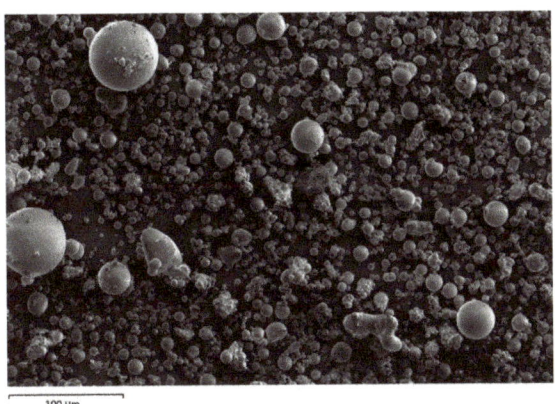

Figure 5. Morphology of fly ash.

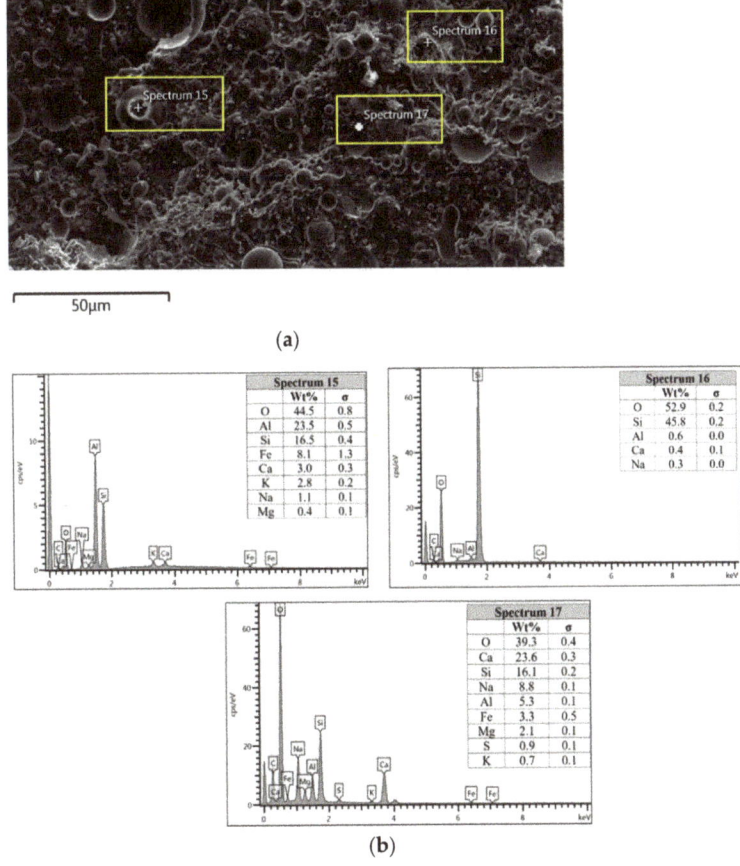

Figure 6. (a) Morphology of AAFA specimens cast in dry condition. (b) EDS for AAFA specimens cast in dry condition.

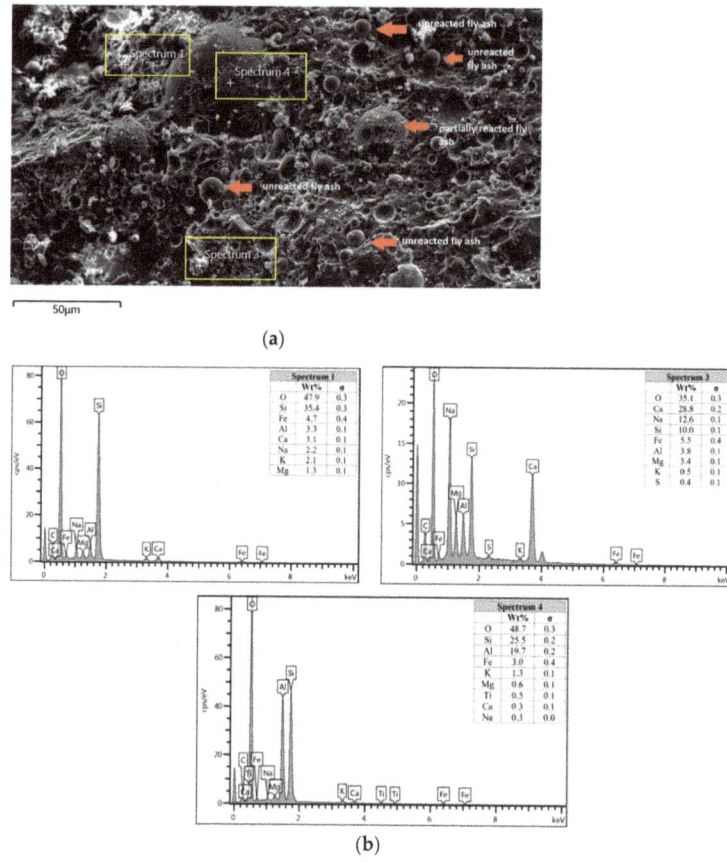

Figure 7. (**a**) Morphology of AAFA specimens cast in seawater. (**b**) EDS for AAFA specimens cast in seawater.

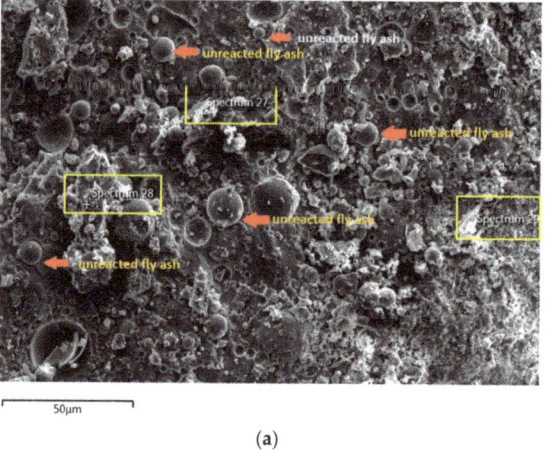

(**a**)

Figure 8. *Cont.*

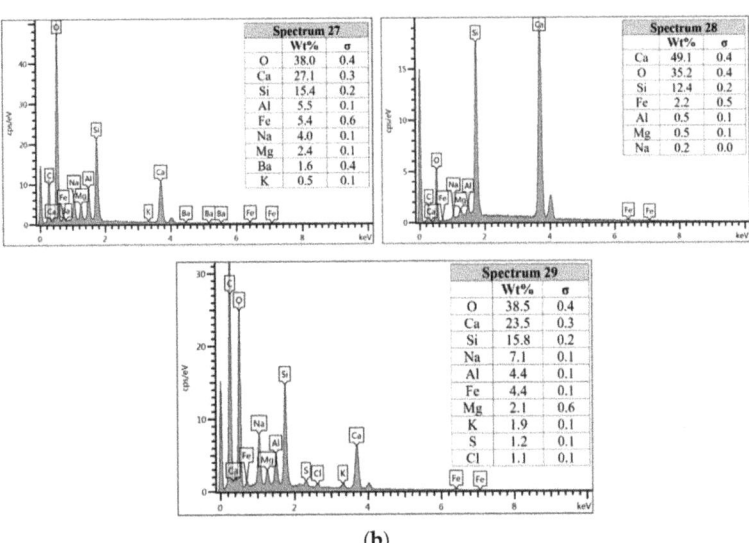

Figure 8. (a) Morphology of AAFA specimens cast in river water. (b) EDS for AAFA cast in river water.

For the AAFA specimens casted in the dry condition as in Figure 6a, the present of unreacted fly ash still detected on the specimen. For the elemental composition of spectrum 15 is occupied by Si, Al, and Fe, with Ca and Na less than 5 wt.%. Referring to the FESEM images, spectrum 15 showed the particle shapes of fly ash. It can therefore be surmised that the unreacted fly ash contributed to the strength increment with respect to the aging period due to the complex reaction between the surfaces of the particles via bonding strength [53–57]. Meanwhile, the elemental composition in spectrum 16 majorly consists of Si, but for spectrum 17 is dominated by Si, Ca, Na, and Al as in Figure 6b. It can be hypothesised that these elemental compositions represent the reaction product of C-A-S-H and C-S-H due to the high content of Ca in the source material (fly ash).

Figure 7a shows the specimen cast in seawater where unreacted and partially reacted fly ash were detected. Through EDS analysis spectrum 1 is dominated mostly by Si with Na, Al, Ca, and Fe less than 5 wt.%. Spectrum 3 is dominated by Ca, Na, and Si, which indicate the formation of C-S-H. Additionally, spectrum 4 show high concentration of Si and Al as in Figure 7b which represent unreacted fly ash.

The unreacted fly ash remains present between AAFA matrix as confirmed by the FESEM image in Figure 8a. The AAFA specimen's elemental compositions cast in river water (Figure 8b) are represented by spectrum 27, 28, and 29. From the three different spots, the Ca and Si are predominant indicating the existing of calcium silicate hydrate (C-S-H).

The microstructure of specimens cast in lake water (Figure 9a) showed micro-crack and it is believed to be due to sample preparation for FESEM. For spectrum 54, it is dominated by Ca, Na, and Si, which signifying the formation of C-S-H. However, spectrum 55 is mostly dominated by Ca, Si, Al, Fe and by referring to Figure 9b, the location of this spectrum is on spherical shape of fly ash. The elemental composition of spectrum 56 is predominated by Si, Al, and Fe.

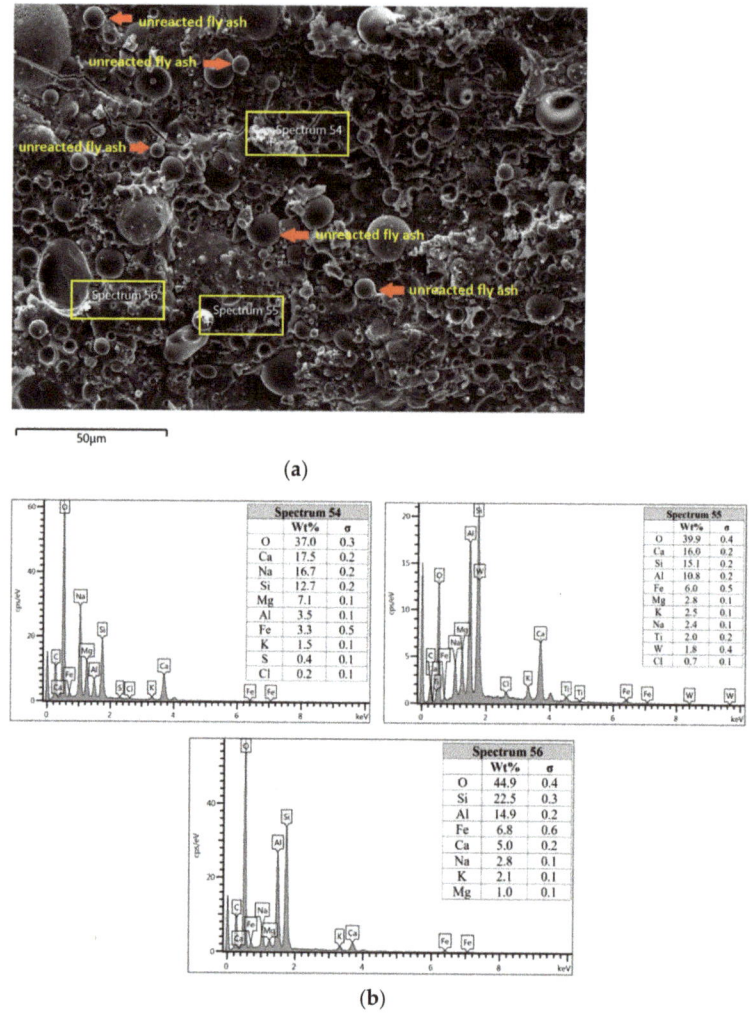

Figure 9. (a) Morphology of AAFA specimens cast in lake water. (b) EDS for AAFA specimens cast in lake water.

In Section 3.4, C-S-H presence is confirmed via XRF analysis due to the increment in Ca and Si content. Additionally, the same finding also noted in EDS analysis where the C-S-H supported the compressive strength by acting as a micro-aggregate in the AAFA which produced denser AAFA matrix. The formation of C-S-H started from dissolution of Ca from the source material where some of the Ca will precipitate in the form of calcium hydroxide ($Ca(OH)_2$) and C-S-H. Likewise, Si species also favourably to react with dissolved Ca rather than polymerise with soluble Al [58,59]. Hence, the presence of excessive Al will force out from Ca-rich area into the AAFA network. Past research also found that C-S-H gel contribute to the strength development at later age such as at 28th days [57,58].

Through EDS analysis, the existence of Fe was noticeable from all specimens and was reconfirmed by the XRF result. The high percentage of Fe in AAFA network due to involvement as substitution for Al, which leads to formation ferro-sialate-siloxo and ferro-sialate-disiloxo poly binders where Ca^{2+} and Na^+ act as charge-balancing cations [57].

4. Conclusions

The use of concrete for underwater placement is a significant challenge due to washout effect as well as the presence of various ions in the water which can influence the properties of concrete. The present study analysed the strength, changes in water pH, and AAFA setting time when go through underwater placement in seawater and freshwater. The chemical composition of AAFA paste is analysed using XRF and EDS. The AAFA can be used as binder for underwater concrete without the addition of anti-washout admixture (AWA). The maximum compressive strength of 71.10 MPa was obtained from the specimens cast in seawater on 28th days. It demonstrates 12.11% of strength reduction compared to specimens cast in dry condition and according to JSCE standard, the AAFA specimens are qualified to use for underwater casting. It was found that the presence of Cl- ions in seawater leads to formation of calcium chloride ($CaCl_2$) which acts as accelerator for early setting time and strength development.

Author Contributions: Conceptualization, investigation, formal analysis, and visualization, Z.Y.; Supervision, funding acquisition, and project administration, M.M.A.B.A.; writing—review and editing, Funding acquisition and project administration, L.-y.L.; formal analysis and data curation, D.D.B.N.; Conceptualization, methodology, writing—original draft preparation and resources, M.A.A.Z.H.; validation and writing—review and editing, A.V.S.; interpretation, visualization and validation (P.V.); data curation and visualization, R.A.R. All authors have read and agreed to the published version of the manuscript.

Funding: This research was funded by ROYAL SOCIETY NEWTON-UNGKU OMAR FUND (grant number NI170199), EUROPEAN UNION (grant number H2020-MSCA-RISE-2015-689857-PRIGeoC) and publications grant of the Gheorghe Asachi Technical University of Iasi (TUIASI) project number GI/P38/2021.

Institutional Review Board Statement: Not applicable.

Informed Consent Statement: Not applicable.

Conflicts of Interest: The authors declare no conflict of interest.

References

1. Assaad, J.J.; Daou, Y.; Khayat, K.H. Simulation of water pressure on washout of underwater concrete repair. *ACI Mater. J.* **2009**, *106*, 1–8.
2. Yousri, K.M. Self-flowing underwater concrete mixture. *Mag. Concr. Res.* **2008**, *60*, 1–10. [CrossRef]
3. Grzeszcyk, S.; Jurowski, K.; Bosowska, K.; Grzymek, M. The role of nanoparticles in decreased washout of underwater concrete. *Constr. Build. Mater.* **2019**, *203*, 670–678. [CrossRef]
4. Khayat, K.H.; Sonebi, M. Effect of mixture composition on washout resistance of highly flowable underwater concrete. *ACI Mater. J.* **2001**, *98*, 289–295.
5. Jung-Jun, P.; Jae-Heum, M.; Jun-Hyoung, P.; Sung-Wook, K. An estimation on the performance of high fluidity anti-washout underwater concrete. *Key Eng. Mater.* **2014**, *577-578*, 501–504.
6. Sam, X.Y.; Berner, D.E.; Gerwick, B.C. *Assessment of Underwater Concrete Technologies for in-the-Wet Construction of Navigation Structure*; US Army Corps of Engineers: Washington, DC, USA, 1999.
7. Heniegal, A.M. Developing underwater concrete properties with and without anti-washout admixtures. *Int. J. Sci. Eng. Res.* **2017**, *7*, 97–106.
8. Wang, J.; Xie, J.; Wang, Y.; Lui, Y.; Ding, Y. Rheological properties, compressive strength, hydration products and microstructure of seawater-mixed cement pastes. *Cem. Conc. Comp.* **2020**, *114*, 1–15. [CrossRef]
9. Mohammed, T.U.; Hamada, H.; Yamaji, T. Performance of seawater-mixed concrete in the tidal environment. *Cem Conc. Res.* **2004**, *34*, 593–601. [CrossRef]
10. Japan Society of Civil Engineers. Recommendations for Design and Construction of Anti-Washout Underwater Concrete. *Concr. Libr. JSCE* **1992**, *19*, 1–61.
11. Zhang, J.; Shi, C.; Zhang, Z.; Ou, Z. Durability of alkali-activated materials in aggressive environments: A review on recent studies. *Constr. Build. Mater.* **2017**, *152*, 598–613. [CrossRef]
12. Mcleod, R.S. Ordinary portland cement with extraordinary high CO_2 emission. What can be done to reduce them? *BFF Autumn.* **2005**, *15*, 30–33.
13. Singh, B.; Ishwarya, G.; Gupta, M.; Bhattacharyya, S.K. Geopolymer concrete: A review of some recent developments. *Constr. Build. Mater.* **2015**, *85*, 78–90. [CrossRef]

14. Wardhona, A.; Gunasekara, C.; Law, D.W.; Setunge, S. Comparison of long-term performance between alkali activated slag and fly ash geopolymer concretes. *Constr. Build. Mater.* **2017**, *143*, 272–279. [CrossRef]
15. Rakngan, W.; Williamson, T.; Ferron, R.D.; Sant, G.; Juenger, M.C.G. Controlling workability in alkali-activated class C fly ash. *Constr. Build. Mater.* **2018**, *183*, 226–233. [CrossRef]
16. OECD/IEA. *World Energy Outlook 2008*; International Energy Agency (IEA) Publications: Paris, France, 2008; p. 578.
17. Alvarez-Ayuso, E.; Querol, X.; Plana, F.; Alastuey, A.; Moreno, N.; Izquierdo, M.; Font, O.; Moreno, T.; Diez, S.; Vázquez, E.; et al. Environmental, physical, and structural characterisation of geopolymer matrixes synthesised from coal (co-) combustion fly ashes. *J. Hazard. Mater.* **2008**, *154*, 175–183. [CrossRef] [PubMed]
18. Fernandez-Jimenez, A.; de la Torre, A.G.; Palomo, A.; Lopez-Olmo, G.; Alonso, M.M.; Aranda, M.A. G Quantitative determination of phases in the alkali activation of fly ash. Part 1. Potential ash reactivity. *Fuel* **2006**, *85*, 625–634. [CrossRef]
19. Siddique, S.; Jang, J.G. Acid and sulfate resistance of seawater based alkali activated fly ash: A sustainable and durable approach. *Constr. Build. Mater.* **2021**, *281*, 122601. [CrossRef]
20. Ryu, G.S.; Lee, Y.B.; Koh, K.T.; Chung, Y.S. The mechanical properties of fly ash-based geopolymer concrete wit alkaline activators. *Constr. Build. Mater* **2013**, *47*, 409–418. [CrossRef]
21. Rattanasak, U.; Chindaprasirt, P. Influence of NaOH solution on the synthesis of fly ash geopolymer. *Miner. Eng.* **2009**, *22*, 1073–1078. [CrossRef]
22. Mustafa Al Bakri, A.M.; Kamarudin, H.; Khairul Nizam, I.; Bnhussain, M.; Zarina, Y.; Rafiza, A.R. Correlation between Na_2SiO_3/NaOH ratio and fly ash/alkaline activator ratio to the strength of geopolymer. *Adv. Mater. Res.* **2012**, *341-342*, 189–193. [CrossRef]
23. Palomo, A.; Grutzeck, M.; Blanco, M. Alkali-activated fly ashes. A cement for the future. *Cem. Concr. Res.* **1999**, *29*, 1323–1329. [CrossRef]
24. van Jaarsveld, J.G.S.; van Deventer, J.S.J.; Lukey, G.C. The effect of composition and temperature on the properties of fly-ash and kaolinite-based geopolymers. *Chem. Eng. J.* **2002**, *89*, 63–73. [CrossRef]
25. Abdullah, M.M.A.; Kamarudin, H.; Mohammed, H.; Khairul Nizam, I.; Rafiza, A.R.; Zarina, Y. The relationship of NaOH molarity, Na_2SiO_3/NaOH ratio, fly ash/alkaline activator ratio, and curing temperature to the strength of fly ash-based geopolymer. *Adv. Mater. Res.* **2011**, *328–330*, 1482–1485.
26. Kupwade-Patil, K.; Allouche, E.N. Examination of chloride-induced corrosion in reinforced geopolymer concretes. *J. Mater. Civ. Eng.* **2012**, *25*, 1465–1476. [CrossRef]
27. Shaikh, F.U. Effects of alkali solutions on corrosion durability of geopolymer concrete. *Adv. Concr. Constr.* **2014**, *2*, 109–123. [CrossRef]
28. Chindaprasirt, P.; Chalee, W. Effect of sodium hydroxide concentration on chloride penetration and steel corrosion of fly ash-based geopolymer concrete under marine site. *Constr. Build. Mater.* **2014**, *63*, 303–310. [CrossRef]
29. Sakkas, K.; Panias, D.; Nomikos, P.P.; Sofianos, A.I. Potassium based geopolymer for passive fire protection of concreter tunnels linings. *Tunnel. Under. Space Tech.* **2014**, *43*, 148–156. [CrossRef]
30. Yun-Ming, L.; Cheng-Yong, H.; Long-Yuan, L.; Jaya, N.A.; Abdullah, M.M.A.; Soo-Jin, T.; Hussin, K. Formation of one-part mixing geopolymers and geopolymer ceramics from geopolymer powder. *Constr. Build. Mater.* **2017**, *156*, 9–18.
31. ASTM C618-12a. Standard specification for coal fly ash and raw or calcined natural pozzolan for use in concrete. In *Annual Book of ASTM Standards*; ASTM International: West Conshohocken, PA, USA, 2013.
32. Mustafa Al Bakri, A.M.; Kamarudin, H.; Bnhussain, M.; Khairul Nizar, I.; Rafiza, A.R.; Zarina, Y. Microstructure of different NaOH molarity of fly ash-based green polymeric cement. *J. Eng. Tech. Res.* **2011**, *3*, 44–49.
33. ASTM C191-01. *Standard Test Method for Time of Setting of Hydraulic Cement by Vicat Needle*; ASTM International: West Conshohocken, PA, USA, 2001.
34. ASTM C109 / C109M-16a. *Standard Test Method for Compressive Strength of Hydraulic Cement Mortars (Using 2 in. or [50 mm] Cube Specimens)*; ASTM International: West Conshohocken, PA, USA, 2016.
35. Chindaprasirt, P.; Jaturapitakkul, C.; Chalee, W.; Rattanasak, U. Comparative study on the characteristic of fly ash and bottom ash geopolymers. *Waste Mngmnt.* **2009**, *29*, 539–543. [CrossRef] [PubMed]
36. Weng, L.; Sagoe-Crentsil, K. Dissolution process, hydrolisis and condensation reactions during geopolymer synthesis: Part I—Low Si/Al ratio systems. *J. Mater. Sci.* **2007**, *42*, 2997–3006. [CrossRef]
37. Skvara, F.; Kopecky, L.; Smilauer, V.; Bittnar, Z. Material and structural characterization of alkali activated low-calcium brown coal fly ash. *J. Hazard. Mater.* **2009**, *168*, 711–720. [CrossRef] [PubMed]
38. Rattanasak, U.; Pankhet, K.; Chindaprasirt, P. Effect of chemical admixture on properties of high-calcium fly ash geopolymer. *Int. J. Miner. Metal. Mater.* **2011**, *18*, 364–369. [CrossRef]
39. Mohamed, R.; Abd Razak, R.; Abdullah, M.M.A.B.; Shuib, R.K.; Subaer, C.J. Geopolymerization of class C fly ash:reaction kinetics, microstructure properties and compressive strength of early age. *J. Non-Crys. Sol.* **2021**, *553*, 1–15. [CrossRef]
40. Puligilla, S.; Mondal, P. Role of slag in microstructural development and hardening of fly ash-slag geopolymer. *Cem. Concr. Res.* **2013**, *43*, 70–80. [CrossRef]
41. Yang, S.; Xu, J.; Zang, C.; Li, R.; Yang, Q.; Sun, S. Mechanical properties of alkali-activated slag concrete mixed by seawater and sea sand. *Constr. Build. Mater.* **2019**, *196*, 395–410. [CrossRef]

42. Duxson, P.; Fernandez-Jimenez, A.; Provis, J.L.; Lukey, G.C.; Palomo, A.; van Deventer, J.S.J. Geopolymer technology: The current state of the art. *J. Mater. Sci.* **2007**, *42*, 2917–2933. [CrossRef]
43. Kumar, S.; Kumar, R.; Mehrotra, S.P. Influence of granulated blast furnace slag on the reaction structure and properties of fly ash based geopolymer. *J. Mater. Sci.* **2010**, *45*, 607–615. [CrossRef]
44. De Silva, P.; Sagoe-Crentsil, K.; Sirivivatnanon, V. Kinetics of geopolymerisation: Role of Al_2O_3 and SiO_2. *Cem. Conc. Res.* **2007**, *37*, 512–518. [CrossRef]
45. Weng, L.; Sagoe-Crentsil, K.; Brown, T.; Song, S. Effects of aluminates on the formation of geopolymers. *Mater. Sci. Eng. B.* **2005**, *117*, 163–168. [CrossRef]
46. Perera, D.S.; Cashion, J.D.; Blackford, M.G.; Zhang, Z.; Vance, E.R. Fe specification in geopolymers with Si/Al molar ratio of 2. *J. Eur. Ceram. Soc.* **2007**, *27*, 2697–2703. [CrossRef]
47. Rossano, S.; Behrens, H.; Wilke, M. Advanced analyses of ^{57}Fe Mossbauer data of alumino-silicate glasses. *Phys. Chem. Miner.* **2008**, *35*, 77–93. [CrossRef]
48. Xu, H. Geopolymerisation of Aluminosilicate Minerals. Ph.D. Thesis, University of Melbourne, Melbourne, Australia, 2001.
49. Van Jaarsveld, J.G.S.; Van Deventer, J.S.J.; Lukey, G.C. The characterisation of source materials in fly ash-based geopolymers. *Mater. Lett.* **2003**, *57*, 1272–1280. [CrossRef]
50. Timakul, P.; Rattanaprasirt, W.; Aungkavattana, P. Improving compressive strength of fly ash-based geopolymer composites by basalt fibers addition. *Ceram. Int.* **2016**, *42*, 6288–6295. [CrossRef]
51. Gani, M.S.J. *Cement and Concrete*, 1st ed.; Chapman and Hall: London, UK, 1997.
52. Richardson, I.G. The nature of C-S-H in hardened cements. *Cem. Concr. Res.* **1999**, *29*, 1131–1147. [CrossRef]
53. Xu, H.; van Deventer, J.S.J. The geopolymerisation of aluminosilicate minerals. *Int. J. Miner. Process* **2000**, *59*, 247–266. [CrossRef]
54. Kumar, R.; Kumar, S.; Mehrotra, S.P. Towards sustainable for fly ash through mechanical activation. *Resour. Conser. Recycl.* **2007**, *52*, 157–179. [CrossRef]
55. Yip, C.K.; Lukey, G.C.; van Deventer, J.S.J. The coexistence of geopolymeric gel and calcium silicate hydrate at the early stage of alkali activation. *Cem. Concr. Res.* **2005**, *35*, 1688–1697. [CrossRef]
56. Wang, Y.; He, F.; Wanf, J.; Hu, Q. Comparison of effects of sodium bicarbonate and sodium carbonate on the hydration and properties of Portland cement paste. *Materials* **2019**, *12*, 1033. [CrossRef]
57. Wang, Y.; He, F.; Wang, J.; Wang, C.; Xiong, Z. Effects of calcium bicarbonate on the properties of ordinary Portland cement paste. *Constr. Build. Mater.* **2019**, *225*, 591–600. [CrossRef]
58. Kumar, S.; Djobo, J.N.Y.; Kumar, A.; Kumar, S. Geopolymerization behaviour of fine iron-rich fraction of brown fly ash. *J. Build. Eng.* **2016**, *8*, 172–178. [CrossRef]
59. Nergis, D.D.B.; Vizureanu, P.; Ardelean, I.; Sandu, A.V.; Corbu, O.C.; Matei, E. Revealing the Influence of Microparticles on Geopolymers' Synthesis and Porosity. *Materials* **2020**, *13*, 3211. [CrossRef] [PubMed]

MDPI
St. Alban-Anlage 66
4052 Basel
Switzerland
Tel. +41 61 683 77 34
Fax +41 61 302 89 18
www.mdpi.com

Materials Editorial Office
E-mail: materials@mdpi.com
www.mdpi.com/journal/materials

www.ingramcontent.com/pod-product-compliance
Lightning Source LLC
LaVergne TN
LVHW070226100526
838202LV00015B/2095